Science of Fungi in Grapevine

Olivier Viret • Katia Gindro

Science of Fungi
in Grapevine

 Springer

Olivier Viret
General Direction of Agriculture
Viticulture and Veterinary Affairs
Morges, Vaud, Switzerland

Katia Gindro
Mycology
Agroscope
Nyon, Vaud, Switzerland

ISBN 978-3-031-68662-7   ISBN 978-3-031-68663-4   (eBook)
https://doi.org/10.1007/978-3-031-68663-4

Agroscope

This Springer imprint is published by the registered company Springer Nature Switzerland AG
The registered company address is: Gewerbestrasse 11, 6330 Cham, Switzerland

If disposing of this product, please recycle the paper.

Foreword

One of the world's most important cultivated plants, the grapevine was domesticated from the wild vine over 11,000 years ago. The fungi associated with it are likely as old as the plant itself. Despite their co-evolution with the vine over the centuries, it was only with the invention of the microscope in the seventeenth century that precise descriptions of fungi began to become available.

The arrival in Europe of powdery and downy mildew in the second half of the nineteenth century brought about a complete change in vine cultivation practices worldwide, as systematic protection from both of these fungi was required throughout their growing cycle. Over 150 years after their introduction, they are still the focus of scientific research and in-depth techniques. Around 20 other fungal diseases, including black rot, are present in grapevine, making plant protection an essential and highly complex endeavour.

Olivier Viret and Katia Gindro's work illustrates in an exemplary manner the scientific knowledge acquired over the course of history on fungi themselves and on their multiple interactions with the plant and its environment. The initial focus of the research lies in the biology and epidemiology of fungi, as well as the identification of symptoms and effective means of protection. Thereafter, it is extended to the complete study of the pathogen–host plant system, which forms a bioecological unit now referred to as the holobiont.

An entire, richly illustrated chapter of this book is devoted to the vine itself—its development, phenology and the structure and anatomy of its various organs, from root to berry via the trunk, shoots, buds, leaves and inflorescences.

The breeding of fungal disease-resistant varieties is dealt with in detail, based particularly on the authors' scientific work on vine defence mechanisms and selection using biochemical and genetic markers. The recent creation of high-quality varieties obtained through natural crossing and equipped with a number of resistance genes points to revolutionary prospects for protecting grapevine against the main fungal diseases.

The study of the vine's microbiota is comprehensively described by the authors in their work. They show that the number of fungal species associated with vine diseases is on the increase, and that, among the totality of microorganisms living on the vine and in the soil, certain ones are likely to boost the resistance of the plant to diseases and stress. The role of fungal endophytes, their biodiversity, interactions with plants, way of life and possible evolutionary pathways are analysed objectively on the basis of the most recent scientific findings.

Three main chapters, sumptuously illustrated, describe the quasi-totality of cryptogamic diseases of the vine, starting with those developing on the green parts of the plant. The symptoms, biology, life cycles, modes of infection, forecasting methods and protection strategies are dealt with in detail, particularly for the four most important diseases, i.e. downy mildew, powdery mildew, grey mould and black rot. The other types of rot, anthracnose, red fire disease and the toxins present on the harvest are also addressed.

Wood diseases such as esca, black dead arm, Eutypa dieback, black spot disease and black foot are on the rise in vineyards, and the authors give an account of the latest knowledge on this important issue. Lastly, the diseases that attack the roots such as honey fungus and root rot caused by *Dematophora, Roesleria* or *Phytophthora* conclude the chapters dedicated to the different fungal diseases of the vine.

Grafting and the production of grapevine plants can be sources of infection by fungi. The different graft and rootstock assembly techniques and the risks of contamination at the different stages of multiplying the young plants are analysed in addition to the mycobiome present in nursery plants and disinfection techniques.

The final chapter of this book tackles the issue of methods of protection against fungal diseases of grapevine. The background to the introduction of fungicides, the production of synthetic organic fungicides and the regular appearance of pathogen fungicide resistance phenomena are addressed by the authors, as are the alternatives to plant-protection products emerging from combinatorial chemistry and the use of antagonistic microorganisms and elicitors.

Integrated, organic and biodynamic protection strategies are presented. An important success factor is the proper management of protection against diseases through the implementation of prophylactic measures, suitable choice and dosage of active substances and the fine-tuning of spraying equipment.

In conclusion, *The Science of Fungi in Grapevine*, written and coordinated by the scientists Olivier Viret and Katia Gindro, is a remarkable reference work on the fungal diseases of grapevine. Based on the latest scientific findings, it takes a holistic approach to the subject of the protection of grapevine from fungi by placing pathogen–vine interactions squarely at the centre of its reflections and analyses. Aimed both at researchers, educators, students and technicians as well as those working in the wine industry, it is written to be accessible to both scientists and practitioners alike. The rich illustrative material allows for easy comprehension of the different subjects addressed.

Viticulture Committee of the International   François Murisier
Organisation of Vine and Wine (OIV), Dijon, France

Head of Viticulture and Oenology Research Group at
Agroscope until 2008, Pully, Switzerland

Contributors

The authors would like to thank:

Dr. Valérie Hofstetter internationally recognised Mycologist and Phylogeneticist Emerita, Staff Research Scientist in Agroscope's 'Mycology' Research Group, for writing the chapters on wood and root diseases, as well as the subchapters 3.2.4.7 'Why Do Fungal Names Change? and 4.7 'Sour Rot (Acid Rot)'

Dr. Pierre-Henri Dubuis expert in phytopathology and in the use and application of fungicides in viticulture, Staff Research Scientist in Agroscope's 'Mycology' Research Group, for his editorial contribution on plant-protection products and control strategies, and authoring Sects. 8.1.2 and 8.1.3.

Dr. Leonie Pellissier expert in chemistry and natural products, Staff Research Scientist within Agroscope's 'Mycology' Research Group and at the University of Geneva, responsible for authoring Sects. 3.3 and 3.4.

Acknowledgements

The authors would like to extend their sincerest gratitude to the following colleagues for their contributions to this work:

Alain Gaume, Head of the 'Plant Protection' Research Division at Agroscope, Switzerland, for his support in developing agronomic research;

Carole Parodi, professional photographer, Agroscope, Switzerland, for creating high-quality illustrations of the vine;

Debra Nicol and Carol Finch, Scotland-based scientific and technical translators, for translating the manuscript from French into English;

Emilie Michellod, Agroscope, Switzerland, for enabling the histological imaging to be performed;

François Murisier, Switzerland, author of the foreword to this work;

Mario del Curto, professional photographer, Switzerland, for creating several illustrations of the fungal kingdom;

Regula Wolz-Gysi, Agroscope, Switzerland, for supporting this project and coordinating the translation of the manuscript;

Robin Huber, University of Geneva, Switzerland, for creating illustrations on specific chemical topics;

Samuel Wüst, Agroscope, Switzerland, for writing the subchapter on 'Recessive Resistances and Loss-of-Susceptibility Genes';

Virginie Duquette, graphic designer, Switzerland, for creating all the epidemiological cycles in this work.

Contents

About the Authors

Olivier Viret   currently Head of Vitiviniculture for the Canton of Vaud (Switzerland), has also pursued his scientific career at Agroscope, where he has in turn been Head of Mycology, Plant Protection and Viticulture-Oenology at national level. Prior to this, he served an apprenticeship as a winemaker-cellarer and completed a Master's in Agronomic Engineering and a PhD in Mycology at the ETH Zurich (Switzerland), where he currently lectures in vine pathology. His scientific expertise and practical experience cemented his international reputation, following a stint at Penfolds in Australia and collaborations with prestigious vineyards in Bordeaux, Champagne, and Tuscany, the University of Melbourne in Australia; he was also involved in a project for the development of viticulture in Ukraine. He has served as a national expert in the International Organisation of Vine and Wine (OIV) in the 'Genetic Resources and Vine Breeding' and 'Vine Protection and Viticultural Techniques' groups for over 15 years. O. Viret is the author of hundreds of scientific and extension publications as well as book chapters and spearheaded the Agrometeo and Vitimeteo plant-protection and vineyard risk forecasting systems developed in partnership with the Freiburg Weinbauinstitut in Breisgau (Germany).

Katia Gindro   is the Head of the 'Mycology and Pathology' Research Group at the Swiss Federal Agronomic Research Station Agroscope, where she also completed her Master's thesis in Biology and her PhD (Dr. ès Science) on *Botrytis cinerea* at the University of Lausanne (Switzerland) between 1994 and 2000. K. Gindro's extensive experience in mycology and vine-fungi interactions, along with her pioneering research on secondary metabolites and vine phytoalexins, has significantly enhanced her reputation in the field. Her work, which includes the use of optical and electron microscopy to monitor the epidemiology of fungal diseases, has led to numerous research projects in collaboration with both public and private sector partners. Notably, she is currently working with startups such as Agrosustain, Biorem Engineering, or Invaio. As a lecturer in mycology at the University of Geneva and a Board member of the Swiss Mycology Society, K. Gindro has authored over 200 scientific, technical, and extension publications, as well as several books and book chapters on phytopathology, biochemistry, and mycology.

Mindful of the practical application of scientific research findings while publishing their work internationally, the two authors have assembled in this book over 30 years' work on all aspects linking grapevine to fungi, whether the latter are pathogenic or simply form part of the plant microbiome as endophytes.

Chapter 1
Introduction

1.1 Origins and Plasticity of Grapevine

Grapevine is one of the most ancient of cultivated plants. The origins of the wild form of the vine can be traced back to the Paleolithic era over 200,000 years ago, when humans were hunter-gatherers. The cultivated vine *Vitis vinifera* is descended from the wild vine *Vitis vinifera* subsp. *sylvestris* which was domesticated in the Near East (Western Asia) and the South Caucasus (Georgia, Armenia, Azerbaijan) from two distinct populations separated during the last ice age, 11,700 years ago. The population from the Caucasus is thought to be associated with the earliest wine production, while the European population is thought to have originated from introgression between wild European populations and domesticated vines of the Near East (Dong et al. 2023). Today, grapevine is one of the most economically important cultivated plants (Töpfer and Trapp 2022). In 2021, the area under vine worldwide was nearly 7.3 million hectares for wine production (34.1 million tonnes or 262 million hl), grape juice (3.1 million tonnes or 26 million hl), table grapes (30.1 million tonnes) and raisins (5.4 million tonnes or 1.35 million tonnes dry weight). Grapevine has remarkable plasticity in terms of its area of adaptation and the genetic diversity of the genus *Vitis*. Developing primarily in temperate regions, it is also found in colder, wetter regions. Its climatic requirements depend mainly on temperature, sunshine and precipitation—factors which are influenced by geographic data such as latitude, elevation, exposure, slope, topography or the proximity of water bodies. Due to the global warming effect which has been observed on a planetary scale since the 1980s, the geographic boundaries for grapevine cultivation have gradually changed. In both the northern and southern hemispheres, they range between the thirtieth parallel south and slightly beyond the fiftieth parallel north, with an extension into southern England, southern Denmark and Norway.

Air temperature is the biggest limiting factor for the vine; the annual average must be higher than 9 °C, with the optimum ranging between 11 and 16 °C. This figure decreases by 0.6 °C for each increasing degree of latitude and by 0.5 °C per

O. Viret, K. Gindro, *Science of Fungi in Grapevine*,
https://doi.org/10.1007/978-3-031-68663-4_1

100 m of elevation. The elevation limit for growing grapevine is around 650–700 m, with exceptions associated with specific microclimatic conditions. In the Swiss Alps, for example, the continental conditions of the Valais allow vines to reach their peak up to an elevation of 900 m. The highest vineyards in Europe are at 1200 m elevation in Morgex in the Val d'Aosta in Italy and at 1150 m elevation in Visperterminen in the Valais (Switzerland). In China, vines are grown at up to 1500 m elevation on the high plateaus of Yunnan Province, and northern Argentina holds the world record of 3100 m for the highest elevation at which grapevine is cultivated. In these extreme conditions, sizeable production costs are incurred by the battle against frost (which has given rise to coping strategies such as ridging and the burying of grapevine stumps) and drought. Although vines can develop vegetatively in relatively extreme conditions provided that certain protective measures are implemented, the physiological equilibrium of the plant is only achieved in very specific conditions.

In Europe, the vine is particularly well adapted in the Mediterranean basin, where a significant proportion of the world's winegrowing area is to be found (Table 1.1). In the southern hemisphere, the vine cultivation area is mainly limited to New Zealand, Australia, South Africa, Brazil, Argentina and Chile. In these countries as well as on the West Coast of the United States (California, Washington, Oregon) or in certain arid regions in China (Gansu and Hebei provinces, the Xinjiang region situated at −154 m elevation with an average 20 mm of rainfall per year), the limiting factor is the water supply, which must be ensured by irrigation taken from groundwater or watercourses. Globally, 85% of New World vineyards are irrigated, in contrast with less than 10% of the area under vines in Europe. The optimal rainfall pattern for vines is around 600 mm during the growing season, ideally

Table 1.1 Viticultural area of the world's main wine-producing countries (>100,000 ha) in 2022 (in bold: Mediterranean countries; data from the International Organisation of Vine and Wine OIV)

Country	Area under vines (ha)	%
Spain	**968,668**	**13.0**
China	875,000	11.8
France	**792,528**	**10.6**
Italy	**704,738**	**9.5**
Turkey	**448,319**	**6**
United States	438,858	5.9
Argentina	218,233	2.9
Chile	202,638	2.7
Portugal	**192,287**	**2.6**
Romania	191,310	2.6
Iran	152,558	2.0
Australia	146,128	1.9
South Africa	125,989	1.7
Greece	**105,732**	**1.4**
Germany	102,873	1.4
Others (74 countries)	2,356,000	32
World	7,254,512	100

spread out to induce moderate water stress during ripening, which positively influences the quality of the grapes. Soil water availability depends on soil water storage capacity, expressed by soil water reserves, the slope of the vineyard, soil type, plant cover and its upkeep (Zufferey et al. 2022). The minimum number of hours of sunshine required by the vine is around 1500–1600 h per annum, with at least 1200 h during the growing season.

1.2 Historical Background of Fungal Diseases in Grapevine

1.2.1 Origin and Evolution of Diseases in Grapevine

From their origins through to antiquity and the Middle Ages, both wild and domesticated vines most likely had to contend with fungal pathogens. The civilisations of these eras were autocracies and subject to the constraints of their natural environment. The actual origins of the most virulent fungal diseases date back to a relatively recent period nearly two centuries ago, which is not to say that the vine was previously free from pathogens such as black rot, grey mould, wood diseases or anthracnose. Greek and Roman authors mentioned diseases such as grapevine leaf rust, a term coined by analogy with cereal rusts, which doubtless did not correspond to the rusts known nowadays in crops. Old grapevine literature (Reymondin 1798; Chaptal and Parmentier 1806) mentions 'grape rot' and 'hail disease'—the latter probably corresponding to grapevine white rot *(Coniella diplodiella)*—without touching on their importance or economic impact.

In the late nineteenth century, the winegrowing world was plunged into despair, vineyards were laid waste and growing practices were considerably complicated by the successive arrival in Europe of powdery mildew in 1845, phylloxera in 1863 and downy mildew in 1878 (Viala 1885), followed in the early twentieth century by the growth of pests of European origin such as grape leaf rust mite, spider mites and European grapevine moths. At this time, viticulture was contending with a new constraint obliging winemakers to 'sulphate' vines (i.e. to apply copper sulphate to them via a sprayer) at regular intervals to ensure yields and quality.

As late as the 1950s, the regular use of sulphur and copper against powdery and downy mildew as well as the grafting of vines onto American rootstock to control phylloxera were widespread practices that brought peace of mind to the vineyards. European grapevine and grape berry moths (*Lobesia botrana* and *Eupoecilia ambiguella*) and the vine leafroller tortrix (*Sparganothis pilleriana*) were controlled with highly toxic arsenics salts, gradually replaced by plant extracts such as pyrethrum and nicotine, trialled between 1890 and 1910 (Linder et al. 2016). At this time, plant protection products were considered to be 'remedies', by analogy with human medicine.

The history of phytopathology in general and vine pathology in particular date back to the Greek philosopher Theophrastus (370–286 BC), to whom we owe the earliest descriptions of fungi on cereals, trees and vegetables. From the Roman era to antiquity, humanity was in thrall to fungal pathogens, suffering harvest losses of varying sizes, and appealing to the divinities through ignorance and their own powerlessness to control these pathogens. Greco-Latin authors refer to numerous fungi in association with plants, classifying them as part of the plant kingdom. For over 2000 years, knowledge on pathogens was lacking, aside from the description provided by historians of their negative impacts on crops. The invention of the microscope in the seventeenth century led to the discovery of the invisible parts of fungi. From the mid-seventeenth century, Pier Antonio Micheli (1679–1737), considered the father of modern mycology, observed the spores and filaments of fungi and described a great number of microscopic species, including *Botrytis cinerea*, the causal agent of grey mould in grapevine.

The history of phytopathology is punctuated by dramatic episodes such as the historic Irish Famine of 1845–46 which killed hundreds of thousands of people and compelled one-and-a-half million individuals to emigrate to the United States following the devastation of the potato harvests by a then-unknown causal agent. It was not until 1847 that Miles Joseph Berkeley experimentally elucidated the destructive role played by *Phytophthora infestans*, the agent responsible for potato late blight which to this day remains one of the most virulent pathogens of this crop, and which is morphologically similar to downy mildew of grapevine (*Plasmopara viticola*).

1.2.2 Control of Fungal Diseases

Over the course of time, grapevine co-evolved from wild vines as these became established throughout the world and from genotypes of the *Vitis* genus, contending with obligatory pathogenic biotrophic fungi (such as the downy and powdery mildews) with highly variable levels of resistance (Morales-Cruz et al. 2021). The species *V. vinifera*, literally "the wine-bearing vine", has made its mark throughout the world, and today holds a quasi-monopoly. The number of varietals, varieties or cultivars described around the globe varies according to bibliographic source but can be reasonably estimated at 6000. Nevertheless, the varietal concentration, taking France as an example, with 20 varieties covering nearly 90% of the area under vines despite 317 varieties being permitted by the regulations, shows that only a fraction of viticultural biodiversity is being exploited (Lacombe 2012). The ubiquity of *V. vinifera* and its susceptibility to fungal diseases make cultivated vines dependent on plant protection products. To avoid the use of the latter, large-scale research projects on new, resistant grapevine varieties using other genotypes than those of *V. vinifera* was launched in the late nineteenth century (Mian et al. 2023). Numerous interspecific hybrids were created, mainly in France, by crossing *V. vinifera* with species of *Vitis* that were particularly resistant to powdery mildew. Hundreds of these varieties were obtained through crossings, with obvious success in terms of pathogen resistance but less success in terms of their oenological potential. In Switzerland, for instance, the varieties 'Seibel 1000', 'Seibel 5455' (syn. 'Plantet') and 'Oberlin 604' were planted for commercial wine production, but quickly

Table 1.2 Charges in vineyard area planted with interspecific grape varieties in France

Year	Total area [ha]	Hybrid area [ha]	%
1927	1,485,670[a]	216,197[a]	14.5
1947	1,550,000[a]	370,000[a]	23.8
1960	1,290,000	400,000	31.0
2007	835,805	6285	0.75
2017	779,500	5492	0.70

Until they were banned in the 1960s these varietals accounted for nearly one-third of the area under vine; this figure has now fallen to less than 1%
[a]Data from Mathieu (1949)

disappeared with the development of plant protection products. In France, hybrids met with great success until into the 1960s (Table 1.2) and accounted for nearly 30% of the area under vine, despite the banning in 1935 of six *V. labrusca* crossings ('Jacquez', 'Noah', 'Herbemont', 'Clinton', 'Isabelle', 'Othello') due to their foxy note and human toxicity of the high content of methanol and malvidin (= malvidol diglucoside) in the wines (De la Fuente Lloreda 2018; Tampaktsi et al. 2023). These ultimately had their AOC (*Appellation d'Origine Contrôlée* or 'Controlled Designation of Origin' label) revoked in 1953.

1.3 Over Two Centuries of Research on Fungi in Grapevine

Practically all plants are infected by pathogens. Fungi are among the most mysterious and harmful microorganisms, responsible for around 80% of known diseases to date. Fungal biodiversity is immense, with over 148,000 species and counting described so far. Mycologists estimate that the number of species worldwide ranges between 3.5 and six million (Wu et al. 2019). Able to develop on living or dead plant tissues, fungi contain specific structures allowing them to survive in a latent state until conditions are conducive to their further development. They can penetrate plant tissues or grow solely on the surface of the organs of the plant by anchoring specialised mycelial structures in its epidermis. The fungal spores which spread the diseases are essentially transported by the wind, water, and the insects or microfauna of the soil, and can be found frozen in the upper layers of the atmosphere with their germinative power intact (Gindro and Pezet 2001).

Knowledge on the fungal diseases of plants in general and grapevine in particular developed with the invention of the microscope in the seventeenth century and the advent of fungicides in the late nineteenth century. Countless scientific writings on this subject worldwide and the fact that it has occupied scientists for over two centuries highlights not only the economic importance of fungal diseases, but also the impossibility of permanently resolving plant-protection issues. Due to their specific biological properties, fungi have an extraordinary ability to adapt to their environment through regular genome mutations and through their sexual phase, which allows for genetic recombinations. *Botrytis cinerea*, for example—the causal agent of grey mould—has multinucleate unicellular conidia endowing it with great adaptive potential. Moreover, it develops sclerotia which are extremely resistant to environmental constraints, ensuring its survival in the most extreme conditions, and making it one of the most difficult pathogens to control (Walker 2016).

The regular use of specific fungicides with unisite modes of action to avoid undesirable effects on the environment has in general led to the adaptation of fungal pathogens to the point of destroying the efficacy of these products. By contrast, multisite fungicides prevent the emergence of resistant fungal isolates, but like copper cause undesirable effects on the environment. As for active substances of natural origin—the definition of which becomes contentious from the moment in which the molecules are synthesised artificially—they are partially effective, and require more frequent application. Hence, the cultivation of *Vitis vinifera*, a species that is susceptible to fungal diseases, involves the active control of fungal pathogens through the application of fungicides at regular intervals, regardless of the technical management approach or production method chosen.

Widely explored from the late nineteenth to the early twentieth century, the hybrids obtained by crossing *V. vinifera* with downy and powdery mildew-resistant Vitaceae offered new prospects for viticulture in that they did not require treatment with sulphates. The hybrids, direct-producer hybrids or direct, first-generation plants, described as "new vines" (Mathieu 1949) were controversial owing to their flavour notes, which could not compare with those of the traditional *V. vinifera* varieties. In plant-protection terms they offered clear benefits, which in the case of the most resistant hybrids could extend to not requiring fungicide treatments.

More recently, scientists in Germany pioneered the creation and breeding of fungal disease-resistant grape varieties with the obtaining of the 'Regent' variety from the crossing of 'Diana' and 'Chambourcin' in 1967 at Geilweilerhof in Rhineland-Palatinate, which, despite its monogenic resistance genes against downy mildew (Rpv 3.1) and polygenic resistance genes against powdery mildew (Ren 3, Ren 9) was classified in 1996 as *V. vinifera* (Maul and Topfer 2015). The classification of hybrids as *V. vinifera* despite the proportion of genomes alien to this species remains a contentious subject (De la Fuente Lloreda 2018). This resistant grape variety is among the most commonly planted in Germany, occupying nearly 2200 ha between 2006 and 2009. From this point onwards the area under 'Regent' vines steadily contracted to 1700 ha in 2020 when their resistance to downy mildew was overcome by *Plasmopara viticola*, necessitating nearly the same plant-protection treatments as the traditional grape varieties.

Since the beginning of the twenty-first century, selection for resistance has once more been at the top of the 'sustainable viticulture' agenda, a result of the questioning of the impact of plant-protection products on public health and the environment. Numerous institutes are working on interspecific crossings, in pursuit of grape cultivars that are as close as possible to the traditional varieties, by combining resis-

Table 1.3 Most commonly planted interspecific varieties worldwide (2017) irrespective of end use (OIV 2017)

Variety	Colour	Area (ha)	Country
Kyoho	Black	365,000	China
Concord	Black	34,000	USA
		2000	Brazil
Isabella	Black	13,000	Brazil
Couderc noir	Black	2000	Brazil
Jacquez	Black	1000	Brazil

tance genes against the fungal pathogens. Molecular biology allows for a more targeted approach in the introgression of resistance genes while avoiding transgenic approaches, but still requires agronomic evaluation under vineyard and vinification conditions to determine wine quality.

A great many hybrids have been obtained over the course of time, some of which have met with great success, like the grape variety 'Kyoho', the most commonly planted in China for the production of table grapes and occupying the largest area in global terms (Table 1.3, OIV 2017).

For wine grapes, the choice of grape variety remains difficult in a traditional context influenced by market globalisation, with several dozen varieties dominating worldwide. According to statistics from the OIV (International Organisation of Vine and Wine) (2017), the world's most commonly planted wine-grape varieties are, in descending order of area, Cabernet Sauvignon, Merlot, Tempranillo, Syrah, Grenache Noir or Garnacha Tinta and Pinot Noir for red wines, and Airen, Chardonnay, Sauvignon Blanc, Ugni Blanc or Trebbiano Toscano for white wines. Changing vineyard composition should be examined in light of fungal disease resistance being a key element in reducing plant-protection product use. In this respect, the traits of resistant grape varieties must be painstakingly described, in particular the resistance genes for each fungal disease in question as well as the specific agronomic and oenological features of the varieties. Resistance levels can vary considerably between different varieties and may be polygenic or monogenic in nature. Monogenic resistances are more easily overcome by pathogenic fungi. This information is crucial for a vineyard's management approach and dictates the necessary control strategies. Put differently, at a global level, vineyards will probably never be able to completely dispense with the use of plant protection products. Sadly, all pathogen self-regulation approaches based on environmental and cultivation conditions that are favourable for the pathogen's antagonists or which contribute to the improved physiological equilibrium of the vine and which aim to dispense completely with the use of fungicides have so far failed (Pertot et al. 2017). The control of pathogenic fungi in crops in general and in grapevine in particular offers worthwhile alternatives which nevertheless have their limits when climatic conditions are favourable for harmful microorganisms.

References

Chaptal R, Parmentier D (1806) Abrégé du traité théorique et pratique sur la culture de la vigne. JL Roard, Paris, p 283

De la Fuente Lloreda M (2018) Use of hybrids in viticulture. A challenge for the OIV. OENO One 52(3):231–234. https://doi.org/10.20870/oeno-one.2018.52.3.2312

Dong Y, Duan S, Xia Q et al (2023) Dual domestication and origin of traits in grapevine evolution. Science 379:892–901. https://doi.org/10.1126/science.add8655

Gindro K, Pezet R (2001) Effect of long-term storage at different temperatures on conidia of *Botrytis cinerea* Pers.: Fr. FEMS Microbiol Lett 204:101–104

Lacombe T (2012) Contribution l'étude de l'histoire évolutive de la vigne cultivée (*Vitis vinifera* L.) par l'analyse de la diversité génétique neutre et de gènes d'intérêt. Thèse de doctorat, SupAgro Montpellier, p 328

Linder C, Kehrli P, Viret O (2016) La Vigne: vol. 2, ravageurs et auxiliaires (Ed. AMTRA, Nyon, Suisse), p 394

Mathieu J (1949) La vigne nouvelle, intérêt et importance des hybrides producteurs en face de la vigne française. Imprimerie Brunel et Cie Montpellier, p 157

Maul E, Topfer R (2015) Vitis International Variety Catalogue (VIVC) A cultivar database referenced by genetic profiles and morphology. BIO Web Conf 5:01009

Mian G, Nassivera F, Sillani S, Iseppi L (2023) Grapevine resistant cultivars: a story review and the importance on the related wine consumption inclination. Sustain For 15:390

Morales-Cruz A, Aguirre-Liguori JA, Zhou Y, Minio A, Riaz S, Walker AM, Cantu D, Gaut BS (2021) Introgression among North American wild grapes (*Vitis*) fuels biotic and abiotic adaptation. Genome Biol 22:254. https://doi.org/10.1186/s13059-021-02467-z

OIV (2017) Distribution of the world's grapevine varieties. Focus OIV 2017:55

Pertot I, Caffi T, Rossi V, Mugnai L, Hoffmann C, Grando MS, Gary C, Lafond D, Duso C, Thiery D, Mazzoni V, Anfora G (2017) A critical review of plant protection tools for reducing pesticide use on grapevine and new perspectives for the implementation of IPM in viticulture. Crop Prot 97:70–84. https://doi.org/10.1016/j.cropro.2016.11.025

Reymondin P (1798) L'art du vigneron (Lausanne), p 406

Tampaktsi C, Gancel AL, Escudier J-L, Samson A, Ojeda H, Pic L, Rousseau J, Gauthier P, Viguier D, Furet MI, Teissedre P-L (2023) Phenolic potential of new red hybrid grape varieties to produce quality wines and identification by the malvin. BIO Web Conf 56:02012. 43rd World Congress of Vine and Wine. https://doi.org/10.1051/bioconf/20235602012

Töpfer R, Trapp OA (2022) A cool climate perspective on grapevine breeding: climate change and sustainability are driving forces for changing varieties in a traditional market. Theor Appl Genet 135:3947–3960

Viala P (1885) Les maladies de la vigne (*Peronospora*, oïdium, anthracnose, pourridié, cottis, *Cladosporium*, etc.). In: Bibliothèque du Progrès Agricole et Viticole, Montpellier, Paris, p 240

Walker AS (2016) Diversity within and between species of *Botrytis*. In: Fillinger S, Elad Y (eds) *Botrytis*—the fungus, the pathogen and its management in agricultural systems. Springer, Berlin, pp 91–125

Wu B, Hussain M, Zhang W, Stadler M, Liu X, Xiang M (2019) Current insights into fungal species diversity and perspective on naming the environmental DNA sequences of fungi. Mycology 10(3):127–140. https://doi.org/10.1080/21501203.2019.1614106

Zufferey V, Verdenal T, Gindro K, Murisier F, Viret O (2022) La Vigne: vol. 4, Anatomie et physiologie, alimentation et carences, accidents physiologiques et climatiques. Ed. AMTRA, Nyon, p 564

Chapter 2
Grapevine

2.1 Systematics of the Genus *Vitis*

The Vitaceae are a family of around 950 species belonging to 16 different genera (Ma et al. 2021), including creepers, found in tropical and temperate zones (Süssengut 1953; Wen 2007). The Vitaceae, whose oldest recorded fossils thus far date from the Late Cretaceous period (66 Mya) and are relatively common in the Tertiary period (65–2.5 Mya) (Manchester et al. 2013; Rozefelds and Pace 2018), are known for the production of table grapes, raisins and wine grapes, and are one of the most important crops in economic terms (Gerrath et al. 2015). Domestication of the grapevine occurred around 11,000 years ago in Western Asia and the Caucasus, enabling the production of grapes for eating or winemaking. The domestic species of Western Asia spread throughout Europe with the first farmers, introgressed into ancient wild Western ecotypes and subsequently diversified along human migratory paths to give rise to Muscat and the unique wine grapes of the West in the Late Neolithic (Dong et al. 2023). According to recent studies by Wen (2018), the Vitaceae can be classified into five tribes corresponding to the five main clades identified in the phylogenetic studies of the family (Zhang et al. 2015; Lu et al. 2018), defined by specific morphological criteria (type of inflorescence, thickness of floral disc, position and structure of the ovary or structure of the seeds). These are as follows:

- **Ampelopsidae** J. Wen and Z.L. Nie, containing the genera *Ampelopsis* Michx, *Nekemias* Raf., *Rhocissus* Planch. and *Clematicissus* Planch.
- **Cissae** Rchb., containing the genus *Cissus* L.
- **Cayratieae** J. Wen and L. M. Lu, with the genera *Cayratia* Juss., *Cyphostemma* (Planch.) Alston, *Causonis* Raf., *Pseudocayratia* J. Wen, L.M. Lu and Z.D. Chen, *Acareosperma* Gagnep., '*Afrocayratia*' and *Tetrastigma* (Miq.) Planch.
- **Parthenocissae** J. Wen and Z.D. Chen, with the genera *Parthenocissus* Planch. and *Yua* C. L. Li.
- **Viteae** Dumort., containing the genera *Vitis* L., *Ampelocissus* Planch. including *Nothocissus* and *Pterisanthes*.

O. Viret, K. Gindro, *Science of Fungi in Grapevine*,
https://doi.org/10.1007/978-3-031-68663-4_2

The genus *Vitis* is subdivided into two sub-genera, viz., the sub-genus *Vitis* Planch. (2n = 38) (anc. *Euvitis*) containing 82 described species, and the sub-genus *Muscadinia* (Planch.) Rheder (2n = 40) with 3 described species, including *Vitis rotundifolia* Michaux, 'Muscadine', also called 'Soco' (Table 2.1). These two sub-genera are thought to have diverged around 18 million years ago (Wan et al. 2013). The sub-genus *Muscadinia* is found in the warm and temperate zones of the south-eastern United States, in northeastern Mexico, and in Belize, Guatemala and the Caribbean, whilst the sub-genus *Vitis* (Wan et al. 2013, 2018) is spread throughout the globe, including northern South America, North and Central America, Europe and Asia (Planchon 1887; Galet 1988; Zecca et al. 2012). The group originating in Europe and the Near East includes only the species *V. vinifera*, subdivided into two subspecies encompassing the cultivated varieties (subsp. *vinifera*) and wild grape-vines (subsp. *sylvestris*). The group originating in North and Central America encompasses over 24 species, including *V. riparia*, *V. rupestris* and *V. aestivalis*, as well as *V. labrusca* L and *V. rotundifolia*.

With few exceptions, the grape varieties cultivated for winemaking or eating belong to the only European species, *Vitis vinifera* subsp. *sativa* L., domesticated from the wild parent *V. vinifera* subsp. *silvestris* around 6000–8000 years ago in the Near East (Myles et al. 2011; McGovern 2019). Nevertheless, *V. vinifera* is just one of numerous *Vitis* species known across the globe.

The systematics of the group of Asian *Vitis* species have evolved a great deal over the last few years with the recent discovery of numerous species. Nearly 30 species are native to a broad swathe of East Asia, China, Japan and Java (Wan et al. 2013) covering wide geographic areas with a subtropical-to-continental climate. The best-known representatives are, among others, *V. coignetiae*, *V. quinquangularis*, or *V. amurensis*, whose natural range (China, southern Siberia) explains their high tolerance of winter cold. Besides the European grapevine *V. vinifera* ssp. *vinifera*, which includes the bulk of the cultivated varieties renowned for their quality, certain winemaking or eating varieties of grapes were directly bred from other species such as *V. labrusca* L. and *V. rotundifolia*. These varieties generally exhibit very particular aromatic qualities described as "foxed", associated with the synthesis of certain molecules such as methyl anthranilate in *V. labrusca* (Galet 1988), an ester of anthranilic acid often used in perfumery for its notes of wild strawberry or fruits of the forest, which is found in wine and is on the whole not popular with consumers.

Over the years, grapevine cultivation has brought about a significant increase in genetic diversity through sexual reproduction. This high number of varietals or grape varieties is the result of various processes such as the domestication of wild vines (*Vitis vinifera* subsp. *silvestris*) from a number of different sites, controlled crosses between domesticated and local wild vines, the ancient practice of cultivating plants produced by spontaneous crossings and, to a lesser extent over the past century, controlled breeding. Interspecies hybridisation, i.e. between the species of the genus *Vitis*, widely developed from the end of the nineteenth century to deal with the phytosanitary crisis (phylloxera, downy and powdery mildew), led to the creation of direct-producer hybrids (DPHs) and rootstocks, thus contributing to a greater diversification of plant material. Biodiversity also increased due to natural 'somatic'

Table 2.1 Alphabetical list of the geographical distribution of the validated species (https://powo. science.kew.org/ Royal Botanic Garden, Kew) of the genus *Vitis*

Species	Species
Vitis acerifolia Raf.	*Vitis luochengensis* W.T. Wang
Vitis aestivalis Michx.	*Vitis menghaiensis* C.L. Li
Vitis amurensis Rupr.	*Vitis mengziensis* C.L. Li
Vitis arizonica Engelm.	*Vitis metziana* Miq.
Vitis baihuashanensis M.S. Kang & D.Z. Lu	*Vitis monticola* Buckley
Vitis balansana Planch.	*Vitis mustangensis* Buckley
Vitis bashanica P.C. He	*Vitis nesbittiana* Comeaux
Vitis bellula (Redher) W.T. Wang	*Vitis x novae-angliae* Fernald
Vitis berlandieri Planch.	*Vitis novogranatensis* Moldenke
Vitis betulifolia Diels & Gilg	*Vitis nuristanica* Vassilcz.
Vitis biformis Rose	*Vitis palmata* Vahl
Vitis blancoi Munson	*Vitis pedicellata* M.A. Lawson
Vitis bloodworthiana Comeaux	*Vitis peninsularis* M.E. Jones
Vitis bourgaeana Planch.	*Vitis piasezkii* Maxim.
Vitis bryoniifolia Bunge	*Vitis pilosonervia* F.P. Metcalf
Vitis californica Benth.	*Vitis popenoei* J.L. Fennell
Vitis x champinii Planch.	*Vitis pseudoreticulata* W.T. Wang
Vitis chunganensis Hu	*Vitis qinlingensis* P.C. He
Vitis chungii F.P. Metcalf	*Vitis retordii* Rom. Caill. ex Planch.
Vitis cinerea (Engelm.) Engelm. Ex Millardet	*Vitis riparia* Michx.
Vitis coignetiae Pulliat ex Planch.	*Vitis romanetii* Rom. Caill.
Vitis davidiana (Carrière) N.E.Br	*Vitis rotundifolia* Michx.
Vitis davidii (Rom. Caill.) Föex	*Vitis rupestris* Scheele
Vitis x doaniana Munson ex Viala	*Vitis ruyuanensis* C.L. Li
Vitis erythrophylla W.T. Wang	*Vitis saccharifera* Makino
Vitis fengqinensis C.L. Li	*Vitis shenxiensis* C.L. Li
Vitis ficifolia Bunge	*Vitis shuttleworthii* House
Vitis flavicosta Mickel & Beitel	*Vitis silvestrii* Pamp.
Vitis flexuosa Thunb.	*Vitis sinocinerea* W.T. Wang
Vitis giradiana Munson	*Vitis sinoternata* W.T. Wang
Vitis hancockii Hance	*Vitis tiliifolia* Humb. & Bonpl. ex Schult.
Vitis heyneana Schult.	*Vitis tsoi* Merr.
Vitis hissarica Vassilcz.	*Vitis vinifera* L.
Vitis hui W.C. Cheng	*Vitis vulpina* L.
Vitis jaegeriana Comeaux	*Vitis wenchowensis* C. Ling
Vitis jinggangensis W.T. Wang	Vitis wenxianensis W.T. Wang
Vitis kiusiana Momiy.	*Vitis wilsoniae* H.J. Veitch
Vitis jacquemontii R. Parker	*Vitis wuhanensis* C.L. Li
Vitis labrusca L.	*Vitis xunyanensis* P.C. He
Vitis lanceolatifoliosa C.L. Li	*Vitis yunnanensis* C.L. Li
Vitis longquanensis P.L. Chiu	*Vitis Zhejiang-adstricta* P.L. Chiu

Blue: American continent; green: Asia; yellow: Europe

genetic mutations, a common phenomenon in grapevines and fixed in a temporarily stable manner through vegetative multiplication. This type of multiplication uses cutting, layering or grafting to obtain copies that are genetically identical to the original parent, known as clones, and remains true only if no somatic mutation occurs in the undifferentiated meristematic cells. Today, the genus *Vitis* boasts a rich genetic diversity of several thousand grape varieties. The international catalogue of vine grape varieties identifies 21,045 different names, 12,250 of which are for *V. vinifera* alone. This figure should be taken with a pinch of salt, however, as it includes a substantial number of synonyms and homonyms. The true number of recognised *V. vinifera* grapevine varieties is estimated at over 8000 (data obtained from information provided by the International Organisation of Vine and Wine, www.oiv.int).

Somatic Mutations
(Torregrosa et al. 2019)

Grapevines are generally propagated by cuttings or grafting, which produces genetically identical individuals for a given clone. Nevertheless, spontaneous or 'somatic' mutations may occur. These consist of a local modification of the nucleotide sequence of the grapevine genome due to a change, an insertion or a deletion of one or more nucleotides. This process of mutation is natural and is one of the main reasons for the evolution of species which produces genetic diversity and serves as a basis for natural or directed selection. Somatic mutations appear in a somatic cell, i.e. a cell that is not directly involved in the sexual reproduction of the plant. When a mutation occurs in a cell it can be transmitted to the offspring. These mutations can then be propagated and preserved through vegetative multiplication (cuttings or grafting).

The consequences of genetic mutations in the grapevine can vary greatly. Some mutations such as the colour of the berries (Pinot Noir, Pinot Blanc and Pinot Gris) or leaf shape have a visible effect on phenotype, while 'silent' mutations do not modify the appearance of the plant. It is sometimes difficult to determine visually whether a particular phenotype (for example, leaf deformation) is caused by an infectious agent, or is the result of a somatic mutation. To differentiate between these two cases, graft transmission tests must be conducted. Unlike modifications caused by somatic mutations, those due to viral infection are transmissible by grafting. In some cases only a portion of the cells carry the mutation. We then speak of a chimera. This is the case for Pinot Meunier, a grape variety composed of two types of cells. Some of these cells are identical to Pinot Noir cells, while the outer cells of the hairy epidermis are different owing to a somatic mutation. This example, which has been propagated for centuries, highlights the stability of some of these mutations. Certain *in vitro* culture techniques enable a grapevine plant to be reconstituted from an initial cell. The two types of Pinot Meunier cells give rise to whole plants with very different phenotypes. One cell population will yield grapevine plants that are morphologically identical to the Pinot Noir variety, while the other will produce a Pinot Noir mutant called a 'microvine' (Rienth et al. 2016). A microvine is thus a Pinot Noir variety with a mutation in the gene controlling the vine's response to a phytohormone, gibberellic acid. In this instance, the mutation concerns a change in a single nucleotide that affects the active site of the protein, rendering it non-functional (Boss and Thomas 2002). In addition to the significant hairiness of the leaves, this mutation is responsible for a reduction in the vegetative organs (leaves, internodes) and the transformation of the tendrils into inflorescences, which leads to a reduction in the size of the plant and the continuous production of grapes.

In vitro culture represents a stress for the grapevine that can potentially increase the frequency of somatic mutations. Consequently, the use of this technique can lead in certain situations to morphological and physiological modifications in the plants. This phenomenon results in somaclonal variations that can be researched to create genetic variability within plant breeding and improvement programmes.

2.2 Development of the Vine in Its Environment and Phenology

Throughout its history, the grapevine has had to adapt to evolving environmental conditions. The numerous interactions between the environment and biological processes determine its development and the composition of its products. The great genetic diversity present in the genus *Vitis* and in the species *V. vinifera* in particular have led to its cultivation in most of the world's countries, in very different climates ranging from zones of extreme drought to others of great humidity and heat, or, conversely, extreme cold in winter (Fig. 2.1).

Grapevines are also capable of developing in different types of soils—sandy or clay soil, deep or shallow, acidic or highly alkaline, with all possible combinations in between. With the near-ubiquitous arrival of phylloxera at the end of the nineteenth century, their genetic component became even more complex with the grafting of *V. vinifera* varieties onto other *Vitis* species. Humans have always had to use their ingenuity to develop growing techniques adapted to the different environmental conditions. At first they relied on empirical knowledge essentially acquired by observation and transmitted from generation to generation. The great variety of vine training systems that exist throughout the world perfectly illustrates this knowledge and ability to adapt to different natural conditions. One might mention as an example two extremes: vine cultivation on the island of Lanzarote in the Canaries in pits dug in the volcanic ash, where the shoots are left on the ground to protect them from the wind and to capture the low atmospheric humidity (Fig. 2.2); and at the other end of the spectrum, the ridging or hilling of the vines with earth in winter, or the protection of the vines with duct tape equipped with heating wires to counteract the extreme winter cold in Quebec and other Canadian provinces.

Fig. 2.1 Manual grapevine
cultivation in foothill-to-
montane altitude zones in
northern climates with cold
winters

Fig. 2.2 Grapevine cultivation on the island of Lanzarote (Canaries), with its particularly arid
climate (**a**), in pits dug in the volcanic ash. Protected from the wind by low stone walls (**b**)

Increasingly refined scientific knowledge on the ecophysiology of the vine and environmental factors has been steadily acquired and provides producers with invaluable information for deciding on the most appropriate technical approaches. The new findings on the structure and anatomy of its different organs, from the root to the berry, via the trunk, shoots, buds, leaves and inflorescences, help explain certain functional disorders. The progress made on the mechanisms and factors governing photosynthesis, respiration, transpiration and gaseous exchanges in general help guide the choice of growing techniques in terms of the training and feeding. The findings on the metabolism, transport, migration and storage of assimilation products are invaluable for explaining functional deficiencies and suggesting the means of remedying them. Like all perennial plants, grapevines are more complex in their behaviour than annual plants. In addition to guaranteeing balanced growth and optimal development of the fruit in productive and qualitative terms, they must ensure the storage in the roots and wood of reserves that will be mobilised the following year, and even over the longer term.

The role of these reserves—carbon and nitrogen reserves in particular—is vital for the functioning of the grapevine (Zufferey et al. 2015). Too heavy a load of grapes can lead to a reduction in carbohydrates, especially starch, stored in the perennial parts of the plant. Among others, the carbohydrates accumulated in the roots will furnish the energy necessary for the enzymatic reactions responsible for reducing iron via the acidification of the environment, thus allowing the absorption and transport of iron in the plant to reduce the risks of chlorosis (Fig. 2.3).

Research involving the stable isotope of nitrogen (^{15}N) has brought to light the accumulation, in the roots and old wood, of nitrogen reserves that are used in the spring, at the outset of vine growth. Much like the vascular system in the animal kingdom, plants possess a network of vessels enabling continuous flow exchanges via the xylem and phloem ducts, between the aerial and underground parts, forming the basis of growth and the annual cycle of development.

Fig. 2.3 Yellowing of leaves due to an iron deficiency (chlorosis) associated with an overload of grapes the previous year

As the level of knowledge advances, the complexity of the plant world is revealed. Detailed analysis of vine physiology highlights the fact that one factor is often not enough on its own to explain a malfunction. In most cases there are several constantly interacting physiological, cultural and environmental causes at work. It also applies to water supply and the grapevine's response to drought, to carbon and mineral nutrition, to physiological and weather-related events. For example, a mineral deficiency often cannot be explained merely by the lack of this element in the soil or the plant. Blockage, antagonism or synergism phenomena and disorders associated with the absorption and transport of the elements may be at the root of the symptoms observed in the plant. From bud burst in the spring to bunch maturity in the autumn, the vine develops in successive stages. This annual cycle—from the first leaf to lignification of the shoots—is wholly governed by the weather conditions of the current year, and in particular by temperature and precipitation. Baggiolini (1952) defined and illustrated 16 phenological stages for the vine. The subsequently developed Eichhorn code comprising 22 stages from 01 to 47 (Lorenz et al. 1994) was included in a decimal code (from 00 to 100) in use since 1990; known by the name of the BBCH scale (an acronym formed from the initial letters of the coordinating institutions the Federal Biological Institute (*Biologische Bundesanstalt*), the Federal Variety Office (*Bundessortenamt*) and the **Ch**emical industry), this code is applied worldwide to all cultivated plants (Lancashire et al. 1991; Hack et al. 1992). In this system, the main developmental stages of plants are listed in ten points from 0 to 9. In the case of the grapevine, stages 2, 3 and 4, which are specific to monocotyledons, do not exist:

0   Bud burst, bud development
1   Leaf development
5   Inflorescence development
6   Flowering
7   Fruit development
8   Maturation of berries
9   Senescence and beginning of dormancy

The relationship between the BBCH scale and the Baggiolini scale, which is still very widespread in practice, is given in Fig. 2.4.

The climate is warming (Fig. 2.5), and according to predictions on a planetary scale, warming of the order of 2–5 °C could occur in the coming decades, depending on the measures for limiting greenhouse gases taken by all the world's countries. In the temperate regions a moderate and steady increase in temperatures will initially produce positive effects, whilst the phenomenon is undoubtedly more problematic in the zones where temperatures are already high.

Through the monitoring of its phenological stages, the grapevine is a good marker of climate change, and of temperatures in particular. Over the past 60 years, the dates of bud burst, flowering, veraison (berry ripening) and the start of harvests have advanced in numerous wine-growing regions by several days or even several weeks on average (Duchêne and Schneider 2005; van Leeuwen et al. 2019). This is a general trend, although seasonal or interannual variations remain considerable, especially in temperate zones. The growing season has generally become shorter (Table 2.2).

Fig. 2.4 Phenological stages of the vine according to the BBCH code (in black) and the equivalent stages of Baggiolini (1952) (in red)

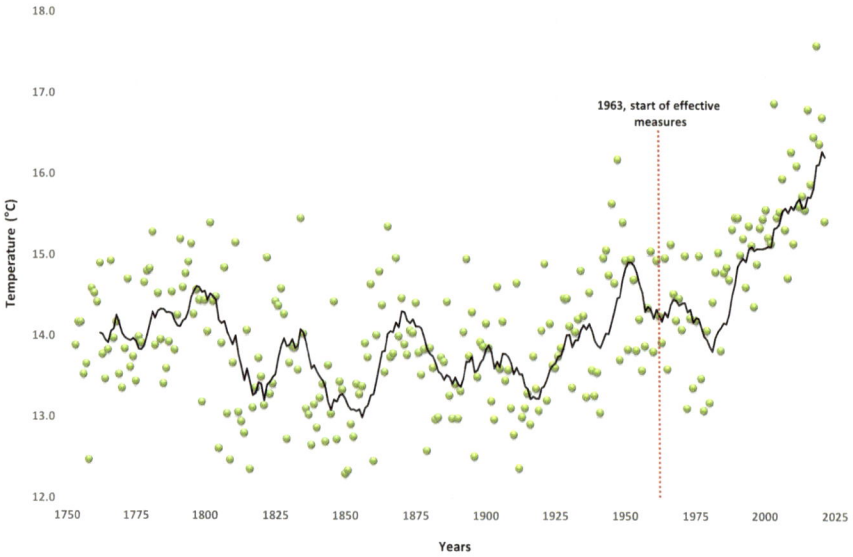

Fig. 2.5 Average annual changes in temperature from 1 April to 31 October in Changins, Canton of Vaud, Switzerland, from 1753 to 2023. Data extrapolated from the Geneva Observatory and Geneva Airport stations until 1963, the start of effective measures. The solid line represents the moving average over 10 years

Table 2.2 Monitoring the phenological stages of *Vitis vinifera* cv. Chasselas from 1925 to 2021 on the Pully site (Lake Geneva basin, Canton of Vaud, Switzerland)

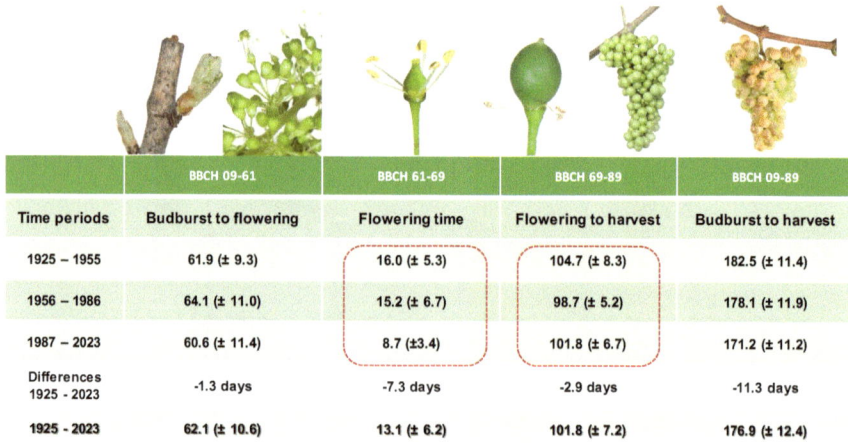

	BBCH 09-61	BBCH 61-69	BBCH 69-89	BBCH 09-89
Time periods	**Budburst to flowering**	**Flowering time**	**Flowering to harvest**	**Budburst to harvest**
1925 – 1955	61.9 (± 9.3)	16.0 (± 5.3)	104.7 (± 8.3)	182.5 (± 11.4)
1956 – 1986	64.1 (± 11.0)	15.2 (± 6.7)	98.7 (± 5.2)	178.1 (± 11.9)
1987 – 2023	60.6 (± 11.4)	8.7 (±3.4)	101.8 (± 6.7)	171.2 (± 11.2)
Differences 1925 - 2023	-1.3 days	-7.3 days	-2.9 days	-11.3 days
1925 - 2023	62.1 (± 10.6)	13.1 (± 6.2)	101.8 (± 7.2)	176.9 (± 12.4)

Average duration and standard deviations in days from bud burst (BBCH 09) to flowering (BBCH 61), of the flowering period (BBCH 61–69), from flowering to harvest (BBCH 69–89) and of the whole of the growing season (BBCH 09–89) for 30-year periods (1925–1955, 1956–1986 and 1987–2021). Over 85 years, climate change has caused a shortening of the Chasselas growing season of the order of 13 days. The length of the period between the end of flowering and harvest remains of the order of 100 days

The early onset of budburst can expose the grapevine to greater risk of frost. In warm zones, the breaking of dormancy could be disrupted if winter temperatures remain too high. Exposure of the grapes to excessively high temperatures can modify their composition and increase the risk of scorching. The wine alcohol content tends to increase and acidity tends to decrease. Climate change not only concerns temperatures but also the amount and distribution of precipitation, with forecasts varying significantly from region to region. Aggravated problems of drought or excessive rainfall could arise. The specificity of a wine produced in a given terroir is often based on a subtle balance between the variety, soil, climate and human know-how. A major change in climatic conditions can upset this balance and call into question the historical value of certain terroirs. Grapevine cultivation throughout the world has kept on evolving, as shown e.g. by the establishment of vineyards in new zones, with the installation of irrigation where there is not sufficient rainfall. This evolution will continue into the future, in the expectation that global warming will be contained so as to allow changes to be made gradually.

The grapevine has a great capacity for adapting to changes in the climate, first and foremost in terms of plant material. There are differences in earliness of up to 40–50 days between the cultivated varieties. Global planting focuses on a limited number of so-called international varieties. According to the International Organisation of Vine and Wine, just 13 grape varieties occupy one-third of the world's vineyard surface area, and 33 account for half of this surface area (OIV 2017). The recovery of ancestral grape varieties and the creation of new ones can be of great interest in terms of adaptation potential. The rootstocks are also capable of influencing the length of the growth cycle, but their effects remain limited to several days. Some of them have drought-tolerance levels enabling the vine to better withstand water shortages, but they are not usable in all pedoclimatic situations. Breeding new rootstocks appears necessary in order to create cultivars that are better at regulating transpiration and hydraulic conductivity, but this long-term task is plagued with numerous uncertainties (Cukierman et al. 2021). In addition to the suitability of the rootstock itself, the process must take into account the interactions between the rootstock and the graft. Consequently, few research institutes are interested in this approach. Water management and measures for preventing climate extremes will doubtless be a priority for many of the world's vineyards that are highly exposed to drought and heat waves.

One of the ways to adapt to climate change—in particular to rising temperatures—is to change the siting of vineyards by opting to establish vines at more suitable latitudes and/or altitudes or with less favourable exposures. Such a trend is already emerging, especially in northern European countries. The average temperature decreases by around 0.6 °C for each increasing degree of latitude, or for a 100 m increase in altitude. These are average worldwide values which do not take account of local meso- and microclimatic variations or the fact that temperature-change forecasts are not uniform and vary according to region. Bioclimatic indices can provide indications for choosing the most suitable zones for vine cultivation and the production of high-quality grapes, but they still need to be fine-tuned by incorporating current physiological knowledge (Zufferey et al. 2022).

2.3   Structure and Anatomy

The grapevine is a very robust perennial deciduous creeper. The wild grapevine, *Vitis sylvestris L.*—forebear of the cultivated grapevine *Vitis vinifera* L.—is established in forests or groves. Thanks to its tendrils and stem it can grow on trees, using their trunk as a stake and deploying its foliage at the height of the tree canopy. *V. sylvestris* is polygamodioecious, meaning that it has unisexual flowers concentrated on different vines: The male flowers with longer stamens and aborted ovaries are isolated on specific vines and the female flowers, through almost-complete stamen abortion, are grouped on other vines (Fig. 2.6). Nevertheless, grapevines in the wild have become a rarity, owing to attacks by different pests and pathogens as well as the near-systematic elimination of creepers during forest maintenance. Different species of domesticated vine that belong to the genus *Vitis* and are native to America, Asia or Europe are cultivated throughout the world. The species *Vitis vinifera*, also called the European grapevine, is the most frequently planted. The cultivated grapevine is regularly pruned and clipped, thereby losing its creeping habit. In this state, the vine exhibits the classic organisation of a woody plant, comprising vegetative organs such

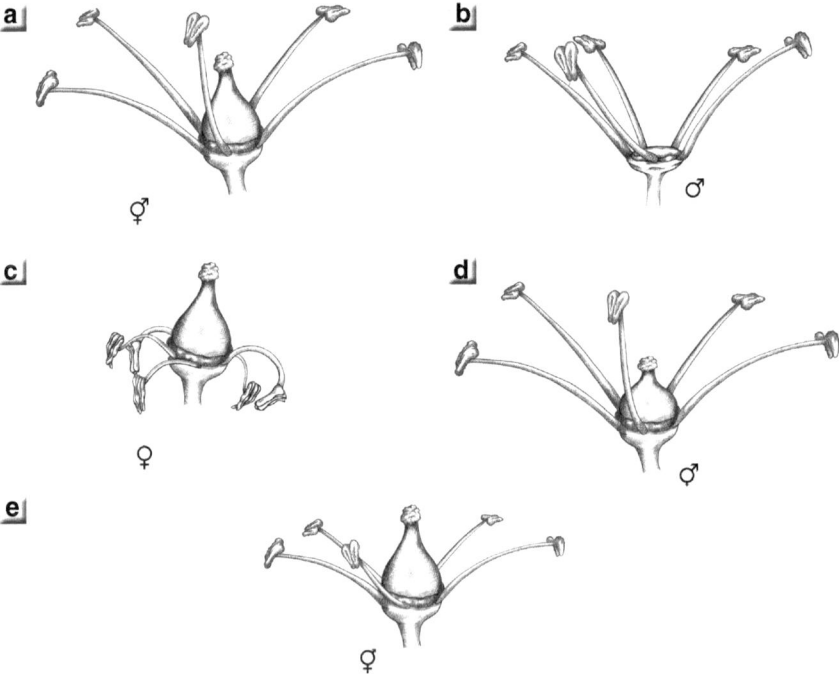

Fig. 2.6 Different flower morphotypes within the Vitaceae family. (**a**) Hermaphrodite flower of *V. vinifera*. (**b**) Strictly male flower. (**c**) Female flower resulting from the almost complete abortion of the stamens. (**d**) and (**e**), intermediate forms of which either the stamens (**d**) or the pistil (**e**) are more developed. (Illustration © Virginie Duquette, Un Monde d'illustrations, Switzerland)

as roots, trunk, stalks, leaves and tendrils, as well as reproductive organs in the form of flowers and fruits (berries) arranged in bunches. A vinestock is distinguished by an underground part (roots) and an aerial part (trunk and shoots forming the vine). Although the basic shape of a vine is predefined, its general architecture may vary according to how it is trained and pruned. Because of the phylloxera crisis affecting Europe in the nineteenth century, European varieties were grafted onto American rootstock resistant to the root-dwelling form of the aphid responsible for phylloxera (*Daktulosphaira vitifoliae*). In terms of morphology, the aerial part of grafted vines differs genetically from the roots. The graft point is marked by the formation of a more-or-less apparent callus which gradually disappears as the vines age.

This chapter describes the anatomy and morphology of the different organs of the cultivated vine in broad brushstrokes without considering their functions, which are addressed in detail in the following chapters. Knowledge of anatomy and morphology helps us better understand how water and minerals are absorbed by the roots and conveyed to all the aerial organs of the grapevine, as well as how fungi interact with the various parts of the vine at the microscopic level.

2.3.1 Root System

The root system is the plant-soil interface. It ensures physical anchoring in the soil and the absorption of water and minerals by the plant. The roots also serve as a storage structure of e.g. carbohydrates in the form of starch and of minerals or phytohormones (cytokinins, etc.). In the case of the cultivated grapevine, root formation generally starts from the heel of the woody cuttings of the rootstock (Fig. 2.7), via differentiation of the cambium.

These adventitious roots are the main roots of the vinestock, from which secondary, tertiary, etc. lateral roots will form to create a complex root system, not particularly dense but often very extensive and deep, depending on the nature of the soil and the availability of water and mineral salts in it. Seldom exceeding a diameter of 3–4 cm and developing both horizontally and vertically in the soil, the woody roots serve to anchor the vine to the soil and to transport and stock nutrients. In spring, at the start of the growing season, fine roots (rootlets) continually differentiate from the main roots, enabling the absorption of water and mineral salts (Fig. 2.8).

2.3.1.1 Rootlets

A rootlet is distinguished by the presence of an apical root cap—protective tissue facilitating soil penetration—followed immediately by the rootlet itself with a zone of absorbent hairs arising from the differentiation of certain cells of the piliferous layer (peripheral epidermal cells) (Fig. 2.9). Due to their very thin, hydrophilic walls, these absorbent hairs are the main sites for the absorption of water and mineral salts. In grapevines, they are distributed quite evenly and form a continuous

Fig. 2.7 Root system of a woody cutting of a grapevine. (**a**) General view. (**b**) Detail

Fig. 2.8 Root system of a young grafted seedling (**a**). (**b**) Detail with long woody roots and fine rootlets (arrow) for absorbing water and minerals

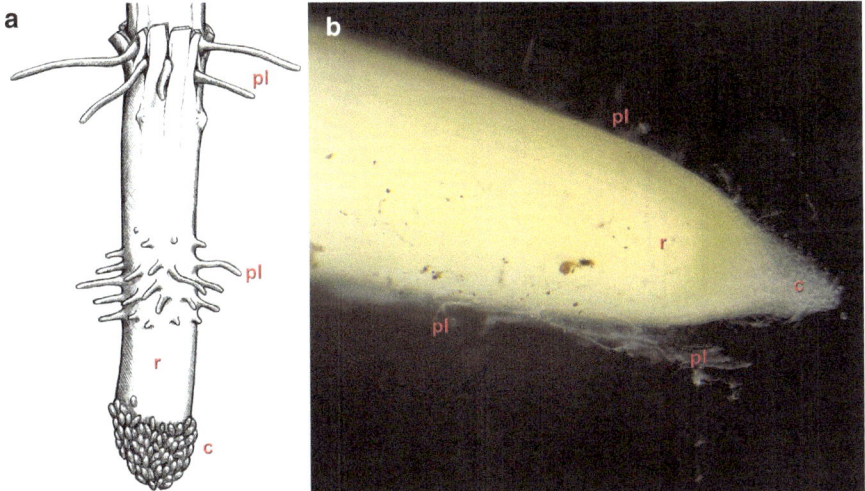

Fig. 2.9 Tip of a rootlet (r) with cap (c) as well as numerous very fine absorbent hairs forming the piliferous layer (pl)

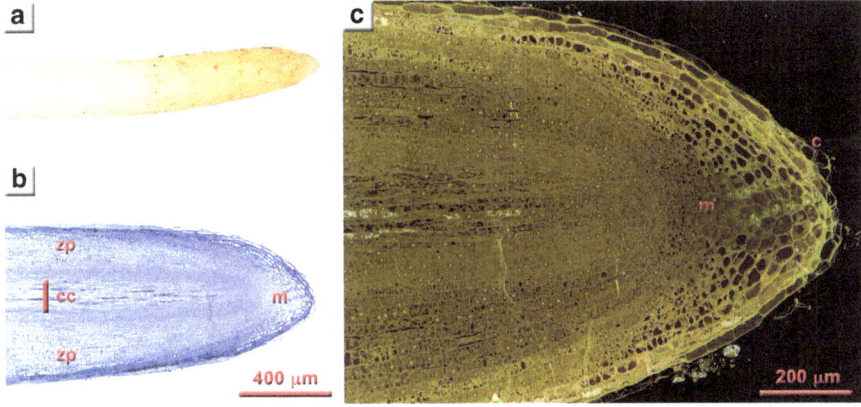

Fig. 2.10 Anatomy of a vine rootlet. (**a**) General view. (**b**) Longitudinal section of the tip of a rootlet, showing the arrangement of the conducting tissues (*cc* central cylinder, *m* meristem) and peripheral cell layers called the cortex or peripheral zone (zp). (**c**) Structure of the root cap (c) with the meristem (m)

piliferous zone. The absorbent hairs regenerate constantly over the course of the season. Viewed in longitudinal section, following the root cap, we can distinguish a small cushion of cells—the primary meristem—enabling the production of the root cap in one direction and the genesis of all the primary tissues of the rootlet in the other direction (Fig. 2.10).

In cross section (Fig. 2.11), the rootlets exhibit two distinct zones—a peripheral cell cortex consisting of rigid cells and limited on the outside by a layer of

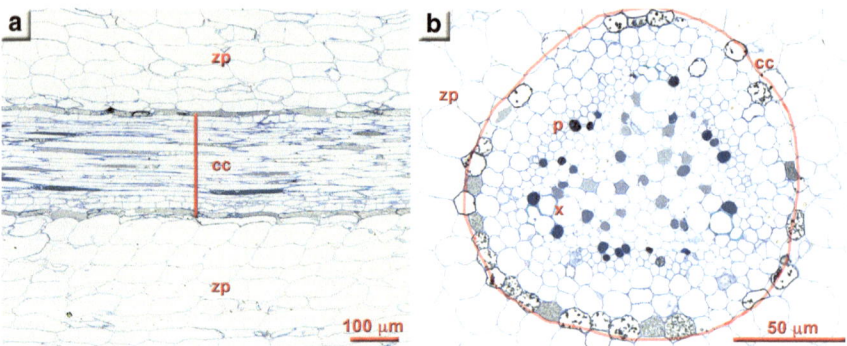

Fig. 2.11 Anatomy of a vine rootlet in longitudinal and cross section. (**a**) Longitudinal section showing the central cylinder (cc) consisting of conducting tissues (in red) and cells of the peripheral zone (zp). (**b**) Conducting tissues not differentiated into steles, comprising xylem (x) and phloem (p) vessels

suberised cells, and a central zone, called the central cylinder, bounded on the outside by a layer of rigid cells called the endoderm. Following the endoderm are two to three layers of support cells of variable size, the pericycle, which is followed in turn by conducting tissue, i.e. the xylem conducting the raw sap (water and mineral salts), and the phloem conducting the processed sap (the products of photosynthesis). In the rootlet, the conducting tissues are arranged in clumps surrounding a regularly shaped core of cells forming the medullary parenchyma or pith. All rootlet structures are transient primary structures which will be supplemented more-or-less quickly by secondary structures during the root lignification process after 3–4 weeks.

2.3.1.2 Sclerified or Lignified Roots

The structure of the lignified roots differs from that of the rootlets. On the outside, several layers of suberised cells create a protective envelope for the root (Fig. 2.12). A very thick brown cap covers them at their apex. The primary meristem gives way to secondary meristem which serves to establish secondary tissues. The first secondary meristem is the cambium, consisting of several layers of more-or-less elongated generating cells and located between the primary xylem (wood) and the primary phloem (soft bast) and enabling the formation of centripetal secondary xylem, called secondary wood (towards the inside), as well as centrifugal secondary phloem, called hard bast (towards the outside), accompanied by a number of bast fibres (bast support cells) arranged in an irregular pattern. The second secondary meristem is the subero-phellodermal layer, forming phelloderm on the inside and suber on the outside. The suber (or cork) is composed of rectangular dead cells

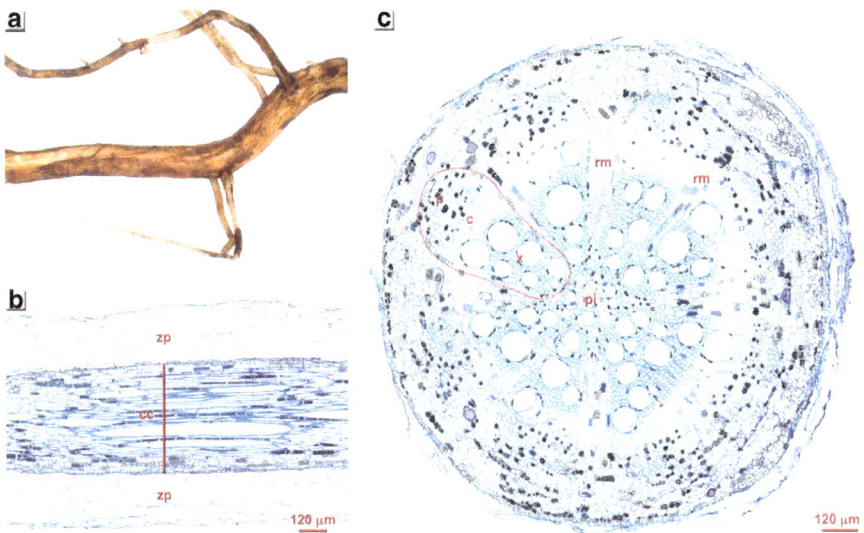

Fig. 2.12 Anatomy of a sclerified root in longitudinal and cross section. (**a**) General view. (**b**) Longitudinal section showing the central cylinder (cc) consisting of conducting tissues and peripheral cell layers called the cortex or peripheral zone (zp) comprising the phelloderm and suber. (**c**) Cross section showing the organisation of the conducting bundles into several steles (surrounded in red) comprising the phloem (p), cambium (c) and xylem (x). A central cell mass is composed of pith (pi) cells. The steles are separated from one another by rows of elongated cells, the medullary rays (rm)

whose walls are covered with suberin, a hydrophobic polymer. The suber mainly provides thermal protection for the root. In the case of *Vitis vinifera*, the cork is composed of five to six layers of cells. The phelloderm is a starchy supporting parenchyma forming the cortical layers of the root. The phelloderm is succeeded by the vascular tissues (or vascular system) grouped into elongated bundles, separated from one another by medullary rays consisting of starchy parenchymal cells (Fig. 2.12).

These bundles—also called steles—consist, from the outside to the inside, of primary functional phloem (soft bast), secondary phloem (hard bast), cambium, primary xylem (wood) and secondary xylem (secondary wood). The central core of the root is formed of pith cells.

From this structure the roots will develop year after year, thus forming growth rings that are difficult to identify. From the following spring, the cambial layers will enable the inward formation of secondary wood surrounding the secondary wood of the previous year, as well as the outward formation of secondary hard bast which contains more peribast fibres arranged in regular layers, lining that of the previous year. Development thus continues year after year.

The Non-Conductng Tissues of the Vine
- **Parenchyma**: Despite their fairly simple cytological organisation, parenchymal cells perform essential functions such as photosynthesis and the
  storage of reserves. Each parenchymal cell has a large vacuole. Their cellular walls are thin, and traversed by numerous plasmodemata. Among the
  parenchyma, we distinguish (1) the chlorophyllous parenchyma, located in
  the aerial parts of the plant (essentially in the leaf blades), containing
  numerous bulky chloroplasts; (2) the starchy parenchyma or reserve parenchyma, which abound in the underground organs (roots) as well as in the
  cells of the pith. The latter store numerous starch granules from the products of photosynthesis. Depending on the stimuli present, parenchymal
  cells can easily modify their initial specialisation and transform into
  another type of cell, or revert to a non-differentiated state.
- **Epidermis**: The epidermis is a layer of cells covering stems and leaves and
  acting as a barrier against external assault from e.g. pathogens and the climate while at the same time enabling gaseous exchanges with the atmosphere. It allows the differentiation of stomatal cells and hairs. Found on
  the underside of the vine leaves, the stomata can reach a density of 200–300
  stomata per mm^2 of leaf area. Epidermal cells possess rudimentary chloroplasts without chlorophyll, and secrete a covering or cuticle composed of
  waxes and cutin to the outside.
- **Collenchyma**: The collenchyma is the support tissue for growing organs.
  It forms rapidly under the epidermis. Cytologically, the walls of the collenchyma are composed of successive layers of cellulose whose fibres are
  alternately parallel or perpendicular to the axis of the cell, endowing it with
  significant strength. Despite this strength, the collenchyma remains an
  elastic tissue, allowing the elongation of the plant organ.
- **Sclerenchyma**: The sclerenchyma is a support tissue composed of a diversified set of cells whose common characteristic is the creation of a very
  hard cell wall. Once the cell has stopped growing, the primary wall lines
  itself with a secondary wall consisting of very tightly packed cellulose
  fibres that render the walls elastic but not plastic. Subsequently, the whole
  sclerifies through the addition of lignin, becoming non-extensible and
  very hard.

The Conducting Tissues: Xylem and Phloem
The vascular tissues consist of two main conducting systems: the xylem trans-
ports water and ions from the root to the rest of the plant, and the phloem
distributes the products of photosynthesis as well as a wide variety of other
solutes throughout the plant. In vines, the conducting tissues are arranged in a
more-or-less regular circle of conducting bundles (or steles) comprising

phloem vessels (bast) on the outside and xylem vessels (wood) towards the inside. These vessels consist of long, narrow tubular cells whose structure is different in the xylem and phloem (Fig. 2.13). The tracheids and the vessel elements constitute the conducting cells of the <u>xylem</u>. The walls of these two types of cells have secondary thickenings (lignin deposits) forming patterns (spiral, annular, etc.) to prevent the collapse of the vessels. Once mature, the cells of the xylem lose their cytoplasm, i.e. they are dead once they are functional for the transport of water and minerals. The vessel elements (tracheids) are arranged end-to-end to form a larger unit called a vessel (2.14). Tracheids are linked to one another at their ends, but water can circulate from one to another through openings called 'punctuations' (Fig. 2.13). Other types of cells are present in the xylem, including sclerenchyma fibres and parenchymal cells, important for the storage of energy-rich molecules (starch) and phenolic compounds. The <u>phloem</u> is composed of long, thin-celled walls stacked in vertical columns, the whole forming a long unit called a sieve tube. The sieve-tube cells have perforated transverse walls enabling the processed sap to pass from one tube to another. Unlike the xylem they are living tissue and contain cytoplasm and mitochondria, but no ribosomes or cell nuclei. The central part of the cell contains circulating processed sap rather than cytoplasm. The sieve tubes are closely linked to small, dense parenchymal cells called companion cells, which contain all the cellular machinery enabling certain metabolic processes of the sieve cells (Fig. 2.14). In addition, the phloem frequently contains storage parenchyma and supporting fibres.

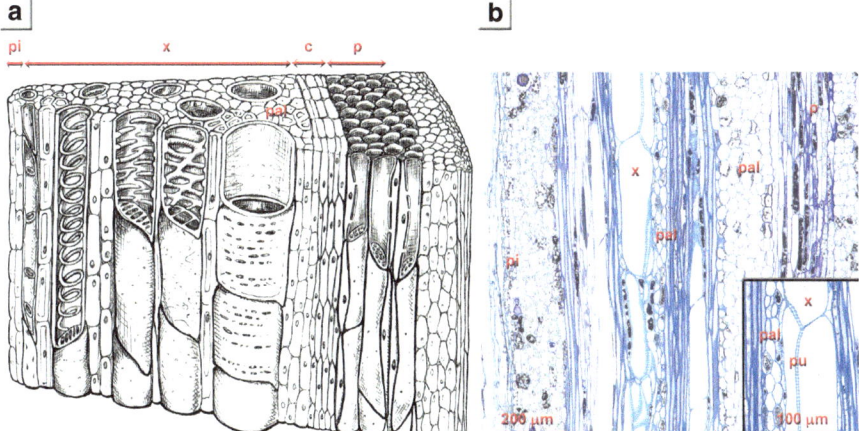

Fig. 2.13 Arrangement and structure of conducting vessels. (**a**) Schematic representation of a stele with conducting bundles showing, from the outside to the inside, the sieve tubes of the phloem (p), the cambium (c), the xylem (x) supported by lignified parenchyma (pal) and the cells of the pith (pi). (**b**) Longitudinal section of a stele with conducting bundles with the various elements described in A. Inset: Detail of the tracheids (xylem) with the characteristic wall punctuations (pu) allowing circulation of the raw sap

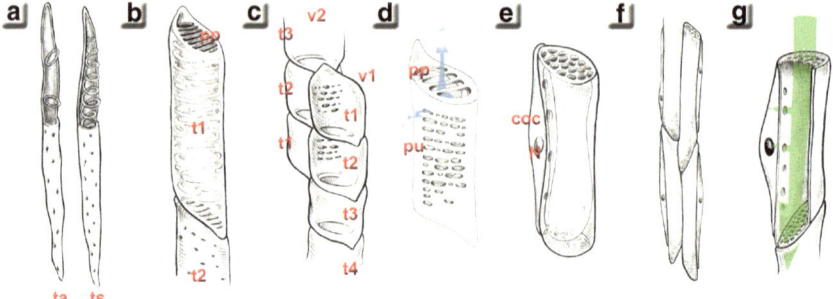

Fig. 2.14 Schematic representation of the conducting-vessel structures of the xylem (**a–d**) and phloem (**e–g**). (**a**) annular (ta) and spiral (ts) tracheids. (**b**) Conducting vessel consisting of 2 tracheids (t1 and t2) ending in a perforation plate (pp). (**c**) Arrangement of the vessels (v1 and v2), between them consisting of 3 and 4 tracheids (t1 to t4). (**d**) Ascending circulation of the raw sap through the perforation plate (pp) and the punctuations (pu). (**e**) Phloem cell with a companion cell (coc) having a nucleus (N). (**f**) Stack of cells forming a sieve tube. (**g**) Descending circulation of the processed sap. Illustration © Virginie Duquette, Un Monde d'illustrations, Switzerland

Fig. 2.15 Rhytidome of a grapevine trunk (**a**) peeling off in strips (**b**)

2.3.2   The Trunk and (Woody) Shoots

The aerial part of a vinestock comprises a trunk that is more-or-less developed depending on its age and conducting system (a short trunk for a goblet-trained vine or a long one for a high Lenz Moser or Sylvoz-type training, for example) and the woody shoots (that year's shoots) with their buds. The trunks of cultivated grapevines are flexuous, supported by artificial supports on which they develop. A cultivated grapevine remains a creeper that maintains a defined shape and architecture thanks to pruning. This rather twisted trunk is covered by a rhytidome (or outer bark) whose thickness depends on the age of the plant, and which is shed in strips (Fig. 2.15). The rhytidome consists of dead tissue formed from the different layers of cork or suber (replacing the juvenile bark, or cortex).

In the case of espaliered grapevines, the trunk may extend over a horizontal trellis wire to enable the formation of one or two permanent cordons (arms), or else stop at the height of the trellis wire for Guyot (cane) pruning, in which the fruit branches bearing the latent buds giving rise to the young shoots (the year's fruiting herbaceous shoots) are renewed annually. Grapevines are characterised by the production of two types of shoots, viz., the (long) primary shoots and the shorter secondary shoots or lateral shoots. Morphologically the shoots are distinguished by the production of numerous alternate distichous leaves having in their axis a main latent bud and a secondary bud which grows that same year, giving rise to the lateral shoots. The leaves are attached to the stalk from successive nodes separated from one another by portions of the shoot called internodes.

The nodes in the lower rows do not have tendrils. The organisational regularity of a nonfruiting shoot described by Viala and Vermorel (1910) consists of two nodes in which leaves and tendrils alternate with a tendril-less leaf. The same arrangement is then repeated up to the end of the shoot (Fig. 2.16). This regularly intermittent arrangement of the majority of *Vitis* species can, however, diverge with continuous arrangements where each node has a tendril, as with *V. labrusca* or with subcontinuous arrangements, where three nodes with tendrils alternate with a tendril-less node (some descendants of *V. labrusca*, Noah, Clincton, Othello). In fruiting shoots, the tendrils of the lower rows are generally replaced from the third node onwards by inflorescences on one, two or three successive nodes, terminating with alternating leaves and tendrils up to the apex.

In cross section, a green shoot sampled before lignification exhibits the same structure as the roots, with an outer cortical layer followed by a central cylinder composed of conducting tissues, then by central cells forming the pith (Fig. 2.17). The shoot is therefore bounded on the outside by a layer of epidermal cells (short, polygonal cells) arranged in a single longitudinal row, followed by several layers of hypodermal cells bordering a zone of cortical cells forming the cortical parenchyma. Within this cortical parenchyma, thicker-walled supporting cell masses called collenchyma become individualised.

This collenchyma generally compresses into bundles in the radial extension of the libero-ligneous bundles (Fig. 2.18). Likewise, the position of the collenchyma corresponds to the presence of more-or-less salient ribs visible on the shoots. The centre of the shoot is composed of large cells that are icosahedral in appearance, the cells of the pith. The physiological role of the bundles is to ensure the mechanical transport of fluids. The steles (libero-ligneous bundles) of conductive tissue are arranged in more-or-less regular circles and are separated from one another by primary medullary rays composed of two to five layers of cells with lignfied walls. Secondary medullary rays are sometimes found inside the steles when the latter are particularly large. From the outside to the inside, the structure of a stele (Fig. 2.18) consists of a semicircular mass of cells with a more-or-less thick pecto-cellulosic wall, the sclerenchyma, followed by the primary phloem (soft bast), the cambium, composed of several rows of tabular cells, then the primary xylem (wood). The cambium will then centrifugally generate hard-bast strata with lignified walls (secondary phloem, Fig. 2.19) serving as a support, alternating with soft bast which is responsible for circulating the sap produced (descending flux of the products of photosynthesis).

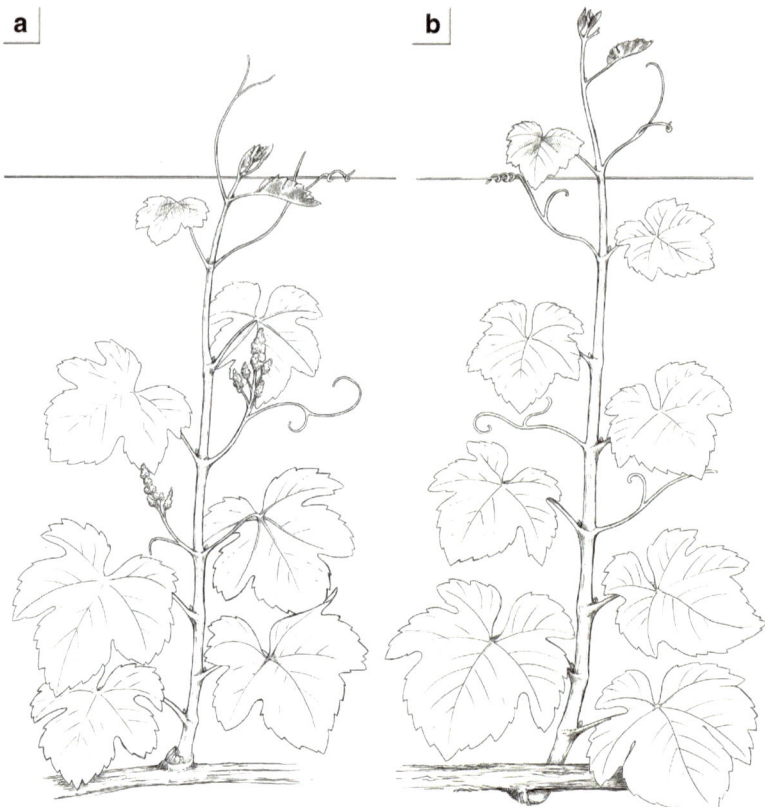

Fig. 2.16 Arrangement of the vegetative organs (leaves, tendrils and bunches) of a grapevine shoot. (**a**) Fruiting shoot from a main bud on two-year-old wood. (**b**) Non-fruiting shoot usually sprouting from a dormant bud on old wood. (Illustration © Virginie Duquette, Un Monde d'illustrations, Switzerland)

Fig. 2.17 Cross section of a green shoot. *ep* epidermis, *pi* pith, *p* phloem, *x* xylem

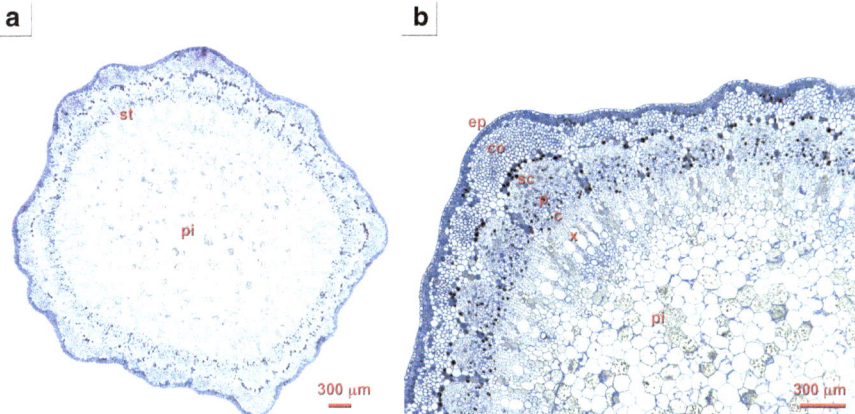

Fig. 2.18 Cross section of a green shoot at the growing tip. (**a**) General structure highlighting the zone of the conducting bundles organised into steles (st) and the cells of the pith (pi). (**b**) Detail of the structural anatomy of the green shoot: (**c**) cambial zone or cambium; *co* collenchyma, *ep* epidermis, *pi* pith, *p* phloem, *sc* sclerenchyma, *x* xylem

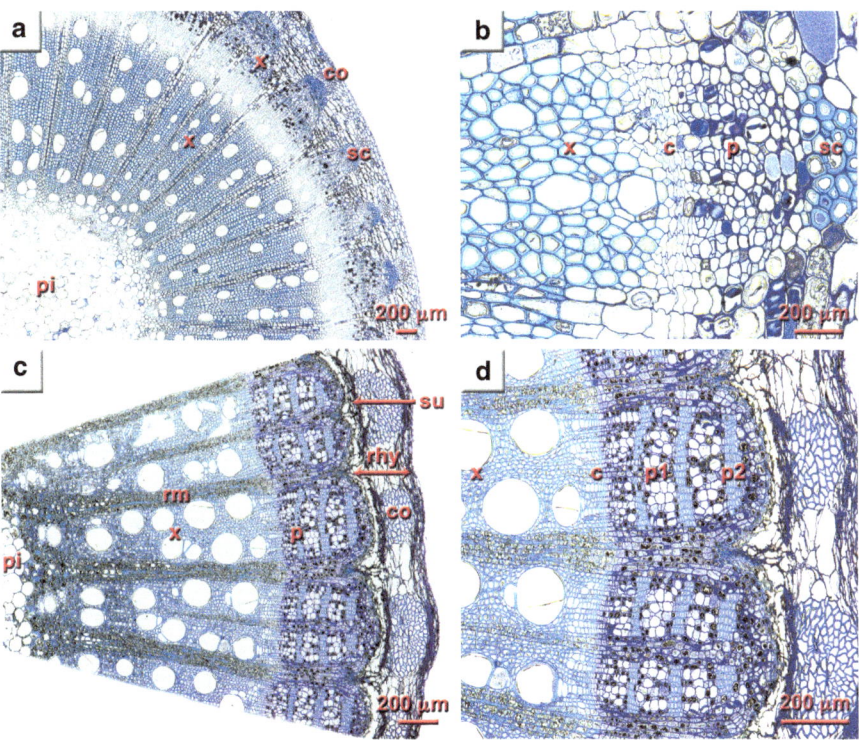

Fig. 2.19 Cross section of a green shoot (**a, b**) compared to a lignified shoot (**c, d**). (**a**) Several steles of conducting bundles, each separated by a few layers of cells forming the medullary rays. (**b**) Detail. (**c**) Lignified shoot consisting of five steles of conducting bundles, each separated by a few layers of cells forming the medullary rays. *Co* collenchyma, *c* cambium, *pi* pith, *p* phloem, *p1* primary phloem, *p2* secondary phloem, *rm* medullary ray, *rhy* rhytidome, *sc* sclerenchynma, *su* suber, *x* xylem

Likewise, in the other direction (ascending flux from the roots towards the green organs), the cambium will generate secondary wood (secondary xylem) with vessels of very large diameter ensuring the transport of the raw sap, followed by the primary wood (secondary xylem) (Fig. 2.19). These fairly early differentiations will be followed by the tissue lignification process as a prelude to the hardening off of the woody shoot. The hardening-off per se is characterised by the formation of several layers of cortical cells with suberised walls, called suber (cork), located between the sclerenchyma and the phloem. The cortical zone running from the epidermal cells to the suber constitutes the bark of the woody shoot of the grapevine, commonly called the rhytidome.

Cambial activity ceases during the winter and recommences the following spring, leading to the formation of stratified bast (alternating soft and hard bast) as well as secondary xylem on the outside of the secondary xylem of the previous year. Similarly, a new subero-phellodermic cell layer will lead to the production of the secondary bark elements on the same model as the previous year, taking with it the outermost part of the secondary phloem which is included in the formation of the secondary bark (secondary rhytidome). From there, the rhytidome will begin to peel off in strips. In woody plants not belonging to the Vitaceae family the annual growth rings are far more distinct than in grapevines since both the cambium and the conducting vessels of the former are continuous over the whole of the trunk circumference, whilst in the grapevine they are interrupted by the medullary rays. Unlike other woody plants, grapevines do not develop true callus tissue. Because of this, successive annual pruning wounds leave dead wood on the grapevine (Fig. 2.20) which can have a significant influence on the development of wood diseases (Viret and Gindro 2014).

2.3.3   Buds

With grapevines, we distinguish several types of buds that are generally defined as embryos of the shoots consisting of a vegetative cone ending in a meristem and with stumps of leaves and inflorescences. At the end of the stem is the apex, which ensures the lengthways growth of the shoot by means of cellular multiplication and the differentiation of new internodes, nodes, leaves, buds and tendrils. The apex degenerates once the herbaceous growth of the shoot has finished. The leaves are attached at node level and at their axil there are two buds—an axillary bud which on developing yields the lateral shoots or laterals, and a latent bud (or 'latent eye') (Fig. 2.21) which will remain dormant on the woody shoot in winter and will only develop the following year, giving rise to that year's shoot and to the leaves, tendrils and inflorescences (Fig. 2.22).

This latent eye consists of a main bud framed by one or two smaller secondary buds which can either grow simultaneously or take over from the primary bud if the latter is damaged. The buds of the same eye are protected by at least two brown, overlapping, lignified bud scales (Fig. 2.23). Latent buds may also be located below the rhytidome. Once vegetative growth resumes, the young shoots arising from these cortical buds, referred to as 'gourmands', are usually eliminated, as they never

Fig. 2.20 Cross section of a two-year-old vine shoot (**a** and **c**) compared with another woody plant (*Ficus benjamina*) (**b** and **d**). (**c**) Steles of conducting bundles intersected by the medullary rays (rm) showing the alternation of the secondary phloem (p2) with the primary phloem (p1) as well as the primary xylem (x1) with the secondary xylem (x2). Because of this, the vine (**e**) does not overlay pruning wounds, unlike other woody plants (**f**, apple tree). (**d**) Cuttings from an *F. benjamina* shoot, with uninterrupted phloem, cambium and xylem

Fig. 2.21 Latent vine bud located at each node. *bs* bud scale

bear grapes. In individual cases they are kept to renew the vinestock—for example, where cutting-back is performed in the case of diseases of the wood, modification of the training system, or the formation of a new trunk.

The main bud consists of a several-millimetre-long vegetative cone representing the stump of a stem with a certain number of preformed nodes, also bearing the stumps of leaves and inflorescences. Two different types of thermal protection structures protect these basic vegetative elements, viz., more-or-less thin bud scales called cataphylls, arranged in strata, and the tomentum or woolly down, a dense tangle of fine hairs working its way into all the empty spaces of the bud in the form of a dense cotton wool.

In spring, the end of latent-bud dormancy is manifested in the extrusion of the tomentum (*bourre* in French) giving rise to the French term *débourrement* (literally, the shedding of the tomentum), associated with the vegetative recovery of the grapevine, leading to the fall of the two tough brown bud scales. From this stage onwards, the grapevine develops through the growth of vegetative tissues. In macroscopic cross section, an elongation of the vegetative cone and a dedifferentiation of the leaves, tendrils and future bunches can be seen (Fig. 2.24). The emergence of the leaf primordia at the apex of the bud is the sign of intense metabolic activity and marks the beginning of the phenological development of the grapevine.

2.3.4 Leaf, Petiole and Tendril

2.3.4.1 Leaf

The general leaf morphology of *Vitis vinifera* is characterised by a more-or-less dissected, five-lobed blade (palmatilobate leaves) with a serrated edge, borne on a petiole covered by more-or-less salient ribs, with a deeper depression (groove) on the shoot side. The base of the petiole is attached at the shoot node (Fig. 2.25).

Fig. 2.22 Phases of development of the latent bud giving rise to that year's shoot and to the leaves, tendrils and inflorescences. (**a**) Swelling of latent bud (BBCH stage 01). (**b** and **c**) Bud in wool (BBCH 05). (**d**) Green tip (BBCH 09). (**e**) Leaves primordia appear. (**f**) Leaves appear (BBCH 10). (**g**) Leaf development; three leaves unfolded (BBCH 13). (**h**) Bunches clearly visible (BBCH 53)

The leaf blade has five primary veins lying fairly flat on the adaxial face (upper surface) individualising in the same place, then splitting off into secondary and tertiary veins and so on, covering the entire surface of the blade (Fig. 2.26). A large number of stomata line the abaxial face (lower surface) only (see insert),

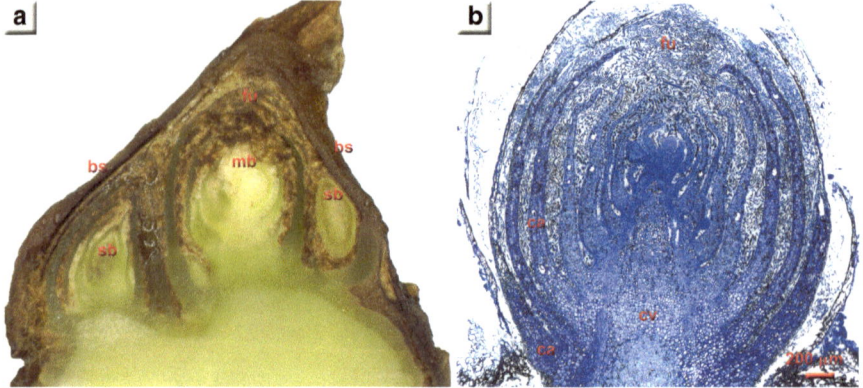

Fig. 2.23 Cross section of a latent bud. (**a**) Main bud (mb) and vegetative cone (cv), surrounded by two secondary buds (bs), protected by a thick cotton wool or fuzz (fu). (**b**) Semi-thin section of the main bud showing the arrangement of the cataphylls (ca), the vegetative cone (cv) with leaf stumps and fuzz (fu)

ensuring gaseous exchanges (photosynthesis, respiration and transpiration). The leaf has a flat, more-or-less embossed, hairy or blistered surface. Its general appearance may vary greatly from one grape variety to another, i.e. from one clone to another. Somatic mutations associated with the shape of the leaves can take on very unusual appearances, as in the case of the marbled leaves of the Chasselas cv. Cioutat variety (Fig. 2.26). Leaf traits form the basis of ampelographic identification (Fig. 2.21), which, according to the list of descriptors of the International Organisation of Vine and Wine, takes *inter alia* the following criteria into consideration (Dupraz and Spring 2010; OIV 2022):

- The hairiness, pigmentation and openness of the tip of the shoots, the colouring of the ventral and dorsal side of the shoots;
- The hairiness and colouring of the young leaves, the shape of the leaf blade, the number of lobes, the colour of the upper face, the shape of the leaf serration, the opening and shape of the petiolar sinus;
- the shape, length, compactness, size and number of wings of the bunches
- the shape, length and colour of the berries.

The Stomata and Lenticels

Stomata are characteristic openings in the epidermis of plants, present essentially on the lower face of leaves (bifacial leaves), as well as more irregularly on all green organs such as stalks, leaf petioles or herbaceous stalks, inflorescences and green berries. They are formed by two specialised, elongated, photosynthetic epidermal cells called guard cells, connected only at their ends, creating an intercellular space, the ostiole. The guard cells control the diameter of the ostiole. Through the epidermis and the cuticle, the ostiole creates a direct connection between the outside and a particularly important intercellular space of the mesophyll, the substomatal chamber (Fig. 2.27). The width of the ostiole is controlled by the opening and closing of the guard cells: the more turgid the guard cells, the more open the ostiole. Although the lower face of a leaf generally has between 100 and 500 stomata per mm^2, the pore area only represents 0.5%–2% of the leaf area, even when the spaces are completely open. Stomata are distributed in a specific manner within each epidermis depending on the plant species or grape variety in question.

The stomatal cells control gaseous exchanges between the outside environment and the substomatal chamber by modifying the size of the ostiole. This movement is caused by variations in cell osmotic pressure. High pressure causes maximum tension and deformation of the walls and retracts the guard cells of the ostiole (spacing of the lips). An internal drop in pressure brings them closer together. This opening and closing movement is controlled by several factors. Light and low CO_2 levels favour opening, whilst drought and a water deficit will cause closure. With normal weather, a daily opening (diurnal) and closing (nocturnal) rhythm is created, in all cases modulated by the environmental conditions.

The green organs of the grapevine are also composed of lenticels. The latter are channels or pores linking the atmosphere with the suberised parenchyma. Lenticels vary in shape and allow exchanges of oxygen, water vapour, and volatile chemical compounds, among others.

2.3.4.2 Petiole

The structure of the petiole is very similar to that of a green shoot. In cross section, the outer zone of the petiole is bounded by a layer of epidermal cells punctuated by stomata, followed by a parenchyma stiffened by the presence of collenchyma, then by the conducting bundles. The latter are organised in a circle or ellipse and are followed by a zone of icosahedral cells forming the pith. In addition to the ring of conducting bundles there are two steles in the two depression zones of the petiole (deeper groove). From the outside to the inside of the phloem, the conducting bundles (steles), which differ in size, consist of a zone of aligned cambial cells followed by secondary and primary xylem (Fig. 2.28). The petiole continues into the leaf blade proper, connecting at the junction of the main veins. From the adaxial face (upper side facing the light) to the abaxial face (lower side), the leaf blade in cross section (Fig. 2.29) consists of a layer of epicuticular waxes, a layer of cutin of varying thickness (see inset), followed by a single row of epidermal cells, then by long, regularly aligned 'palisade' cells, in turn followed by a lacunar parenchyma (containing numerous empty lacunae) composed of polygonal cells.

These lacunae often correspond to substomatal zones of the stomata dotted about the lower epidermis, which is composed of smaller cells than the upper epidermis.

Fig. 2.24 Phases of development of the latent bud in cross section, highlighting the outlines of leaves (ol) and inflorescences (oi), as well as the internodes (in) on the vegetative cone (cv). (**a** and **b**) 'Bud in wool' stage (BBCH 05) with the bud scales (bs), the first leaf primordia (l), the cataphylls (ca) of the primary (main) bud (mb) borne on the vegetative cone (cv), the fuzz (fu) and the secondary bud (sb). (**c** and **d**) 'Green tip' stage (BBCH 09), in which it is possible to distinguish the beginnings of inflorescences (oi) and the internodes (in) on the vegetative cone. (**e** and **f**) At the 'leaves appear' stage (l) (BBCH 10), the future inflorescences which will become the bunches (bu) are bordered by a bract (br)

Fig. 2.25 Petiole of grapevine leaf with more-or-less prominent ribs. A deeper depression (groove, arrow) is visible on the side of the shoot. To the left, tendril base adjacent to the leaf petiole

Fig. 2.26 Grapevine leaf. (**a**) Adaxial (upper) face of the leaf blade with fairly flat veins. (**b**) Abaxial (lower) face of the leaf blade showing the prominent veins (arrow) and a beaded gland whose function remains hypothetical (transpiration, gum). (**c**) 'Marbled' leaf of Chasselas Cioutat variety

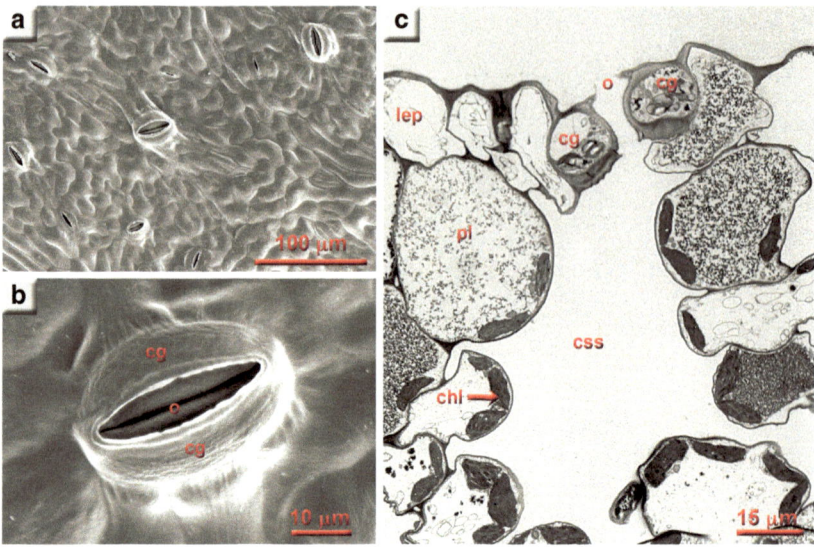

Fig. 2.27 Structure of the stomata present on the lower face of the grapevine leaves, observed by electron microscopy. (**a**) Random distribution of the stomata of the abaxial (lower) surface of a grapevine leaf viewed under a scanning electron microscope. (**b**) Detail of a stoma viewed under a scanning electron microscope, with the two guard cells (cg) and ostiole (o). (**c**) Detail of structure of leaf blade with the two guard cells (cg) and the ostiole (o), followed by the substomatal chamber (css) integrated into the lacunar parenchyma (pl) bounded by the lower epidermis (lep). The cells of the lacunar parenchyma contain a large number of chloroplasts (chl)

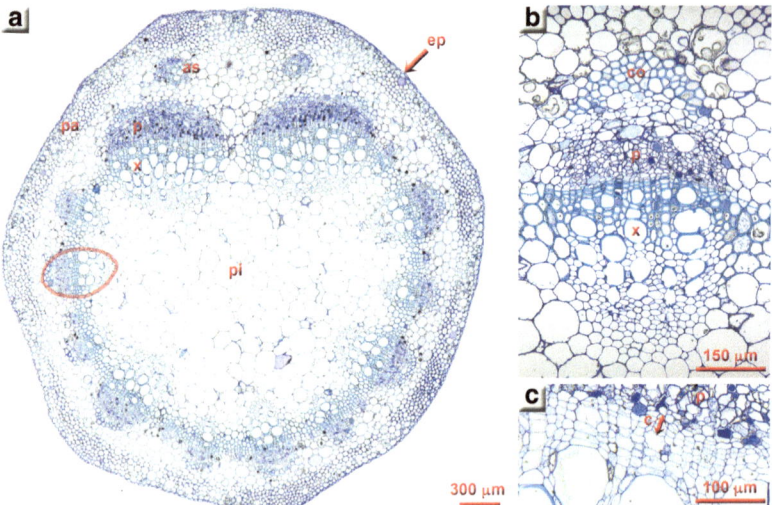

Fig. 2.28 Cross section of a leaf petiole. (**a**) General view showing the organisation of the petiole, its steles of conducting bundles arranged more-or-less in a circle, as well as two additional steles present at the groove. (**b**) Detail of the organisation of a stele of conducting bundles arising from (**a**), surrounded by red dashes. (**c**) Detail of transition zone between phloem and xylem. *c* cambium, *co* collenchyma, *ep* epidermis, *pi* pith, *p* phloem, *pa* parenchyma, *as* additional stele, *x* xylem

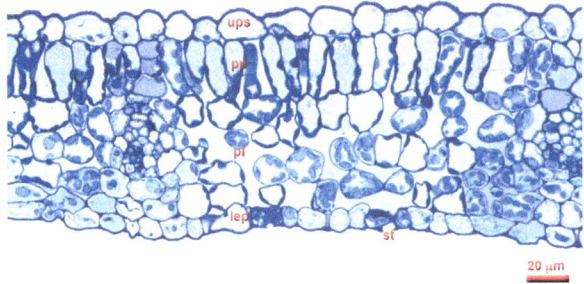

Fig. 2.29 Cross section of the leaf blade from the upper (adaxial) face to the lower (abaxial) face, with the upper epidermis (ups) followed by the palisade parenchyma (pp), the lacunar parenchyma (pl) and the lower epidermis (lep) with a stomata (st)

On the upper side there are a few lenticels, usually arranged along main veins. The leaf blade is punctuated by numerous veins. At the main and secondary veins, the palisade parenchyma is replaced by collenchyma, a support tissue forming a quasiring, followed by the lacunar parenchyma and generally by six conducting bundles or steles, whose layout is identical to that of the petioles. From outside to inside we distinguish the phloem, followed by a zone consisting of several rows of cambial cells, then the xylem. As the lower veins become smaller and smaller, the steles are increasingly less organised and their numbers dwindle to a single stele in the thinnest veins (Fig. 2.30).

Epicuticular Waxes, Cuticle and Cutin

The interface between the plant tissues and the environment is formed by the cuticle, a waterproof extracellular protective layer covering all aerial plant organs. The cuticle is topped by free waxes forming a specific crystalline structure depending on its chemical composition. The cuticle proper is of heterogenous chemical composition. Viewed under an electron microscope, the cuticle consists of two distinct parts (Fig. 2.31):

(a) the cuticular waxes, consisting of a complex mixture of long-chain carbon compounds (paraffins, esters, ketones, alcohols and acids) directly exposed to the atmosphere;

(b) the cutin, consisting of fatty acids forming a polymeric layer with some incrustations of waxes and other chemical compounds such as flavones or terpenes. Cutin is defined as a saturated and unsaturated fatty-acid polyester with a chain length of 16–18 carbon atoms synthesised in the epidermal cells. It may contain fungicidal compounds whose concentration may increase with the age of the plant organ, which may explain in part its resistance to certain fungal pathogens (Schnee 2008; Özer et al. 2017).

The thickness of the cuticle and its chemical composition as a whole vary from one plant and grape variety to another. Broadly speaking, plant epicuticular waxes combine a high degree of crystallisation and a strongly hydrophobic nature with low chemical reactivity. The cuticle thus plays the important role of physiological barrier on the surface of plant organs, enhancing the plant's resistance to a wide range of biotic (fungal pathogens, phytophagous insects, etc.) and abiotic stressors (drought, UV radiation, frost, etc.). In grapevines, cuticle thickness and wax content are key factors for the ontogenic resistance of the berries to powdery mildew (*Erysiphe necator*) or grey mould (*Botrytis cinerea*), linked to the genotype-specific crystallisation pattern of epicuticular waxes.

2.3.4.3 Tendrils

Tendrils are herbaceous and may become ligneous over the course of the season. In the majority of cases tendrils are bifurcated (i.e. have two branches), although on very rare occasions we may observe non-ramified tendrils, or ones with three or more branches. A bifid tendril is distinguished by a basal part, the hypoclade, and by two different-sized branches, a (larger) major branch facing downwards and a minor branch facing upwards (Fig. 2.32). Owing to their extreme flexibility, the young green tendrils can attach themselves to a support and spiral around it (Fig. 2.33), enabling the continuous aerial acropetal development of the shoot, which can achieve impressive dimensions.

Tendrils and Movement
(Lampsidis and Simon 1963; Jackson 2008; Keller 2020)

Tendrils are modified stalks of inflorescences whose differentiation into flowers is inhibited by gibberellin, the hormone responsible for cellular elongation. Tendril morphogenesis depends on the balance between gibberellin and cytokinin, the cellular division hormone. The absence of production of gibberellin or the suppression of its activity at the point of initiation of the organ causes the formation of an inflorescence, whilst the presence of gibberellin causes the production of a tendril. By contrast, cytokinin favours the formation of inflorescences rather than tendrils. It seems likely that the response to these hormones is quantitative, as there is a continuity of intermediate structures between tendrils and inflorescences. Bunch inflorescences are themselves considered to be modified stalks. It is no surprise, therefore, that tendrils have certain features in common with stalks. Unlike the growth of vegetative stalks, tendril growth is determinate, i.e. strictly limited. Tendrils go through three stages of development and functioning. First, they develop natural openings at their tips called hydathodes, which secrete water. Next, the hydathodes degenerate and pressure-sensitive cells in their epidermis develop all along the tendril. Upon contact with solid objects, these special-ised cells activate the elongation and cellular growth of the opposite side of the tendril, causing the tendril to twine around the object touched. Tendrils allow wild vines growing in a forest or cultivated vines planted among trees which serve as their support to access sunlight above the tree canopy. The tips of the tendrils search for surrounding objects by making sweeping and rota-tional movements during oscillatory growth; this pattern of development is called convolution or circumnutation. When one of the tips detects a support (thanks to the epidermal cells), the branches twine quickly around the sup-port. Following this movement—described as 'thigmotropic'—the entire ten-dril lignifies and stiffens to prevent its untwining. According to Lampsidis and Simon (1963) "the strength of cirumnutation depends on the age of the ten-dril; initially weak, it rises during the period of strong growth, then steadily decreases to nothing." With the differentiation of the collenchyma in the cor-tex and of the xylem and the lignification of the medullary ray cells, tendrils become woody and rigid at maturity. Tendrils that fail to find support die off after abscission at their point of attachment to the shoot.

According to Viala and Vermorel (1910), the anatomy of a grapevine's tendrils is indisputably reminiscent of that of modified flower stalks, as evidenced by the fact that it is not unusual to find one or more lateral or apical flowers on them (Fig. 2.34).

Anatomically the tendril is organised similarly to the veins of a leaf blade, viz., a layer of outer epidermal cells followed by several annular layers of continuous collenchyma, a cortical parenchyma, then a ring of steles of conducting tissue, culminating in a central cylinder of icosahedral cells forming the pith. From out-side to inside, the conducting bundles consist of primary phloem, a few rows of cambial cells, then the primary xylem (Fig. 2.35). The structure of the lignified tendrils is similar to that of the hardened-off shoots, and is characterised by the appearance of secondary phloem (hard bast) and secondary xylem (wood), as well as by the formation of bunches of sclerenchyma bordering the upper part of the steles (Fig. 2.36).

2.3.5   Inflorescence and Flowers

The vine is characterised by inflorescences in compound bunches (Fig. 2.37), i.e., a group of flowers arranged around a common axis (or rachis), the stem, extended by a peduncle attaching it to the shoot. The different elements making up the bunch are the peduncle, which attaches it to the fruiting shoot (that year's green shoot) at a node; the stalk (rachis), the main axis of the bunch to which the flower-bearing pedi-cels are attached at different points; and then the berries, at a pedicel extension zone, the bulge (Fig. 2.37).

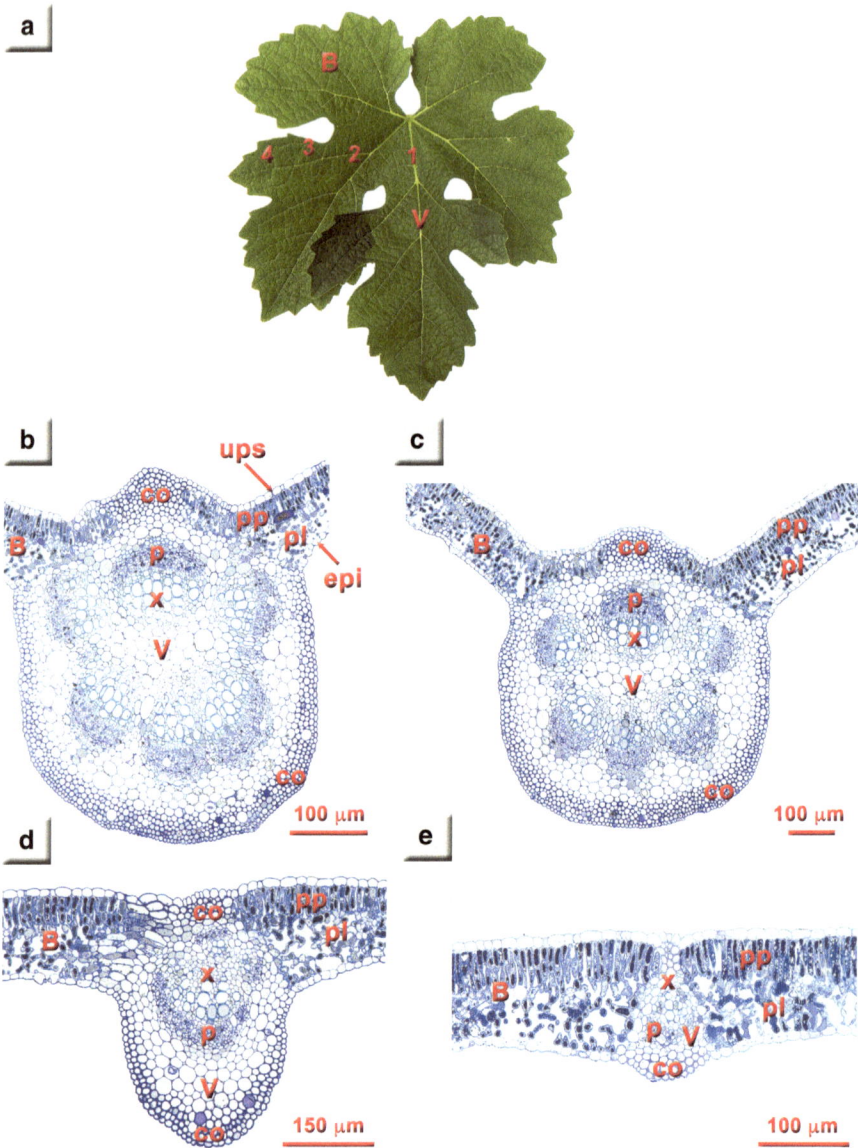

Fig. 2.30 Anatomy of the main-to-quaternary veins (V) of the leaf blade (B). (**a**) Ramification of the main veins (1) into secondary, tertiary and quaternary veins (2–4). (**b**) Cross section of a primary vein, transition between blade and vein with replacement of the palisade parenchyma by the collenchyma. The steles of conducting bundles are arranged more-or-less in a circle. (**c**) Cross section of a secondary vein; the steles of conducting bundles are again arranged more-or-less in a circle. (**d**) Cross section of a tertiary vein. Just one well developed stele remains. (**e**) Cross section of a quaternary vein. Just one rudimentary stele is still present. *co* collenchyma, *ups* upper epidermis, *epi* lower epidermis, *p* phloem, *pl* lacunar parenchyma, *pp* palisade parenchyma, *x* xylem

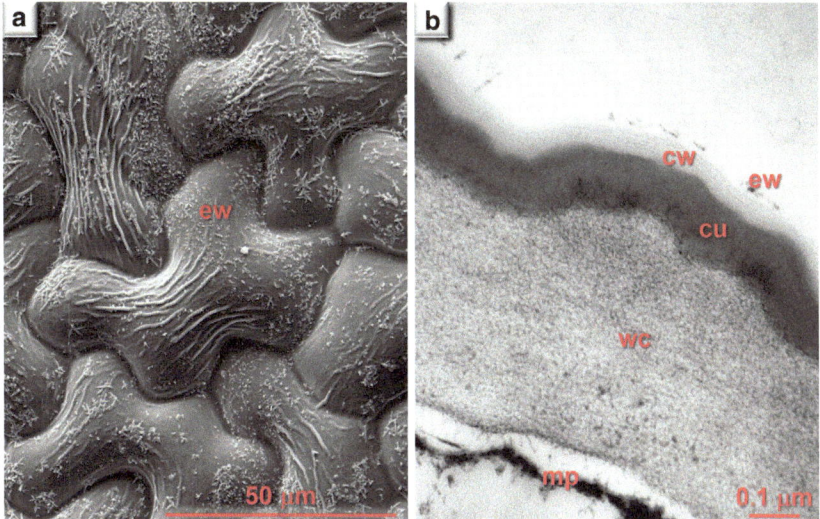

Fig. 2.31 Structure of the interface between the grapevine leaf and the external environment. (**a**) Cell structure of the adaxial (upper) surface of a grapevine leaf showing the crystallisation patterns of the epicuticular waxes (ew) seen by scanning electron microscopy. (**b**) Cross section of the leaf viewed by transmission electron microscopy with, from outside inwards, the epicuticular waxes (ew), the cuticular waxes (cw), the cutin (cu) the cell wall (wc) and the cell plasma membrane (mp)

Fig. 2.32 Two-armed, bifurcated herbaceous grapevine tendril (**a**) which can twin around trellis wire (**b**) and will later be lignified (arrow). *Hy* hypoclade (basal part), *Bma* major branch, *Bmi* minor branch

Fig. 2.33 Examples of herbaceous tendrils twined around supporting wires (**a** and **c**) or parts of the vine (**b** and **d**)

Fig. 2.34 Herbaceous grapevine tendril producing several flowers in an apical (**a**) or lateral (**b**) position

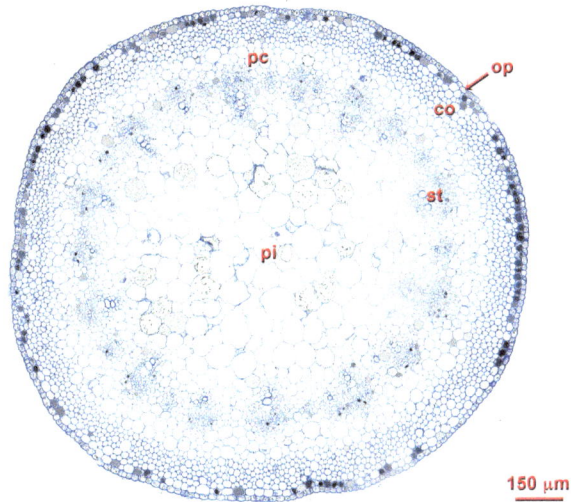

Fig. 2.35 Cross section of a herbaceous tendril. The outer epidermis (op) is followed by several layers of annular collenchyma (co), cortical parenchyma (pc) and steles of conducting bundles (st) arranged in a circle, as well as pith (pi) cells

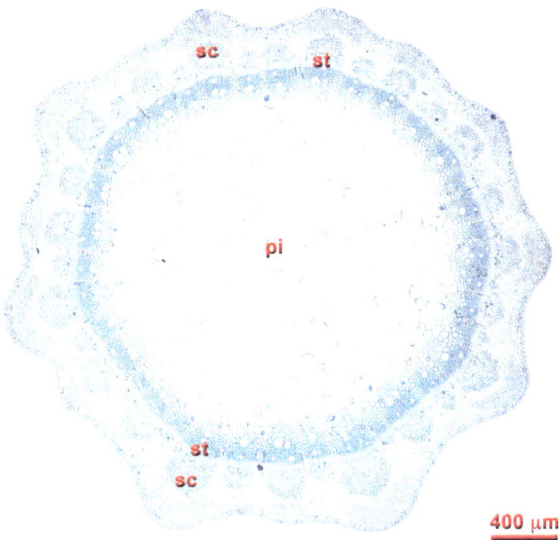

Fig. 2.36 Cross section of a lignified tendril. Unlike an herbaceous tendril, clusters of sclerenchyma (sc) appear along the upper part of the steles (st). *pi* pith

Like the tendrils, the inflorescences are lateral to the shoot, opposite a leaf. In the cultivated grapevine and more generally in the genus *Vitis*, inflorescences appear from budburst onwards, usually on the first two to three nodes counting from the base of the shoot. The level of insertion of the bunches on the shoot and the bunch/

Fig. 2.37 Inflorescence in compound bunches (**a**). The flowers are borne (**b**) on thin pedicels (pi) inserted on a common axis, the rachis or stalk (R)

Fig. 2.38 Young grapevine inflorescences appearing shortly after budburst, within bundles of young leaves (BBCH 51–53). At this stage, they appear as small compact flower glomerules (gl) bounded by a thin bract (b)

tendril succession vary according to grape variety, year, and number of bunches present on the shoot (Bouard 1986). From the third or fourth node the tendrils give way to inflorescences and allow the shoot to attach itself to a support, thus ensuring its lianescent development. Shortly after budburst, inflorescences appear in the form of small compact cones, distinctly visible among the young leaves. At this stage (BBCH 51–53), the flowers are grouped into several glomerules covered by thin bracts that will degenerate as the flowers mature (Fig. 2.38). The overwhelming majority of the flowers of the cultivated grapevine (*Vitis vinifera*) are bisexual (hermaphroditic), and possess five free deciduous petals, in exceptional cases four, generally fused at their apex into a calyptra or corolla (flower hood), as well as (four or) five stamens placed opposite the petals.

Each stamen consists of a filament and an anther with two pollen sacs, each of which contains two pollen locules). The ovoid pollen grain is around 10 μm in diameter and contains over 1000 pollen grains, each consisting of a vegetative cell and two generative cells. The highly reduced calyx (also termed a 'receptacle') consists of (four or) five rudimentary fused sepals forming a slight excrescence at the top of the flower receptacle. The ovary is superior and bilocular, i.e. divided into two chambers, each of which contains two ovules (Fig. 2.39).

The grapevine flower contains a maximum of four ovules, often two, which give rise after fertilisation to 2–4 seeds. Anatomically, the flower hoods and pistil are distinguished by the presence of numerous functional stomata (10–20 mm²) enabling

Fig. 2.39 Structure of grapevine flowers. (**a**) 'Closed flower bud' stage (BBCH 57); the sexual organs are sheathed in the calyptra or hood (c). (**b**) Section of a closed flower bud showing the stamens (st) and bilocular ovary (ov) containing two visible ovules (ovu). (**c**) 'Open bud' stage (BBCH 65) showing the five stamens, each consisting of a filament (f) and an anther (a) as well as the discoid stigma (sti). (**d**) Detail of the nectaries (ne) alternating with the stamens (only the filament (f) is visible), the ovary (ov) surmounted by a single stylus (sty) then the discoid stigma (sti)

them to ensure their own gaseous exchange. In the gynoecium (where all the female organs are located), five glands (nectaries or osmophores) alternate with the stamens, forming a disc surrounding the lower part of the ovary. The style or pistil is simple, ending in a bulging (capitate) or disc-shaped (discoid) stigma (Fig. 2.40).

On flowering, the calyptra detaches from the flower receptacle from the bottom upwards, shrivels quickly and falls off, allowing the androecium (stamens, the totality of male organs) and the gynoecium (nectaries, style and stigma) to expand. The role of the nectaries is to attract pollinating insects. Pollination is not dependent on the insects, however, as most varieties of cultivated grapevines are hermaphroditic. Several days before the calyptra falls off, the stamens release pollen grains which spread and can germinate. The pollen tube, the speed of whose growth depends on the temperature, grows through the tissues of the stigma and style until it reaches and fertilises the mature ovule. Not all ovules are fertilised, and some degenerate. This phenomenon can be exacerbated by environmental factors (rainfall and low temperatures) which cause a higher number of aborted ovules or decrease the pollen germination rate (problems with shot berries and millerandage). The ovule is characterised by the presence of two integuments:

– the outer integument, limited by a cellular epidermis delimiting a parenchyma with two to four cell layers.
– the inner integument, consisting of an epidermal layer, a thin parenchyma, and then an inner epidermis marking the boundary with the nucellus and the various embryonic elements such as the oospheres or the embryo sac (Fig. 2.40).

A grape berry with two locules can have a maximum of two seeds, and usually berries have four locules. Because the grapevine is strongly heterozygous, the multiple genetic recombinations associated with the occasional somatic mutations cause each cell of the nucellus and each cell of the seed to be genetically different. *Vitis vinifera* grape berries usually have one to two seeds, less commonly three to four. Because of the multiple genetic recombinations and occasional somatic mutations, each nucellus cell and each pollen cell on the same plant are genetically different, which explains why each grape seed within the same berry has a unique genome. This huge potential for genetic recombination underlies all the new grape varieties which the breeder can influence to an extent by selecting parents via targeted artificial pollination.

2.3.6 *Berries*

After fertilisation, the stamens degenerate and dry up and the ovary swells. Phenologically, this stage—referred to as fruit set (BBCH 71)—marks the start of herbaceous growth of the berry up to the stage of veraison or ripening (BBCH 81). The latter stage marks the end of growth and the onset of senescence of the berry, i.e. the change in colour according to variety (disappearance of chlorophyll) (Fig. 2.41), metabolic and structural modifications such as a reduction in skin

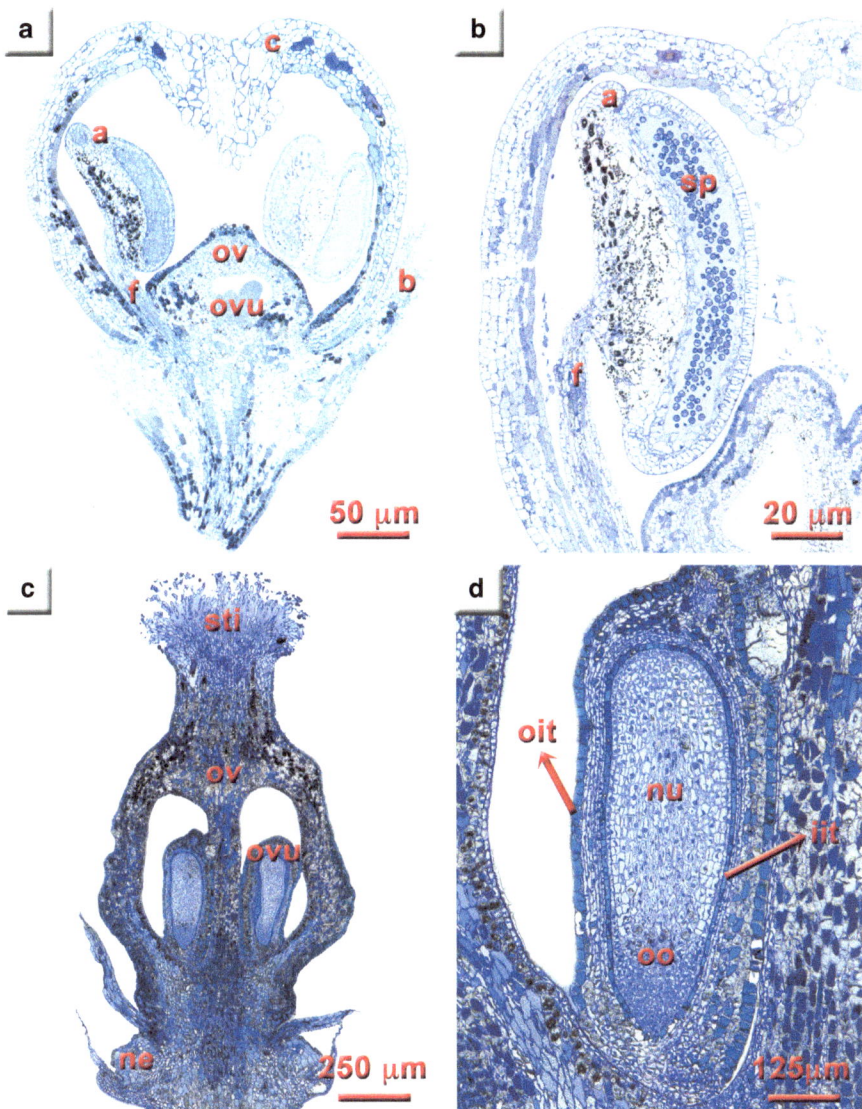

Fig. 2.40 Cross section of grapevine flowers. (**a**) 'Separate flower buds' stage (BBCH 57); the sex organs, viz., the stamens, each consisting of a filament (f) and an anther (a) and the bilocular ovary (ov) with a visible ovule (ovu) are enclosed in the calyptra or hood (c) and bounded by a thin bract (b). (**b**) Detail of a stamen with filament (f) and anther (a) bearing the pollen sac (sp) full of numerous pollen grains. (**c**) 'Full flowering' stage (BBCH 65); the two locules of the ovary (ov) contain two very visible ovules (ovu), as well as the globular stigma (sti), and the nectaries (ne). (**d**) Detail of the ovule bounded on the outside by the outer integument (oit) and the inner integument (iit). The oospheres (oo) are differentiated at the base of the nucellus (nu)

Fig. 2.41 Phenological stages of the grapevine from flowering to full ripeness. (**a**) In full flower (BBCH 65). (**b**) End of flowering (BBCH 67-69). (**c**) Fruit-set (BBCH 71). (**d**) Development of the berries (BBCH 73). (**e**) 'Pea stage' development of the berries (BBCH 75). (**f**) Bunch closure (BBCH 77). (**g**) Onset of veraison (BBCH 81). (**h**, **i**) Veraison (BBCH83-85). (**j**) Full ripeness (BBCH 89)

thickness and inactivation of constitutive defence mechanisms (phenolic compounds: proanthocyanidins, phenolic acid, glycolic acid, tannins) against *Botrytis* (Pezet et al. 2004), and induced defence mechanisms (stilbene phytoalexins: resveratrol, pterostilbene, viniferins, etc.) against *Botrytis* and downy mildew (Gindro et al. 2012).

Grape berries come in all sorts of shapes from more-or-less spherical to oblong, depending on the variety, and are attached to the rachis by the pedicel, which terminates in an annular bulge (receptacle) to which the berry is attached (Fig. 2.42). The

Fig. 2.42 General anatomy of a young grape berry after fruit-set (BBCH 71). (**a**) The berries are attached to the rachis (R) by a thin pedicel (pd) terminating in the receptacle (re) in which the berry is fixed. The berry is surmounted by the remainder of the stigma (sti) and speckled with small, prominent dots, the lenticels (le). (**b**) Cross section of a berry showing two locules (lo) in which the seeds develop

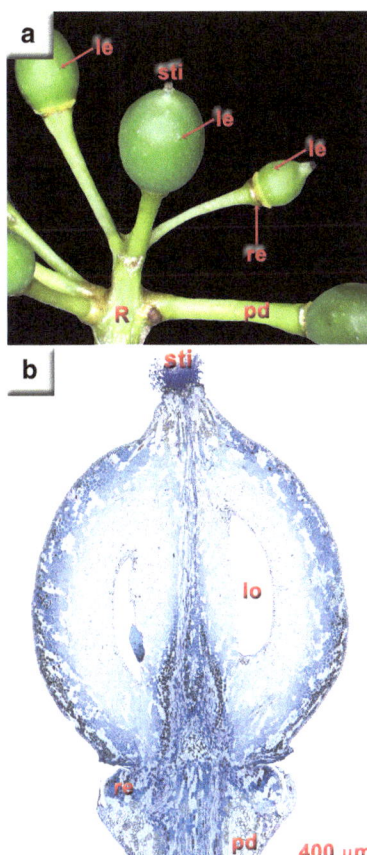

intact surface of a young berry is covered with a waxy, powdery substance called 'bloom' (in Latin, *pruina* ('hoarfrost')) composed of epicuticular waxes that can be eliminated by simple rubbing and which play a protective role.

The surface of the berry is dotted with a number of stomata, and in particular lenticels visible as small brownish spots in relief (Fig. 2.42). The remains of the stigma are visible as a tiny brush at the top of the berry. The berry itself consists of three different layers of cells succeeding one another up to the seeds and together forming the pericarp (Fig. 2.43). The skin corresponds to the epicarp, composed of a layer of relatively small epidermal cells covered by a layer of epicuticular waxes. The epicarp is followed by several layers of tangential cells which play a supporting role: the tangential layer (collenchyma).

The number of layers of tangential cells and the thickness of their walls depend on the grape variety and determine the greater or lesser rigidity of the skins of the berries regardless of their colour (Fig. 2.44).

The cells of the epicarp house all of the berry's cellular defence mechanisms against pathogens such as *Botrytis cinerea* (botrytis blight). This phenotypical

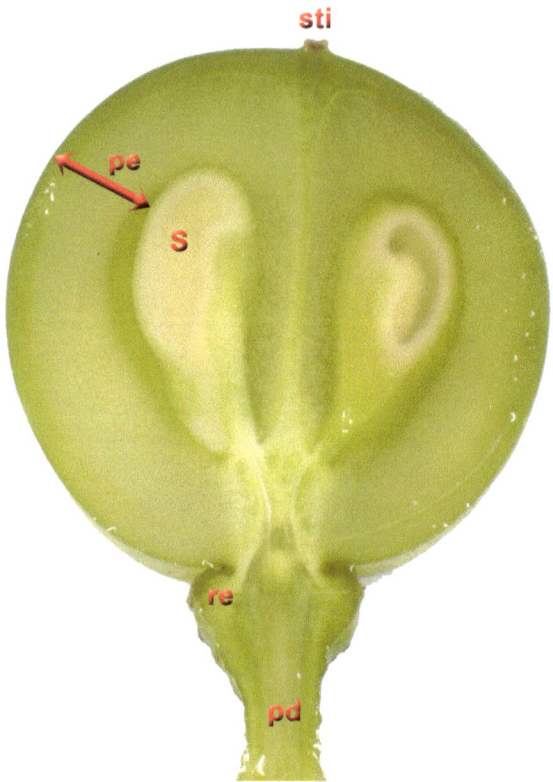

Fig. 2.43 General anatomy of a grape berry in cross section at the 'bunch closure' stage (BBCH 77). *pe* pericarp, *pd* pedicel, *S* seed, *re* receptacle, *sti* stigma

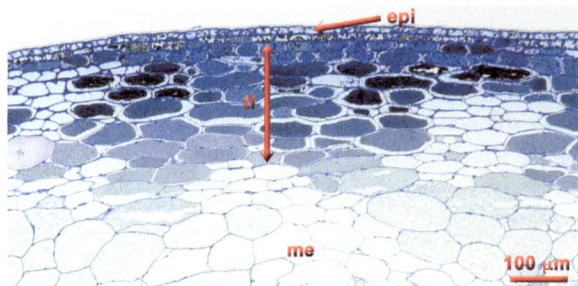

Fig. 2.44 Structure of the pericarp of a grape berry. *tl* tangential layer (collenchyma); *epi* epicarp composed of one to two layers of epidermal cells, *me* mesocarp (flesh cells)

parameter is solely associated with genotype and is almost impervious to the effects of plant-protection treatments or other viticultural practices. The more cell layers there are, the greater the berry's resistance to the pathogen. This element partially explains the resistance to botrytis of the Gamaret (Pezet et al. 2004), Galotta,

Carminoir, Gamarello, Cornarello, Nerolo, Cabernello and Merello grape varieties obtained in Switzerland (Spring et al. 2017). The flesh, termed the mesocarp, constitutes the essential tissue of the berry. Consisting of large, thin-walled cells, it contains the different compounds obtained when the grape juice is extracted—essentially, water, sugars, organic acids and mineral salts. The seed locules are bounded by a thin cellular layer, the endocarp, which lies below the mesocarp. From the outside to the inside, the structural elements of the seeds consist of the outer integument, the inner integument and the albumen. The outer integument consists of the outer epidermis (a layer of small cells) followed by several layers of outer parenchymal cells, followed by the inner epidermis which is composed of sclerified cells arranged longitudinally. These are followed by the inner integument, consisting of an outer epidermis and a layer of parenchymal cells which precedes a layer of cells forming the inner epidermis before the actual albumen (Fig. 2.45). There are a maximum of four seeds, given that a grapevine possesses no more than four fertilisable ovules. Frequently, however, there are only one to three seeds, or even none, or mere seed traces depending on the type of seedlessness (stenospermocarpic or parthenocarpic).

The whole of the berry is vascularised by a group of conducting bundles—the libero-ligneous bundles extending from the pedicel, the whole of which is called the brush (from the name of the 'brush-like' elements remaining attached to the pedicel when the berry is detached from it), the peripheral bundles vascularising the circumference of the berry just below the skin, the central bundles extending up to the stigma and the bundles vascularising the seeds (Fig. 2.46). In the structure of the base of the berry, we find the conducting bundles of the grapevine organised into steles, consisting of the collenchyma followed by the phloem, a cambial layer, and the xylem. Fine rootlets at the apex of the shoots, which pass through the wood of the trunk, the inflorescences, bunches, tendrils and leaves, and the structure of the conducting tissues of the grapevine, are organised in a similar manner, allowing the transport of the raw and processed sap throughout the living parts of the plant.

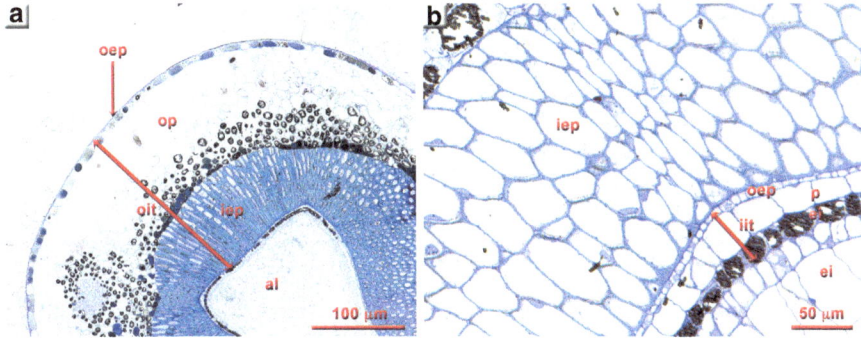

Fig. 2.45 Structural elements of grape seeds, from the outer integument (oit) to the albumen (al). (**a**) Cross section of outer integument (oit): from outside to inside, the outer epidermis (oep), outer parenchyma (op) and inner epidermis (iep). (**b**) Detail of inner integument (iit), with, from outside to inside, the outer epidermis (oep), parenchyma (p) and inner epidermis (iep) bounding the albumen (al)

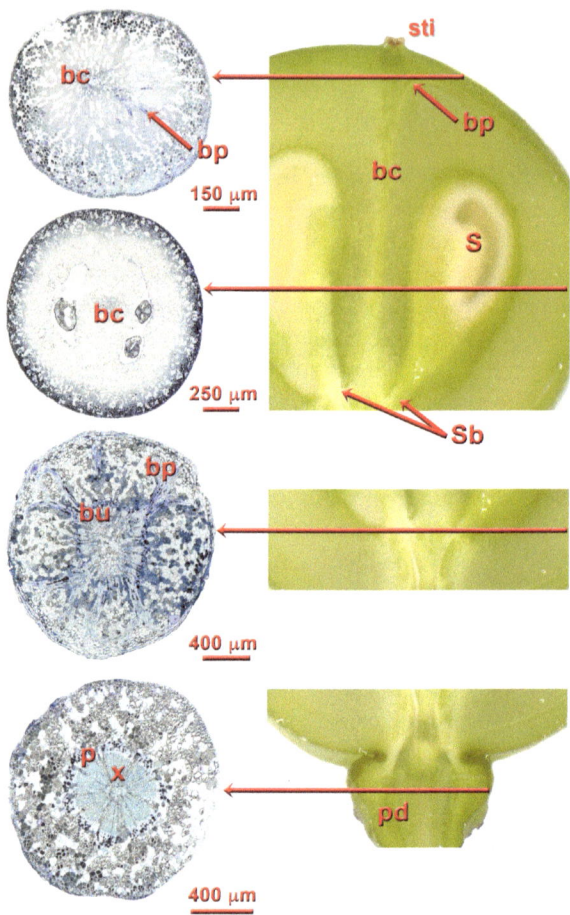

Fig. 2.46 Grape berry vascularisation system from the base of the pedicel (pd) up to the stigma (sti). The cross sections to the left highlight the conducting bundles consisting of steles of xylem (x) and phloem (p) arranged in libero-ligneous bundles or a brush (bu), in central bundles (bc), in peripheral bundles (bp), and in bundles (Sb) vascularising the seeds (S)

2.4   Grape Varieties Resistant to Fungal Diseases

2.4.1   History of Resistant Vines

Grapevine resistance to diseases and pests only became a topic for consideration towards the mid-nineteenth century, when the great scourges—powdery and downy mildew, black rot and phylloxera—swept through European vineyards from North America. Spatial separation had confronted the three groups of *Vitis*—European, American and Asian—with very different natural conditions and biotic stresses, including fungal diseases and pests, which had resulted in the selection of species adapted to their habitats (Rousseau et al. 2013). The consecutive introduction into Europe of powdery mildew in 1845 and downy mildew in 1878 struck with full force in European grapevines, whose ancient natural selection had not been shaped by these pathogens. Grape varieties resulting from natural or directed crossings between *V. labrusca* L. and European varieties discovered in the United States, such as the variety 'Isabella', proved resistant to powdery mildew when cultivated in Europe alongside the severely affected European varieties. The introduction of such varieties (Clinton, Noah, Othello, etc.) was thus strongly encouraged, which doubtless also contributed to the introduction in Europe of phylloxera, caused by the aphid-like insect *Daktulosphairia vitifoliae* (Fig. 2.47), and of downy mildew (Rousseau et al. 2013).

Nevertheless, the path was paved and the idea of introducing disease-resistance or pest-resistance genes through hybridisation with resistant species or varieties was quickly exploited starting in the late nineteenth century. The trauma caused by the appearance and rapid expansion of phylloxera raised awareness of the importance of studying and making use of the American species' sources of resistance (Planchon mission to America in 1873). This was the starting signal for hybridiser activity, especially in France in the first instance, for the creation of phylloxera-resistant rootstocks intended to reconstitute European wine production and grape varieties with low susceptibility to fungal diseases (direct-producer hybrids). Interspecific hybridisation had its glory days in France from 1880 to 1940: during this period,

Fig. 2.47 Grapevine phylloxera, caused by the aphid-like insect *Daktulosphairia vitifoliae*. (**a**) Leaf damage. (**b** and **c**) Closed phylloxera galls on the underside of a grape leaf, containing a yellowish-orange female and numerous eggs (**d**)

famous hybridisers (Seibel, Seyve-Villard, Joannes Seyve, Castel, Couderc, Baco, Kuhlmann, Rességuier et al.) bred numerous rootstocks still used nowadays, as well as hundreds of interspecific hybrids. Certain of these were so instrumental in the regeneration of the wine sector that they covered 400,000 ha (30% of vineyards) in France in 1958 (Rousseau et al. 2013). However, the inadequate quality of the wines supplied by most of these initial hybrids, and those obtained from *V. labrusca* L. in particular, resulted in their rejection and exclusion, made possible by the introduction of legislation for the protection of designations of origin from the 1930s onwards. This witch hunt ended in the banning of interspecific hybrids from the French Designation of Origin regions over the course of the 1950s, and to the drastic curbing of their expansion. At the same time, hybridisation decreased substantially in France whilst undergoing an upsurge in numerous countries like Germany, Austria, Bulgaria, Spain, Portugal, Hungary, Italy, Moldavia, Turkey, the Czech Republic, the ex-USSR, Romania, Serbia, China, India, South Korea and Japan. Among other things, research conducted in the Eastern countries has prospected the genetic heritage of the Asian *Vitis* group, and in particular *V. amurensis* Rupr., often in order to find winter-cold-resistant genotypes. The utilisation of varied sources of resistance, by which we mean the utilisation of different species of the genus *Vitis*, enables us to obtain polygenic or 'pyramidised' varieties, i.e. those with several resistance genes that are probably associated with several different resistance mechanisms for confronting fungal pathogens. The advantage of combining resistance genes is that it makes it harder for a fungus to bypass the overall resistance of the variety.

2.4.2   Natural Resistance of Grapevine to Fungal Diseases and Breeding of Resistant Cultivars

Among the ten or so species of fungi considered to be the main pathogens of grapevine, three are responsible for the most serious damage in vineyards: botrytis bunch rot or grey mould (*Botrytis cinerea* Pers.), downy mildew (*Plasmopara viticola* (Berk. & M.A. Curtis) Berl. & De Toni) and powdery mildew (*Erysiphe necator* Schwein.). More recently in Europe, owing to a reduction in the use of plant protection products and the introduction of resistant hybrids, black rot (*Phyllosticta ampelicida* (Engelm.)) has made something of a comeback in vineyards (Fig. 2.48).

All of the *V. vinifera* grape varieties cultivated throughout the world are more or less vulnerable to these diseases and must be protected by repeated applications of fungicides. Depending on the weather conditions and the technical itineraries chosen, 8–12 treatments are required on susceptible *V. vinifera* depending on the performance of the fungicides involved. Only less vulnerable varieties enable an evolution towards truly environmentally friendly and sustainable viticulture with a significant reduction in crop protection inputs, including those from sulphur and copper treatments, both of which are indispensable in organic viticulture. Varietal creation represents the most promising approach to achieving this objective. The breeding of

Fig. 2.48 Damage caused by the main fungal diseases of the aerial organs of the vine. (**a**) Downy mildew (*Plasmopara viticola*); (**b**) Powdery mildew (*Erysiphe necator*); (**c**) Botrytis bunch rot or grey mould (*Botrytis cinerea*); (**d**) Black rot (*Phyllosticta ampelicida*)

grapevines resulting from targeted crossings is one of the most promising methods for preserving and utilising the genetic diversity that also enables the selection of specific genetic traits such as resistance to pathogens. Each crossing produces numerous offspring (seedlings from germinated grape seeds) (Fig. 2.49) or germplasm (collection of genetic resources) whose level of resistance must be quickly estimatable to avoid long and tedious on-the-spot evaluations. For downy and powdery mildew, artificial inoculation of the seedlings with *P. viticola* sporangia or *E. necator* conidia and estimation of the development of the disease after an incubation period allows for reliable breeding (Pezet et al. 2004). The production and

Fig. 2.49 Grape seedlings (germplasm) after a phase of crossings by castration of the inflorescences and targeted pollinisation

Fig. 2.50 Assessment of resistance to downy mildew (*Plasmopara viticola*) after artificial inoculation on leaf discs (**a**) or whole leaves (**b**)

density of the sporangia produced by these artificial inoculations are widely accepted as a basis for estimating the resistance of the vine to fungal pathogens (Delp 1954). Climate chamber or greenhouse resistance tests on vine cuttings are not always representative of the true level of resistance observed in the field, however (Brown et al. 1999). Other resistance criteria must be rigorously tested on the seedlings to enable the correlation of resistance in the greenhouse or in *in vitro* tests with resistance in the vineyard.

Evaluation of the level of resistance by traditional, so-called empirical methods is based on the visual assessment of symptoms in the field over several years with a view to ensuring adequate disease pressure and the reproducibility of results. This is particularly true for botrytis bunch rot, resistance to which must be evaluated mainly on the flowers and bunches. In the greenhouse, resistance to downy and powdery mildew can be assessed after artificial inoculation of the grape seedlings resulting from the crossings, or on detached whole leaves or leaf discs (Fig. 2.50). The development of both powdery and downy mildew is then monitored and symptoms are visually assessed after several days. This methodology has the merit of shortening the breeding scheme by eliminating the tedious evaluation of resistance in the field. Resistance in the greenhouse and laboratory is not always representative of actual resistance in the vineyard, however, and therefore requires additional molecular (genotype) and/or biochemical (production of secondary defensive metabolites, phytoalexins) and histological (study of the structure of the tissues) analyses pertaining to the phenotype and natural defence mechanisms of the plant.

2.4.2.1   Resistance to Grey Mould

Resistance to *Botrytis cinerea* can be selected within the European grapevine *V. vinifera* itself (intraspecific crossings, hybridisation). For over 40 years now, the Agroscope research station in Switzerland has been conducting grapevine breeding programmes with a view to improving the diversity and quality of new grey-mould-resistant grape varieties. In 1970, the breeder André Jaquinet obtained the grape variety Gamaret through intraspecific crossing (Gamay × Reichensteiner) (Fig. 2.51), which was approved in 1990. Over the last two decades, its cultivation has gained increasing importance owing to its oenological qualities, its production potential and its remarkable resistance to grey mould, both in Switzerland and France, where it has been included in the national catalogue ('Official National List of Varieties (Grapevine Plants), List A'). From this programme, 11 new red grape varieties (Fig. 2.51) have been launched and disseminated in Switzerland (Table 2.3).

Gamaret, as opposed to the susceptible Gamay, has become a model for research on *B. cinerea* resistance mechanisms. Saprophytic and necrogenic (i.e. it kills to feed itself), this fungus exhibits a very low sensitivity to resveratrol and to the other stilbene derivatives, except for pterostilbene (Pezet and Pont 1990). Nevertheless, Gamaret does not synthesise this molecule in concentrations that would affect *B. cinerea*. Its resistance must therefore be explained by other mechanisms, even if a synergy can be observed between glycolic acid, a natural constituent of the grape berries, and pterostilbene, the mixture of these metabolites having proven highly toxic to the conidia of *B. cinerea*. The *Vitis-Botrytis* pathosystem is complex, bringing into play induced and constitutive defence mechanisms as well as elements linked to the structure and morphology of the berries and bunches. The structural defences are linked to the primary infection process (i.e. the formation of the infection cushion and the penetration of the plant tissues through the production of a wide range of hydrolytic enzymes) (Fig. 2.52), whilst the inducible responses are associated with the subsequent infection processes.

Grapevines can secrete an array of chitinases and other hydrolytic enzymes to degrade the cell wall of the fungus (Chong et al. 2008). Nevertheless, *B. cinerea* is able to catalyse the conversion of chitin into chitosan, thus preventing the degradation of its own cell wall (El Gueddari et al. 2002). Some studies have shown that following infection by a necrotrophic fungus such as *Botrytis cinerea*, the grapevine was able to activate the jasmonic acid/ethylene pathway, causing the induction of genes linked to the biosynthesis of phytoalexins, such as phenylalanine ammonia-lyase (PAL) and stilbene synthase (STS) (Girault et al. 2008; Wang et al. 2015). The synthesis and secretion of pathogenesis-related (PR) proteins also form part of the inducible defence mechanism deployed by *V. vinifera*. Giacomelli et al. (2012) demonstrated an increase in the transcripts of five defensin-like genes in grape tissues infected by *B. cinerea*. Three of the new defensins are capable of inhibiting the germination of the conidia of the fungus, suggesting that these genes play a role in the defence against *B. cinerea*. Likewise, it should be noted that the resistance of the berries is also closely linked to their stage of development. At the veraison stage, for example, an attempt at infection often results in the accumulation of *ROS (reactive oxygen species)*, which enable the activation of the salicylate-dependent defence pathway, the synthesis of resveratrol, the reinforcement of the cell wall and the formation of papillae underneath the infection cushions, which may stop the fungus in its tracks. By contrast, the much more susceptible infected ripe berries activate the jasmonate pathway, which is incapable of stopping this sort of nectrotrophic

Fig. 2.51 Bunch morphology of the intraspecific red grape varieties bred in Switzerland for their resistance to grey mould (*Botrytis cinerea*). (**a**) Cabernello; (**b**) Carminoir; (**c**) Cornarello; (**d**) Diolinoir; (**e**) Galotta; (**f**) Gamarello; (**g**) Gamaret; (**h**) Garanoir; (**i**) Mara; (**j**) Merello; (**k**) Nerolo

Fig. 2.51 (continued)

Table 2.3 Created and disseminated intraspecific grape varieties with resistance to grey mould (*Botrytis cinerea*) (Spring et al. 2017; Spring and Dupraz 2019)

Grape variety	Colour	Intraspecific crossings	Year obtained	Level of resistance
Cabernello	Red	Cabernet franc × Gamaret	1995	High
Carminoir	Red	Pinot noir × Cabernet Sauvignon	1982	High
Cornarello	Red	Humagne Rouge × Gamaret	1995	High
Diolinoir	Red	Robin noir × Pinot noir	1970	Fairly high
Galotta	Red	Ancellotta × Gamay	1981	High
Gamarello	Red	Merlot × Gamaret	1995	High
Gamaret	Red	Gamay × Reichensteiner	1970	Very high
Garanoir	Red	Gamay × Reichensteiner	1970	High
Mara	Red	Gamay × Reichensteiner	1970	High
Merello	Red	Merlot × Gamaret	1995	High
Nerolo	Red	Nebbiolo × Gamaret	1995	Very high

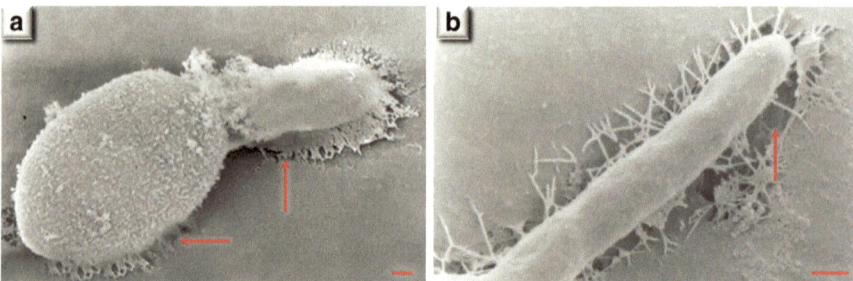

Fig. 2.52 Digestion of the cuticle of the vine leaf (arrows) by *Botrytis cinerea*, which penetrates into the plant tissues owing to the production of hydrolytic enzymes from the esterase family secreted by the conidia and germination tubes of the fungus (**a**) as well as by the growing hyphae (**b**) Scale bar represents 1 μm

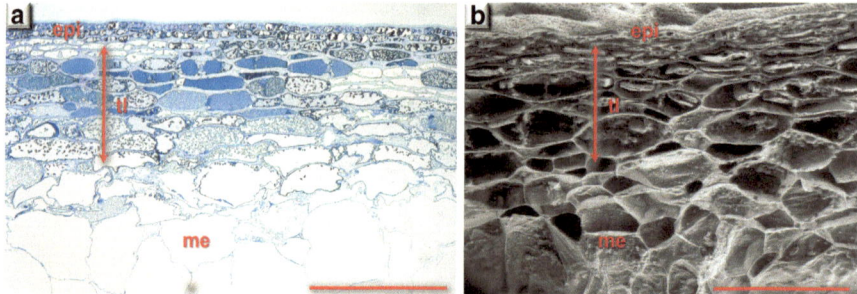

Fig. 2.53 Structure of the epidermis of the berries of the intraspecific grape variety Gamaret, observed by optical microscopy (**a**) and scanning electron microscopy (SEM) (**b**). *epi* epidermis, *me* mesocarp, flesh cells of the pulp, *tl* tangential layer. Scale bar represents 200 μm

infection (Kelloniemi et al. 2015). The most likely biochemical resistance mechanism is linked to the phenolic and polyphenolic compounds composing the berries (polymeric proanthocyanidins), known for their ability to inhibit stilbene oxydase and other hydrolytic enzymes of this fungus—key enzymes involved in the detoxification of these defence components (Perret et al. 2003). The development of grey mould is closely linked to the detoxifying activity of the phenolic compounds, which means that the inhibition of the activity of these enzymes prevents the development of the fungus. In structural terms, Gamaret berries possess a thick skin, itself consisting of an epidermis (small cells) below which lie several layers of tangentially stretched cells (tangential layers of the skin), and lastly the isodiametric, very large, very thin-walled cells of the pulp (Fig. 2.53).

Depending on the grape variety, the structure of the skin can differ in two points: the number of cell layers and the thickness of the cell walls, which distinguish varieties with a hard, thick skin from those with a tender, thin skin. The tangential layers are also where the phenolic compounds and the tannins are found, the quantity and composition of which determine the plant's resistance to grey mould (Perret 2001). The bigger the tangential layer, the more active metabolites it contains and the more the induced defence mechanisms can be expressed long-term during the ripening of the grapes (Fig. 2.54), bearing in mind that ripening is a phenomenon of senescence or cellular aging, and hence of losses in reactivity when faced with biotic and abiotic aggressions. For this reason, pre-existing defences make an active contribution to the defence of *Vitis vinifera* against *B. cinerea*.

2.4.2.2   Resistance to Downy and Powdery Mildews

Because the traditional *V. vinifera* grape varieties are not resistant to downy mildew, powdery mildew or black rot, the main sources of resistance to these pathogens are provided by *Vitis* species from Asia or America which have been subject to natural selection pressure for centuries. This is why interspecific crossing programmes have been launched in various research institutes to develop the grapevine genotypes with the greatest possible resistance. Historically, it was the American and Asian species that were used in the first interspecific crossings with the European grapevine to obtain resistant genotypes. The great majority of these species from the

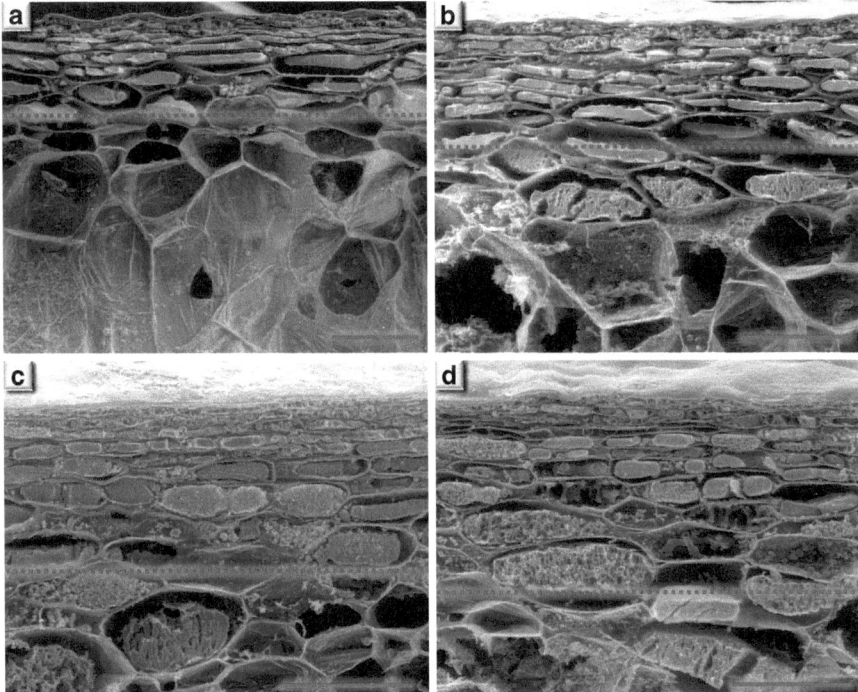

Fig. 2.54 Structure of the epidermis of the berries of four grape species at ripeness, exhibiting increasing resistance to grey mould (*Botrytis cinerea*), observed by scanning electron microscopy (SEM). The dotted red line defines the boundary between the tangential layer of the epidermis and the pulp. (**a**) Gamay (susceptible); (**b**) Chasselas (susceptible); (**c**) Divico (resistant); (**d**) Gamaret (resistant). Scale bar represents 100 μm

subgenus *Vitis* (2n = 38) do not pose a fertility problem in an interspecific crossing with the European grapevine. The situation is slightly different with the subgenus *Muscadinia* (2n = 40) which has a different number of chromosomes. Despite this obstacle, in 1972 the French researcher Alain Bouquet initiated a series of back-crosses of this genotype with European grape varieties, based on a first crossing between the European vine and *V. rotundifolia* Michaux performed in 1904. The representatives of the last generation of these backcrosses are of major significance in recent varietal improvement programmes.

2.4.2.3 Marker-Assisted Breeding

Molecular marker-assisted breeding enables the expression of the phenotype to be linked to the genotype, provided that the link between the markers and the genes in question is sufficiently strong. Certain characteristics are defined by a single site (locus) within the genome, while other traits such as resistance to biotic or abiotic stresses depend on different loci, in addition to environmental influences. Moreover, such 'quantitative' traits may play a more important role than others for the same trait. When closely associated with a quantitative trait, these loci are called QTLs (quantitative trait loci). Nevertheless, the exact location of the QTLs within the genome is not usually known. To get round this, markers located near the QTLs are determined based on their binding strength, so that they can be used for assisted selection. The closer the marker is to the QTL, the greater the chance that it will always be associated and that the link between marker and QTL will not be broken from one generation to the next. Good markers will allow the individuals of a population to be segregated according to resistance traits. The offspring are divided into groups according to their inherited genotypes and their phenotypes are compared to identify the significant associations between the resistance traits and the allelic variants.

   Although time-consuming and expensive, this approach permits the development of reliable markers for selecting the seedlings with the most resistant genotypes through the progressive combination of the different QTLs via backcrosses. To date, over 50 populations have been genotyped and phenotyped to map the resistance QTLs to *P. viticola* and *E. necator* and to create precise genetic maps, *inter alia* through the use of high-throughput sequencing technologies and the availability of the grapevine genome. Thirty-three QTL markers for resistance to downy mildew (Table 2.4) and 13 QTLs for resistance to powdery mildew (Table 2.5) have been identified, located on 14 and 8 different chromosomes, respectively. In addition, there are the resistance QTLs to black rot (*Phyllosticta ampelicida* Engelm.) (Table 2.6). Three major resistance QTLs (*Rgb*1 and *Rgb*3 on chromosome 14 and *Rgb*2 on chromosome 16) have been identified. Rgb1 and Rgb2 were identified in a population back-crossed with the grape variety Börner an interspecific hybrid of *Vitis riparia* and *Vitis cinerea* (Rex et al. 2014) and *Rgb*3 was identified from the cross Merzling (Bettinelli et al. 2023). Some minor QTLs were also detected on chromosomes 3, 4, 8, 10, 12, 13 and 19 (Rex et al. 2014). The term *Rgb* comes from the obsolete name of the fungus which causes black rot, *Guignardia bidwellii* (*Rgb* = Resistance *Guignardia bidwellii*).

Table 2.4 Current list of the 33 main resistance loci to downy mildew of grapevine (*Plasmopara viticola*) identified on American and Asian grapevines

QTL name	Source	Authors	Chromosome	Resistance level[a]
Rpv1	*Vitis rotundifolia*	Merdinoglu et al. 2003; Wiedemann-Merdinoglu et al. 2006	12	Partial
Rpv2	*Vitis rotundifolia*	Merdinoglu et al. 2003; Wiedemann-Merdinoglu et al. 2006	18	Total
Rpv3.1	*Vitis rupestris*	Fischer et al. 2004; Welter et al. 2007; Bellin et al. 2009; Venuti et al. 2013; Divilov et al. 2018; Foria et al. 2020	18	Partial
Rpv3.2	*Vitis rupestris*	Bellin et al. 2009; Venuti et al. 2013; Zyprian et al. 2016	18	Weak partial
Rpv3.3	*Vitis riparia*	Vezzulli et al. 2019	18	Weak partial
Rpv4	American *Vitis* spp.	Fischer et al. 2004; Welter et al. 2007; Bellin et al. 2009; Divilov et al. 2018	4	Limited
Rpv5	*Vitis riparia*	Marguerit et al. 2009; Morcira et al. 2011	9	Weak partial
Rpv6	*Vitis riparia*	Marguerit et al. 2009; Moreira et al. 2011	12	Weak partial
Rpv7	American *Vitis* spp	Fischer et al. 2004; Welter et al. 2007; Bellin et al. 2009; Divilov et al. 2018	7	Limited
Rpv8	*Vitis amurensis*	Blasi et al. 2011; Schwander et al. 2012; Venuti et al. 2013; Song et al. 2018; Lin et al. 2019	14	High partial
Rpv9	*Vitis riparia*	Marguerit et al. 2009; Moreira et al. 2011	7	Limited
Rpv10	*Vitis amurensis*	Blasi et al. 2011; Schwander et al. 2012; Venuti et al. 2013; Song et al. 2018; Lin et al. 2019	9	Partial
Rpv11	American *Vitis* spp	Fischer et al. 2004; Welter et al. 2007; Bellin et al. 2009; Divilov et al. 2018	5	Limited
Rpv12	*Vitis amurensis*	Blasi et al. 2011; Schwander et al. 2012; Venuti et al. 2013; Song et al. 2018; Lin et al. 2019	14	High partial
Rpv13	*Vitis riparia*	Marguerit et al. 2009; Moreira et al. 2011	12	No data
Rpv14	*Vitis cinerea*	Ochssner et al. 2016	5	Limited
Rpv15	*Vitis piazezkii*	Pap et al. in preparation	11	No data
Rpv16	*Vitis piazezkii*	Pap et al. in preparation	12	No data
Rpv17	American *Vitis* spp	Fischer et al. 2004; Welter et al. 2007; Bellin et al. 2009; Divilov et al. 2018	8	Limited
Rpv18	American *Vitis* spp	Fischer et al. 2004; Welter et al. 2007; Bellin et al. 2009; Divilov et al. 2018	11	Limited
Rpv19	*Vitis rupestris*	Welter et al. 2007; Bellin et al. 2009; Divilov et al. 2018; Vezzulli et al. 2019; Foria et al. 2020	14	No data
Rpv20	American *Vitis* spp	Fischer et al. 2004; Welter et al. 2007; Bellin et al. 2009; Divilov et al. 2018	6	Limited

(continued)

Table 2.4 (continued)

QTL name	Source	Authors	Chromosome	Resistance level[a]
Rpv21	American *Vitis* spp	Fischer et al. 2004; Welter et al. 2007; Bellin et al. 2009; Divilov et al. 2018	7	Limited
Rpv22	*Vitis amurensis*	Blasi et al. 2011; Schwander et al. 2012; Venuti et al. 2013; Song et al. 2018; Lin et al. 2019; Fu et al. 2020	2	Weak partial
Rpv23	*Vitis amurensis*	Blasi et al. 2011; Schwander et al. 2012; Venuti et al. 2013; Song et al. 2018; Lin et al. 2019; Fu et al. 2020	15	Weak partial
Rpv24	*Vitis amurensis*	Blasi et al. 2011; Schwander et al. 2012; Venuti et al. 2013; Song et al. 2018; Lin et al. 2019; Fu et al. 2020	18	Weak partial
Rpv25	*Vitis amurensis*	Blasi et al. 2011; Schwander et al. 2012; Venuti et al. 2013; Song et al. 2018; Lin et al. 2019	15	Weak partial
Rpv26	*Vitis amurensis*	Blasi et al. 2011; Schwander et al. 2012; Venuti et al. 2013; Song et al. 2018; Lin et al. 2019	15	Partial
Rpv27	*Vitis aestivalis*	Sapkota et al. 2015, 2019	18	Weak partial
Rpv28	*Vitis rupestris*	Bhattarai et al. 2021	10	Partial
Rpv29	Caucasian *Vitis* spp.	Sargolzaei et al. 2020	14	Limited
Rpv30	Caucasian *Vitis* spp.	Sargolzaei et al. 2020	3	Limited
Rpv31	Caucasian *Vitis* spp.	Sargolzaei et al. 2020	16	Limited

Rpv: resistance *Plasmopara viticola*
[a]Resistance level according to Possamai and Wiedemann-Merdinoglu (2022a, b): Total: no sporulation; High partial: weak pathogen development and very weak sporulation rate; Partial: stunted, limited growth of pathogen; Weak partial: slightly stunted, limited growth; Limited: weak resistance to pathogen

Table 2.5 Current list of the 15 grapevine resistance loci to powdery mildew (*Erysiphe necator*) identified on American and Asian grapevines

QTL name	Source	Authors	Chromosome	Resistance Level[a]
Ren1	*Vitis vinifera* cv. Kismish vatkana	Hoffmann et al. 2008	13	Partial
Ren1.2	*Vitis vinifera*, numerous accessions (Caucasus)	Riaz et al. 2020	13	Partial
Ren2	*Vitis* sp. interspecific hybrids (America)	Dalbò et al. 2001; Feechan et al. 2015	14	Weak partial
Ren3	Regent: *V. vinifera* × American spp.	Welter et al. 2007; Zendler et al. 2020	15	Partial
Ren4	*Vitis romanettii*	Raming et al. 2011	18	Total
Ren5	*Vitis rotundifolia*	Blanc et al. 2012	14	Total

(continued)

Table 2.5 (continued)

QTL name	Source	Authors	Chromosome	Resistance Level[a]
Ren6	*Vitis piasezkii*	Pap et al. 2016	9	Total
Ren7	*Vitis piasezkii*	Pap et al. 2016	19	Partial
Ren8	*V. vinifera* × American spp	Zyprian et al. 2016	18	Limited
Ren9	Regent: *V. vinifera* × American spp	Zendler et al. 2017; Zendler et al. 2020	15	Partial
Ren10	Seyval blanc: *V. vinifera* × American spp	Teh et al. 2017	2	Limited
Ren11	*Vitis aestivalis*	Karn et al. 2021	15	Partial
Run1	*Vitis rotundifolia*	Barker et al. 2005; Feechan et al. 2013; Argurto et al. 2017	12	Total
Run2.1	*Vitis rotundifolia*	Riaz et al. 2011	18	Partial
Run2.2	*Vitis rotundifolia*	Riaz et al. 2011	18	Partial

Ren: resistance *Erysiphe necator*, *Run* resistance *Uncinula necator* (former name of powdery mildew of grapevine)

[a]Resistance level according to Possamai et al. (2021) and Possamai and Wiedemann-Merdinoglu (2022a, b): Total: no sporulation; High partial: weak pathogen development and very weak sporulation rate; Partial: stunted, limited growth of pathogen; Weak partial: slightly stunted, limited growth; Limited: weak resistance to pathogen

Table 2.6 Current list of the three major grapevine resistance loci to black rot (*Phyllosticta ampelicida*)

QTL name	Source	Authors	Chromosome
Rgb1	Börner (*Vitis riparia* × *Vitis cinerea*)	Rex et al. 2014	14
Rgb2	Börner (*Vitis riparia* × *Vitis cinerea*)	Rex et al. 2014	16
Rgb3	Merzling (*V. vinifera* × *V. rupestris* × *V. aestivalis* var. *lincecumii*)	Bettinelli et al. 2023	14

Rgb resistance *Guignardia bidwellii* (former name of black rot of grapevine)

Bettinelli et al. (2023) demonstrated that the Rgb1 locus is enriched for genes pertaining to phloem dynamics and mitochondrial proton transfer, whilst the Rgb3 locus exhibits a group of Germin-like protein genes associated with pathogenesis that trigger programmed cell death. The findings of these authors therefore suggest the strong involvement of mitochondrial oxidative burst and phloem occlusion in Black Rot resistance mechanisms.

In addition to the search for loci for vine resistance to *Plasmopara viticola*, *Erysiphe necator* and *Phyllosticta ampelicida*, resistance QTLs for other fungal diseases of the vine were highlighted:

- Su et al. (2023) were able to map the two stable QTLs for resistance to *Botrytis cinerea* located on chromosomes 2 and 7. In combination with RNA sequencing (RNA-seq), these authors succeeded in identifying a structural gene termed VIEDR2 (vitvi02g00982) and three transcription factors that are clearly involved in the expression of the VIEDR2 gene for its role in resistance to grey mould.

- A QTL associated with resistance to leaf anthracnose (*Elsinoë ampelina*), termed Rea1 (Resistance *Elsinoë ampelina* 1), was located on chromosome 18 in a region containing genes coding for proteins involved in the formation of reactive oxygen species (ROS). A second locus located on chromosome 19 was associated with shoot resistance in a region containing genes coding for proteins responsible for synthesising terpenes and quinones (Modesto et al. 2022).
- A White Rot resistance QTL (*Coniella diplodiella*), Rcd1 (Resistance *Coniella diplodiella*) was identified on chromosome 14, opening up prospects for the breeding of grape varieties resistant to this pathogen (Zhang et al. 2017; Su et al. 2021). The findings of Li et al. (2023) on an interspecies crossing (*Vitis vinifera* L. × *Vitis davidii* Foex.) enabled the identification of a new resistance QTL located on chromosome 3.
- The first QTL reported for grape ripe rot (*Colletotrichum* spp.), termed Cgr1 (*Colletotrichum gloeosporioides* resistance 1), was identified on chromosome 14 of the 'Shuang Hong' vine by Fu et al. (2019). Moreover, a SNP (SNP = Single Nucleotide Polymorphism) closely linked with the peak of Cgr1 termed 'np19345' was identified as a molecular marker for the Cgr1 resistance haplotype.
- High-density genetic maps were utilised to locate new qualitative resistance loci to dead-arm of grapevine (*Phomopsis viticola*), termed Rda1 and Rda2 (Resistance *Diaporthe ampelina*), from *Vitis cinerea* B9 and 'Horizon', respectively (Barba et al. 2018). The linearity between the genetic and physical reference maps enabled us to pinpoint the locus Rda2 on chromosome 7 and the locus Rda1 on chromosome 15, which spans a group of five NB-LRR (nucleotide-binding site-leucine-rich repeat) genes linked to the detection of pathogen-associated proteins, usually the effector molecules responsible for virulence (De Young and Innes 2006).

Selection programmes, whether based on empirical criteria or marker-assisted, are relatively common in Europe (Germany, Switzerland, France, Italy, Austria, Spain, Portugal, Serbia, Czech Republic, Moldavia, Romania, Bulgaria), the Middle East (Turkey), South America (Brazil) and Asia (Japan, South Korea, China, India). Australia and the United States have also developed selection programmes, but based on GM technologies (GMO = genetically modified organism). To cite just a few examples, the following European research institutes have developed marker- (or QTL-) assisted selection programmes:

- INRAE (National Research Institute for Agriculture, Food and the Environment) in France, developing polygenically resistant grape varieties as part of the Resdur programmes ("Résistance Durable");
- Agroscope's Viticulture Research Centre in Pully (Canton of Vaud) in Switzerland (Fig. 2.55);
- The Institutes for grapevine breeding in Germany: Julius Kühn Institute (JKI) Geiweilerhof in Siebeldingen,; Weinbauinstitut Freiburg in Breisgau (WBI), Geisenheim University;
- *Instituto Murciano de Investigación y Desarrollo Agrario y Medioambiental* (IMIDA) in Spain;

Fig. 2.55 Agroscope's Viticulture Research Centre in Pully (Canton of Vaud), Switzerland, has 5 ha of experimental land housing the national grapevine collection and activities for selecting new fungal disease-resistant grapevine varieties

- Pécs and Kecskemét Centres in Hungary;
- Applied Genomics Institute, Udine in Italy.

The progressive spread of the grapevines produced by these selection programmes should enable a reduction in the use of plant protection products, leading gradually to an increasingly sustainable wine sector. Consumers still need to get to grips with the tastes of these new wine or table varieties. Developing new varieties easily takes 15–20 years, and the adoption of these by the sector and the consumer is a matter of several decades. Indeed, the most popular, and hence most commonly cultivated European grapevine varieties have generally been maintained for centuries by clonal propagation (This et al. 2006). The vinification processes for all these varieties have usually been optimised by the producers, and must be re-established for the new resistant varieties. Consequently, the market is often skittish about adopting the new grape varieties created by resistance selection programmes, owing to consumer preference for the traditional varieties. At the same time, the hope is that the use of new, more ecologically sustainable varieties will become widespread with growing consumer awareness of the problems associated with the application of plant-protection products. For example, the grape variety 'Divico', developed by Agroscope (Switzerland), was launched in 2013, 16 years after it was created by crossing; growing international interest has accounted for its steadily rising fortunes.

The current approach to the selection of disease-resistant varieties obtained by classic hybridisation is based on the principle of pyramiding resistance genes in order to minimise the risk of pathogen adaptation, which seems to be considerably higher in cases of monogenic resistance (Cadle-Davidson et al. 2011). Alongside conventional interspecific hybridisation, certain research institutes have not given up on the project of introducing disease-resistance genes into susceptible European varieties via genetic transformation (Feechan et al. 2013).

2.4.2.4 Recessive Resistances and Loss-of-Susceptibility Genes

The traditional breeding approach is to introgress dominantly inherited major-effect resistance genes from wild relatives into elite material. Resistance in these cases is typically conferred by receptor proteins that allow plants to recognise the pathogen or its effectors and mount an immune response to them (Merdinoglu et al. 2018). However, major-effect resistance genes are often relatively quickly overcome by new pathogen strains that have evolved to evade recognition by the plant host (Peresotti et al. 2010; Feechan et al. 2015; Wingerter et al. 2021). The erosion of resistance subsequently requires new genes to be identified and introgressed, resulting in a continuous 'arms race' against the pathogen. Although such 'Red Queen' breeding strategies are successful, they often only work in the short-term. As a result, they are frequently replaced by gene pyramiding, i.e. the use of multiple-resistance genes acting against the same pathogen (Mundt 2018). However, this approach also has its disadvantages (Stam and McDonald 2018), because (1) it relies on the occurrence of multiple-resistance genes; (2) introgression of genes from wild relatives often has a negative effect on wine quality due to linkage drag, a problem that increases with the number of resistances used; (3) it is only possible for resistances for which molecular markers are available, since individuals with a resistance pyramid can only be identified using markers; and (4) depending on the number of resistance loci, ever-larger populations need to be screened for individuals that carry the desired allele combination. Importantly, however, resistances represent a 'public good', and the effectiveness of pyramids can be undermined if the component resistances are utilised and deployed in isolation in other varieties (Wingerter et al. 2021). An important question in (grapevine) breeding, therefore, is whether there are other strategies leading to longer-term resistance against the major pathogens. One possible alternative relies on the removal of so-called susceptibility genes from the grapevine genome.

Fungal pathogens such as grapevine powdery and downy mildew essentially cause disease by manipulating a plant's immune responses and physiology and exploiting its metabolism. Small proteins (typically so-called effectors) secreted by the pathogens often play a key role in this. Removing non-essential targets of these effectors has well-demonstrated benefits for preventing pathogen attack in a variety of crops (Van Schie and Takken 2014). In addition, such loss-of-susceptibility-type resistances have proven to be among the most durable resistances in breeding history (Biffen 1905; Freisleben and Lein 1942; Mishra et al. 2005; Humphry et al. 2006). While speculative, it may therefore be that these resistances represent an extraordinary 'evolutionary challenge' for the pathogen which prevents their rapid breakdown. They also differ from the traditional resistances ('R-genes') in terms of mechanisms and inheritance: for example, R-genes are often dominantly inherited, whilst loss of a susceptibility gene is recessively inherited. A prominent case of such so-called susceptibility factors are Mildew Resistance Locus O (MLO) proteins that are necessary for powdery mildew virulence (Büschges et al. 1997). Certain alleles of MLO in barley therefore provide broad-spectrum, very long-lasting resistance against powdery mildew (Freisleben and Lein 1942; Büschges et al. 1997; Humphry et al. 2006). Moreover, modern breeding tools allow the targeting and knockdown of homologous genes in a varied range of crops, including grapevine, which also results in reduced susceptibility to mildews (Feechan et al. 2008; Pessina et al. 2016; Kusch and Panstruga 2017). It may therefore appear that such resistances represent a 'silver bullet' in plant breeding, but of course there is a catch: since susceptibility genes have a conserved biochemical and physiological function in the plant, their loss normally decreases host fitness (especially in the absence of pathogens). For example, MLO genes can play various roles in the host plant, such as balancing immune functioning and controlling root growth (Chen et al. 2009), or controlling sexual reproduction (Kessler et al. 2010). Trade-offs associated with the loss of a susceptibility gene might therefore be unacceptable for the breeder, but they need not be so. Indeed, despite relatively few papers published explicitly on the topic, susceptibility genes (or recessive resistances) for disease-resistance breeding have been successful exploited for over a century. Moreover, susceptibility genes are not as rare as one might expect: a recent review lists over 150 examples, from a variety of plant species and which confer resistance to viruses, bacteria, fungi or nematodes (Van Schie and Takken 2014). For example, approximately half the known resistances against plant viruses are recessively inherited (Truniger and Aranda 2009; Hashimoto et al. 2016), and a recently discovered recessive resistance provides an important new tool for controlling grapevine leafroll virus (Djennane et al. 2021). Furthermore, recent genetic screening for recessively inherited resistances against grapevine downy and powdery mildew has identified several candidates which appear to achieve major effects. The cloning of these resistances will not only provide more insights into the underlying mechanisms, but also provide new resistances for traditional breeding or new targets for biotechnological applications.

2.4.2.5 Defence Mechanisms of Grapevine and Selection by Biochemical and Microscopical Markers

Like all living organisms, plants are in constant contact with other potentially pathogenic microorganisms (viruses, bacteria, phytoplasmas, fungi). They can detect an attack by sensing conserved signatures called 'microbe-associated molecular patterns' (MAMPs) or via molecular patterns associated with tissue deterioration, termed 'host-derived damage-associated molecular patterns' (DAMPs) (Héloir et al. 2019). These mechanisms are made possible by specific pattern-recognition receptors (plant plasma membrane pattern-recognition receptors or PRRs) that are present in the plasma membrane. MAMPs can be structural components of micro-organisms such as lipopolysaccharides (LPSs), peptidoglycans (PGNs), rhamnolip-ids (RLs), chitin or the β-glucans (glucose polymers) of fungi cell walls, as well as toxins or secreted enzymes such as the xylanases or the fungal endopolygalacturo-nases. DAMPs are endogenous danger signals such as oligogalacturonides (OGs) or cutin monomers, released by the plant cell wall e.g. through the action of cutinases produced by the fungi when penetrating the plant tissues (Gindro and Pezet 1999). The perception of MAMPs/DAMPs triggers a complex cascade of signalling events including ionic fluxes such as Ca^{2+} influx, production of reactive oxygen species (ROS: O^{2-} / H_2O_2) and nitric oxide (NO), or activation of kinase proteins and over-expression of pathogen-related genes (PR genes). These early events cause a mas-sive transcriptional reprogramming of primary and secondary metabolites. The induction of genes specifically associated with defence can lead to the synthesis of:

1. pathogenesis-related proteins (PR proteins), including hydrolytic enzymes such as the β-1,3-glucanases or the chitinases, associated with cell-wall degradation in microorganisms,
2. induced, toxic secondary metabolites such as the phytoalexins,
3. compounds involved in cell-wall reinforcement such as callose,
4. or proteins involved in hypersensitive response (HR), associated with pro-grammed cell death (PCD).

These defence reactions are regulated by phytohormones involving the metabo-lism of salicylic acid (SA), jasmonic acid (JA) and ethylene (ET). The SA path-way is mainly triggered by biotrophic pathogens, while the JA pathway is activated mainly by necrotrophic pathogens such as *Botrytis cinerea*. Like the majority of higher plants, the grapevine possesses a complex constitutive and/or biochemical defense system against fungal aggressions. Depending on the *Vitaceae* species in question, these systems are expressed differently, supplying the plants with more or less effective means of defence. With cultivated vines, this produces different levels of sensitivity to diseases. It should be emphasised that major grapevine pathogens such as grey mould, downy and powdery mildew or black rot do not respond in the same way to the plant's defence molecules (which can be constitu-tive in the green organs and berries, or induced during stress), since their epidemi-ology differs. Likewise, grapevines induce different resistance mechanisms depending on the pathogen in question and its mode of infection (Armijo et al. 2016). For example, downy mildew development can be curbed by induced mech-anisms such as blockage of the stomata by callose, or by stilbenes synthesised in the epidermal cells. Powdery mildew can be kept in check by the physical and chemical structure of the surface waxes (epicuticular waxes) constituting the green organs, while grey mould development is thwarted by the presence of spe-cific constituent tannins in the berry epidermis and by skin thickness. The efficacy of grapevine defences depends on the rapidity of expression of the genes control-ling the different metabolic pathways involved and on the constituent elements characteristic of each *Vitaceae* and each variety of *Vitis vinifera*. The study of these defence systems is a complex matter, since they are conditioned both by the genetic inheritance of the grape varieties and by the biochemical mechanisms of infection of the different pathogens. Research has identified different systems of resistance which when taken separately prove to be of greater or lesser efficacy, but which when combined can inhibit several different pathogens. Knowledge of these mechanisms is a basic requirement for selecting new grapevine varieties which can jointly resist several fungi.

Grapevines are known to have several defence genes which are associated with PAL (phenylalanine ammonia lyase), STS (stilbene synthase), LOX (lipoxygenase), different pathogenesis-related proteins (PRprots) such as chitinases and glucanases, and polygalacturonase-inhibiting proteins (PGIPs) (Adrian et al. 2012). PAL and STSs lead to the synthesis of resveratrol, one of the main stilbenic phytoalexins of grapevine. Other enzymes have the effect of destroying the parietal and membrane structures of fungal pathogens. Certain building blocks of grapevine cells can also play a major role in defence mechanisms: as explained above, the physical and chemical structure of leaf surface waxes in certain powdery mildew-resistant Vitaceae block the development of fungal infection structures (Schnee 2008). Likewise, the degree of polymerisation of certain berry tannins inhibits the lytic enzymes of *Botrytis cinerea* (Perret et al. 2003). The secretion of callose plugging the stomata has been observed during *Plasmopara viticola* attacks (Gindro et al. 2006; Yu et al. 2012). In general, however, it is the induced defence reactions, also termed 'hypersensitive reactions' (HRs), which ensure the complete inhibition of parasitic developments in fungal disease-resistant grapevine varieties.

Stilbenes of Grapevine

The stilbenes are a family of phenolic molecules whose chemical structure, both in the monomer and oligomer states, consists of a *cis*- or *trans*- oriented 1,2-diphenylethylene nucleus (Fig. 2.56) with the formula $C_{14}H_{12}$. When exposed to UV light, they emit an intense blue fluorescence (Fig. 2.57). This characteristic has given rise to the name 'stilbene', derived from the Greek 'στίλβω'» ('stilbo'), 'to shine'.

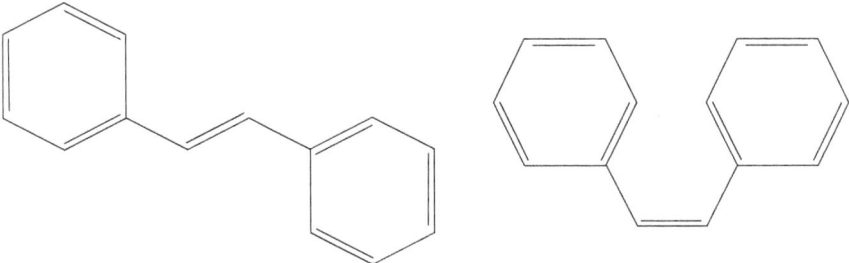

Fig. 2.56 1,2-diphenylethylene nucleus (left: *trans*-oriented [*trans*-1,2-diphenylethylene, (E)-stilbene]; right: *cis*-oriented [*cis*-1,2-diphenylethylene, (Z)-stilbene]

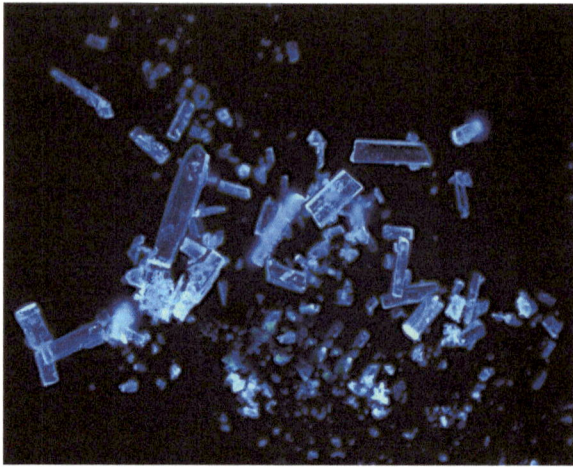

Fig. 2.57 Blue fluorescence emitted by resveratrol crystals exposed to UV light

Stilbenes are secondary metabolites produced via the phenylalanine/polymalo-nate pathway (phenylpropanoid pathway) in which the stilbene synthase (STS; EC 2.3.1.95) catalyses the formation of the stilbene skeleton (simple monomeric stil-benes) (Fig. 2.58) from a CoA ester derived from cinnamic acid (p-coumaroyl-CoA in the case of resveratrol) and 3 units of malonyl-CoA in a single reaction. STS belongs to the type-III polyketide synthase enzyme superfamily and shares a high degree of amino acid sequence homology with chalcone synthase (CHS EC 2.3.1.74), which utilises the same substrates as STS, but is responsible for the for-mation of flavonoid-type compounds. Over 48 STS genes have been identified in plants, 32 of which are functional (Parage et al. 2012).

Simple stilbenes can be hydroxylated, methylated, esterified, glycosylated or prenylated by the action of specific enzymes (Fig. 2.59). They can be isolated from plants in the form of monomers or polymers (dimers, trimers, tetramers and hexam-ers). Resveratrol, the first stilbene ever identified (Takaoka 1939), and its derivatives such as pterostilbene, oxyresveratrol or the viniferins, are the best-known stilbenes owing to their varied biological, *inter alia* biocidal, activity and their medicinal properties.

In the Vitaceae, resveratrol (a monomer) and α, β and ε viniferin (dimers) have been identified as important phytoalexins (Langcake and Pryce 1977), to wit, anti-microbial substances synthesised *de novo* by the grapevine which accumulate rap-idly in the zones infected by various pathogens, in particular fungal pathogens. Pterostilbene, a methylated stilbene, was subsequently identified in grapevine by Langcake et al. (1979). The name 'resveratrol' derives from the abbreviation for the class of molecules to which it belongs, to wit, resorcinol, coupled with the name of the plant from which it was extracted and identified for the first time (*Veratrum*), + -ol, indicating the presence of a hydroxyl group. Pterostilbene, for its part, was first described by Späth and Schläger (1940) in red sandalwood (*Pterocarpus santalinus*

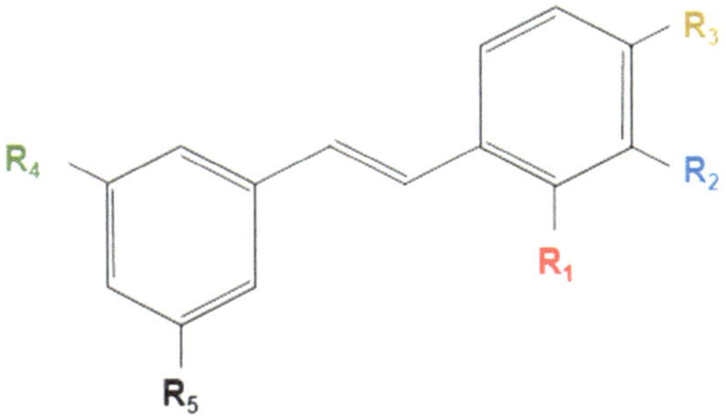

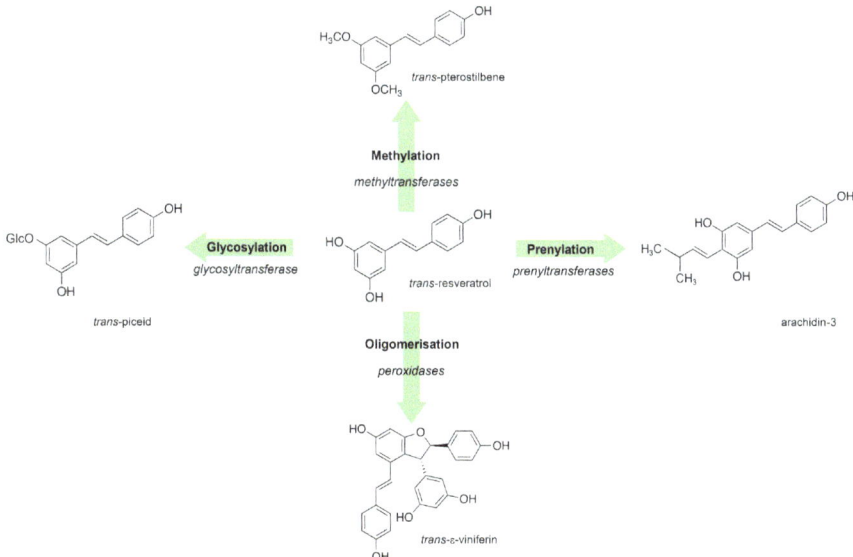

R_1	R_2	R_3	R_4	R_5	Phenolic acid precursor	Stilbene
H	H	OH	OH	OH	p-coumaric	resveratrol
H	H	OH	OCH₃	OCH₃	p-coumaric	pterostilbene
H	H	H	OH	OH	cinnamic	pinosylvine
H	OH	OH	OH	OH	caffeic	piceatannol
H	OH	OCH₃	OH	OH	isoferulic	rhapontigenin
OH	H	OH	OH	OH	2'-4'-dihydroxycinnamic	hydroxyresveratrol
H	H	OH	OGlu	OH	p-coumaric	piceid
H	OH	OH	OGlu	OH	caffeic	astringin

Fig. 2.58 Chemical structure of some monomeric stilbenes in the *trans* form and of their precursors

Fig. 2.59 Examples of enzymatic reactions on a monomeric stilbene such as resveratrol and the formation of various derivatives

L.). A total of 459 stilbenes have been identified in over 45 botanical families and 196 plant species, the exhaustive list of which was compiled by Teka et al. (2022). Beyond the plant kingdom, stilbenes have also been identified e.g. in an Antarctic sponge with the name *Kirkpatrickia variolosa* (Garmini et al. 1995), as well as in an entomopathogenic bacterium *Phtorhabdus luminescens* (Richardson et al. 1988). The currently available list of stilbenic compounds is probably not exhaustive, and other molecules will be identified thanks to constant optimisation of the analytical methods; even so, said list highlights the wide distribution of the biochemical pathway of stilbene synthesis in plants and other living organisms.

Over 23 constitutive monomeric stilbenes are present in *Vitis vinifera* out of the 80 currently identified within the Vitaceae family. Nevertheless, some stilbenic compounds are induced in response to a biotic or abiotic stressor and are not necessarily present constitutively in the vegetative organs of the plant (Błaszczyk et al. 2019). Goufo et al. (2020) identified over 80 stilbenes in the vegetative organs of *V. vinifera*, including the 23 aforementioned monomers, 30 dimers, 8 trimers, 16 tetramers, and one hexamer, viniphenol. Table 2.7 lists a sample of the different oligomers of *V. vinifera* taken from the review of Goufo et al. (2020). The chemical structure of these same molecules is illustrated in Fig. 2.60. Pentamers have been identified within the Vitaceae family, such as Amurensin E (Fig. 2.61) in *Vitis amurensis*, but have not yet been detected in *Vitis vinifera*.

Although resveratrol and its derivatives are present in lignified stem tissue (Schnee et al. 2013; Lambert et al. 2013; Guerrero et al. 2016), certain compounds are absent from the leaves, or from the young shoots in the free form. By contrast, numerous derivatives of resveratrol ranging from the monomer to the hexamer, like chunganenol (Fig. 2.61), identified in *Vitis chunganensis*, or methylated, such as

Table 2.7 Example of stilbenic compounds out of the 78 identified in the different vegetative organs of *Vitis vinifera*, taken from the publication of Goufo et al. (2020)

Group of compounds	Stilbenic compound	Vegetative Organs				
		Leaves	Stems	Canes	Wood	Roots
Monomers	*trans*-piceid	Green	Green	Green	Green	Green
Monomers	*trans*-resveratrol	Green	Green	Green	Green	Green
Monomers	*trans*-pterostilbene	Green	Green	Red	Red	Red
Monomers	*trans*-piceatannol	Green	Green	Green	Green	Green
Monomers	*trans*-pinostilbene	Green	Red	Red	Red	Red
Dimers	Leachianol G	Red	Green	Green	Green	Red
Dimers	Ampelopsin A	Red	Green	Green	Green	Green
Dimers	Pallidol	Green	Green	Green	Green	Green
Dimers	*trans*-δ-viniferin	Green	Green	Green	Red	Green
Dimers	*trans*-ε-viniferin	Green	Green	Green	Green	Green
Dimers	Vitisinol C	Red	Green	Red	Red	Green
Trimers	*trans*-miyabenol C	Green	Green	Green	Green	Green
Trimers	α-viniferin	Green	Green	Red	Green	Red
Trimers	Viniferol D	Red	Green	Green	Green	Green
Tetramers	Hopeaphenol	Green	Green	Green	Green	Green
Tetramer	Vitisin A	Green	Green	Red	Green	Green
Tetramers	Viniferol A	Red	Green	Green	Green	Green
Hexamers	Viniphenol A	Red	Green	Red	Red	Red

Green: detected in the vegetative organ; Red: not detected in the vegetative organ

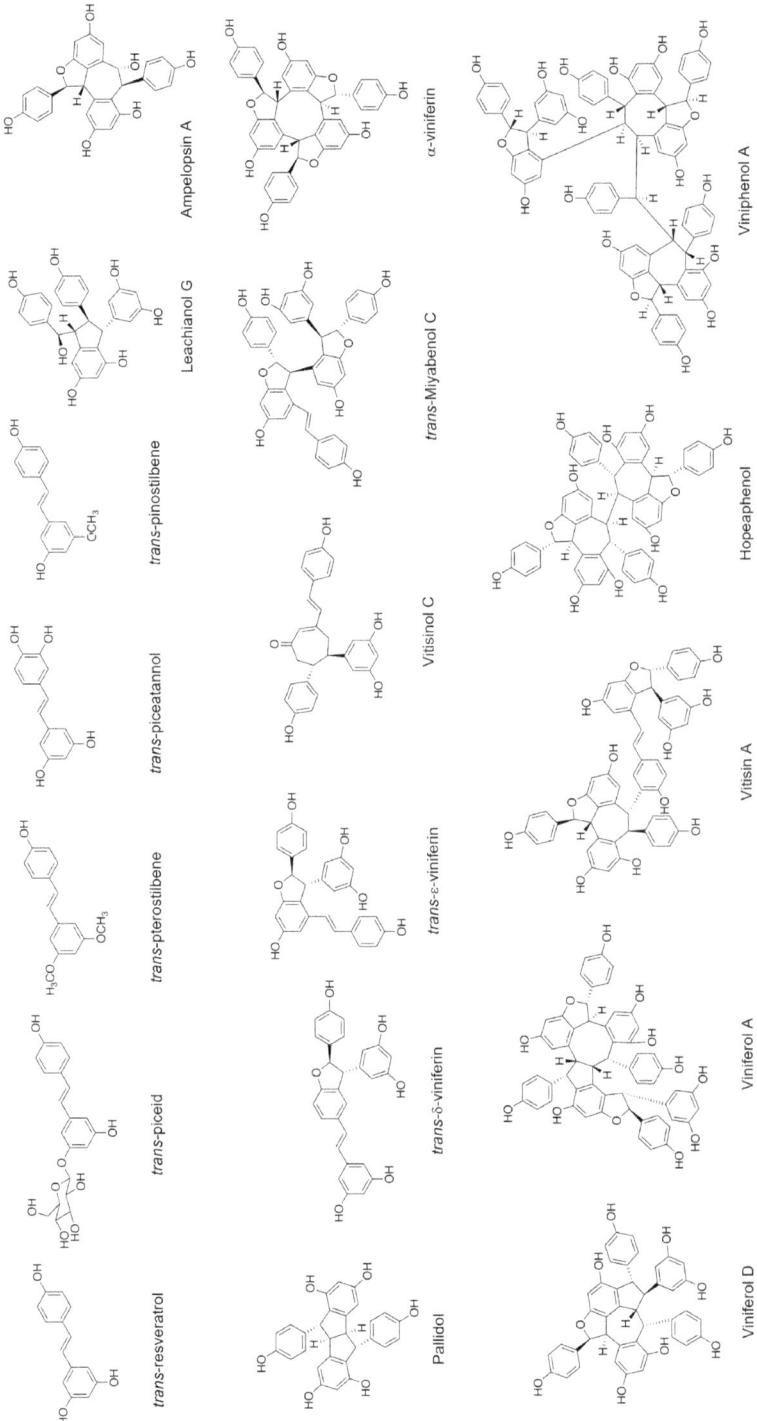

Fig. 2.60 Chemical structure of stilbenic monomers and oligomers of *Vitis vinifera* listed in Table 2.7

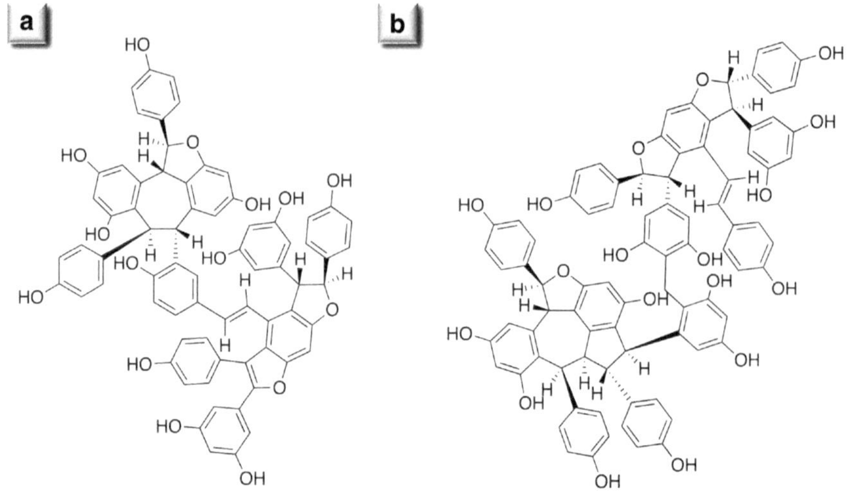

Fig. 2.61 Chemical structure of amuresin E (**a**), a resveratrol pentamer found in *Vitis amurensis*, and of chunganenol (**b**), a stilbene hexamer found in the berries of *Vitis chunganensis*

pterostilbene, are present in the berries of *V. vinifera* (Pezet and Pont 1988; Błaszczyk et al. 2019).

Pezet and Pont (1990) demonstrated the toxicity of pterostilbene to the asexual spores (conidia) of *Botrytis cinerea*, the said compound interfering with cellular respiration leading to the rapid destruction of cell content including the ribosomes, the endoplasmic reticulum and the mitochondrial and nuclear membranes, as well as the complete structural breakdown of the fungal cell membranes (Fig. 2.62).

The resveratrol synthesis rate after induction by a stress depends on the grape variety and enables the evaluation of the grapevine's resistance to powdery mildew, grey rot and downy mildew. Resveratrol and some of its derivatives are present in the grape berries and the mature woody shoots as well as in the wine. When Renaud and de Lorgeril (1992) and Frankel et al. (1993) demonstrated the cardiovascular-protective effects of resveratrol, the 'French paradox' was born—a theory, contested to this day, that a Mediterranean diet, including the moderate consumption of red wine, is beneficial for cardiovascular ailments. Since then, numerous research projects have touched on the medical benefits of resveratrol and pterostilbene, such as their anticancer, anti-ageing or anti-inflammatory activity (Li et al. 2017). Since then, furthermore, research groups throughout the world have studied the composition of the stilbenes found in the wine and berries (Pezet and Cuenat 1996; Błaszczyk et al. 2019). The synthesis of pure pterostilbene and resveratrol has enabled the study of the toxic effect of these stilbenes on the *B. cinerea* pathogen. The enzymatic synthesis of the resveratrol dimer δ-viniferin, produced by the peroxidases of *B. cinerea*, as well as the purification of ε-viniferin from lignified shoots, has enabled the evaluation of their toxicity to downy mildew (*Plasmopara viticola*) and powdery mildew (*Erysiphe necator*). In the same vein, several research projects have focused on the composition of the stilbenes naturally present in the woody

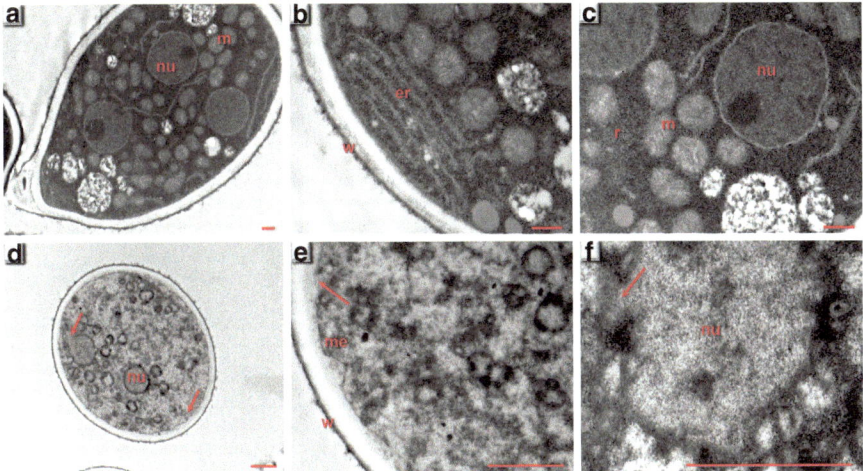

Fig. 2.62 Effect of pterostilbene on the conidia (asexual spores) of *Botrytis cinerea* observed by transmission electron microscopy. From **a** to **c**: non-treated control conidia; from **d** to **f**: conidia treated for 2 h with 1 mM of pterostilbene. The red arrows show the structural breakdown of the cellular elements, including the plasma and nuclear membranes, as well as the destruction of the integrity of cellular content compared to non-treated conidia. *er* endoplasmic reticulum, *m* mitochondria, *me* plasma membrane, *nu* nucleus, *r* ribosomes, *w* wall. Scale bar represents 2 μm

shoots of grapevines, in which a wide range of stilbenic compounds are immobilised. Vine shoots are thus an important source of stilbenes that can be exploited (Schnee et al. 2013; Aliano-Gonzales et al. 2020) both as antifungal molecules in agronomy and as medically beneficial compounds, owing to their antioxidant, antimicrobial, anti-ageing, antiviral and anticancer activity. Given the benefits of these compounds, it is hardly surprising that certain methodologies have been developed to create a wide new range of bioactive chemicals from known stilbenic compounds. For example, the use of the *B. cinerea* secretome has allowed the generation via biotransformation of chemically new compounds from resveratrol and/or pterostilbene (Fig. 2.63) (Gindro et al. 2017)—compounds exhibiting fungicidal (Schnee et al. 2022), bactericidal (Righi et al. 2020) or Wnt-inhibiting activity in the case of triple-negative breast cancer (Huber et al. 2022).

Selection Tools

Powdery mildew (*Erysiphe necator)* and grapevine downy mildew (*Plasmopara viticola)* are obligate biotrophic organisms that are entirely dependent on the host cells of photosynthetically active tissues to complete their life cycle. Plants use two main resistance strategies to limit the invasion and growth of biotrophic pathogenic fungi: penetration resistance and programmed cell death (PCD). Penetration resistance blocks the intrusion of the cell wall and the membrane by the germination tube of the germinated spores, thereby preventing the formation of haustoria.

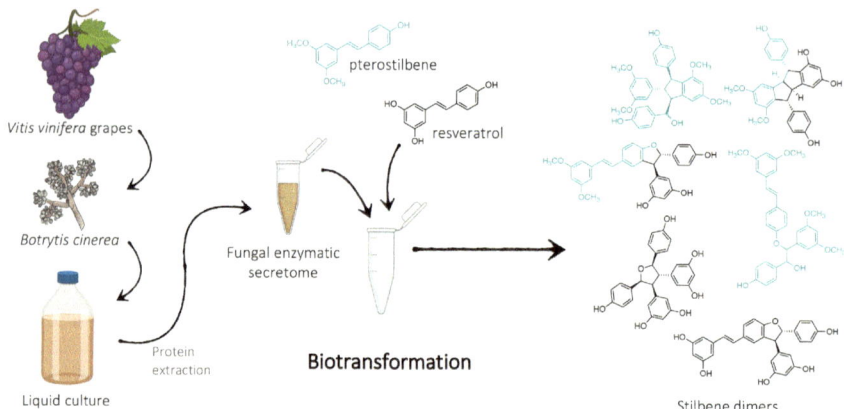

Fig. 2.63 Schematic view of a biotransformation approach for the production of stilbene dimers. *Botrytis cinerea* is isolated from grapes and grown in a liquid medium. Extracellular proteins are extracted, resulting in a fungal enzymatic secretome, which is mixed with stilbenes (*trans*-pterostilbene in blue, *trans*-resveratrol in black). This type of reaction produces many stilbene dimers with different structures, some of which are shown here. (Illustration © Robin Huber, University of Geneva, Switzerland)

Programmed cell death, for its part, takes place inside the epidermal cells after penetration by the pathogen and causes the degeneration of the invaded cell, thereby cutting off the supply of nutrients needed by the biotrophic fungus to continue growing and developing.

In the plant cell, the various defence responses, all of which are interconnected, are immediately triggered according to two processes:

1. After recognition of specific molecules released by the pathogen and activation of a response cascade (pathogen-associated molecular patterns, 'PAMPs'), for example the chitin released from the fungal cell wall. In this case, the released molecules are recognised by plasma membrane receptors of the host cell which activate response cascades, inducing various defence mechanisms such as ROS production, defence gene activation, stress hormone and phytoalexin synthesis, or cell wall reinforcement. This strategy is the first possible line of defence.
2. Through the action of effectors, which are generally proteins secreted by the pathogen into the plant cell whose purpose is to interfere with the first line of defence. Depending on the individual case, the plant develops R proteins capable of recognising these effectors and then initiating the aforementioned defence response cascades.

Powdery Mildew (*Erysiphe necator*)

Powdery mildew is an ectophytic or ectoparasitic biotrophic fungus that develops on the surface of the green organs of the vine and sends out rootlike structures (haustoria) that penetrate into the epidermal cells of the plant, enabling the fungus to obtain nutrients and pursue its surface development (Fig. 2.64). Consequently, *E. necator* does not require specific openings or wounds to penetrate into the plant's tissues.

In terms of resistance, Table 2.5 shows the 15 loci currently identified in several grapevine species native to North America, China and Central Asia which are assumed to contain resistance genes ('*R* genes') conferring a high level of resistance to powdery mildew. *R* genes represent the most important class of genes associated

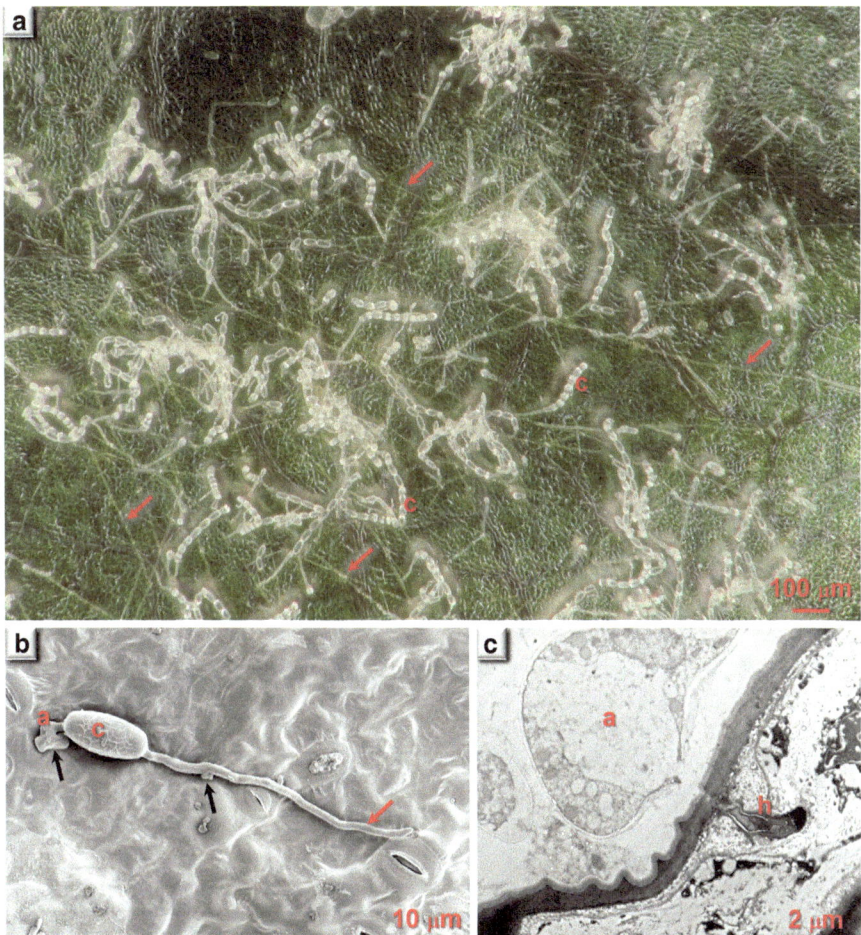

Fig. 2.64 Development of *Erysiphe necator*, an ectophytic biotrophic fungus on the surface of a grapevine leaf. (**a**) Observation under an optical microscope, with the red arrows indicating the filaments of the mycelial network; (**b**) Observation of a germinated conidium under a scanning electron microscope, showing a hypha on the surface of a grapevine leaf with the production of appressoria (infection cushions, black arrows) which represent the sites where the haustoria have penetrated into the epidermal cell layer; (**c**) Observation under a transmission electron microscope of a haustorium that has penetrated into the epidermal cell layer; *a* appressorium, *c* conidium, *h* haustorium

with plant resistance, coding for proteins with nucleotide-binding site leucine-rich repeat domains (Gururani et al. 2012). In Vitaceae, the *R* genes are grouped in tandem repeats of genomic regions. These regions have been genetically mapped, revealing nine loci coding for *R* gene sequences conferring resistance to *E. necator*, such as *Run1*, *Run2*, *Ren1*, *Ren2*, *Ren3*, *Ren4*, *Ren5*, *Ren6* and *Ren7* (Qiu et al. 2015). Among the *R* genes identified in connection with the resistance of certain grapevine varieties to powdery mildew, we can cite in particularly MrRUN1, linked to the activation of programmed cell death in the infected epidermal cells of *Vitis rotundifolia* (Feechan et al. 2013). With infection by *E. necator*, it appears that either the transcription of certain genes may be increased, or that certain genes have a different level of expression in susceptible and resistant varieties (Fung et al. 2008). For example, lipoxygenase genes play an important role in grapevine defence through lipid peroxidation events, through the synthesis of compounds involved in signalling which lead to the production of compounds such as ROS at the infection site, or which exhibit antifungal activity (Guche et al. 2021). Moreover, more-recent new breeding technologies have been used to genetically improve grapevines by eliminating endogenous genetic material (Dalla Costa et al. 2016; Dalla Costa et al. 2020) or generating DNA-free modifications using ribonucleoproteins (Malnoy et al. 2016). Likewise, the use e.g. of "Microvine" (Torregrosa et al. 2019) or "Picovine" (Chaïb et al. 2010) has helped to accelerate the discovery of new target genes for deciphering the resistance of *Vitis* to powdery mildew (Dai et al. 2012).

Besides the molecular and proteomic approach, it is possible to study the events linked to the infection metabolomically and microscopically in order to identify cytological and analytical tools for evaluating the resistance of the grape varieties to *E. necator*. Optical microscopy can highlight resistance events such as the presence of callose in the papillae of certain grape varieties creating mechanical barriers to the penetration of powdery mildew haustoria (Feechan et al. 2011). Scanning electron microscopy of the adaxial surface of grapevine leaves has revealed that the epicuticular wax crystallisation pattern differs between susceptible and resistant grape varieties. The susceptible variety *V. vinifera* cv Chasselas has a relatively smooth surface and a few scattered protuberances, giving it a crusty appearance. By contrast, the surface of *V. candicans*, which is highly resistant to powdery mildew, is densely covered with wafer-like crystals jutting out perpendicularly to the plane of the leaf. The width and crystallisation patterns have been confirmed by transmission electron microscopy (Fig. 2.65). The wafers have thin margins and relatively triangular shapes on which no haustoria have been observed.

Although these findings are interesting, scanning and transmission electron microscopy are very time-consuming, and are therefore not suitable for rapidly evaluating the resistance of several hundred seedlings in a breeding programme. Nevertheless, it remains worthwhile to apply this microscopic approach alongside the chemical determination of the wax composition of these new cultivars in order to obtain a partial explanation of their resistance to powdery mildew. A recent study on the chemical analysis of the epicuticular waxes of four powdery mildew-resistant grapevine genotypes (Italia × Mercan-174, Gürcü, Isabella, Özer Karası) and of two susceptible varieties (Cabernet Sauvignon and Italia) demonstrated that there was

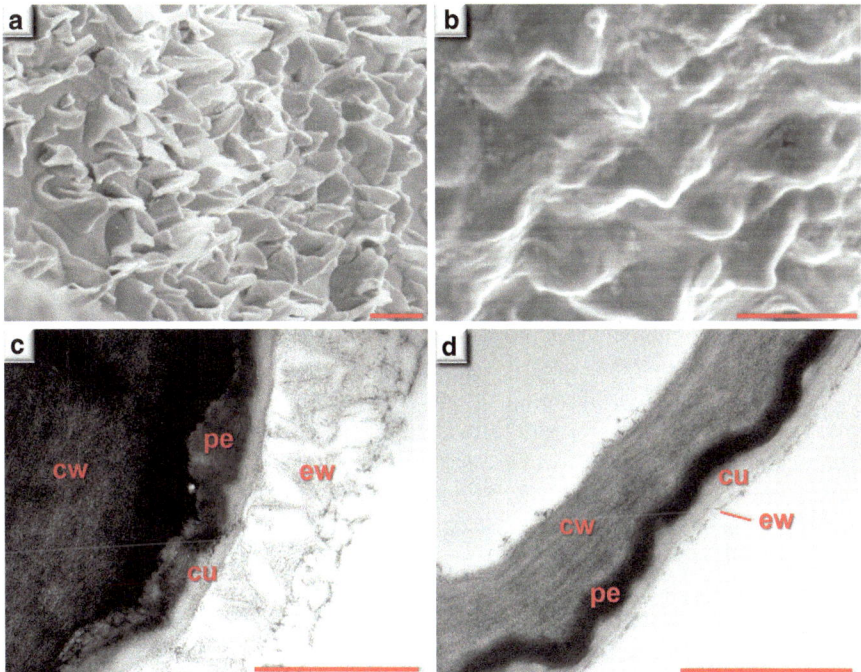

Fig. 2.65 Observation by scanning and transmission electron microscopy of crystallisation patterns of the adaxial surface of a vine leaf of a powdery mildew-resistant (*Vitis candicans*) and powdery mildew-susceptible (*Vitis vinifera* cv. Chasselas) grape variety. (**a** and **c**) Surface covered with a scaly tangle of epicuticular wax crystals in V. *candicans*. (**b** and **d**) Smooth epicuticular wax surface with several small, scattered protuberances. *ew* epicuticular waxes, *cu* cutin, *cw* cell wall, *pe* pectin. Scale bars represent 1 μm

no difference in terms of the total amount of wax on either the leaves or the berries. It is in the chemical composition of the waxes that the most striking differences are to be found, in particular the presence of specific compounds in the resistant varieties such as various fatty acids, alkanes, terpenes and several indole and ketone derivatives exhibiting strong antifungal activity (Özer et al. 2017). Identifying specific compounds with a strong fungicidal activity should enable the development of tools for improving the identification of resistant genotypes in powdery mildew-resistant variety breeding programmes. However, the microscopic approach remains worthwhile, since the crystallisation pattern of the waxes is closely linked to their composition.

Artificial inoculations are generally performed on leaf discs or on whole leaves incubated in optimal conditions for the growth of *E. necator* (Feechan et al. 2011). The observation of the latter's development and the quantification of the conidia germination rate, the extent of appressorium formation, mycelial network density and sporulation level 6 days after inoculation are epidemiological criteria for determining the level of susceptibility of the seedlings to *E. necator*. A phenomenon peculiar to infection by *E. necator* is the production of mycelium strictly on the surface of the host. Consequently, the induction of defence metabolites only increases during the development of infectious structures (appressoria, infection cushion and differentiation of intracellular haustoria). *E. necator* haustoria only infect the first layer of epidermal cells, whilst *P. viticola* develops an intercellular mycelium that invades the mesophyll and forms numerous infectious structures (haustoria) in the cells, enabling it to absorb the plant constituents required for its growth and development.

As described by Sosa-Zuniga et al. (2022), hormones play a key role in plant defence responses, particularly jasmonic acid (JA) and ethylene (Et) for necrotrophic pathogens such as *Botrytis*, and salicylic acid (SA) for biotrophic pathogens such as powdery mildew. Although the best-described hormonal response pathway against powdery mildew attack is that of SA, it has also been shown that Et and JA contribute to grapevine response to infection by powdery mildew. This response mechanism is associated with the induction of a series of defence proteins, and although there is no direct evidence linking the induction of these defence proteins to the phenylpropanoid pathway, a correlation has been observed in the increased biosynthesis of phytoalexins and the upregulation of phenylalanine ammonia-lyase and stilbene synthase genes. These increases were positively correlated with the accumulation of toxic stilbenes (Jacobs et al. 1999; Schnee 2008). Nevertheless, given the highly localised synthesis of stilbene phytoalexins at the infection sites, quantification of the stilbenes induced by powdery mildew infections must be linked to the number of appressoria and infectious structures (Schnee 2008). Recent studies have shown that, following infection by powdery mildew, greater amounts of phenolic and polyphenolic compounds such as catechins, epicatechin and gallic acid were produced in response to the infection in the resistant grape varieties (Atak et al. 2021). It appears that total resistance to powdery mildew, leading to a complete absence of sporulation, has only been observed in certain grape varieties that build up a mechanical barrier (epicuticular wax patterns, callose deposits) that limits penetration and an effective PCD response following penetration into the epidermal cells (ROS, stilbenes).

Downy Mildew (*Plasmopara viticola*)

Unlike powdery mildew, *Plasmopara viticola* only penetrates through the stomata in the green tissues of the grapevine. Active plant structures that can open and close, the stomata are responsible for exchange processes (e.g. evapotranspiration, mineral, gas and hormone exchange) between the environment and the internal tissues of the plant (Fig. 2.66).

The first phase of infection by downy mildew consists in the germination of the zoospore, penetration of the germ tube into the stomatal cavity and the formation of a vesicle in the substomatal chamber which produces one or more hyphae. These filaments then develop by insinuating themselves between the cells of the mesophyll and sending haustoria out into the cells to absorb nutrients, then, depending on environmental conditions, sporulating by releasing a large number of sporangia

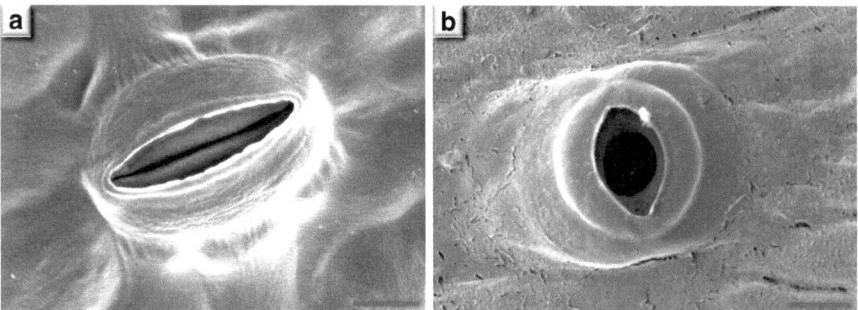

Fig. 2.66 Stomata on a grapevine leaf observed under scanning electron microscopy; closed (**a**) and open (**b**). Scale bars represent 10 μm

through these same stomata (Fig. 2.67). Moreover, the research of Guillier et al. (2015) demonstrated a deregulation of stomatal activity after infection by downy mildew, causing the stomata to remain abnormally open. This phenomenon is most likely due to the induction of two thermostable, glycosylated proteins accumulating in the apoplastic fluids: the first was identified as a kinase with a phototropin conserved domain and the second as a kinase containing a lysophospholipase conserved domain. In susceptible and tolerant grape varieties, infection by *P. viticola* results in an overall reduction in photosynthetic processes leading to altered carbohydrate content, the inappropriate upregulation of genes coding for PR (pathogenesis-related) proteins or phenylpropanoid-pathway enzymes, the slight accumulation of hydrogen peroxide (H_2O_2), and lipid peroxidation. Taken together, these factors are not sufficient to inhibit downy mildew development during the initial stages of the infection (Nascimento et al. 2019).

Certain QTLs are linked to very high levels of grapevine resistance to downy mildew, such as *Rpv1*, or even to total resistance in the case of *Rpv2* (Table 2.4). One of the first resistance genes to be identified and functionally characterised was named *MrRPV1* (Feechan et al. 2013). This gene confers high resistance to downy mildew, strongly inhibiting hyphae development and sporulation whilst being co-expressed with certain stilbene synthase genes or with ROS production (Qu et al. 2021). *P. viticola*'s attempts to infect resistant grape varieties lead to the induction of the expression of various genes linked to the secondary metabolism of the plant, such as the PR genes (e.g. PR2), stilbene synthases (Richter et al. 2006; Malacarne et al. 2011), genes linked to callose synthase expression (Yu et al. 2012), or to the reaction cascade involving the transcription factor VvWRKY33 (a DNA-binding protein) activating the promoter of the VvPR10.1 gene belonging to the PR10 gene groups (Merz et al. 2015). Lenzi et al. (2016) have also demonstrated an induction of *NAC* genes (protein transcription factors involved in the regulation of genetic expression) in microdissected stomata. The modulation of these genes is limited to the stomata cells, indicating that host response is mainly confined to the infection sites and that short-distance signals are produced by the stomatal cells towards the adjacent cells.

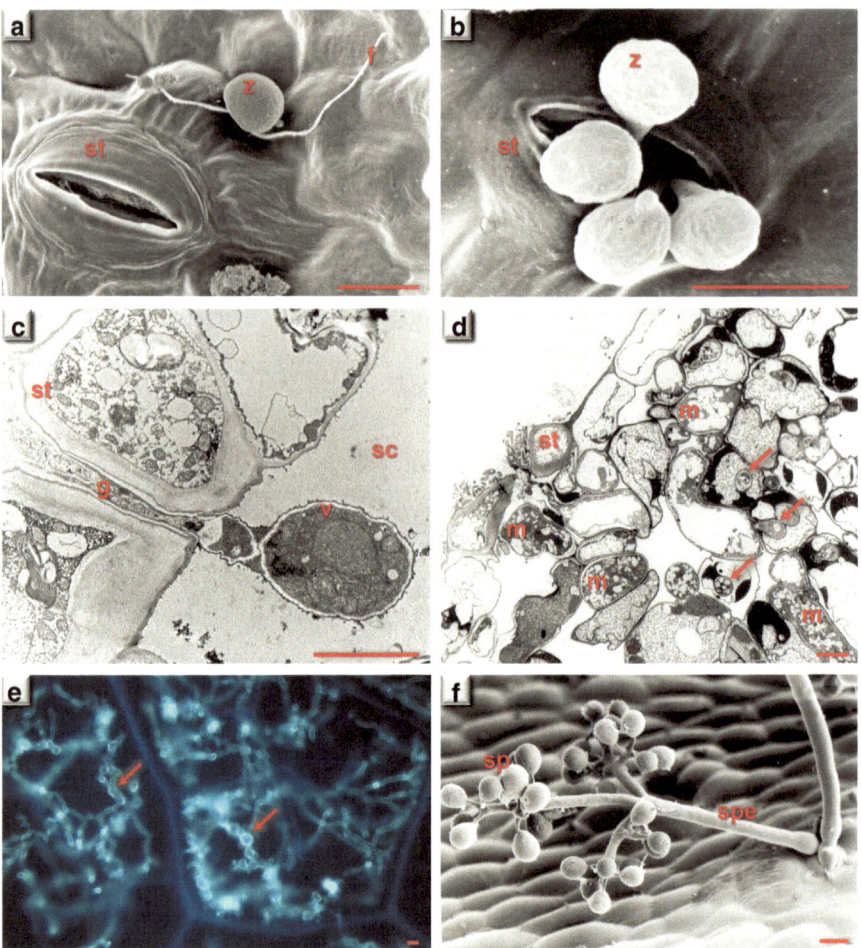

Fig. 2.67 Development of downy mildew in the leaf tissues of the grapevine (*Vitis vinifera* cv. Chasselas) seen by scanning (**a**, **b**, **f**) and transmission electron microscopy (**c**, **d**) as well as fluorescence microscopy (**e**). A biflagellate zoospore approaches a stoma (**a**); The zoospores shed their flagella, encyst and germinate, sending out a germ tube (**b**) crossing the stomatal space and arriving at the substomatal chamber with the formation of a vesicle (**c**). Downy mildew hyphae insinuate themselves between the leaf cells and send out intracellular haustoria (arrows) (**d**). Downy mildew invades the leaf tissues (arrows) (**e**), then releases sporangiophores terminating in sporangia through the stomata (**f**). *f* flagellum, *g* germ tube, *m* downy mildew, *sc* substomatal chamber, *sp* sporangia, *spe* sporangiophore, *st* stomata, *z* zoospore. Scale bars represent 10 µm

Various physiological events are representative of the grapevine's resistance to downy mildew and can be studied via metabolomics and/or microscopy. Together with genomics and transcriptomics, this allows the study of resistance biomarkers and the development of tools that can be used in breeding programmes. Among the vine's responses to downy mildew, Cavaco et al. (2023) have demonstrated that the fatty-acid (FA) content of different grape varieties was significantly higher in the susceptible accessions.

Callose is a polysaccharide in the form of β-1,3-glucan with several β-1,6 branches. This compound is a constituent of the cell walls of a great many higher plants and plays important roles during the development of the plants and/or in response to biotic and abiotic stressors. Produced by enzymes belonging to the callose synthase family, callose is degraded by β-1,3-glucanases. When attacked, plants physically reinforce their cell walls with callose deposits to delay or prevent penetration by pathogens. The rapid synthesis of callose in the stomata of vine leaves and the formation of callose deposits around the haustoria after infection by *P. viticola* are phenomena that prevent or limit downy mildew penetration and development in resistant grapevine varieties (Gindro et al. 2006; Yu et al. 2012; Yin et al. 2017) (Fig. 2.68).

Based on findings from morphological observations and microscopic studies, Yu et al. (2012) identified five levels of grapevine resistance to downy mildew: (1) immune varieties exhibiting callose deposits around the stomata and inhibition of zoospore germination; (2) extremely resistant, callose deposits around and in the stomata, zoospore germination not impacted and possible formation of hyphae; (3) resistant, with callose deposits near the stomata and around the haustoria, but unable to block infection; (4) partially resistant, same as for (3) but with greater development of the hyphae; and (5) susceptible, with pathogen and hyphae development throughout the leaf mesophyll. Classifications of this type, based solely on histological responses and on the analysis of a limited number of varieties, can lead to contradictory results when other criteria come into play. To establish resistance scales, it is preferable to associate and combine several types of grapevine defence response: observation of cytological, morphological (formation of necroses and types of necroses, callose deposits) and metabolomic events such as the production of stilbene phytoalexins like resveratrol, ε or δ-viniferin (Pezet et al. 2003; Pezet et al. 2004; Gindro et al. 2006; Malacarne et al. 2011), whose expression is upregulated when infection occurs.

After penetration of the substomatal chamber a hypersensitive reaction may be triggered quickly and resveratrol is synthesised in the cells. Though itself of low toxicity for downy mildew, this compound generates other active and toxic stilbenes

Fig. 2.68 Callose deposit in a grape leaf stomata seen under a scanning electron microscope. Scale bar represents 10 μm

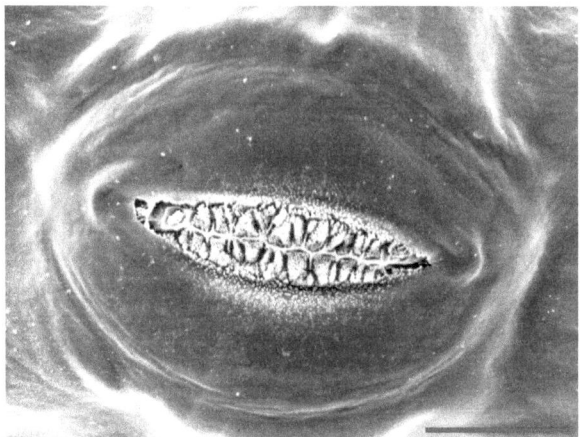

in resistant grape varieties such as viniferins, dimers of resveratrol produced by its oxidation by grapevine peroxidases. Resveratrol can also be methylated into highly toxic pterostilbene or glycosylated into non-toxic piceid (Pezet et al. 2004). The efficacy of this defence reaction is due entirely to the rapid synthesis of stilbenes and their concentration at the actual site of infection, causing the structural break-down of the fungal elements and inhibiting downy mildew growth. Thus, a grape-vine variety capable of resisting downy mildew will detect the parasite at an early stage and will immediately synthesise strong concentrations of resveratrol, itself oxidised into viniferins and/or methylated into pterostilbene. Conversely, suscepti-ble *V. vinifera* grape varieties synthesise resveratrol but this compound is largely glycosylated into piceid, which is harmless to downy mildew (Alonso-Villaverde et al. 2011).

Resveratrol and its oxidised or methylated derivatives are produced in the leaves and bunches of resistant grape varieties at the site of infection after artificial inocu-lation and can be analysed and quantified by HPLC (high-performance liquid chro-matography). Two of the oxidation products of resveratrol, for example, are ε-viniferin and δ-viniferin, described as two of the main stilbenes present in the leaves of stressed grapevines. To date, pterostilbene—the most toxic stilbene for downy mildew, powdery mildew and grey mould—has only been produced in large quantities in specific grapevine genotypes and backcrosses. The use of these geno-types must be prioritised to promote resistance. In susceptible grape varieties, res-veratrol is mainly glycosylated to form piceid. This addition of glucose to resveratrol protects the latter against subsequent oxidation. This is particularly important if we consider the respective toxicity of the different stilbenes facing each fungal patho-gen. The qualitative and quantitative analysis of the stilbenes in the leaves of grape-vine seedlings 48 h post-inoculation is highly predictive of the level of resistance of genotypes to *P. viticola*. Artificial infections are generally induced by inoculating drops of downy mildew propagule suspension. Given that the stilbenes are synthe-sised at the infection sites themselves, sampling the drop zone, which often exhibits more-or-less-extensive necrotic patches, is a good compromise for monitoring the appearance of the stilbenes produced. In immune varieties or those produced by their backcrosses such as *Vitis rotundifolia*, however, the necrotic zones consist of a number of scattered micro-necroses located below the surface of the infection drops (Fig. 2.69). In this case, samples must be taken with great care just around the necrotic zones, to avoid the risk of diluting signal intensity.

In muscadine *(V. rotundifolia)*, the infection process is blocked before the devel-opment of vesicles and infectious structures, which leads to a rapid accumulation of substantial quantities of stilbenes even at the level of the stomata (Fig. 2.70). Twenty-four hours after infection, the most toxic stilbenes, i.e. δ-viniferin and pterostilbene, are present at levels 24 and 42 times higher than their respective ED_{50} (the ED_{50} or median effective dose being the concentration that prevents 50% of pathogen development) (Alonso-Villaverde et al. 2011). Consequently, the most important stage in the inhibition of disease in muscadine could be in connection with the rapid induction of metabolic responses, which are produced before the appearance of haustoria.

Fig. 2.69 Appearance of patches of infection after artificial inoculation of leaves of different vine varieties with downy-mildew suspension drops. (**a**) Chasselas, abundant sporulation with no formation of necrotic patches extending beyond the area of the drop (×30); (**b**) Volturnis (*Rpv12*), fairly dense sporulation over the inoculation surface, presence of necrotic patches (×30); (**c**) Cabernet Volos (*Rpv12–Ren3*), scattered sporulation over the inoculation surface, presence of necrotic patches (×30); (**d**) Divico (*Rpv10–Rpv3.3–Ren3–Ren9*), no sporulation, extensive necrotic punctuations on the inoculation surface (×30); (**e**) Backcrossing within the 'Resdur' programme (*Rpv1–Rpv3–Rpv10–Run1–Ren3*), no sporulation, necrotic punctuations covering the inoculation surface (×30); (**f**) *Vitis rotundifolia* (muscadine), no sporulation, necroses confined to the infected stomata (arrows) (×200)

Fig. 2.70 Symptoms of infection of *Vitis rotundifolia* (muscadine) by *Plasmopara viticola*. (**a**) *V. rotundifolia*, photograph © Vincent Dumas, INRAE; (**b**) Infected stomata observed by fluorescence microscopy, strong induced fluorescence following infection (arrow); (**c**) Infected stomata with downy-mildew structural breakdown (arrow) observed by transmission electron microscopy; (**d**) Infected stomata seen under a binocular magnifier 48 h after inoculation, exhibiting necroses of the stomatal cells (arrow); *sc* substomatal chamber, *st* stomata. Scale bars represent 10 μm

Stilbene production rate is an excellent biomarker of the resistance of grape varieties, as confirmed by Eisenmann et al. (2019) on the correlation between the QTL *Rpv3* and the expression of genes associated with stilbene biosynthesis and programmed cell death. Experiments using protease inhibitors on *V. vinifera* leaves induced by methyl jasmonate have demonstrated that cysteine proteases are the key enzymes of the hypersensitivity reaction. The application of protease inhibitors (more specifically, caspases and cysteines) appears to partially block the jasmonic acid pathway and thus impede the synthesis of stilbenes, produced by the phenylalanine pathway. Both protease inhibitor treatments increased the infection rate in the resistant and immune varieties, reduced the production of toxic stilbenes and modified the plants' level of susceptibility to the pathogen. In particular, the immune grape variety *V. rotundifolia* became resistant (hyphae and haustoria were observed), the resistant variety (Divico) reached the level of a susceptible grape variety (a sporulation was observed) and the susceptible grape variety became even more susceptible, *P. viticola* having colonised the whole of the leaf mesophyll (Gindro et al. 2012) (Fig. 2.71).

Fig. 2.71 Procedure for selecting downy mildew-resistant grapevine varieties developed in Switzerland, consisting of two sorting stages. (**a**) Cotyledon formation after germination of a grape seed; (**b**) Growth and first leaf. (**c**) Stage 10 leaves: sampling of leaves 4 and 5 from the apex; (**d**) and (**e**) Inoculation of stripped leaves with an aqueous suspension of downy mildew zoospores, followed by incubation for 1 week at 100% humidity; (**f**) Downy mildew sporulation on certain susceptible genotypes (arrow), which are eliminated; (**g**) 20 μL drop inoculation of a dense suspension of downy mildew on the whole leaves of the remaining genotypes, incubation for 48 h at 100% humidity; (**h**) Callose production analyses and HPLC profiles of stilbene phytoalexin production; elimination of genotypes only producing non-toxic or low-toxicity stilbenes such as resveratrol and piceid, followed by return of the retained genotypes to the breeder. (**i**) Growth of the selected genotypes to obtain wood for further grafting and planting in the field for the agronomic and organoleptic evaluation

The Resistance Breeding Program at Agroscope in Switzerland Predates the Use of Reliable QTL Markers

Since 1996, varietal creation in Switzerland has been geared to obtaining downy and powdery mildew-resistant grape varieties through conventional interspecific hybridisation (Agroscope Viticulture Research Group). In a first phase, the grape variety Gamaret, a European parental line (*V. vinifera*) chosen for its qualitative potential and its exceptional resistance to *Botrytis cinerea* (Pezet 1993), was crossed with a wide range of disease-resistant grape varieties carrying downy- and powdery-mildew resistance genes from wild American and Asian grapevines. To speed up the selection process and make it more reliable, and based on the criteria described above, a two-stage procedure (Fig. 2.71) was implemented and is currently being used to select the downy mildew-resistant seedlings. The procedure comprises the following stages: (1) artificial inoculation of whole leaves by spraying with an aqueous suspension of downy mildew zoospores, assessment of sporulation after a week of incubation at 100% humidity and elimination of the susceptible plants; and (2) artificial inoculation of whole leaves with 20 mL drops of an aqueous suspension of downy mildew zoospores, application of histological and biochemical criteria (sporulation, callose in the stomata, quantification of stilbenes (ε-, α- and δ-viniferins, resveratrol, piceid, grandiphenol A, hopaphenol, isohopeaphenol and pterostilbene) to categorise the leaves stripped from the remaining plants and eliminate those only producing stilbenes of low toxicity (resveratrol and piceid). Symptoms are visually assessed according to OIV (International Organisation of Vine and Wine) criteria 7 days after infection to validate the results obtained. The resistant seedlings are then transferred to the breeder, propagated, then planted in the vineyard for agronomic and oenological evaluation (Gindro et al. 2013) (Fig. 2.71).

Over 70 crossings have been made to date, producing over 30,000 seedlings, 1200 of which were selected using early testing (an average of 4% of the original plants). These genotypes (30 red and 3 white) were multiplied and 13 genotypes were successfully planted in extended studies. Two new grape varieties stemming from this programme have been approved and disseminated in practice: the red grape variety Divico (Spring et al. 2013) and the white grape variety Divona (Spring et al. 2018) (Fig. 2.72). Divico is the result of a 1997 crossing between Gamaret and

Fig. 2.72 The resistant interspecific grape varieties Divico (left) and Divona (right), carriers of the QTLs *Rpv10*, *Rpv3.3* for downy mildew, *Ren3*, *Ren9* for powdery mildew and *Rgb1* for black rot, obtained by Agroscope in Switzerland

Bronner, a resistant German white grape variety obtained by the Freiburg Institute in Breisgau. Its complex genealogy involves wild grapevines of American (*Vitis rupetris, Vitis lincecumii*) and Asian (*Vitis amurensis*) origin, from which its downy- and powdery mildew resistance traits derive. According to the genetic analyses conducted, Divico is a carrier of the QTLs *Rpv10* and *Rpv3.3* for downy mildew and *Ren3* and *Ren 9* for powdery mildew. It also possesses *Rgb1*, a *Guignaridia bidwellii* resistance QTL conferring low susceptibility to grapevine black rot. Divico also exhibits a very high resistance to grey rot, inherited from Gamaret. Produced from the same crossing, Divona exhibits the same disease-resistance factors and levels as Divico. The new crosses carried out in collaboration with INRAe (National Center for grapevine breeding, Colmar, France) no longer use this procedure and are based exclusively on selection assisted by molecular markers.

References

Adrian M, Trouvelot S, Gamm M, Poinsson B, Héloir MC, Daire X (2012) Activation of grapevine defense mechanisms: theoretical and applied approaches. In: Mérillon JM, Ramawat KG (eds) Plant defence: biological control. Springer, Berlin. https://doi.org/10.1007/978-94-007-1933-0_13

Aliano-Gonzales MJ, Richard T, Cantos-Villar E (2020) Grapevine canes extracts: raw plant material, extraction methods, quantification, and applications. Biomol Ther 10:1195. https://doi.org/10.3390/biom10081195

Alonso-Villaverde V, Voinesco F, Viret O, Spring JL, Gindro K (2011) The effectiveness of stilbenes in resistant Vitaceae: ultrastructural and biochemical events during *Plasmopara viticola* infection process. Plant Physiol Biochem 49(3):265–274. https://doi.org/10.1016/j.plaphy.2010.12.010

Argurto M, Schlechter RO, Armijo G, Solano E, Serrano C, Contreras RA, Zuniga GE, Arce-Johnson P (2017) RUN1 and REN1 pyramiding in grapevine (*Vitis vinifera* cv. Crimson Seedless) displays an improved defense response leading to enhanced resistance to powdery mildew (*Erysiphe necator*). Front Plant Sci 8:758. https://doi.org/10.3389/fpls.2017.00758

Armijo G, Schlechter R, Agurto M, Muñoz D, Nuñez C, Arce-Johnson P (2016) Grapevine pathogenic microorganisms: understanding infection strategies and host response scenarios. Front Plant Sci 30(7):382. https://doi.org/10.3389/fpls.2016.00382

Atak A, Göksel Z, Yilmaz Y (2021) Changes in major phenolic compounds of seeds, skins, and pulps from various *Vitis* spp. and the effect of powdery and downy mildew diseases on their levels in grape leaves. Plants 10:2554. doi: 10.3390/plants10122554

Baggiolini M (1952) Les stades repères dans le développement annuel de la vigne et leur utilisation pratique. Rev Romande Agric Arboric 8:4–6

Barba P, Lillis J, Luce RS, Travadon R, Osier M, Baumgartner K, Wilcox WF, Reisch BI, Cadle-Davidson L (2018) Two dominant loci determine resistance to Phomopsis cane lesions in F1 families of hybrid grapevines. Theor Appl Genet 131(5):1173–1189. https://doi.org/10.1007/s00122-018-3070-1

Barker CL, Donald T, Pauquet J, Ratnaparkhe MB, Bouquet A, Adam-Blondon AF, Thomas MR, Dry I (2005) Genetic and physical mapping of the grapevine powdery mildew resistance gene *Run*1 using a bacterial artificial chromosome library. Theor Appl Genet 111:370–377. https://doi.org/10.1007/s00122-005-2030-8

Bellin D, Peressotti E, Merdinoglu D, Wiedemann-Merdinoglu S, Adam-Blondon AF, Cipriani G, Morgante M, Testolin R, Di Gaspero G (2009) Resistance to *Plasmopara viticola* in grapevine

"Bianca" is controlled by a major dominant gene causing localised necrosis at the infection site. Theor Appl Genet 120:163–176. https://doi.org/10.1007/s00122-009-1167-2

Bettinelli P, Nicolini D, Costantini L, Stefanini M, Hausmann L, Vezzulli S (2023) Towards marker-assisted breeding for black rot bunch resistance: identification of a major QTL in the grapevine cultivar 'Merzling'. Int J Mol Sci 24(4):3568. https://doi.org/10.3390/ijms24043568

Bhattarai G, Fennell A, Londo JP, Coleman C, Kovacs LG (2021) A novel grape downy mildew resistance locus from *Vitis rupestris*. Am J Enol Vitic 72:1–20. https://doi.org/10.5344/ajev.2020.20030

Biffen RH (1905) Mendel's laws of inheritance and wheat breeding. J Agric Sci 1(1):4–48. https://doi.org/10.1017/S0021859600000137

Blanc S, Wiedemann-Merdinoglu S, Dumas V, Mestre P, Merdinoglu D (2012) A reference genetic map of *Muscadinia rotundifolia* and identification of *Ren5*, a new major locus for resistance to grapevine powdery mildew. Theor Appl Genet 125:1663–1675. https://doi.org/10.1007/s00122-012-1942-3

Blasi P, Blanc S, Wiedemann-Merdinoglu S, Prado E, Rühl EH, Mestre P, Merdinoglu D (2011) Construction of a reference linkage map of *Vitis amurensis* and genetic mapping of *Rpv8*, a locus conferring resistance to grapevine downy mildew. Theor Appl Genet 123:43–53. https://doi.org/10.1007/s00122-011-1565-0

Błaszczyk A, Sady S, Sielicka M (2019) The stilbene profile in edible berries. Phytochem Rev 18:37–67. https://doi.org/10.1007/s11101-018-9580-2

Boss PK, Thomas MR (2002) Association of dwarfism and floral induction with a grape 'green revolution' mutation. Nature 416(6883):847–850. https://doi.org/10.1038/416847a. PMID: 11976683

Bouard J (1986) La disposition des grappes sur les rameaux principaux de *Vitis vinifera* L. Connaiss Vigne Vin 20:195–206. https://doi.org/10.20870/oeno-one.1986.20.4.1307

Brown MV, Moore JN, Fenn P, McNew RW (1999) Comparison of leaf disk, greenhouse and field screening procedures for evaluation of grape seedlings for downy mildew resistance. Hort Sci 34:331–333

Büschges R, Hollricher K, Panstruga R, Simons G, Wolter M, Frijters A, van Daelen R, van der Lee T, Diergaarde P, Groenendijk J, Töpsch S, Vos P, Salamini F, Schulze-Lefert P (1997) The barley Mlo gene: a novel control element of plant pathogen resistance. Cell 88(5):695–705. https://doi.org/10.1016/s0092-8674(00)81912-1

Cadle-Davidson LE, Consolie NH, Chicoine DR (2011) Variation within and between *Vitis* species for foliar resistance to the powdery mildew pathogen *Erysiphe necator*. J Phytopathol 95:202–211. https://doi.org/10.1094/PDIS-02-10-0092

Cavaco AR, Laureano G, Duarte B, Marques da Silva J, Gameiro C, Cunha J, Dias JE, Matos AR, Figueiredo A (2023) First assessment of leaf lipids and fatty acids as biomarkers of grapevine tolerance/susceptibility to *Plasmopara viticola*. Physiol. Mol. Plant Pathol 124. https://doi.org/10.1016/j.pmpp.2023.101948

Chaïb J, Torregrosa L, Mackenzie D, Corena P, Bouquet A, Thomas MR (2010) The grape microvine—a model system for rapid forward and reverse genetics of grapevines. Plant J 62(6):1083–1092. https://doi.org/10.1111/j.1365-313X.2010.04219.x

Chen Z, Noir S, Kwaaitaal M, Hartmann HA, Wu MJ, Mudgil Y, Sukumar P, Muday G, Panstruga R, Jonesa AM (2009) Two seven-transmembrane domain mildew resistance locus o proteins cofunction in *Arabidopsis* root thigmomorphogenesis. Plant Cell 21(7):1972–1991. https://doi.org/10.1105/tpc.108.062653

Chong J, Le Henanff G, Bertsch C, Walter B (2008) Identification, expression analysis and characterization of defense and signaling genes in *Vitis vinifera*. Plant Physiol Biochem 46(4):469–481. https://doi.org/10.1016/j.plaphy.2007.09.010

Cukierman J, Quenol H, Bouffard M (2021) Quel vin pour demain? Le vin face au défi climatique. Dunod (ed), Paris. ISBN: 2100824732

Dai R, Ge H, Howard S, Qiu W (2012) Transcriptional expression of stilbene synthase genes are regulated developmentally and differentially in response to powdery mildew in Norton and Cabernet Sauvignon grapevine. Plant Sci 197:70–76. https://doi.org/10.1016/j.plantsci.2012.09.004

Dalbò MA, Ye GN, Weeden NF, Wilcox WF, Reisch BI (2001) Marker-assisted selection for powdery midew resistance in grapes. J Am Soc Hortic Sci 126:83–89. https://doi.org/10.21273/JASHS.126.1.83

Dalla Costa L, Piazza S, Campa M, Flachowsky H, Hanke MV, Malnoy M (2016) Efficient heat-shock removal of the selectable marker gene in genetically modified grapevine. Plant Cell Tissue Organ Cult 124:471–481. https://doi.org/10.1007/s11240-015-0907-z

Dalla Costa L, Piazza S, Pompili V, Salvagnn U, Castaro A, Moffa L, Vittani L, Moser C, Malnoy M (2020) Strategies to produce T-DNA free CRISP red fruit trees via *Agrobacterium tumefaciens* stable gene transfer. Sci Rep 10:20155. https://doi.org/10.1038/s41598-020-77110-1

De Young BJ, Innes RW (2006) Plant NBS-LRR proteins in pathogen sensing and host defense. Nat Immunol 7(12):1243–1249. https://doi.org/10.1038/ni1410

Delp CJ (1954) Effect of temperature and humidity on the grape powdery mildew fungus. Phytopathology 44:615–626. ISSN:0031-949X

Divilov K, Barba P, Cadle L, Bruce D (2018) Single and multiple phenotype QTL analyses of downy mildew resistance in interspecific grapevines. Theor Appl Genet 131:1133–1143. https://doi.org/10.1007/s00122-018-3065-y

Djennane S, Prado E, Dumas V, Demangeat G, Gersch S, Alais A, Gertz C, Beuve M, Lemaire O, Merdinoglu D (2021) A single resistance factor to solve vineyard degeneration due to grapevine fanleaf virus. Commun Biol 4:637. https://doi.org/10.1038/s42003-021-02164-4

Dong Y, Duan S, Xia Q, Liang Z, Dong X, Margaryan K, Musayev M, Goryslavets S, Zdunić G, Bert PF, Lacombe T, Maul E, Nick P, Bitskinashvili K, Bisztray GD, Drori E, De Lorenzis G, Cunha J, Popescu CF, Arroyo-Garcia R, Arnold C, Ergül A, Zhu Y, Ma C, Wang S, Liu S, Tang L, Wang C, Li D, Pan Y, Li J, Yang L, Li X, Xiang G, Yang Z, Chen B, Dai Z, Wang Y, Arakelyan A, Kuliyev V, Spotar G, Girollet N, Delrot S, Ollat N, This P, Marchal C, Sarah G, Laucou V, Bacilieri R, Röckel F, Guan P, Jung A, Riemann M, Ujmajuridze L, Zakalashvili T, Maghradze D, Höhn M, Jahnke G, Kiss E, Deák T, Rahimi O, Hübner S, Grassi F, Mercati F, Sunseri F, Eiras-Dias J, Dumitru AM, Carrasco D, Rodriguez-Izquierdo A, Muñoz G, Uysal T, Özer C, Kazan K, Xu M, Wang Y, Zhu S, Lu J, Zhao M, Wang L, Jiu S, Zhang Y, Sun L, Yang H, Weiss E, Wang S, Zhu Y, Li S, Sheng J, Chen W (2023) Dual domestications and origin of traits in grapevine evolution. Science 379(6635):892–901. https://doi.org/10.1126/science.add8655

Duchêne E, Schneider C (2005) Grapevine and climatic changes: a glance at the situation in Alsace. Agron Sustain Dev 25:93–99. https://doi.org/10.1051/agro:2004057

Dupraz P, Spring JL (2010) Cépages, principales variétés de vigne cultivées en Suisse, glossaire ampélographique selon la liste des descripteurs de l'OIV. AMTRA, Nyon

Eisenmann B, Czemmel S, Ziegler T, Buchholz G, Kortekamp A, Trapp O, Rausch T, Dry I, Bogs J (2019) Rpv3-1 mediated resistance to grapevine downy mildew is associated with specific host transcriptional responses and the accumulation of stilbenes. BMC Plant Biol 19(1):343. https://doi.org/10.1186/s12870-019-1935-3

El Gueddari NE, Rauchhaus U, Moerschbacher BM, Deising HB (2002) Developmentally regulated conversion of surface-exposed chitin to chitosan in cell walls of plant pathogenic fungi. New Phytol 156:103–112. https://doi.org/10.1046/j.1469-8137.2002.00487.x

Feechan A, Jermakow AM, Torregrosa L, Panstruga R, Dry IB (2008) Identification of grapevine MLO gene candidates involved in susceptibility to powdery mildew. Funct Plant Biol 35(12):1255–1266. https://doi.org/10.1071/FP08173

Feechan A, Kabbara S, Dry IB (2011) Mechanisms of powdery mildew resistance in the Vitaceae family. Mol Plant Pathol 12(3):263–274. https://doi.org/10.1111/J.1364-3703.2010.00668.X

Feechan A, Anderson C, Torregrosa L, Jermakow A, Mestre P, Wiedemann-Merdinoglu S, Merdinoglu D, Walker A, Cadle-Davison L, Reisch B, Aubourg S, Bentahar N, Shrestha B, Bouquet A, Adam-Blondon AF, Thomas MR, Dry IB (2013) Genetic dissection of a Tir-NB-LRR locus from the wild North American grapevine species Muscadinia rotundifolia identifies paralogous genes conferring resistance to major fungal and oomycete pathogens in cultivated grapevine. Plant J 76:661–674. https://doi.org/10.1111/tpj.12327

Feechan A, Kocksis M, Riaz S, Zhang W, Gadoury DM, Walker MA, Dry IB, Reisch B, Cadle-Davidson L (2015) Strategies for RUN1 deployment using RUN2 and REN2 to manage grapevine powdery mildew informed by studies of race specificity. Phytopathology 105:1104–1113. https://doi.org/10.1094/PHYTO-09-14-0244-R

Fischer B, Salakhutdinov I, Akkurt M, Eibach R, Edwards K, Töpfer R, Zyprian EM (2004) Quantitative trait locus analysis of fungal disease resistance factors on a molecular map of grapevine. Theor Appl Genet 108:501–515. https://doi.org/10.1007/s00122-003-1445-3

Foria S, Copetti D, Eisenmann B, Magris G, Vidotto M, Scalabrin S, Testolin R, Cipriani G, Wiedemann-Merdinoglu S, Bogs J, Di Gaspero G, Morgante M (2020) Gene duplication and transposition of mobile elements drive evolution of the *Rpv3* resistance locus in grapevine. Plant J 101:529–542. https://doi.org/10.1111/tpj.14551

Frankel EN, German JB, Kinsella JE, Parks E, Kanner J (1993) Inhibition of oxidation of human low-density lipoprotein by phenolic substances in red wine. Lancet 341:454–457. https://doi.org/10.1016/0140-6736(93)90206-V

Freisleben R, Lein A (1942) Über die Auffindung einer mehltauresistenten Mutante nach Röntgenbestrahlung einer anfälligen reinen Linie von Sommergerste. Naturwissenschaften 30:608. https://doi.org/10.1007/BF01488231

Fu P, Tian Q, Lai G, Li R, Song S, Lu J (2019) Cgr1, a ripe rot resistance QTL in Vitis amurensis 'Shuang Hong' grapevine. Hortic Res 6:6. https://doi.org/10.1038/s41438-019-0148-0

Fu P, Wu W, Lai G, Li R, Peng Y, Yang B, Wang B, Yin L, Qu J, Song S, Lu J (2020) Identifying *Plasmopara viticola* resistance loci in grapevine (*Vitis amurensis*) via genotyping-by-sequencing-based QTL mapping. Plant Physiol Biochem 154:75–84. https://doi.org/10.1016/j.plaphy.2020.05.016

Fung R, Gonzalo M, Fekete C, Kovacs L, He Y, Marsh E, McIntyre LM, Schachtman DP, Qiu W (2008) Powdery mildew induces defense-orientated reprogramming of the transcriptome in a susceptible but not in a resistant grapevine. Plant Physiol 146:236–249. https://doi.org/10.1104/pp.107.108712

Galet P (1988) Cépages et Vignobles de France, Tome 1: Les vignes américaines. C. Déhan, Montpellier

Garmini S, Jayahilake BJB, McClintock B (1995) Isolation and identification of a stilbene derivative from the Antarctic sponge *Kirkpatrickia variolosa*. Nat Prod 58(12):1958–1960. https://doi.org/10.1021/np50126a028

Gerrath J, Posluszny U, Melville L (2015) Vitaceae systematics (origin, characteristics and relationships). In: Gerrath J, Posluszny U, Melville L (eds) Taming the wild grape: botany and horticulture in the Vitaceae. Springer, Cham, pp 1–21

Giacomelli L, Nanni V, Lenzi L, Zhuang J, Dalla Serra M, Banfield MJ, Town CD, Silverstein KA, Baraldi E, Moser C (2012) Identification and characterization of the defensin-like gene family of grapevine. Mol Plant Microbe Interact 25(8):1118–1131. https://doi.org/10.1094/MPMI-12-11-0323

Gindro K, Pezet R (1999) Purification and characterization of a 40.8-kDa cutinase in ungerminated conidia of *Botrytis cinerea* Pers.: Fr. FEMS Microbiol Lett 171(2):239–243. https://doi.org/10.1111/j.1574-6968.1999.tb13438.x

Gindro K, Viret O, Pezet R (2006) Histological study of the responses of two *Vitis vinifera* cultivars (resistant and susceptible) to *Plasmopara viticola* infections. Plant Physiol Biochem 41(9):846–853. https://doi.org/10.1016/S0981-9428(03)00124-4

Gindro K, Alonso-Villaverde V, Viret O, Spring JL, Marti G, Wolfender JL, Pezet R (2012) Stilbenes: biomarkers of grapevine resistance to disease of high relevance in agronomy, oenology and human health. In: Mérillon JM, Ramawat KG (eds) Progress in biological control, plant defence: biological control. Springer, Dordrecht/Heidelberg/London/New York, pp 25–54

Gindro K, Spring JL, Viret O (2013) Mécanismes de défense naturelle de la vigne et sélection de cépages résistants. In: Rousseau J, Chanfreau S, Bontemps E (eds) Les cépages résistants aux maladies cryptogamiques, panorama européen. ouvrage réalisé par l'ICV (institut coopératif du vin), Montpellier, pp 41–45

Gindro K, Schnee S, Righi D, Marcourt L, Nejad Ebrahimi S, Codina JM, Voinesco F, Michellod E, Wolfender JL, Queiroz EF (2017) Generation of antifungal stilbenes using the enzymatic secretome of *Botrytis cinerea*. J Nat Prod 280(4):887–898. https://doi.org/10.1021/acs.jnatprod.6b00760

Girault T, François J, Rogniaux H, Pascal S, Delrot S, Coutos-Thévenot P, Gomès E (2008) Exogenous application of a lipid transfer protein-jasmonic acid complex induces protection of grapevine towards infection by *Botrytis cinerea*. Plant Physiol Biochem 46:140–114. https://doi.org/10.1016/j.plaphy.2007.10.005

Goufo P, Singh RK, Cortez I (2020) A reference list of phenolic compounds (including stilbenes) in grapevine (*Vitis vinifera* L.) roots, woods, canes, stems, and leaves. Antioxidants (Basel) 9(5):398. https://doi.org/10.3390/antiox9050398

Guche MD, Dalla Costa L, Malnoy M, Moser C, Pilati S (2021) Functional study of lipoxygenase-mediated resistance against Erysiphe necator in grapevine. In: XIth International Symposium on Grapevine Physiology and Biotechnology 2021, Stellenbosch, South Africa, 31 October 2021–5 November 2021, p 99. http://hdl.handle.net/10449/71616

Guerrero RF, Biais B, Richard T, Puertas B, Waffo-Teguo P, Mérillon JM, Cantos-Villar E (2016) Grapevine cane's waste is a source of bioactive stilbenes. J Ind Crops Prod 94:884–892. https://doi.org/10.1016/j.indcrop.2016.09.055

Guillier C, Gamm M, Lucchi G, Truntzer C, Pecqueur D, Ducoroy P, Adrian M, Héloir MC (2015) Toward the identification of two glycoproteins involved in the stomatal deregulation of downy mildew-infected grapevine leaves. MPMI 28(11):1227–1238. https://doi.org/10.1094/MPMI-05-15-0115-R

Gururani MA, Venkatesh J, Upadhyaya CP, Nookaraju A, Pandey SK, Park SW (2012) Plant disease resistant genes: current status and future directions. Physiol Mol Plant Pathol 19:367–370. https://doi.org/10.1016/j.pmpp.2012.01.002

Hack H, Bleiholder H, Buhr L, Meier U, Schnock-Fricke U, Weber E, Witzenberger A (1992) Einheitliche codierung der phänologischen entwicklungsstadien mono-und dikotyler pflanzen–erweiterte BBCH-Skala, Allgemein. Nachrichtenblatt Dtsch Pflanzenschutzdienstes 44:265–270

Hashimoto M, Neriya Y, Yamaji Y, Namba S (2016) Recessive resistance to plant viruses: potential resistance genes beyond translation initiation factors. Front Microbiol 7:1695. https://doi.org/10.3389/fmicb.2016.01695

Héloir MC, Adrian M, Brulé D, Claverie J, Cordelier S, Daire X, Sorey S, Gauthier A, Lemaître-Guillier C, Negrel J, Trdà L, Trouvelot S, Vandelle E, Poinssot B (2019) Recognition of elicitors in grapevine: from MAMP and DAMP perception to induced resistance. Front Plant Sci 10:1117. https://doi.org/10.3389/fpls.2019.01117

Hoffmann S, Di Gaspero G, Kovacs L, Howard S, Kiss E, Galbacs Z, Testolin R, Kozma P (2008) Resistance to *Erysiphe necator* in the grapevine "Kishmish Vatkana" is controlled by a single locus through restriction of hyphal growth. Theor Appl Genet 116:427–438. https://doi.org/10.1007/s00122-007-0680-4

Huber R, Koval A, Marcourt L, Héritier M, Schnee S, Michellod E, Scapozza L, Katanaev VL, Wolfender JL, Gindro K, Ferreira Queiroz E (2022) Chemoenzymatic synthesis of original stilbene dimers possessing Wnt inhibition activity in triple-negative breast cancer cells using the enzymatic secretome of *Botrytis cinerea* Pers. Front Chem 19(10):881298. https://doi.org/10.3389/fchem.2022.881298

Humphry M, Consonni C, Panstruga R (2006) mlo-based powdery mildew immunity: silver bullet or simply non-host resistance? Mol Plant Pathol 7(6):605–610. https://doi.org/10.1111/j.1364-3703.2006.00362.x

Jackson RS (2008) Wine science: principles and applications, 3rd edn. Academic, New York

Jacobs AL, Dry IB, Robinson SP (1999) Induction of different pathogenesis-related cDNAs in grapevine infected with powdery mildew and treated with ethephon. Plant Pathol 48:325–336. https://doi.org/10.1046/j.1365-3059.1999.00343.x

Karn A, Zou C, Brooks S, Fresnedo-Ramirez J, Gabler F, Sun Q, Ramming D, Naegele R, Ledbetter C, Cadle-Davidson L (2021) Discovery of the REN11 locus from *Vitis aestivalis* for stable resistance to grapevine powdery mildew in a family segregating for several unstable and tissue specific quantitative resistance loci. Front Plant Sci 12:733899. https://doi.org/10.3389/fpls.2021.733899

Keller M (2020) The science of grapevines, 3rd edn. Academic, London

Kelloniemi J, Trouvelot S, Heloir MC, Simon A, Dalmais B, Frettinger P, Cimerman A, Fermaud M, Roudet J, Baulande S, Bruel C, Choquer M, Couvelard L, Duthieuw M, Ferrarini A, Flors V, Le Pêcheur P, Loisel E, Morgant G, Poussereau N, Pradier JM, Rascle C, Trdá L, Poinssot B, Viaud M (2015) Analysis of the molecular dialogue between gray mold (*Botrytis cinerea*) and grapevine (*Vitis vinifera*) reveals a clear shift in defense mechanisms during berry ripening. Mol Plant-Microbe Interact 28:1167–1180. https://doi.org/10.1094/MPMI-02-15-0039-R

Kessler SA, Shimosato-Asano H, Keinath NF, Wuest SE, Ingram G, Panstruga R, Grossniklaus U (2010) Conserved molecular components for pollen tube reception and fungal invasion. Science 330(6006):968–971. https://doi.org/10.1126/science.1195211

Kusch S, Panstruga R (2017) Mlo-based resistance: an apparently universal "weapon" to defeat powdery mildew disease. Mol Plant-Microbe Interact 30(3):179–189. https://doi.org/10.1094/MPMI-12-16-0255-CR

Lambert C, Richard T, Renauf E, Bisson J, Waffo-Teguo P, Bordenave L, Ollat N, Mérillon JM, Cluzet S (2013) Comparative analyses of stilbenoids in canes of major *Vitis vinifera* L. cultivars. J Agric Food Chem 61:11392–11399. https://doi.org/10.1021/jf403716y

Lampsidis E, Simon J-L (1963) Sur le mouvement révolutif des vrilles du *Vitis vinifera*. Bull OIV 385

Lancashire PD, Bleiholder H, Boom TVD, Langelüddeke P, Strauss R, Weber E, Witzenberger A (1991) A uniform decimal code for growth stages of crops and weeds. Ann Appl Biol 119:561–601. https://doi.org/10.1111/j.1744-7348.1991.tb04895.x

Langcake P, Pryce RJ (1977) The production of resveratrol and the viniferins by grapevines in response to ultraviolet irradiation. Phytochemistry 16:1193–1196

Langcake P, Cornford CA, Pryce RJ (1979) Identification of pterostilbene as a phytoalexin from *Vitis vinifera* leaves. Phytochemistry 18(6):1025–1029. https://doi.org/10.1016/S0031-9422(00)91470-5

Lenzi L, Caruso C, Bianchedi PL, Pertot I, Perazzolli M (2016) Laser microdissection of grapevine leaves reveals site-specific regulation of transcriptional response to *Plasmopara viticola*. Plant Cell Physiol 57(1):69–81. https://doi.org/10.1093/pcp/pcv166

Li YR, Li S, Lin CC (2017) Effect of resveratrol and pterostilbene on aging and longevity. Biofactors 44(1):69–82. https://doi.org/10.1002/biof.1400

Li P, Tan X, Liu R, Rahman FU, Jiang J, Sun L, Fan X, Liu J, Liu C, Zhang Y (2023) QTL detection and candidate gene analysis of grape white rot resistance by interspecific grape (*Vitis vinifera* L. × *Vitis davidii* Foex.) crossing. Hortic Res 10(5):uhad063. https://doi.org/10.1093/hr/uhad063

Lin H, Leng H, Guo Y, Kondo S, Zhao Y, Shi G, Guo X (2019) QTLs and candidate genes for downy mildew resistance conferred by interspecific grape (*V. vinifera* L. × *V. Amurensis* Rupr.) crossing. Sci Hortic 244:200–207. https://doi.org/10.1016/j.scienta.2018.09.045

Lorenz DH, Eichhorn KW, Blei-Holder H, Klose R, Meier U, Weber E (1994) Phänologische Entwicklungsstadien der Weinrebe (*Vitis vinifera* L. ssp. *vinifera*). Codierung und Beschreibung nach der erweiterten BBCH-Skala. Vitic Enol Sci 49:66–70

Lu L, Cox CJ, Mathews S, Wang W, Wen J, Chen Z (2018) Optimal data partitioning, multispecies coalescent and Bayesian concordance analyses resolve early divergences of the grape family (Vitaceae). Cladistics 34:57–77. https://doi.org/10.1111/cla.12191

Ma Z-Y, Nie Z-L, Ren C, Liu X-Q, Zimmer EA, Wen J (2021) Phylogenomic relationships and character evolution of the grape family (Vitaceae). Mol Phylogenet Evol 154:106948. https://doi.org/10.1016/j.ympev.2020.106948

Malacarne G, Vrhovsek U, Zulini L, Cestaro A, Stefanini M, Mattivi F, Delledonne M, Velasco R, Moser C (2011) Resistance to *Plasmopara viticola* in a grapevine segregating population is associated with stilbenoid accumulation and with specific host transcriptional responses. BMC Plant Biol 11:114. https://doi.org/10.1186/1471-2229-11-114

Malnoy M, Viola R, Jung MH, Koo OJ, Kim S, Kim JS, Velasco R, Nagamangala Kanchiswamy C (2016) DNA-free genetically edited grapevine and apple protoplast using CRISPR/Cas9 ribonucleoproteins. Front Plant Sci 7:1904. https://doi.org/10.3389/fpls.2016.01904

Manchester SR, Kapgate DK, Wen J (2013) Oldest fruits of the grape family (Vitaceae) from the Late Cretaceous Deccan cherts of India. Am J Bot 100:1849–1859. https://doi.org/10.3732/ajb.1300008

Marguerit E, Boury C, Manicki A, Donnart M, Butterlin G, Némorin A, Wiedemann-Merdinoglu S, Merdinoglu D, Ollat N, Decroocq S (2009) Genetic dissection of sex determinism, inflorescence morphology and downy mildew resistance in grapevine. Theor Appl Genet 118:1261–1278. https://doi.org/10.1007/s00122-009-0979-4

McGovern PE (2019) Ancient Winw: the search for the origins of viniculture. University Press, Princeton, NJ

Merdinoglu D, Wiedemann-Merdinoglu S, Coste P, Dumas V, Haetty S, Butterlin G, Greif C (2003) Genetic analysis of downy mildew resistance derived from *Muscadinia rotundifolia*. Acta Hortic 603:451–456. https://doi.org/10.17660/ActaHortic.2003.603.57

Merdinoglu D, Schneider C, Prado E, Wiedemann-Merdinoglu S, Mestre P (2018) Breeding for durable resistance to downy and powdery mildew in grapevine. OENO One 52(3):203–209. https://doi.org/10.20870/oeno-one.2018.52.3.2116

Merz PR, Moser T, Höll J, Kortekamp A, Buchholz G, Zyprian E, Bogs J (2015) The transcription factor VvWRKY33 is involved in the regulation of grapevine (Vitis vinifera) defense against the oomycete pathogen *Plasmopara viticola*. Physiol Plant 153(3):365–380. https://doi.org/10.1111/ppl.12251

Mishra AN, Kaushal K, Yadav SR, Shirsekar GS, Pandey HN (2005) The linkage between the stem rust resistance gene Sr2 and pseudo-black chaff in wheat can be broken. Plant Breed 124:520–522. https://doi.org/10.1111/j.1439-0523.2005.01136.x

Modesto LR, Schandner A, Nodari RO, Dalbo MA, Welter L, Hausmann L, Töpfer R, Da Silva AL (2022) First quantitative trait loci mapped for resistance to grapevine anthracnose (*Elsinoë ampelina*). In: Anais 11° congresso brasileiro de melhoramento de plantas. ISBN: 978-65-5941-579-3

Moreira FM, Madini A, Marino R, Zulini L, Stefanini M, Velasco R, Kozma P, Grando MS (2011) Genetic linkage maps of two interspecific grape crosses (*Vitis* spp) used to localize quantitative trait loci for downy mildew resistance. Tree Genet Genomes 7:153–167. https://doi.org/10.1007/s11295-010-0322-x

Mundt CC (2018) Pyramiding for resistance durability: theory and practice. Phytopathology 108(7):792–802. https://doi.org/10.1094/PHYTO-12-17-0426-RVW

Myles S, Boyko AR, Owens CL, Brown PJ, Grassi F, Aradhya MK, Prins B, Reynolds A, Chia J-M, Ware D, Bustamante CD, Buckler ES (2011) Genetic structure and domestication history of the grape. Proc Natl Acad Sci USA 108:3530–3535. https://doi.org/10.1073/pnas.1009363108

Nascimento R, Maia M, Ferreira AEN, Silva AB, Freire AP, Cordeiro C, Silva MS, Figueiredo A (2019) Early stage metabolic events associated with the establishment of *Vitis vinifera—Plasmopara viticola* compatible interaction. Plant Physiol Biochem 137:1–13. https://doi.org/10.1016/j.plaphy.2019.01.026

Ochssner I, Hausmann L, Töpfer R (2016) *Rpv14*, a new genetic source for *Plasmopara viticola* resistance conferred by *Vitis cinerea*. Vitis 55:79–81. https://doi.org/10.5073/vitis.2016.55.79-81

OIV (2017) Distribution of the world's grapevine varieties. Focus OIV 2017:55

OIV (2022) The 2nd edition of the OIV Descriptor list for grape varieties and Vitis species. Technical documents 2022:1–232

Özer N, Sadubak T, Özer C, Gindro K, Schnee S, Solak E (2017) Investigations on the role of cuticular wax in resistance to powdery mildew in grapevine. J Gen Plant Pathol 83:316–328. https://doi.org/10.1007/s10327-017-0728-5

Pap D, Riaz S, Dry IB, Jermakow A, Tenscher AC, Cantu D, Olah R, Walker MA (2016) Identification of two novel powdery mildew resistance loci, Ren6 and Ren7, from the wild

Chinese grape species Vitis piasezkii. BMC Plant Biol 16:170. https://doi.org/10.1186/s12870-016-0855-8

https://www.vivc.de/docs/dataonbreeding/20220920_Table%20of"/420Loci%20for%20Traits%20in%20Grapevine.pdf

Parage C, Tavares R, Réty S, Baltenweck-Guyot R, Poutaraud A, Renault L, Heintz D, Lugan R, Marais GAB, Aubourg S, Hugueney P (2012) Structural, functional, and evolutionary analysis of the unusually large stilbene synthase gene family in grapevine. Plant Physiol 160:1407–1419. https://doi.org/10.1104/pp.112.202705

Peressotti E, Wiedemann-Merdinoglu S, Delmotte F, Bellin D, Di Gaspero G, Testolin R, Merdinoglu D, Mestre P (2010) Breakdown of resistance to grapevine downy mildew upon limited deployment of a resistant variety. BMC Plant Biol 10:147. https://doi.org/10.1186/1471-2229-10-147

Perret C (2001) Analyse des tannins inhibiteurs de la stilbène oxydase produite par *Botrytis cinerea*. Thèse de doctorat de l'Université de Neuchâtel, Switzerland

Perret C, Pezet R, Tabacchi R (2003) Qualitative analysis of grapevine tannins by mass spectrometry and their inhibitory effect on stilbene oxidase of *Botrytis cinerea*. Chimia 57(10):607–610. ISSN: 0009-4293

Pessina S, Lenzi L, Perazzolli M, Campa M, Dalla Costa L, Urso S, Valè G, Salamini F, Velasco R, Malnoy M (2016) Knockdown of MLO genes reduces susceptibility to powdery mildew in grapevine. Hortic Res 3:16016. https://doi.org/10.1038/hortres.2016.16

Pezet R (1993) La pourriture grise des raisins. Le complexe plante-parasite. Le Vigneron champenois 114(5):65–83

Pezet R, Cuenat P (1996) Resveratrol in wine: extraction from skin during fermentation and post-fermentation standing of must from Gamay grapes. Am J Enol Vitic 47(3):287–290

Pezet R, Pont V (1988) Mise en évidence de ptérostilbène dans les grappes de *Vitis vinifera*. Plant Physiol Biochem 26:603–607

Pezet R, Pont V (1990) Ultrastructural observations of pterostilbene fungitoxicity in dormant conidia of *Botrytis cinerea* Pers. J Phytopathol 129:19–30

Pezet R, Perret C, Jean-Denis JB, Tabacchi R, Gindro K, Viret O (2003) δ-Viniferin, a resveratrol dehydrodimer: one of the major stilbenes synthesized by stressed grapevine leaves. J Agric Food Chem 51:5488–5492. https://doi.org/10.1021/jf030227o

Pezet R, Viret O, Gindro K (2004) Plant microbe interaction: the botrytis grey mould of grapes. In: Hemantaranjan A (ed) Biology, biochemistry, epidemiology and control management. Advances in plant physiology, Varanasi, India, pp 75–120

Planchon JE (1887) Ampelidae. In: Monographiae phanerogamarum: Prodromi nunc continuatio, nunc revisio, Alphonse de Candolle and Casimir de Candolle. G. Masson & Cie, Paris

Possamai T, Wiedemann-Merdinoglu S (2022a) Phenotyping for grapevine QTL identification. The case of resistance to *Plasmopara viticola* and *Erysiphe necator*. A review. Bio Web Conf 50:02009. https://doi.org/10.1051/bioconf/20225002009

Possamai T, Wiedemann-Merdinoglu S (2022b) Phenotyping for QTL identification: a case study of resistance to *Plasmopara viticola* and *Erysiphe necator* in grapevine. Front Plant Sci 13:930954. https://doi.org/10.3389/fpls.2022.930954

Possamai T, Wiedemann-Merdinoglu S, Merdinoglu D, Migliaro D, De Mori G, Cipriani G, Velasco R, Testolin R (2021) Construction of a high-density genetic map and detection of a major QTL of resistance to powdery mildew (*Erysiphe necator* Sch.) in Caucasian grapes (*Vitis vinifera* L.). BMC Plant Biol 21:528. https://doi.org/10.1186/s12870-021-03174-4

Qiu W, Feechan A, Dry I (2015) Current understanding of grapevine defense mechanisms against the biotrophic fungus (*Erysiphe necator*), the causal agent of powdery mildew disease. Hortic Res 2:15020. https://doi.org/10.1038/hortres.2015.20

Qu J, Dry I, Liu L, Guo Z, Yin L (2021) Transcriptional profiling reveals multiple defense responses in downy mildew-resistant transgenic grapevine expressing a TIR-NBS-LRR gene located at the MrRUN1/MrRPV1 locus. Hortic Res 8(1):161. https://doi.org/10.1038/s41438-021-00597-w

Raming DW, Gabler F, Smilanick J, Cadle-Davidson M, Barba P, Mahanil S, Cadle-Davidson L (2011) A single dominant locus, ren4, confers rapid non-race-specific resistance to grapevine powdery mildew. Phytopathology 101(4):502–508. https://doi.org/10.1094/PHYTO-09-10-0237

Renaud S, de Lorgeril M (1992) Wine, alcohol, platelets, and the French paradox for coronary heart disease. Lancet 339:1523–1526. https://doi.org/10.1016/0140-6736(92)91277-F

Rex F, Fechter I, Hausmann L, Töpfer R (2014) QTPL mapping of black rot (Guignardia bidwellii) resistance in the grapevine rootstock "Börner" (V. riparia Gm183 x V. cinerea Arnold). Theor Appl Genet 127:1667–1677. https://doi.org/10.1007/s00122-014-2329-4

Riaz S, Menéndez CM, Tenscher A, Pap D, Walker MA (2020) Genetic mapping and survey of powdery mildew resistance in the wild Central Asian ancestor of cultivated grapevines in Central Asia. Hortic Res 7:104. https://doi.org/10.1038/s41438-020-0335-z

Riaz S, Tenscher AC, Ramming DW, Walkker MA (2011) Using a limited mapping strategy to identify major QTLs for resistance to grapevine powdery mildew (Erysiphe necator) and their use in marker-assisted breeding. Theor Appl Genet 122:1059–1073. https://doi.org/10.1007/s00122-010-1511-6

Richardson WH, Schmidt TM, Nealson KH (1988) Identification of an anthraquinone pigment and a hydroxystilbene antibiotic from Xenorhabdus luminescens. Appl Environ Microbiol 54(6):1602–1605. https://doi.org/10.1128/aem.54.6.1602-1605.1988

Richter H, Pezet R, Viret O, Gindro K (2006) Characterization of 3 new partial stilbene synthase genes out of over 20 expressed in Vitis vinifera during the interaction with Plasmopara viticola. Physiological and Mol Plant Path 3(5):248–260. https://doi.org/10.1016/j.pmpp.2006.03.001

Rienth M, Torregrosa L, Sarah G, Ardisson M, Brillouet JM, Romieu C (2016) Temperature desynchronizes sugar and organic acid metabolism in ripening grapevine fruits and remodels their transcriptome. BMC Plant Biol 16(164). https://doi.org/10.1186/s12870-016-0850-0

Righi D, Huber R, Koval A, Marcourt L, Schnee S, Le Floch A, Ducret V, Perozzo R, de Ruvo CC, Lecoultre N, Michellod E, Ebrahimi SN, Rivara-Minten E, Katanaev VL, Perron K, Wolfender JL, Gindro K, Queiroz EF (2020) Generation of stilbene antimicrobials against multiresistant strains of Staphylococcus aureus through biotransformation by the enzymatic secretome of Botrytis cinerea. J Nat Prod 83(8):2347–2356. https://doi.org/10.1021/acs.jnatprod.0c00071

Rousseau J, Chanfreau S, Bontemps E (2013) Les cépages résistants aux maladies cryptogamiques: panorama européen. Groupe ICV (ed.), Lattes, p 228

Rozefelds A, Pace M (2018) The first record of fossil Vitaceae wood from the Southern Hemisphere, a new combination for Vitaceoxylon ramunculiformis, and reappraisal of the fossil record of the grape family (Vitaceae) from the Cenozoic of Australia. J Syst Evol 56:283–296. https://doi.org/10.1111/jse.12300

Sapkota S, Chen LL, Schreiner K, Ge H, Hwang CF (2015) A phenotypic study of botrytis bunch rot resistance in Vitis aestivalis-derived 'Norton' grape. Trop Plant Pathol 40:279–282. https://doi.org/10.1007/s40858-015-0028-6

Sapkota S, Chen LL, Yang S, Hyma KE, Cadle-Davidson L, Hwang CF (2019) Construction of a high-density linkage map and QTL detection of downy mildew resistance in Vitis aestivalis-derived Norton. Theor Appl Genet 132:137–147. https://doi.org/10.1007/s00122-018-3203-6

Sargolzaei M, Maddalena G, Bitsadze N, Maghradze D, Bianco PA, Failla O, Toffolatti SL, De Lorenzis G (2020) Rpv29, Rpv30 and Rpv31: three novel genomic loci associated with resistance to Plasmopara viticola in Vitis vinifera. Front Plant Sci 11:562432. https://doi.org/10.3389/fpls.2020.562432

Schnee S (2008) Facteurs de résistance à l'oïdium (Erysiphe necator Schwein) chez la vigne (Vitis vinifera L.). Thesis, University of Neuchâtel (Switzerland)

Schnee S, Queiroz EF, Voinesco F, Marcourt L, Dubuis PH, Wolfender JL, Gindro K (2013) Vitis vinifera canes, a new source of antifungal compounds against Plasmopara viticola, Erysiphe necator, and Botrytis cinerea. J Agric Food Chem 61:5459–5467. https://doi.org/10.1021/jf4010252

Schnee S, Huber R, Marcourt L, Michellod E, Wolfender JL, Gindro K, Queiroz EF (2022) Generation of antifungal stilbenes derivatives towards grapevine downy mildew using enzymatic secretome of *Botrytis cinerea*. Bio Web Conf 50:03007. https://doi.org/10.1051/bioconf/20225003007

Schwander F, Eibach R, Fechter I, Hausmann L, Zyprian E, Töpfer R (2012) *Rpv10*: a new locus from the Asian Vitis gene pool for pyramiding downy mildew resistance loci in grapevine. Theor Appl Genet 124:163–176. https://doi.org/10.1007/s00122-011-1695-4

Song S, Fu P, Lu L (2018) Downy mildew resistant QTLs in *Vitis amurensis* 'Shuang Hong' grapevine. In: Proceedings of the XIIth International Grapevine Breeding and Genetics Conference, Abstract Book GBG, Bordeaux, France

Sosa-Zuniga V, Valenzuela AV, Barba P, Espinoza Cantino C, Romero-Romero JL, Arce-Johnson P (2022) Powdery mildew resistance genes in vines: an opportunity to achieve a more sustainable viticulture. Pathogens 11:703. https://doi.org/10.3390/pathogens11060703

Späth E, Schläger J (1940) Über die Inhaltsstoffe des roten Sandelholzes. II. Mitteil.: Die Konstitution des Pterostilbens. Eur J Inorg Chem 73:881–883

Spring JL, Dupraz PH (2019) La sélection de la vigne à Agroscope. Publication spéciale, Amtra (ed), Suisse

Spring JL, Gindro K, Voinesco F, Jermini M, Ferretti M, Viret O (2013) Divico, premier cépage résistant aux principales maladies fongiques de la vigne sélectionné par Agroscope. Rev Suisse Vitic Arboric Hortic 45(5):292–303

Spring JJ, Zufferey V, Verdenal T, Duruz P, May S, Barmes E, Bailly S, Bonvin Y, Reymond R, Ferretti M, Rigoni R, Roesti J, Lorenzini F, Reynard JS, Gindro K, Viret O (2017) Nouveaux cépages Agroscope: les saveurs du Sud. Rev Suisse Vitic Arboric Hortic 49(6):328–336

Spring JL, Gindro K, Laprand F, Zufferey V, Verdenal T, Rösti J, Amiet L, Lorenzini F, Duruz P, Barmes E, Bailly S, May S, Bonvin Y, Reymond R, Viret O, Carlen C (2018) Divona, nouveau cépage blanc résistant aux principales maladies de la vigne sélectionné à Agroscope. Rev Suisse Vitic Arboric Hortic 50(5):286–296

Stam R, McDonald BA (2018) When resistance gene pyramids are not durable-the role of pathogen diversity. Mol Plant Pathol 19(3):521–524. https://doi.org/10.1111/mpp.12636

Su K, Guo Y, Zhong W, Lin H, Liu Z, Li K et al (2021) High-density genetic linkage map construction and white rot resistance quantitative trait loci mapping for genus Vitis based on restriction site-associated DNA sequencing. Phytopathology 111(4):659–670. https://doi.org/10.1094/phyto-12-19-0480-r

Su K, Zhao W, Lin H, Jiang C, Zhao Y, Guo Y (2023) Candidate gene discovery of *Botrytis cinerea* resistance in grapevine based on QTL mapping and RNA-seq. Front Plant Sci 14:1127206. https://doi.org/10.3389/fpls.2023.1127206

Süssenguth K (1953) Vitaceae. In: Die Natürlichen Pflanzenfamiien. Engler A and Prantl K. Dunker & Humblot, Berlin, pp 174–333

Takaoka M (1939) Resveratrol, a new phenolic compound, from *Veratrum grandiflorum*. J Chem Soc Jpn 60:1090–1100

Teh SL, Fresnedo-Ramirez J, Clark MD, Gadoury DM, Sun Q, Cadle-Davidson L, Luby JJ (2017) Genetic dissection of powdery mildew resistance in interspecific half-sib grapevine families using SNP-based maps. Mol Breed 31:1. https://doi.org/10.1007/s11032-016-0586-4

Teka T, Zhang L, Ge X, Li Y, Han L, Yan X (2022) Stilbenes: Source plants, chemistry, biosynthesis, pharmacology, application and problems related to their clinical Application-A comprehensive review. Phytochemistry 197:113128. https://doi.org/10.1016/j.phytochem.2022.113128

This P, Lacombe T, Thomas MR (2006) Historical origins and genetic diversity of wine grapes. Trends Genet 22:511–519. https://doi.org/10.1016/j.tig.2006.07.008

Torregrosa L, Rienth M, Romieu C, Pellegrino A (2019) The microvine, a model for studies in grapevine physiology and genetics. OENO One 53(3):3. https://doi.org/10.20870/oeno-one.2019.53.3.2409

Truniger V, Aranda MA (2009) Recessive resistance to plant viruses. Adv Virus Res 75:119–159. https://doi.org/10.1016/S0065-3527(09)07504-6

Van Leeuwen C, Destrac Irvine A, Duchêne E, Gowdy M, Marguerit E, Pieri P, Rességuier L, Ollat N (2019) An update on the impact of climate change in viticulture and potential adaptations. Agronomy 9:514. https://doi.org/10.3390/agronomy9090514

Van Schie CCN, Takken FLW (2014) Susceptibility genes 101: how to be a good host. Annu Rev Phytopathol 52(1):551–581. https://doi.org/10.1146/annurev-phyto-102313-045854

Venuti S, Copetti D, Foria S, Falginella L, Hoffmann S, Bellin D, Cindrić P, Kozma P, Scalabrin S, Morgante M, Testolin R, Di Gaspero G (2013) Historical introgression of the downy mildew resistance gene *Rpv12* from the asian species *Vitis amurensis* into grapevine varieties. PLoS One 8:e0061228. https://doi.org/10.1371/journal.pone.0061228

Vezzulli S, Malacarne G, Masuero D, Vecchione A, Dolzani C, Goremykin V, Mehari ZH, Banchi E, Velasco R, Stefanini M, Vrhovsek U, Zulini L, Franceschi P, Moser C (2019) The *Rpv3-3* haplotype and stilbenoid induction mediate downy mildew resistance in a grapevine interspecific population. Front Plant Sci 10:234. https://doi.org/10.3389/fpls.2019.00234

Viala P, Vermorel V (1910) Traité général de Viticulture: Ampélographie. Masson & Cie éditeurs, Paris

Viret O, Gindro K (2014) La Vigne: volume 1, Maladies fongiques. AMTRA (ed), Suisse. ISBN:385928097X

Wan Y, Schwaninger HR, Baldo AM, Labate JA, Zhong G-Y, Simon CJ (2013) A phylogenetic analysis of the grape genus (Vitis L.) reveals broad reticulation and concurrent diversification during neogene and quaternary climate change. BMC Evol Biol 13:141. https://doi.org/10.1186/1471-2148-13-141

Wang K, Liao Y, Kan J, Han L, Zheng Y (2015) Response of direct or priming defense against *Botrytis cinerea* to methyl jasmonate treatment at different concentrations in grape berries. Int J Food Microbiol 194:32–39. https://doi.org/10.1016/j.ijfoodmicro.2014.11.006

Welter LJ, Göktürk-Bayadar N, Akkurt M, Maul E, Eibach R, Töpfer R, Zyprian EM (2007) Genetic mapping and localization of quantitative trait loci affecting fungal disease resistance and leaf morphology in grapevine (*Vitis vinifera* L.). Mol Breed 20:359–374. https://doi.org/10.1007/s11032-007-9097-7

Wen J (2007) Vitaceae. In: Kubitzki K (ed) Flowering plants eudicots: Berberidopsidales, Buxales, Crossosomatales, Fabales p.p., Geraniales, Gunnerales, Myrtales p.p., Proteales, Saxifragales, Vitales, Zygophyllales, Clusiaceae Alliance, Passifloraceae Alliance, Dilleniaceae, Huaceae, Picramniaceae, Sabiaceae. Springer, Berlin/Heidelberg, pp 467–479

Wen J, Lu L-M, Nie Z-L, Liu X-Q, Zhang N, Ickert-Bond S, Gerrath J, Manchester SR, Boggan J, Chen Z-D (2018) A new phylogenetic tribal classification of the grape family (Vitaceae). J Syst Evol 56:262–272. https://doi.org/10.1111/jse.12427

Wiedemann-Merdinoglu S, Prado E, Coste P, Dumas V, Butterlin G, Bouquet A, Merdinoglu D (2006) Genetic analysis of resistance to downy mildew derived from *Muscadinia rotundifolia*. In: Ninth International Conference on grape genetics and breeding, Udine, Italy 2–6 July

Wingerter C, Eisenmann B, Weber P, Dry I, Bogs J (2021) Grapevine Rpv3-, Rpv10- and Rpv12-mediated defense responses against *Plasmopara viticola* and the impact of their deployment on fungicide use in viticulture. BMC Plant Biol 21:470. https://doi.org/10.1186/s12870-021-03228-7

Yin X, Liu RQ, Su H, Su L, Guo YR, Wang ZJ, Du W, Li MJ, Zhan X, Wang YJ, Liu GT, Xu Y (2017) Pathogen development and host responses to *Plasmopara viticola* in resistant and susceptible grapevines: an ultrastructural study. Hortic Res 4:17033. https://doi.org/10.1038/hortres.2017.33

Yu Y, Zhang Y, Yin Y, Lu J (2012) The mode of host resistance to *plasmopara viticola* infection of grapevines. Phytopathology 102(11):1094–1101. https://doi.org/10.1094/PHYTO-02-12-0028-R

Zecca G, Abbott JR, Sun W-B, Spada A, Sala F, Grassi F (2012) The timing and the mode of evolution of wild grapes (Vitis). Mol Phylogenet Evol 62:736–747. https://doi.org/10.1016/j.ympev.2011.11.015

Zendler D, Schneider P, Töpfer R, Zyprian E (2017) Fine mapping of Ren3 reveals two loci mediating hypersensitive response against *Erysiphe necator* in grapevine. Euphytica 213:68. https://doi.org/10.1007/s10681-017-1857-9

Zendler D, Töpfer R, Zyprian E (2020) Confirmation and fine mapping of the resistance locus Ren9 from the grapevine cultivar "Regent". Plan Theory 10:24. https://doi.org/10.3390/plants10010024

Zhang N, Wen J, Zimmer EA (2015) Congruent deep relationships in the grape family (Vitaceae) based on sequences of chloroplast genomes and mitochondrial genes via genome skimming. PLoS One 10:e0144701. https://doi.org/10.1371/journal.pone.0144701

Zhang Y, Fan X, Sun H, Jiang J, Liu C (2017) Identification and evaluation of resistance to white rot in grape resource. J Fruit Sci 34:105–1105

Zufferey V, Murisier F, Belcher S, Lorenzini F, Vivin P, Spring JL, Viret O (2015) Nitrogen and carbohydrate reserves in the grapevine (*Vitis vinifera* L. 'Chasselas'): the influence of the leaf to fruit ratio. VITIS J Grapevine Res 54:183–188. https://doi.org/10.5073/VITIS.2015.54.183-188

Zufferey V, Verdenal T, Gindro K, Murisier F, Viret O (2022) La Vigne: volume 4, anatomie et physiologie, alimentation et carences, accidents physiologiques et climatiques. Ed. AMTRA, Nyon, Suisse, p 564

Zyprian E, Oechssner I, Schwander F, Simon S, Hausmann L, Bonow-Rex M, Moreno-Sanz P, Grando MS, Wiedemann-Merdinoglu S, Merdinoglu D, Eibach R, Töpfer R (2016) Quantitative trait loci affecting pathogen resistance and ripening of grapevines. Mol Gen Genomics 291:1573–1594. https://doi.org/10.1007/s00438-016-1200-5

Chapter 3
Fungi and Grapevine Mycobiota

3.1 Mycology Throughout the Ages

History suggests that, anthropomorphically and anthropologically speaking, mushrooms have long been part of our daily lives. This is borne out by Ötzi, the neolithic 'iceman' who lived around 5300 years ago. His mummified body was found near the Tisen Pass in Val Senales in the Italian Alps, along with various items including a cloak, a coat, leggings, an axe, a knife and a pouch containing two species of mushroom: a tinder amadou for starting a fire and a birch polypore for its medicinal and antiparasitic properties (Fig. 3.1).

A source of sustenance, a poison, a spiritual aid, for domestic use or medicinal purposes, the use of mushrooms has been documented since humans were able to express themselves through the written word or any other form of iconographic representation. Yet despite this very ancient knowledge of mushrooms, in antiquity we find little specific writing, and only very vague notions about these organisms. In reality, people in the ancient world paid scant attention to mushrooms, which they regarded as a strange and mysterious order of plants. However, they consumed them with enthusiasm, and sometimes warned of the dangers of certain species. Knowledge of the toxicity of mushrooms was only acquired later. Authors in the ancient world believed that vicinity to infected sites or contact with certain putrid bodies or plants where they were found transferred harmful and malignant qualities to mushrooms. In this era, mushrooms were considered to be fortuitous creations of the earth, with neither roots nor seeds. Some philosophers and poets called them 'children of the Gods' or 'children of the Earth' to signify that they seemed to appear by spontaneous generation. In 300 BCE Theophraste proposed what is thought to be the first attempt to classify mushrooms, grouping all species in four divisions: round mushrooms growing underground, those growing at the surface and attached to a stalk, those with neither stalk nor cap, and round ones resembling a human head. Later, Dioscorides, a Greek physician born in the first century CE, was the first author

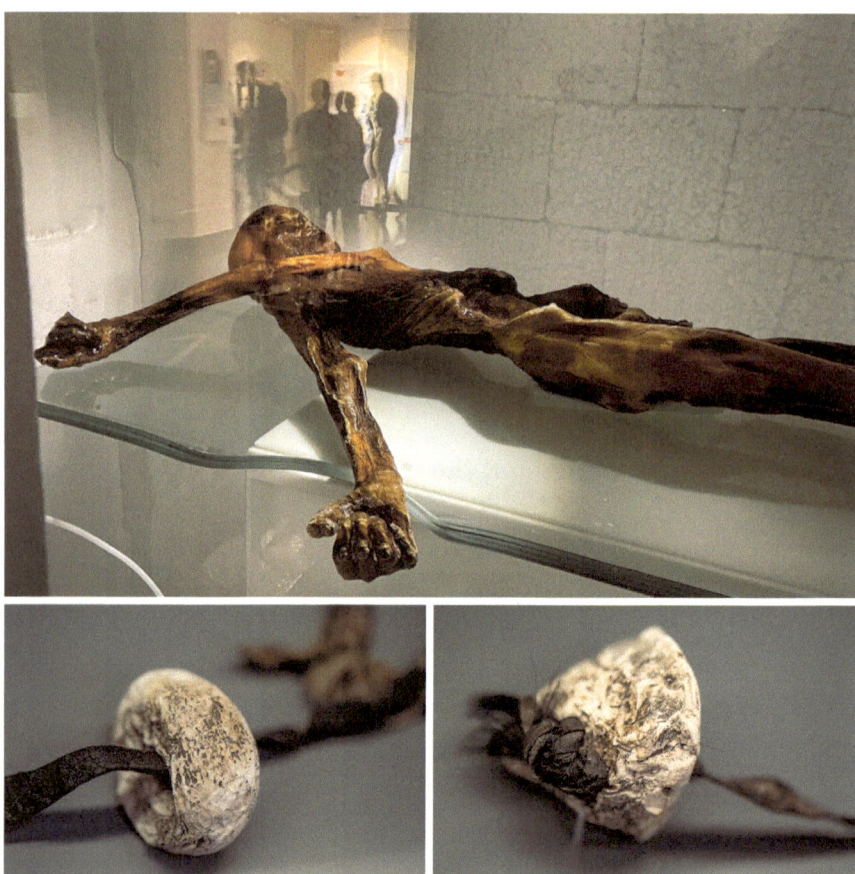

Fig. 3.1 Mummy of Ötzi, the neolithic man found near the Tisen Pass in Italy, preserved at the South Tyrol Museum of Archeology in Bolzano (Italy). As well as tools and equipment, Ötzi carried with him two mushrooms, an amadou and a birch polypore (bottom). (Photographs © Mario del Curto, Switzerland)

to name a specific species and describe its medicinal use. The mushroom in question was the white agaric, which he discovered growing on larch trees in a region called Agaria (between Romania and Kazakstan) and named accordingly. The works of Pliny the Elder around 75 CE are no less remarkable: he was a pioneer of truffle cultivation, describing how they could be multiplied by watering the ground with water from streams that crossed areas where truffles grew. Several authors followed in his footsteps, refining certain notions already described by their predecessors but adding nothing new to the basic body of knowledge that existed about mushrooms. So mushrooms long remained the underdogs of Science, which prioritised the study of animals, flowering plants, medicine and alchemy, as well as the basic elements that surround us.

It was not until the middle of the fifteenth century that the first description of a mushroom appeared, in a book written by Johann Wonnecke von Kaub (Johannes de Cuba). This work, Ortus Sanitatis, was published in Germany in 1485 by Peter Schöffer. Then a Latin translation, *Hortus Sanitatis*, was published on the printing press of Jacob von Meydenbach in Mainz in 1491 (Von Kaub 1491); this book presents a large number of plants and their therapeutic uses alongside descriptions of the animal world both real and mythical. The long list of elements dealt with in the book includes an engraving of a mushroom followed by a brief description of its virtues and dangers (Fig. 3.2).

In 1560, in his work *Commentarii in libros sex pedacii discoriis*, Pietro Andrea Matthioli, an Italian botanist, illustrated and described all the medicinal and domestic uses of the agaric first described by Dioscorides, which he found growing on larches in Italy (Fig. 3.3). Matthioli also continued Pliny's work on truffles and refined the knowledge of their cultivation.

The scientists of the fifteenth and sixteenth century—naturalists, physicians, monks and other observers of life—soon took up the baton, rivalling one another with their engravings and descriptions of the species of mushroom they encountered on their excursions. The greatest advances in mycological knowledge came at the end of the sixteenth century thanks to the works of several botanists and mycologists such as l'Ecluse, Peña, Lobel, Daleschamps, Dodonée, Columbia, Imperrato and the Bauhin Brothers. Charles l'Ecluse (Carolus Clusius), in 1601, wrote the first monograph on mushrooms in his work *Rariorum Plantarum historia*: containing more than 30 pages of detailed and richly illustrated descriptions (Fig. 3.4), it was a far cry from the more obscure and therapeutic accounts of the fifteenth and sixteenth century authors.

Two major revolutions were to take place in the second half of the seventeenth century; microscopy, which revealed the intimate structures of fungi, and nomenclature, returning them to their rightful place in the classification of the living world as organisms in their own right. In 1665, Sir Robert Hooke, father of the scientific revolution, published a seminal work thanks to the development of the microscope, which could magnify objects 30 times or more. Using this innovation, he was able to observe elements that were invisible to the naked eye and discover the microscopic structure of mineral, plant, animal and fungal material. In his scientific treatise *Micrographia*, Hooke described a tiny organism growing on the sheepskin cover of a book in his library which he likened to a miniscule fungus and named 'mould' (Fig. 3.5). Despite the pioneering nature of his research, Hooke still regarded mushrooms as mysterious plants arising by spontaneous generation without means of dissemination or reproduction.

Despite Hooke's work, mycology was to remain a macroscopic science for several years to come. Several researchers followed in his wake, producing increasingly detailed works on fungi and their classification, such as *Theatrum fungorum* by Johannes Fransiscus van Sterbeeck (1675), *Historia Plantarum* by John Ray (1688), and *Eléments de Botanique* by Joseph Pitton de Tournefort (1694).

Pietro Antonio Micheli, professor of botany in Florence, conducted an in-depth study of fungi. Regarded as the father of modern mycology, he published a

Fig. 3.2 First engraving and description of a mushroom in Johannes de Cuba's *Hortus Sanitatis* written in 1491. (Photographs © reproduced by kind permission of the Syndics of Cambridge University Library)

ground-breaking work in 1729 entitled *Nova Plantarum Genera* (Fig. 3.6). He was the first to explore the mode of reproduction and dissemination of fungi and introduced the notion of spores. He was convinced that fungi had to contain seeds to enable them to propagate. In an effort to prove this, he conducted a vast number of experiments and observations which involved flattening the fungal structures under

371

LES
COMMENTAIRES
DE M. PIERRE ANDRE MATTHIOLI
MEDECIN SENOIS, SVR LE III. LIVRE DE
PEDACE DIOSCORIDE ANAZARBEEN, DE LA
MATIERE MEDECINALE.

PREFACE.

Vs deux liures precedens, trefcher Aree, nous auons parlé des drogues aromatiques, des vnguens, des huiles, des arbres, fruis, & larmes d'iceus : puis aprés des animaus, des blés, des herbes qu'on mange ordinairemét, & de celles qui font fortes & acres. En ce troifiéme nous écrirons des racines, des fucs, des femences, des herbes tant domeftiques, que entr'elles femblables, & qui font fort propres en l'vfage de la medecine.

DE L'AGARIC.   CHAP. I.

ON dit que l'agaric eft une racine femblable au laferpitium, moins referree, & plus lache par le deffus, rare & fpongieufe par tout. Il y en a de deus fortes. La femelle, qui eft la meilleure, a au dedans fes ueines & trais longs, & tous drois : le mafle eft rond, & plus maffif & ferré. Tous deus fe prefentent dous au gouft, s'épandans par la bouche on les fent bien amers. Il croift en la region de Sarmatie, nommee Agaria. Aucuns difent l'agaric eftre la racine d'une plante. Les autres qu'il croift fur les troncs des arbres de certaine pourriture, comme champignons. En Galatie d'Afie, & en Cilicie il en uient fur les cedres, mais il n'eft de grande uertu, & s'émie aifement. Le naturel de l'agaric eft de retraindre & d'échauffer. Il eft bon contre les tranchees, humeurs crues & indigeftes, ruptions, contufions, & à ceus qui font tombés d'enhaut. On en donne en bruuage deus oboles auec du uin miellé, à ceus qui font fans fieure, à ceus qui ont fieure auec eau miellee, à ceus qui ont la iauniffe, qui ont l'aleine courte, qui ont mal au foye & aus reins, & aus dyfenteries. On en ordonne une drachme à la difficulté d'urine, à la fuffocation de matrice, à la palleur & mauuaife couleur de toute la perfone, aus tabides auec du uin faiét de raifins cuis en la uigne, au mal de ratelle auec uinaigre miellé : aus deuoiremens d'eftomac tels, que il ne peut tenir la uiande. On mange l'agaric feul, fans qu'il foit detrampé auec aucun' humeur, quand auffi on fait des rots aigres. Si on en prend trois oboles auec de l'eau, il arrefte les crachemens de fang. Il eft bon aus fciatiques, douleurs de iointures, & haut mal, fi on en prend auec uinaigre miellé le pois fufdit : il prouoque les menftrues, il fert contre les uentofités de la matrice, & déliure des friffons, fi on en baille deuant l'accés des fieures, il lafche le uentre pris du pois d'une drachme ou deus auec eau miellee : une drachme d'agaric buë en uin trampé fert de contrepoifon : beu du pois de trois oboles auec du uin, eft un fouuerain remede contre les morfures & piqueures des ferpens. En fomme l'agaric eft bon à toutes maladies interieures, donne felon l'age & la force des patiens, maintenant auec de l'eau, maintenant auec du uin, quelquefois auec uinaigre miellé, aucunefois auec eau miellee.

L'Agaric eft comm'vn potion qui croift fur les arbres. Nous en auons fuffifammét parlé au premier liure traitans de la meleze. On en trouue de fort bon es montaignes de Trente, où i'en ai fouuent coupé auec vne farpe. Pline au liu. 16.chap.8. a écrit tous arbres portans glan produire de l'agaric, toutesfois en tous les bois de Trente, & d'autres endroits d'Italie, que i'aie veu, l'agaric vient fur la feule meleze. Diofc. dit qu'il vient fur le cedre en Galatie d'Afie, & en Cilicie, fans faire

Fig. 3.3 Pietro Andrea Matthioli (1560) published *Commentarii in libros sex pedacii discoriis* in 1560, in which he illustrates and describes all the medicinal and domestic uses of the larch agaric. (Photographs © reproduced by kind permission of the *Bibliothèque des Conservatoire et Jardin botaniques de Genève*, Switzerland)

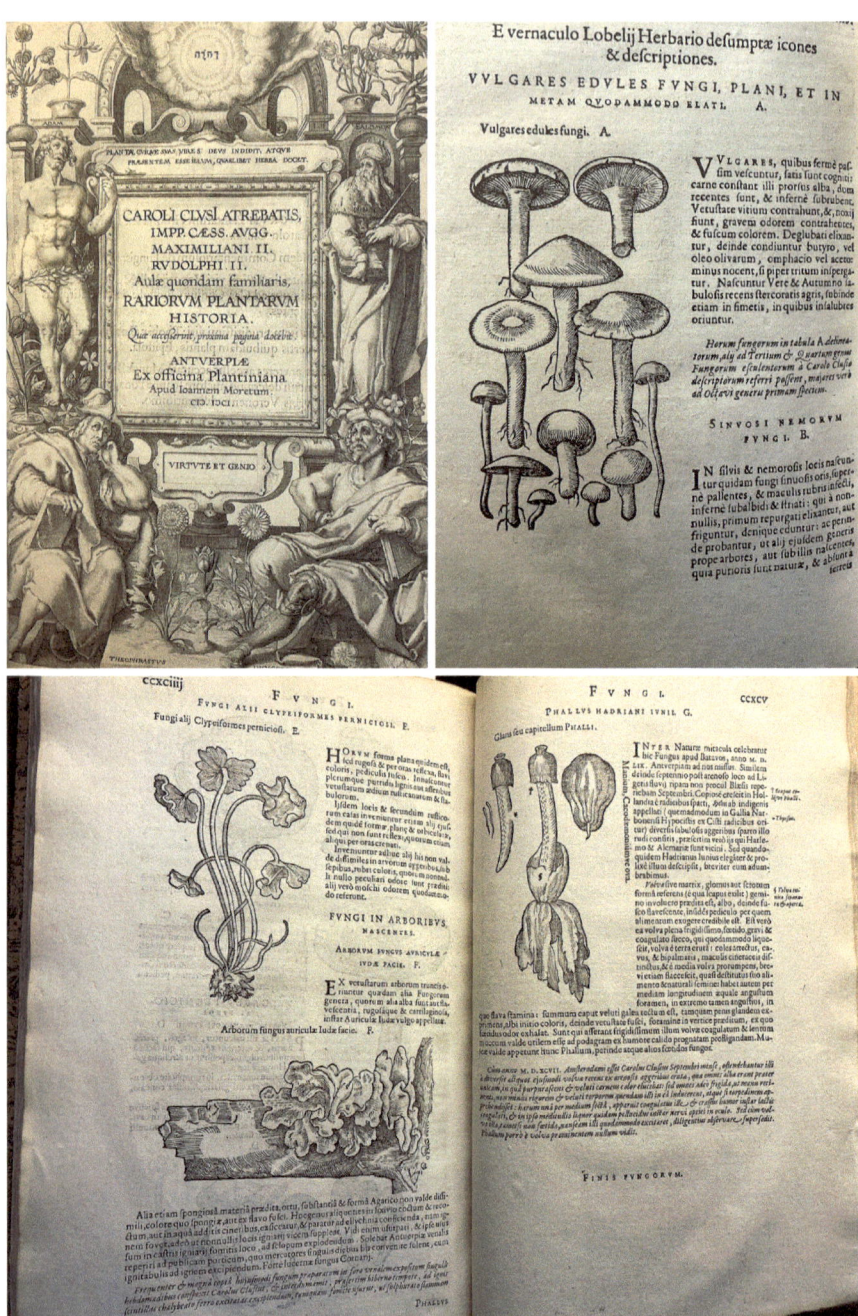

Fig. 3.4 In 1601 Charles l'Ecluse published his work *Rariorum Plantarum historia*, which contains a chapter more than 30 pages long dedicated to mushrooms. (Photographs © reproduced by kind permission of the *Bibliothèque des Conservatoire et Jardin botaniques de Genève*, Switzerland)

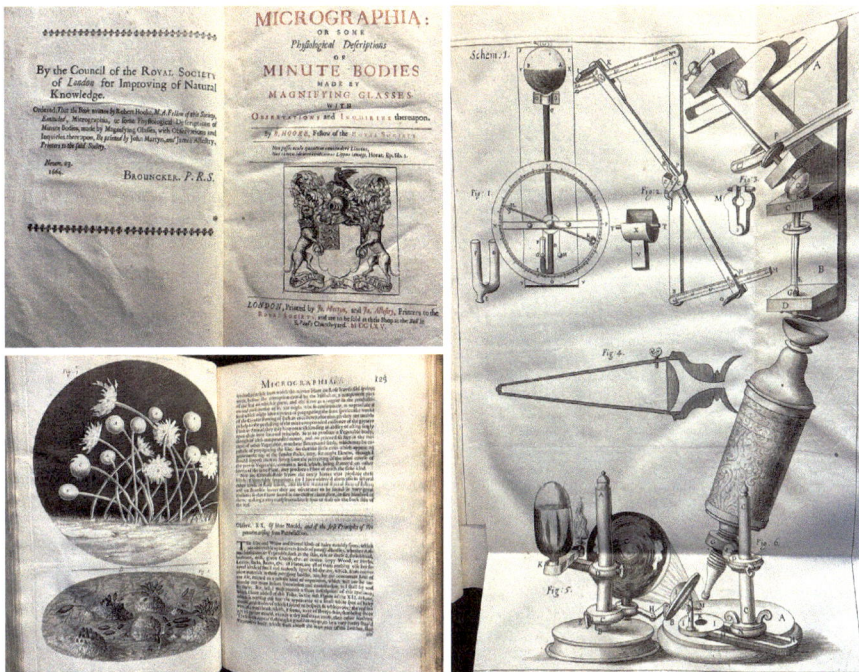

Fig. 3.5 Robert Hooke published *Micrographia* in 1665, in which he described the first mould with the aid of the microscope he developed (right). (Photographs © reproduced by kind permission of the *Bibliothèque des Conservatoire et Jardin botaniques de Genève*, Switzerland)

the lens of his microscope. He also sowed spores and observed and analysed their stages of germination and growth. He described more than 900 species of fungus micro- and macroscopically, including the genera *Botrytis* and *Aspergillus*, as well as numerous lichens.

Thanks to Micheli's work, the eighteenth century saw a massive expansion in mycological knowledge. Microscopy coupled with experimentation enabled major advances in our understanding of their life cycle and means of reproduction, leading to the development of new principles of classification. This sensitive issue was influenced by the author's subjectivity, although the basic principles they wished to adopt confirmed the relevance of their choices. This is easy to illustrate from the way in which the characteristics deemed relevant for the classification of fungi evolved, alternating between the principles of Carl Linnaeus and other eminent scientists of this century. The Swedish naturalist Linnaeus laid the foundations for the binomial system which was to revolutionise the world of classification and taxonomy. Initially in *Systema Naturae* published in Von Linné 1735, he regarded fungi as plants and created a 24th class called *Cryptogamia* encompassing *Flores absconditi*, namely plants with concealed flowers (Fig. 3.7). Then in 1753, Linnaeus published *Species Plantarum,* keeping fungi in *Cryptogamia*, but drastically reducing the number of species within this class (Von Linné 1753).

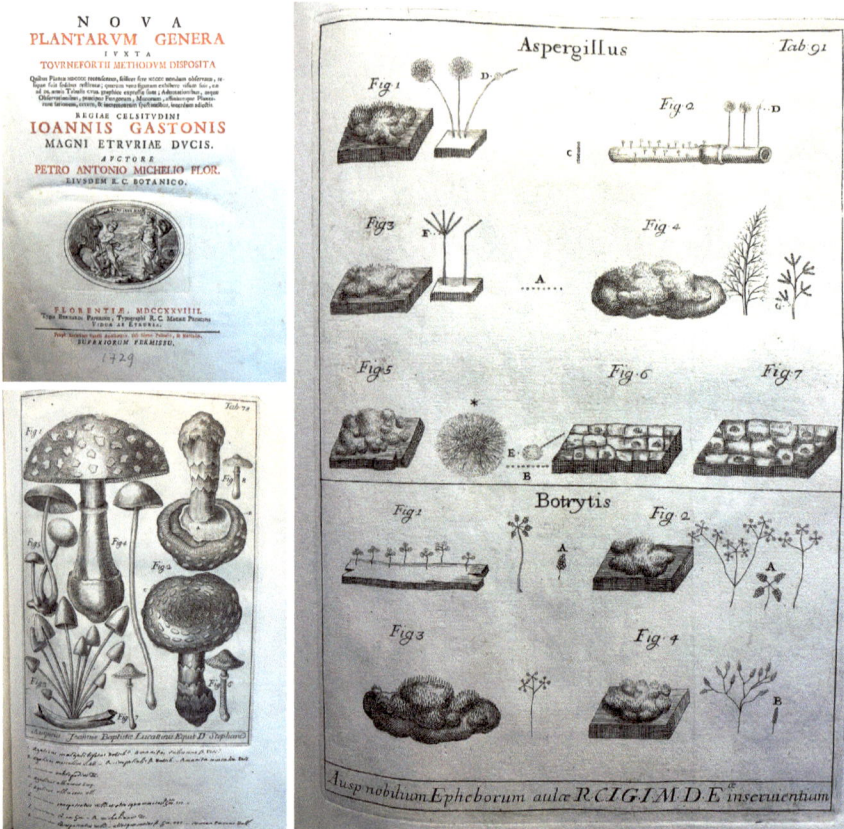

Fig. 3.6 In 1729, Micheli published *Nova Plantarum genera*, the forerunner of modern mycology. The right-hand page shows the iconography of the genera *Aspergillus* and *Botrytis*. (© Photographs reproduced by kind permission of the Bibliothèque des Conservatoire et Jardin botaniques de Genève, Switzerland)

In the same year, German naturalist Johann Gottlieb Gleditsch introduced principles based on the establishment of natural classes in his work *Methodus Fungorum*. He applied these principles to genera and species, exposing the shortcomings of certain classification methods he judged to be artificial. He was one of the first authors to create a classification system based on the location of the fungal 'seeds' (zone of fructification), paving the way for a modern mycology rich in descriptive works. From the eighteenth century, prestigious authors, including Jean-Baptiste François Bulliard, Jean-Jacques Paulet, James Sowerby, Christian Hendrik Persoon, Elias Magnus Fries, Jacob Christian Schäffer and August Johann Georg Karl Batsch (Fig. 3.8) began to reveal the secrets of fungi.

In the nineteenth century James Boltons, Vincent Julius Edler von Krombholz, Léopold Trattinick, Louis Secrétan, Jacob Strumm, Augustin-Pyramus de Candolle, Lewis David von Schweinitz and James Cooke published their observations on

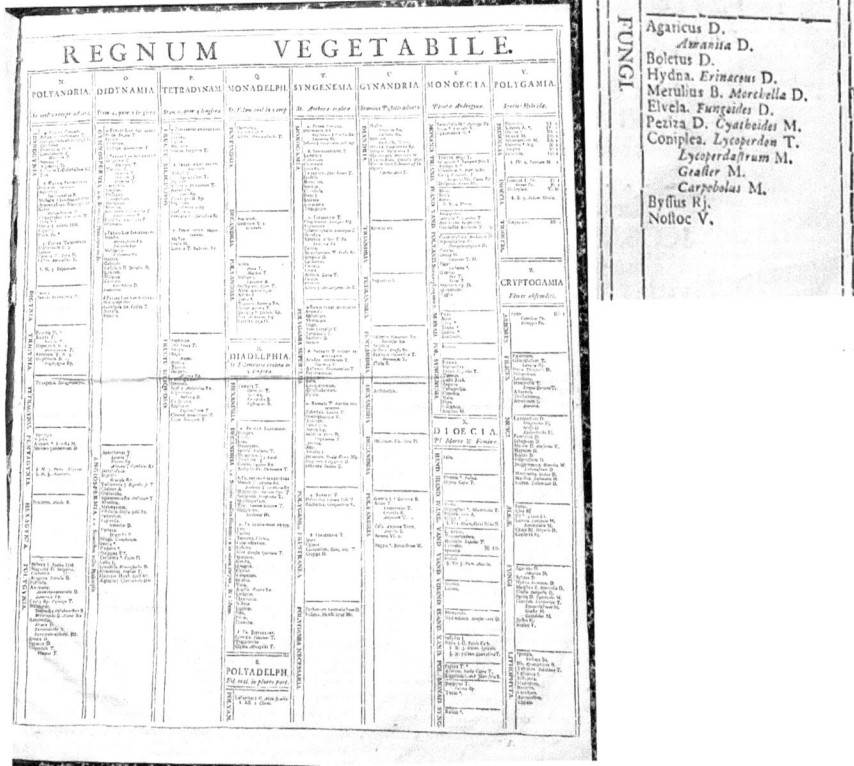

Fig. 3.7 In 1735 Linnaeus published *Systema Naturae,* in which he classified fungi in the *Cryptogamia* (plants with concealed reproductive parts). (Photographs © reproduced by kind permission of the *Bibliothèque des Conservatoire et Jardin botaniques de Genève*, Switzerland)

regional mycoflora and re-examined certain groups of fungi. Their richly illustrated works contained a wealth of new information such as the toxicology of the fungal species presented, the size and structure of the spores and other fungal propagules. The science of mycology crossed the paths of scholars who would go on to revolutionise the study of the fungal world and abolish the principle of spontaneous generation. These authors studied the fine structures of fungi, their mode of reproduction and feeding, developmental factors, relationships with other living organisms and interactions with the environment.

One of them, Heinrich Anton de Bary, German surgeon and botanist, was to turn this science on its head. He showed that a fungus could manifest itself in diverse structures and morphologies which evolved during its development, and that a macroscopic fungus, such as *Amanita phalloides*, is merely the fleeting product of mating between two compatible individuals. He recognised that fungi were ubiquitous underground, where they created a dense network of filaments criss-crossing in all directions. This essential discovery shows that fungi can adapt to environmental conditions and possess considerable adaptive flexibility. De Bary developed the

Fig. 3.8 In 1783, August Johann Georg Carl Batsch published *Elenchus Fungorum*, in which he described a vast number of new species of fungus accompanied by detailed illustrations. (Photographs © reproduced by kind permission of the *Bibliothèque des Conservatoire et Jardin botaniques de Genève*, Switzerland)

concept of epidemiology in parallel with his colleague Miles Joseph Berkeley. The two are regarded as pioneers of plant pathology, or phytopathology.

In 1847, after the Irish famine resulting from the loss of the potato crop, Berkeley showed that a fungus was in fact responsible for the devasting potato blight and not merely a symptom of it. He provided valuable insights into the conditions favouring the development of the pathogen *Phytophthora infestans*.

In the nineteenth century, scientists began to realise that fungi did not belong to the plant kingdom. These advances were made possible through morphological, chemical and developmental analysis. However, it was not until the mid-twentieth century, in 1959, that Robert Harding Whittaker proposed a separate fungi kingdom, in *On the Broad Classification of Organisms*. He divided living organisms into five kingdoms based on mode of nutrition: animals (ingestion), plants (photosynthesis), fungi (absorption), single-celled organisms with nucleus, and bacteria (monera), lacking true nuclei (Fig. 3.9). In Whittaker's classification system, the fungi kingdom was divided into two sub-kingdoms: the Myxomycota, including the Myxomycetes (slime moulds), and the Eumycota (true fungi), including the Phycomycetes, (Archimycetes and Chytrids), the Ascomycetes and the Basidiomycetes. Technological developments and advances in molecular biology, notably DNA techniques such as deep and genome sequencing, led to the emergence of new fields of science which enabled the evolutionary history of living organisms to be reconstructed (phylogenetics). As is often the case in science, knowledge gained through new technologies turned current certainties upside down.

Fig. 3.9 Representation of the classification of organisms proposed by Robert Harding Whittaker (1959) in which living organisms are divided into five kingdoms based on mode of nutrition: animals, plants, fungi, single-celled organisms with nucleus, and bacteria (Monera), which are not shown on the diagram as they come under the protists. (Illustration produced by Whittaker (1959) © Dr. Léonie Pélissier, Switzerland)

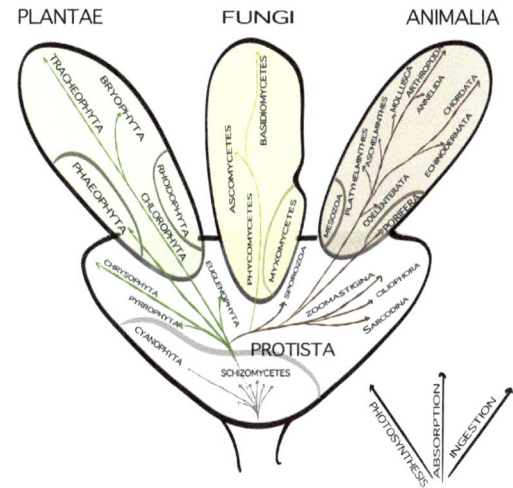

Fig. 3.10 Between 1905 and 1910, Emile Boudier published *Icones Mycologicae*, a reference book containing more than 600 plates of fungi; watercolour illustrations painted by Boudier himself. (Photographs © reproduced by kind permission of the *Bibliothèque des Conservatoire et Jardin botaniques de Genève*, Switzerland)

The fungal classification system is under constant major revision. Fungi constitute a kingdom very far removed from plants and far closer to animals than one might imagine. Distinguished mycologists such as Gillet, Cooke, Tulasne, Patouillard, Quélet and Saccardo published a proliferation of descriptive and illustrated works. Nowadays, thanks to the works of Boudier (Fig. 3.10), Singer, Heim, Kühner, Romagnesi, Webster, Clémençon, Rayner, Boddy, Blackwell, Hawksworth, Ainsworth and Hibbett, to name just a few authors among the thousands of mycologists through the ages who have shed light on this subject, we have an ever-growing understanding of the multiple facets of fungi.

3.2   What Are Fungi?

3.2.1   Structure of Fungi

Most fungi are filamentous organisms (Fig. 3.11), though in some cases they grow in the form of single-celled yeasts by means of fission or more generally by budding (Fig. 3.12). Certain dimorphic species can switch between the filamentous and the yeast stage, while others remain stable in the form of yeasts, such as *Saccharomyces cerevisiae,* the yeast used to make bread, beer and wine. The thallus (the vegetative body) of a filamentous fungus is generally eucarpic, meaning that it is differentiated into a vegetative part for the absorption of nutrients and a reproductive part, which may be sexual or asexual. In some cases, the thallus is undifferentiated but transforms as a whole into propagules. This form of thallus is termed holocarpic, for example *Synchytrium,* a genus belonging to the Chytridiomycetes.

Each filament (hypha) develops from a spore which germinates to form a germ tube which extends by growing from the tip. The term 'hypha' was introduced by

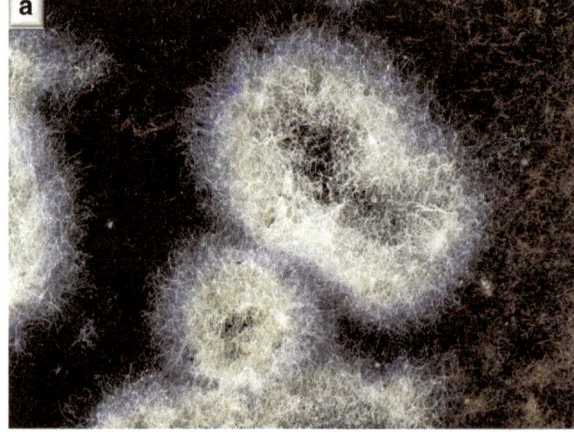
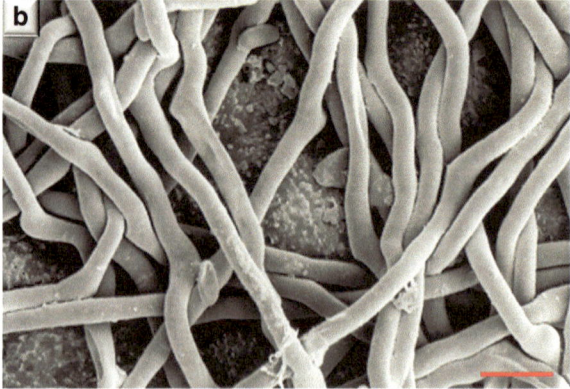

Fig. 3.11 Filamentous fungal growth in the form of a network of varying density, the mycelium (**a**). This network consists of a tangle of tubular filaments (hyphae) (**b**). Scale bar represents 10 μm. (Top photograph © Mario del Curto, Switzerland)

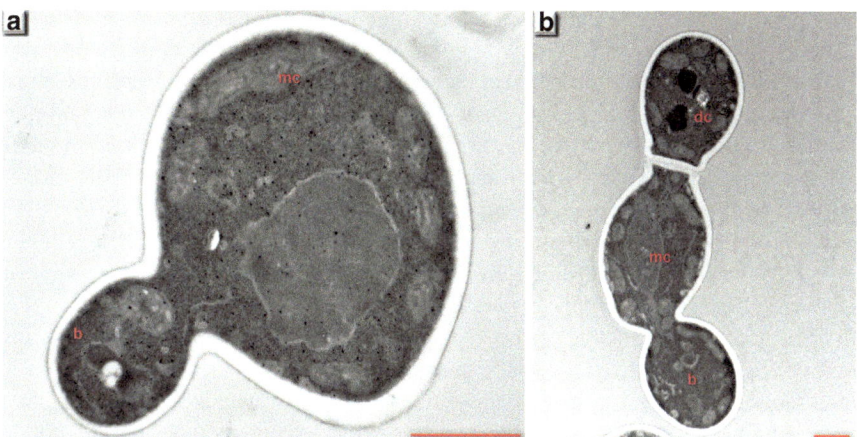

Fig. 3.12 Fungal growth in the form of yeast (*Candida albicans*). Asexual multiplication occurs by budding (**a**), giving rise to a daughter cell which is often slightly smaller than the mother cell. b: bud, dc: daughter cell; mc: mother cell. Scale bar represents 1 μm

Willdenow (1810) but the notion of fungal filaments had already been described by Hooke (1665) and Micheli (1729), although they did not name them specifically. Hyphae can form branches. Mycelium is the term given to the filaments and thallus as a whole. Hyphae are fungal structures protected from the external environment by an outer wall composed of chitin (polymer of N-acetylglucosamine), cellulose (polymer of D-glucose) and, in some taxonomic groups, other glucans such as chitosan derived from a modification of the initial chitin.

Some fungi are able to encase themselves in a sheath reinforced with melanin, a black complex polymer which is extremely resistant and waterproof. During hyphal growth, the materials forming the wall are continually synthesised though the activity of numerous enzymes and added to the growing filaments. Fungal growth is described as 'apical', meaning that elongation always occurs at the apex or tip of the hypha due to the concentration of vesicles transporting the constituents of the cell wall in this area. These vesicles discharge their contents at the tip of the hypha to enable the construction and elongation of the hypha. In some groups of fungi, these vesicles are concentrated in a circular fashion around a micro-corpuscle which, when viewed under transmission electron microscope, reveals itself to be rich in electrons. The vesicles and corpuscle together go by the unusual name of *Spitzenkörper* conferred by Brunswik (1924). The name translates as 'apical body', but the German term is used whatever the language of communication. This *Spitzenkörper* creates a micro-electric field similar to a magnet which enables the vesicles to drain. The growth rate varies according to the species, and it is not unusual for some individuals to grow at speeds of more than 0.8 mm per hour, for example *Botryosphaeria obtusa*. During rapid growth, it is not uncommon to observe several hundred vesicles per minute fusing with the plasma membrane at the apex. When branching occurs, a new *Spitzenkörper* is generated. The direction of hyphal growth is controlled by the position of the *Spitzenkörper* at the apex of the hypha.

The structure of hyphae is generally homogeneous across the different taxonomic groups of fungi. The only significant morphological difference is the presence or absence of cross-walls, called septa, which, unlike septa in plant cells, do not delimit autonomous cells. Instead, they divide the hypha into compartments which communicate with one another via a continuous flow of cytoplasm through a central pore. Thus hyphae can be septate, presenting with defined cellular compartments, or non-septate, in which case they are referred to as coenocytic (Fig. 3.13). The presence or absence of septa is one of the taxonomic criteria enabling Zygomycetes (hyphae generally coenocytic) to be distinguished from Ascomycetes and Basidiomycetes (hyphae generally septate).

Similarly, the structure of the septum may differ according to the taxonomic group. Septa delimiting reproductive structures have no pores (complete). In the Ascomycetes, vegetative hyphae are generally septate. Septa are usually perforated by a single pore which enables the flow of cytoplasm and organelles (mitochondria and nuclei). In the Pezizomycotina, a subphylum of the Ascomycota comprising more than 82,000 species, these pores are organised in an unusual way. They have rounded rims and are frequently accompanied on either side by dense corpuscles called Woronin bodies (Fig. 3.14). These corpuscles were initially described by Woronin (1864) and named in his honour by Buller (1933). Woronin bodies are peroxisomes of 150–500 nm in diameter which perform an important function by plugging the septal pores when the hyphae are damaged to prevent the loss of cell content (Markham and Collinge 1987).

In most species in the Basidiomycetes the septa are perforated by a dolipore; a complex type of pore comprising a thick ring and a central channel (dolium = barrel) (Fig. 3.15). Viewed under electron microscope, the dolipore appears to be surrounded by a structure resembling a pair of parentheses, giving rise to the term parenthesome.

Each defined segment contains one or two, or possibly several cell nuclei—hence mycelium is described as monokaryotic, dikaryotic or plurikaryotic. If the nuclei within the same compartment are genetically identical, the mycelium is said to be

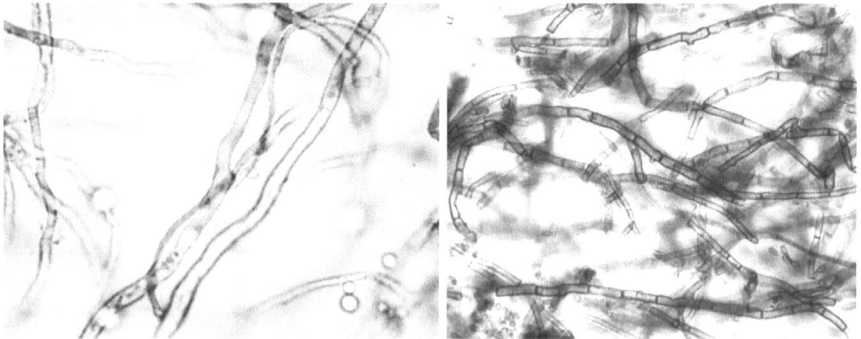

Fig. 3.13 Left, coenocytic hyphae (without septa) of *Mucor racemosus* and right, septate hyphae of *Fomitiporia mediterranea*

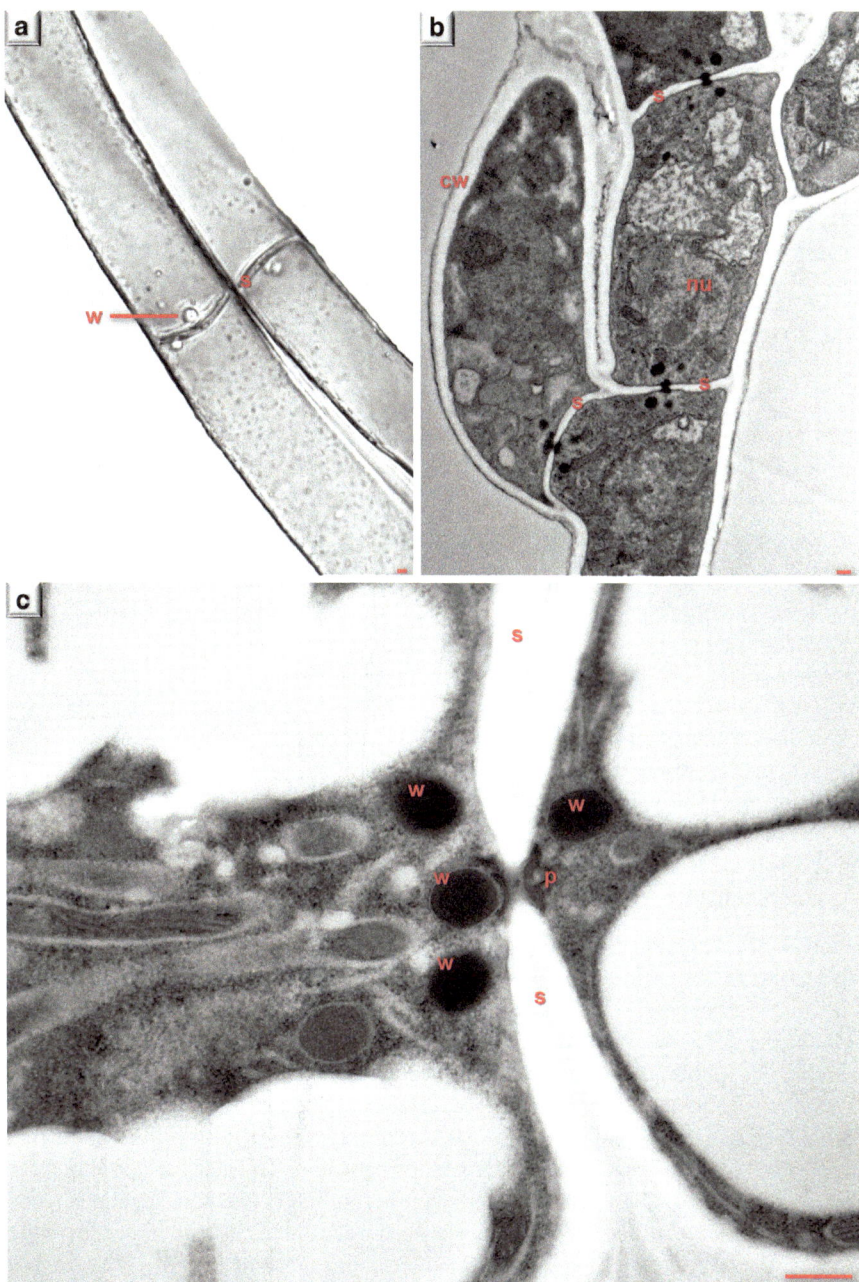

Fig. 3.14 Septa and pores on Pezizomycotina. (**a**) Hyphae of *Helvella coccinea* Scop observed under optical microscope (© Photos R. Dougoud, Switzerland). (**b**) Septa and pores with Woronin body in the hyphae of *Botrytis cinerea* Pers. observed under transmission electron microscope. (**c**) Detail of **b**. *cw* cell wall, *nu* nucleus, *p* pore, *s* septa, *W* Woronin bodies. Scale bar represents 200 μm

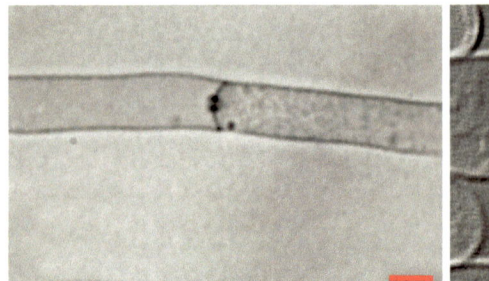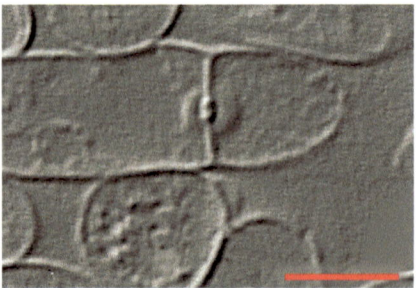

Fig. 3.15 Dolipore in the Basidiomycetes observed under optical microscope. Scale bar represents 5 μm. (Photographs © Heinz Clémençon, Switzerland)

homokaryotic; if the nuclei are genetically different (possibility of anastomosis or fusion of hyphae), the mycelium is said to be heterokaryotic. The two states can be combined in various ways:

- monokaryotic homokaryons (1 nucleus per cellular compartment, all nuclei genetically identical)
- dikaryotic/plurikaryotic homokaryons (2 or several genetically identical nuclei per cellular compartment)
- coenocytic homokaryons (genetically identical nuclei in a non-compartmentalised hypha)
- monokaryotic heterokaryons (1 nucleus per cellular compartment, all nuclei genetically different)
- dikaryotic/plurikaryotic heterokaryons (2 or several genetically different nuclei per cellular compartment)
- coenocytic heterokaryons (genetically different nuclei in a non-compartmentalised hypha).

Fungal cells consist of the organelles typically found in the Eukaroytes, with the exception of the chloroplasts: nuclei, mitochondria with flattened cristae, Golgi apparatus, ribosomes, cytoplasmic microtubules, dictyosomes, lipid inclusions and glycogen inclusions (the storage component in fungi). The nucleus comprises a double phospholipid membrane continuous with the smooth or rough endoplasmic reticulum. Mitosis is intranuclear, so the nuclear membrane does not rupture. In monoflagellated forms of fungi, mitosis is typically centric, taking place in the presence of a centrosome comprising a pair of centrioles (cylindrical structures normally arranged in nine groups of three microtubules); in non-flagellate forms, however, mitosis is non-centric. In this case, a microtubule organising centre called the spindle pole body (SPB) is present, but no centrioles. The SPB is a multi-layered organelle comprising proteins and polymers. In functional terms, it is equivalent to the centrosomes but without centrioles. The SPB organises the microtubule cytoskeleton which plays an important role in organising the spindle and thus in cell division. Another special feature of fungal mitosis is the absence of metaphase plate formation during chromosome alignment. The visible structures of fungi are not composed of true tissue, but an interwoven mass of hyphae arranged in different structures. Called plectenchyma, this 'false tissue' facilitates the formation of macroscopic structures called fruiting bodies containing or bearing the sexual or asexual spores (Fig. 3.16).

3.2.2 Reproduction

Reproduction in fungi is complex, reflecting the heterogeneity of their lifestyle. It may be sexual, asexual or alternate between these two modes of multiplication. The spore is the unit of dissemination in most cases. Spore morphology is exceptionally diverse, with great variation in shape, structure, colour, size and ornamentation (texture and surface patterns). Some spores are septate, having one or more compartments which can germinate independently of one another (Fig. 3.17). The shape and texture of spores are generally linked to their mode of dispersion and dissemination. Dry spores are generally anemochorous and airborne (dispersed by the wind, insects and birds). Some stick together in clumps and spread in the form of projectiles, others are consumed by animals and dispersed through their faeces.

Spores function like seeds but without a preformed embryo. They contain a cytoplasmic mass comprising organelles and other cell structures, for example ribosomes, Golgi apparatus, mitochondria, nucleus(ei), vacuoles and lipid globules.

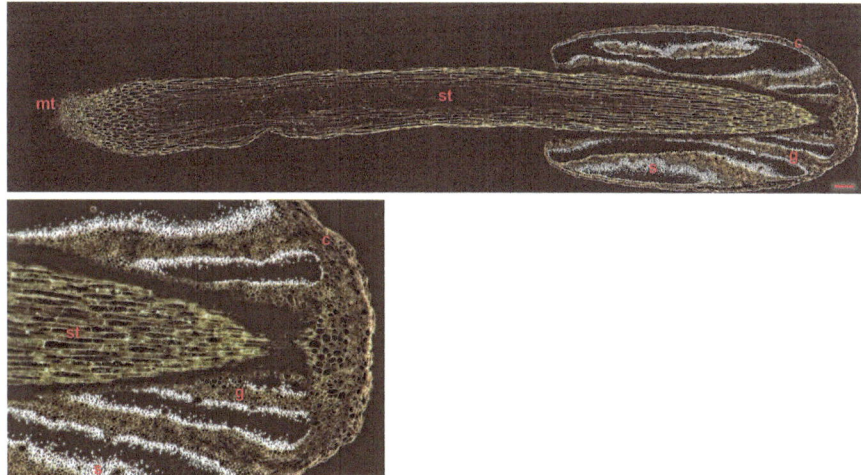

Fig. 3.16 Fruiting body (basidiocarp) of a coprin (*Coprinus* sp.) observed in semi-thin section under optical microscope. (**a**) Interwoven filaments forming the base and cap, resulting from the arrangement of different types of hyphae. The interwoven mass forms a 'false tissue' called plectenchyma. (**b**) Detail of cap with several gills (hymenium, fertile part) having produced a multitude of sexual spores (basidiospores). *c* cap, *g* gills, *mt* mycelial threads, *s* spores, *st* stalk or stipe. Scale bar represents 200 μm

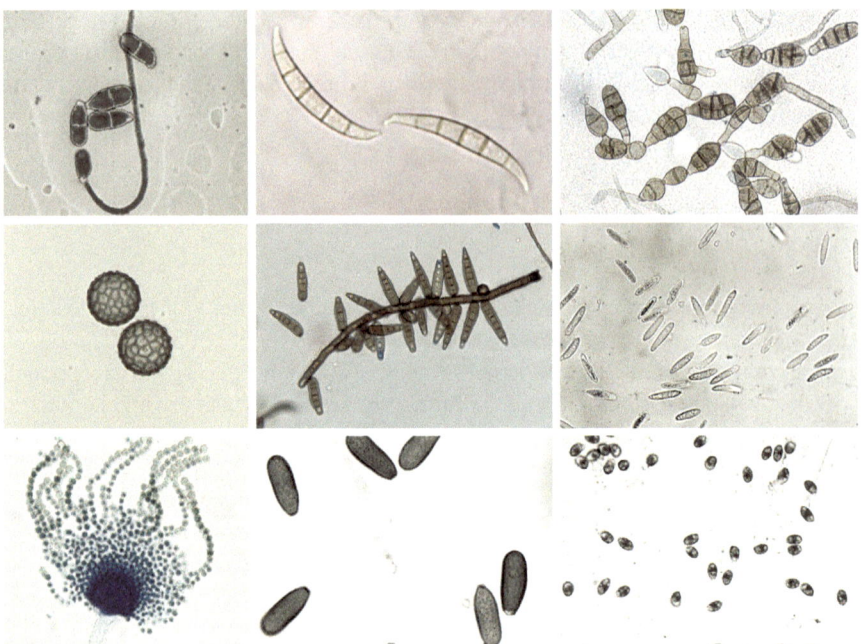

Fig. 3.17 Examples of morphological variation in fungal spores. Some spores are septate, having two or more compartments. From left to right and top to bottom: *Trichothecium* spp., *Fusarium* spp., *Alternaria* spp., *Tilletia caries*, *Helminthosporium* spp., *Coprinellus*, *Aspergillus* spp., *Botryosphaeria* spp., *Phoma* spp

These spores have walls of varying thickness composed of multiple layers, enabling them to survive relatively long periods. Spores are typically produced either by mitotic division without fusion of the nuclei or by sexual reproduction, in which case the nuclei are derived from meiosis. Spores produced by asexual reproduction are called conidia or mitospores; spores produced by sexual reproduction (meiospores) have specific names depending on the taxonomic group of the fungus they are derived from, for example basidiospores, ascospores or zygospores from Basidiomycetes, Ascomycetes or Zygomycetes. Whichever the mode of reproduction, these spores are formed in or on special structures whose arrangement and morphology is also determined by the taxonomic group of the fungus in question.

A third mechanism specific to fungi can be added to these two modes of reproduction: parasexuality. This phenomenon was described by Pontecorvo in *Aspergillus nidulans* (Pontecorvo 1956). Strictly speaking, sexuality in fungi involves two compatible individuals and a mixing of genetic material. However, genetic mixing can also be achieved via the mechanism of parasexuality within a hypha, where several haploid nuclei (N) can cohabit with diploid (2N) or triploid (3N) nuclei (which are short-lived). These nuclei can fuse and separate after exchanging genetic material, resulting in a continuous rearrangement of the genome within the hypha. This genetic exchange can also take place without awaiting a succession of generations. This phenomenon enables certain species of fungus to continually adapt to their environment and to the diverse biotic and abiotic stresses they face and partly explains the exceptional adaptive plasticity of these organisms.

3.2.2.1   Sexual Reproduction

Sexual reproduction in fungi is the result of meiosis, namely the fusion of two cell nuclei carrying two unique sets of chromosomes, which divide and recombine their genetic material. Sexual reproduction is the basis of an individual's genetic diversity and a fundamental requirement for evolution. In fungi, sexual compatibility is ensured by thousands of mating types, including self-fertile sexual reproduction (homothallism), which takes place on the same thallus (auto-compatible thallus), and heterothallic sexual reproduction requiring two separate compatible individuals.

There are three main points to consider regarding sex in fungi: (a) homothallic versus heterothallic reproduction; (b) the sex determinants encoded by the locus of the mating type; (c) sexual systems with a single biallelic locus (bipolar) versus systems with two unlinked, multiallelic sex loci (tetrapolar) (Ni et al. 2011).

A key step in sexual reproduction is mate recognition. Both yeasts and filamentous fungi have evolved systems to detect compatible or incompatible mating partners via specific peptide pheromones and receptors (Jones and Bennett 2011). Pheromone precursor genes have been identified throughout the fungi kingdom in both heterothallic and homothallic species. The presence of precursor genes in homothallic fungi suggests that their products may have functions other than attracting a mate, and may also play a role in post-fusion events (karyogamy and meiosis) as well as in cell-cell recognition and fusion (Kim et al. 2002). In very general terms, sexual reproduction in fungi involves three sequential stages:

- **Plasmogamy**—the fusion of two protoplasts, bringing together two compatible haploid nuclei. Recognition is achieved by volatile peptide signalling.
- **Karyogamy**—the fusion of two haploid nuclei to form one diploid nucleus. In fungi this dikaryotic state can remain stable for a long time.
- **Meiosis**—cell division resulting in a reduction in the chromosome number to one per cell, a mixing of genetic material which restores the haploid state and enables the production of meliospores. Haploid meliospores are formed on or in sporocarps (fruiting bodies), the nature of which varies depending on the taxonomic group, for example:
  - the Ascomycetes form ascocarps containing asci (reproductive cells) in which meliospores called ascospores form. The asci actively open or disintegrate to release the ascospores.
  - the Basidiomycetes form basidiocarps (or basidiomata). The meliospores, called basidiospores, form externally on fertile cells (basidia) and are borne on short sterigmata.

3.2.2.2 Asexual Reproduction

In contrast to sexual reproduction, which can be a slow and complex process, in asexual reproduction a single individual gives rise to a clone. This reproductive mode is a rapid and effective mass multiplication strategy, enabling a fungus to disseminate genetically identical spores and colonise larger areas. Asexual reproduction involves the production of conidia (asexual spores) through the process of thallic conidiogenesis (segmentation, budding or fission) or blastic conidiogenesis. The first mode is typical of some taxa and enables the clonal dissemination of an individual through the formation of spores after separation. In the case of segmentation, fragments of the thallus break away to form new individuals. This is the simplest mode of reproduction because the propagules—referred to as arthroconidia—are formed by the disarticulation of existing hyphae.

Yeasts can reproduce by:

- **Budding**: a bud forms on the surface of the mother cell. The nucleus splits in two, one remaining in the mother cell and the other migrating in the continuous cytoplasm to the daughter cell. The bud forms a wall enabling it to separate from the mother cell and is then ready to bud itself. In some cases, the daughter cell can bud while still attached, creating a chain of budding cells.
- **Fission**: two nuclei produced by mitotic division migrate to each pole of the mother cell. A dividing wall forms equidistant to the nuclei, enabling the two identically sized daughter cells to separate.

In blastic conidiogenesis, conidia are produced on simple or branched structures called conidiophores (non-vegetative hyphae) which may be free or protected in conidiomata or asexual fruiting bodies which vary in size and shape depending on the fungal group in question.

Various ontogenic processes determine whether these conidia are produced on simple or branched conidiophores. In this book we focus only on the development of blastoconidia. There are two principal means by which conidia are produced at the hyphal tip of the conidiophore; both the inner and outer wall of the conidiogenous cell are involved in the formation of the conidium, until it separates (holoblastic ontogeny), or the daughter cell is produced from within the conidiogenous cell involving the inner wall alone (enteroblastic ontogeny). If the external wall of the conidiogenous cell forms a pore through which the internal wall emerges, it is referred to as enteroblastic tretic ontogeny. Conidia can also be formed by the synthesis of a new cell wall which is separate from the conidiogenous cell. In this case, the specialised flask-shaped conidiogenous cells are called phialides (Cole and Samson 1979). Holoblastic development is characteristic of grapevine bunch rot (*Botrytis cinerea*). Each fertile branch of the conidiophore ends in a bulbous swelling which produces holoblastic conidia borne on short denticles in synchronous fashion (Fig. 3.18). Tretic enteroblastic development can be observed in species in the genera *Helminthosporium*, for example, and phialidic enteroblastic development, which is very widespread, in species in the genera *Aspergillus*, *Penicillium* or *Fusarium*.

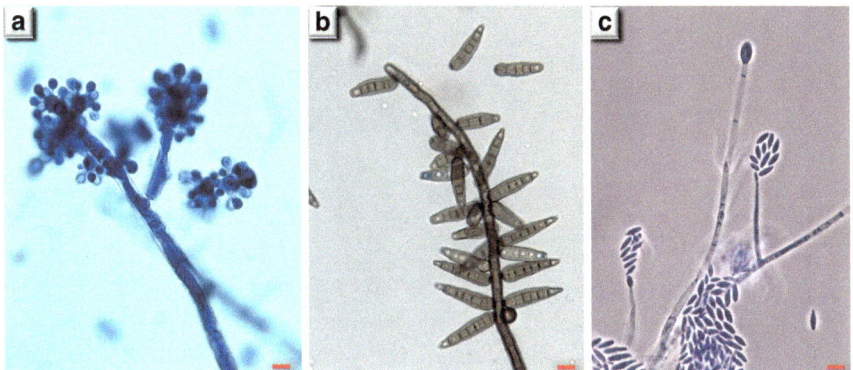

Fig. 3.18 Blastoconidia. (**a**) Holoblastic development of conidia in *Botrytis cinerea* Pers. (**b**) Enteroblastic tretic development of conidia in *Helmintohsporium solani* Durieux & Mont. (**c**) Phialidic enteroblastic development of microconidia in *Fusarium verticillioides* (Sacc.) Nirenberg. Scale bar represents 10 μm

Conidia can also be generated within asexual fruiting bodies (conidiomata); acervuli and pycnidia for example (Fig. 3.19).

- Acervuli are saucer-shaped fruiting bodies which form inside the tissue of the host plant or sub-epidermally (superficially), covered by the plant's cuticle. Conidia are formed by an alignment of conidiogenous cells covering the base of the saucer-like structure. They are released when the cuticle dries out. Acervuli that have not yet opened are characterised by the presence of setae, brown bristle-like filaments with thick walls. Anthracnose caused by species in the genus Colletotrichum forms this type of structure.
- Pycnidia are urn-shaped structures, often ovoid to piriform (pear-shaped), that form within the host tissue or sometimes superficially. Conidia are produced on the conidiogenous cells covering the inside of the structure. Pycnidia have a special pore or opening called an ostiole through which the conidia are released. Release is triggered by the effect of splashing raindrops or simply by the presence of a film of water on the surface of the substrate or the plant organ.

3.2.3   Trophic Modes and Lifestyles

Fungi are cosmopolitan and ubiquitous. They have colonised the world over millions of years, enabling them to evolve strategies of development and survival on and in all the natural and synthetic organic materials on our planet. They are present in Arctic ice, desert sand, soil, sea- and freshwater, plants, animals, rain, clouds, the stratosphere; in all the materials we encounter in our daily lives. They are found at every altitude and latitude. The only limiting factor in their expansion is their acquisition of nutrients. Humans ingest food and gradually digest it in the stomach and

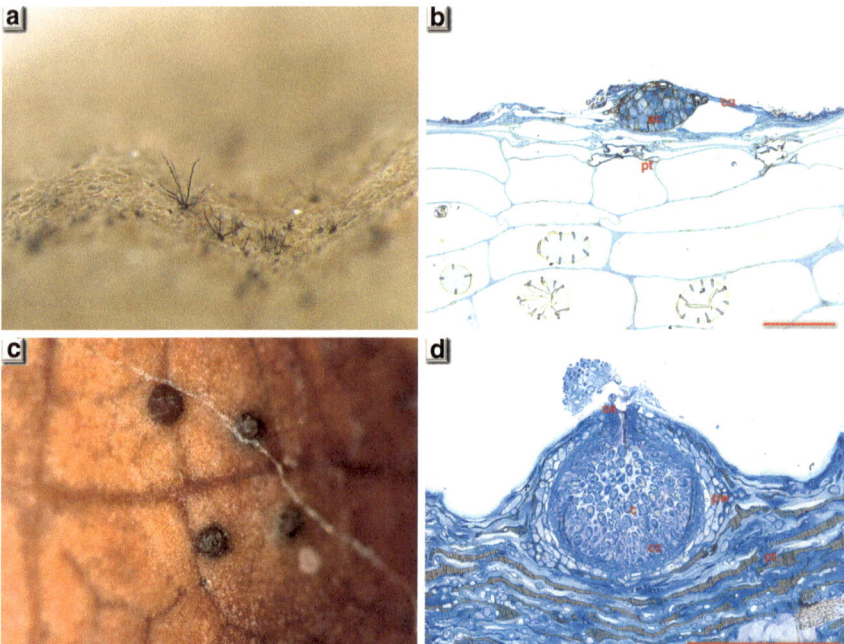

Fig. 3.19 Conidiomata. (**a**) Black acervulus of *Colletotrichum coccodes* (Wallr.) S. Hughes (potato black dot disease) on the surface of the epidermis presenting dark, septate setae. (**b**) Closed sub-epidermal acervulus in the periderm of a potato tuber, covered by the cutin, semi-thin section view. (**c**) Black pycnidia of *Phyllosticta ampelicida* (grapevine black rot) on the surface of a vine leaf. (**d**) Pycnidium of *Phyllosticta ampelicida* (engleman) Aa in the vine leaf tissue, semi-thin section view. Conidia are produced on the conidiogenous cells covering the inside of the structure. *ac* acervuli, *c* conidia, *cc* conidiogenous cells, *cu* cutin, *pt* plant tissue, *pw* pycnidia wall, *os* ostiole. Scale bar represents 100 μm

intestines, where it is broken down chemically through enzyme activity. In this way we obtain the cellular energy needed for growth. Fungi have a different nutritional strategy. They are heterotrophic organisms, meaning that they secrete enzymes extracellularly which enable them to break down organic matter into simple compounds which they ingest by absorption. This process of external digestion results in the decomposition of organic matter. The wide variety of enzymes they produce enables them to break down complex materials such as lignin, chitin, polymeric sugars (glycogen and starch) and synthetic organic polymers such as plastics. The digestion of wood polymers releases carbon locked in cellulose, hemicellulose and lignin. Carbon returned to the cycle in this way is available to plants for photosynthesis, transforming carbon dioxide (CO_2) into sugars and oxygen (O_2). In this way, fungi play a vital role in the recycling of organic matter and thus the carbon cycle. Only a minority of organisms, including some species of fungi, are capable of degrading lignin. This is because this complex and chemically highly diverse polymer releases toxic phenolic and polyphenolic compounds during its enzymatic degradation. Certain specialised species of fungi (white rots) have developed enzymatic detoxification capabilities which enable them to fully degrade lignin.

As a whole, fungi have evolved a variety of strategies to obtain their nutrients, enabling them to degrade both rotting matter and living matter capable of defending itself.

3.2.3.1   Saprophytes and Saprotrophs

Fungi are opportunists capable of adapting to any situation to evolve and obtain nutrients. Human habitat and lifestyle provide these fungi with optimum conditions for growth. They can easily break down processed and unprocessed foodstuffs (fruit, vegetables, jams, sauces), textiles (cotton), books, construction materials (silicone joints, wood, paintwork, plaster) and even synthetic polymers and plastics such as polypropylene, polyethylene, polystyrene, polyurethane, polyvinyl chloride (PVC) and polyethylene terephthalate (PET).

Saprotrophic or saprophytic fungi are capable of degrading all plant- or animal-derived organic substrates (*sapros* means decaying and troph is derived from *trophos*, meaning nourishment). Rather than killing living organisms to obtain their nutrient, saprophytes consume dead and decaying matter. Some species have a limited pool of enzymes. Without the enzymes needed to degrade more complex polymers, they grow only on substrates rich in simple sugars, such as ripe fruit. For this reason, their growth is rapid. They are often ephemeral (short-lived) and rapidly outcompeted and eliminated by more competitive individuals.

Other species produce a wider range of enzymes which enables them to grow on materials made from complex polymers. This type of substrate requires progressive degradation in parallel with detoxification processes. As a result, these fungi grow more slowly. This is the case with lignivorous fungi, which have a complex pool of enzymes: soft rots are capable of depolymerising cellulose, brown rots degrade cellulose and hemicellulose, leaving behind a characteristic pattern of cuboidal cracks (caused by the disassociation of undegraded lignin from the cellulose and hemicellulose) (Fig. 3.20), and white rots degrade all constituents of the wood, including

Fig. 3.20 Characteristic cuboidal structure of wood following enzymatic degradation of the cellulose and hemicellulose, but not the lignin, by brown rot fungi

the lignin which gives the wood its rigidity. Thus white rots are essential for recycling leaf litter and dead trees.

Some opportunists, such as *Serpula lacrimans*, one of the fungi responsible for dry rot, grow on construction timber. This xylophagous fungi is one of the most problematic causes of rot in homes and buildings. It attacks both hard- and softwoods, but since softwoods are more often used in construction, it is on these woods that the fungus is most frequently reported. In the natural environment, dry rot can adapt to a broad range of temperatures from −2 °C to an optimum 23–26 °C. This explains why it has been collected from the trunks of spruce trees in the Himalayas (Balasundaram et al. 2018), Northern Europe, Siberia and Northern California as well as the temperate regions of the world. Human activities in the past, for example timber ship building, are thought to have favoured its spread. In 1937, John Ramsbottom showed that wooden ships were frequently infested with this fungus due to careless use of undried timber combined with poorly ventilated holds. Only timber with humidity levels above 20%–25% are susceptible to attack by this fungus. Dry timber (15%–18% humidity) and proper ventilation prevent its initial growth. If the timber becomes damp due to contact with the ground, damp masonry, defective construction or inadequate ventilation, it may become infected with airborne spores. These spores germinate on the surface of damp wood and then form an abundant mycelial network. Within the wood, the fungus grows mainly at the expense of the cellulose, while the lignin remains untouched. The rotting wood contracts, giving rise to transverse cracks and developing a dry, friable texture. The water produced by progressive degradation of the cellulose may be sufficient for the fungus to continue growing even if moisture levels fall below the threshold for spore germination. The epithet *lacrimans* (crying, weeping) refers to the yellow to orange-coloured beads of liquid produced by the mycelial mass, which resemble tears. The mycelium, white or creamy yellow initially, becomes brownish yellow to rust-coloured as the fungus develops (Fig. 3.21). These sheets of mycelium can spread to adjacent timber and masonry. The fungus can also spread over several metres via mycelial threads (rhizomorphs) which can reach a diameter of 5 mm. These cord-like structures can penetrate walls through mortar and masonry and colonise an entire building if sufficient wood is available as a food source.

Many saprophytes are present in the organs of herbaceous and woody plants, from the roots to the tips of the leaves. A tree may host several species of fungi living opportunistically within its tissue. The trunks of living trees constitute a confined territory. In this environment, when a fungus encounters another, each individual synthesises fungitoxic compounds in an effort to defend their territory, however small it may be. This chemical warfare enables each to mark out their territory by forming blackish lines to restrict intrusion (Fig. 3.22). These battle lines are rich in melanin and polyphenolic compounds. Researchers exploit the confrontation between fungi to discover new bioactive compounds, notably fungicides and bactericides (Bertrand et al. 2013; Costa et al. 2019). Extreme weather events (climate change), prolonged periods of drought or wide temperature variations may be the cause of sap flow disruptions. This phenomenon provides an opportunity for fungi to spread more rapidly by 'borrowing' empty conducting vessels. Repetition of these physiological events can eventually lead to the death of the tree.

Fig. 3.21 Different stages of development of dry rot (*Serpula lacrimans* (Wulfen) J. Schröt.) on construction timber. The mycelium, white or creamy yellow initially, becomes brownish yellow to rust-coloured as the fungus develops. (Photographs © Paraxyl Sàrl, Switzerland)

An animal or insect carcass provides an important source of nutrients for saprotrophic fungi. A carcass is generally attacked by larvae or adult insects, bacteria (rots) and fungi (moulds), which play a vital role in the decomposition of dead and decaying animal matter.

3.2.3.2 Necrotrophs and Hemibiotrophs

Some specialised fungi are capable of killing living plant or animal matter and then deriving their nutrients saprophytically from the dead tissues. These fungi are called necrotrophs (which kill their host and feed on the dead matter). This group contains the largest proportion of phytopathogenic fungi on wild and cultivated plants, responsible for numerous diseases. The plants themselves possess natural means of combatting these fungal pathogens, such as physical barriers and active defence mechanisms, although they are generally unable to withstand repeated attacks. Monocultures greatly favour the spread of diseases caused by necrotrophic fungi.

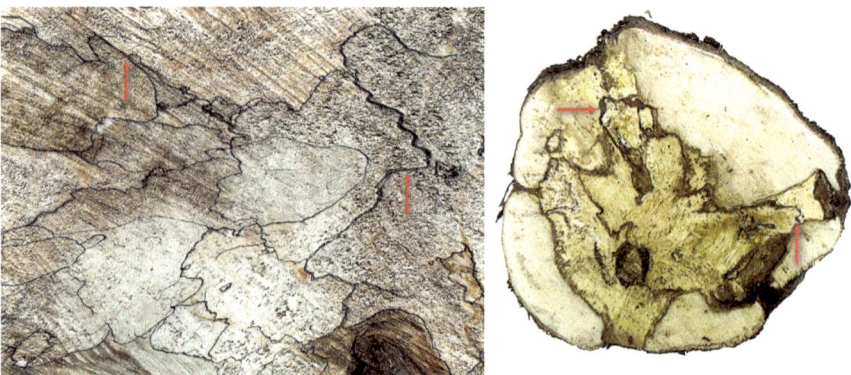

Fig. 3.22 Development of battle lines (arrows) marking the territories of different species of fungus. Left, section of a spruce trunk. Right, section of a vine

This is why it is impossible to produce high-quality food in sufficient quantity without implementing appropriate cultivation and phytosanitary measures.

Every part of a plant may come under attack: trunk, stem, leaves, fruit and flowers. A necrotrophic fungus can penetrate host tissues both mechanically and through the targeted production of enzymes which break down the plant's successive defence barriers. This type of fungus can penetrate the layers of epicuticular wax and degrade the cutin barrier (a mix of free and polymerised fatty acids and sugar polymer), the cell wall composed mainly of cellulose, and finally, the cell membrane composed of long chain fatty acids and proteins. At this stage, many necrotrophs can grow asymptomatically during the early stage of infection. This stage of development is associated with a cryptic biotrophic phase (defined as latency) coinciding with the production of oxalic acid, for example, which suppresses the host's defence mechanisms (Rajarammohan 2021). Fungi capable of undergoing biotrophic growth initially before transitioning to a necrotrophic lifestyle are called hemibiotrophs. *Botrytis cinerea*, for example, can persist asymptomatically in grapevine flower tissues (Van Kan et al. 2014) until the fruit ripens, leading to rapid depreciation of the harvest (Rigotti et al. 2002; Keller et al. 2003). When a necrotroph invades the host cells, it produces a cocktail of enzymes to rapidly macerate the tissues before degrading and consuming them.

Some fungi have developed this type of behaviour on animal tissues. They can infect the skin or internal organs, sometimes leading to the death of the individual. In humans and animals, dermatophyte infections are characterised by the development of necrotic lesions of varying size. These fungi specialise in digesting keratin, enabling them to penetrate the epidermis. They then colonise cells and progressively digest the tissues. The spores they produce spread the infection to neighbouring tissues and can be transported in the bloodstream. Their spread can be inhibited using topical treatments which must be applied for prolonged periods of time. Other fungi can cause the degeneration of internal organs, especially in individuals whose immune system is compromised. The WHO (World Health Organisation) has compiled a list of 19 fungal priority pathogens which require further research. The criteria used to establish the priority listing takes into account resistance to clinical fungicides, mortality rate and access to treatment and accurate diagnosis. *Aspergillus fumigatus* is included in the most critical group. This filamentous fungus is present everywhere in the environment—in soil, rainwater, leaf litter, outdoor air, and even indoor air within our homes. It generally has a saprophytic lifestyle but can transition to a necrotrophic phase and infect living hosts, notably plants, insects, birds and mammals.

A. *fumigatus* is the most widespread opportunistic airborne fungal pathogen and the most dangerous for immunocompromised patients. It produces tiny spores which are easily dispersed in the environment, originating largely from decomposing organic matter. Inhalation of spores into the lungs can cause a multitude of diseases, depending on the host's immune status (Arastehfar et al. 2021). These diseases include invasive pulmonary aspergillosis, chronic pulmonary aspergillosis and various hypersensitivity conditions such as allergic asthma. Other organs can be affected, such as the sinuses, with possible extension to the brain. A. *fumigatus* is responsible for hundreds of millions of cases of invasive aspergillosis worldwide every year. This fungus is also known to cause diseases in animals, especially fish and poultry.

Although the allergic forms of aspergillosis and sinus aspergillosis are not generally life-threatening, the same cannot be said if immunocompromised patients (organ transplants, chemotherapy, cystic fibrosis) are infected by the invasive form. Currently, the therapeutic arsenal available to treat A. *fumigatus* infections is limited. The class of azole antifungal agents is the most widespread option for the treatment and prevention of acute and chronic aspergillosis. However, their use in medicine and agriculture has led to the emergence of resistances (Schürch et al. 2023) which have spread very rapidly at global level. For this reason, there is an urgent need to discover new active substances to treat these diseases.

Some highly specialised necrotrophic fungi attack insects. Various species of fungi belonging to several systematic groups are able to penetrate gaps in the insect's exoskeleton and grow while keeping the insect alive. After several days, the insect dies and fungal structures appear on its body. Entomopathogenic fungi attack a wide range of hosts, such as ants, wasps, butterflies, beetles, grasshoppers and spiders, which can become infected at any stage of development (adult, larva or pupa). Some highly specialised fungi can change the insect's behaviour to maximise their competitive advantage before killing it. This is the case with *Entomophthora muscae*, a parasitic fungus of flies and other diptera (Fig. 3.23). When a spore lands on the body of an insect, it produces a small adhesive cushion which sticks firmly to the exoskeleton. Once the spore has germinated, thanks to the combined action of mechanical pressure and the production of specific enzymes which break down chitin in the exoskeleton, the fungus invades the insect body and initially grows by consuming the host's fatty tissues. The hyphae break into small fragments which are transported by the circulatory system, progressively invading the insect's entire body and tissues. As it grows, the fungus produces compounds that modify the insect's behaviour. These behavioural changes affect sexuality, causing males to reproduce only with infected females and

Fig. 3.23 Left, wasp mummified by a fungus in the genus *Cordyceps*. White fungal structures emerging from the insect corps. Right, fly infected by *Entomophthora muscae*. A few days after infection, the fly dies. Structures emerge from its very extended abdomen, releasing dozens of fungal spores. Bottom, bee mummified by a fungus in the genus *Cordyceps*. (Photographs of wasp and bee © Mario del Curto, Switzerland; Photograph of fly © René Dougoud, Switzerland)

in turn infecting themselves. Secondly, before dying, flies are compelled to crawl up toward the light; they may cling tightly to the tip of a stem, flower or leaf, or glue themselves by their proboscis to windows or walls. A few days after infection, the fly dies. Structures emerge from its very extended abdomen, releasing dozens of fungal spores. This is why it is not unusual to find a dead fly stuck to a window surrounded by a small white halo; these are the spores, poised to infect a new fly. The action of crawling towards the light to a certain extent ensures that the fungal spores are released in areas likely to be visited by other insects.

Other entomopathogenic fungi adopt a different behaviour, transforming the insect into a mummified mass of entangled filaments from which emerge fusiform structures called stromata on the ravaged body of the dead insect (Fig. 3.23). These stromata, which contain the fungal spores, are often brightly coloured. Death is caused by the production of toxic metabolites which spread through the insect's body. These fungi belong to the Cordycipitales, a different systematic group to the Entomophthorales described above. The genus *Cordyceps* contains more than 500 different species, with the epicentre of diversity located in Japan and Northeast Asia. Only 18 species are found on the European continent. Furthermore, these fungi are very well studied and used in the biological control of diverse insect predators. Promising results have been obtained in a recent study investigating the use of entomopathogenic fungi belonging to the genera *Metarhizium* and *Beauveria* to control the development and spread of the Japanese beetle *Popillia japonica*, which attacks more than 400 botanical species, including the cultivated vine (Graf et al. 2023) (Fig. 3.24).

3.2.3.3   Symbiotic Fungi

Over the course of evolution, a huge number of fungi have developed stable associations with diverse organisms. These associations, called symbioses, may involve one fungus and a second protagonist, several protagonists, or even several fungi and

Fig. 3.24 Vine ravaged by the Japanese beetle *Popillia japonica* (top). Photograph © Agroscope, Carole Parodi. Biological control using entomopathogenic fungi kills the pest and provides lasting disease control. (Photographs © Agroscope, Christian Schweizer)

several different partners forming a self-sustaining community. The evolution of molecular techniques has enabled to gain an ever-greater understanding of these complex symbiotic relationships involving the concept of microbiomes (Grimm et al. 2021).

Lichens

Lichens are among the most widespread terrestrial symbioses; the first individuals having evolved a little over 400 million years ago. Lichens can grow on stone, bark (Fig. 3.25), concrete, boulders and other inert and inhospitable substrates such as desert sand, or in the extreme conditions of alpine peaks at altitudes of over 7000 m.

These symbioses colonise virgin substrates and create organic biomass. For a long time, lichens were thought to be stable associations between a fungus and a green alga or a photosynthetic bacterium (cyanobacterium). However, recent studies conducted at microbiome scale have shed light on the extreme complexity of a

Fig. 3.25 Lichens grow on numerous different substrates such as vines, (top left), tree bark (top right), stone (bottom left), construction timber (bottom centre) and lava (bottom right)

lichen, as well as its fragility in response to air pollution. In functional terms, the fungus obtains water and minerals and gives the lichen its specific morphology; in return the alga and/or cyanobacterium provide the products of photosynthesis. As well as photosynthesising, cyanobacteria can fix atmospheric nitrogen. The vast majority of the 200,000 species of lichen described today are associated with a green alga, whereas only 10% live in symbiosis with cyanobacteria.

A study by Grimm et al. (2021) showed the multidimensional structure of *Lobaria pulmonaria* (also called tree lungwort and traditionally used as cough suppressant). This large lichen, found in all regions of the world, produces leaf-like structures on the trunks of trees. Unlike many other species, it is less tolerant of desiccation and highly sensitive to air pollution. The fungus creates a protective layer covering a mass of green algae (*Dictyochloropsis reticulata)* interspersed with clusters of cyanobacteria in the genus *Nostoc*. The non-photosynthetic bacteria (core microbial community) in the form of biofilm are integrated into this quadripartite symbiotic relationship and involved in the production of diverse vitamins needed for fungal growth.

Mutualism Between Fungi, Termites and Ants

The farming of fungi by termites and ants is an example of a mutualistic relationship. This phenomenon was first observed by König (1779), a German botanist who described termite mounds in an area situated in the south-east of Thanjavur (India). Termites are the architects of the animal kingdom. In human terms, the relative height of a termite mound is equivalent to a 180-floor skyscraper. These termite mounds are constructed from grains of soil mixed with saliva (Fig. 3.26).

The workers play an important role in maintaining 'gardens' where they cultivate and nurture fungi on special substrates (combs) built from plant debris which provide an ideal medium for fungal growth. More than 330 species of termite have been reported to farm more than 16 species of fungi in the genus *Termitomyces*. These fungi cover the combs with their mycelium and form tiny pear-shaped

Fig. 3.26 Termite mound constructed by *Macrotermes michaelseni*. (Photograph © Dr. Jan Šobotník, Czech University of Life Sciences)

white nodules 1–2 mm in diameter. These nodules (noduli) are composed of a tangle of hyphae which produce a vast number of spores on their external surface (Fig. 3.27). The nodules represent the early stage of mushroom production, and if not consumed by the termites, they will eventually produce gilled mushrooms of considerable size.

The termites consume numerous gut-resistant spores which are then reseeded via their droppings, where they rapidly germinate and colonise other combs. When the gardens are no longer maintained, the nodules gradually transform into an elongated structure called a pseudopod. These projections have a very pointed tip called a perforator which directs the structure as it grows through the termite mound, eventually emerging above-ground in the form of an edible gilled cap which may reach a considerable size; more than 1 m in diameter in the case of *Termitomyces titanicus*.

Fungus-growing termites derive many benefits from domesticating the mushroom. Firstly, they eat them; the fungi are a valuable source of minerals such as calcium, potassium, magnesium, phosphorus, zinc, copper, selenium, iron, sulphur and sodium as well as proteins, sugars and lipids. The fungi also predigest plant matter foraged by the termites, making it easier for the termites to digest as they lack the enzymes needed to digest cellulose and lignin in their digestive tract.

Fig. 3.27 Termite-fungal relationship. (**a**) The fungus is farmed on combs inside the termite mound: the nodules (first stage of mushroom formation, arrow) are consumed by the termites (here: *Macrotermes bellicosus*). (**b**) When the termite mound is abandoned, the fungus develops from the comb (white arrow), sending out a pseudopod or false foot (red arrow) which works its way through the termite mound and/or earth to form a basidioma on the outside of the mound. (Photographs © Dr. Jan Šobotník, Czech University of Life Sciences)

Similarly, more than 200 species of ants are able to cohabit with and cultivate gilled mushrooms.

Mycorrhizas

Mycorrhizas are mutualistic associations between a biotrophic fungus (which feeds on living organisms) and the roots of a herbaceous or woody plant. This beneficial relationship provides the fungi with simple carbohydrates (sugars, fatty acids) and in exchange the plant receives water through the increased surface area between the soil and the roots; minerals (especially phosphorus and nitrogen); localised detoxification of soils (accumulation of heavy metals) through the transfer and absorption of certain elements (zinc, iron, copper, potassium), and increased tolerance to biotic and abiotic stress (Bruisson et al. 2016; Shi et al. 2023). Mycorrhizas can be divided into two categories: ectomycorrhizas and endomycorrhizas, depending on whether the fungus colonises the intercellular spaces in the root or grows within the cells. The main types of endomycorrhizas are the orchid mycorrhizas associated with the Orchidaceae family (OMs), the ericoids specific to the Ericaceae family (EMs) and the arbuscular mycorrhizas (AMs) (Favre-Godal et al. 2020). Arbuscules are highly branched microscopic fungal structures with a bushy appearance. Around 70%–90% of vascular plants form AMs, 1.5%–2% EMs and 10% OMs (Brundrett and Tedersoo 2018). The AMs are associated with the phylum Glomeromycota which encompasses a little over 240 species of fungi. In the AMs, vesicles and/or arbuscules form inside the cortical cells of the root (Fig. 3.28) and are the sites of symbiotic exchange.

Ectomycorrhizal symbioses are established only with woody plant families. The fungi involved in this type of association belong to very varied taxonomic groups within the Basidiomycota, Ascomycota and Zygomycota. In ectomycorrhizas, the

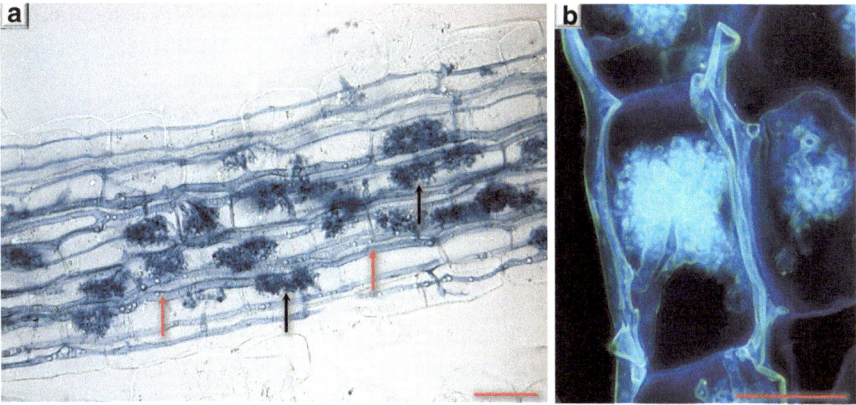

Fig. 3.28 Arbuscular mycorrhizas (AMs) in the root of a vine (*Vitis vinifera*). (**a**) Network of intercellular hyphae (red arrows) and intracellular arbuscules (black arrows). (**b**) Detail of an arbuscule. Scale bar represents 100 μm

fungal hyphae do not penetrate the cortical cells of the roots, but instead form a network of varying density in the intercellular space of the root, called a Hartig net (Brundrett and Tedersoo 2018), where exchanges between plant and fungi take place.

Mycorrhiza in Grapevine

Symbiosis with arbuscular endomycorrhizas can improve plant nutrition. The mycorrhization of young vine plants increases levels of nitrogen, resistance to abiotic stress (water stress, soil salinity, ferric chlorosis and heavy metal poisoning), and metabolites such as nitrate reductase, chlorophyll, phenolic compounds and proline (Trouvelot et al. 2015). Mycorrhized vines in association with AM exhibit increased tolerance to the fungal root pathogens *Armillaria mellea* and *Dactylonectria macrodidyma* (anc. *Cylindrocarpon macrodidymum*), ectoparasitic nematodes (*Meloidogyne incognita* and *Xiphinema index*) and pathogens of green plant parts such as *Plasmopara viticola* and *Botrytis cinerea*. Following the mycorrhization of young vine plants with *Rhizophagus irregularis*, analysis showed that the expression of certain target genes involved in the stilbenoid synthesis pathway (phenylalanine ammonia-lyase PAL, stilbene synthase STS and Resveratrol O-methyltransferase ROMT), was potentiated after infection with downy mildew and grey mould (Bruisson et al. 2016).

3.2.3.4 Obligate Biotrophs

By definition, a biotrophic organism feeds on living tissue without necessarily damaging its host. This is the case with mycorrhizas and endophytic fungi, for example. They draw nutrients and energy from their host plant's living cells and survive in the interstitial space between the cells. However, there is a taxonomically heterogeneous group of fungi which are defined by their inability to survive without their host: the obligate biotrophs. Some species of these fungi can have a wide range of hosts, while others are associated with a single host or a single botanical family. This is the case with grapevine downy mildew, *Plasmopara viticola*, an obligate biotrophic Oomycete associated with the Vitaceae family. Several obligate biotrophs cause diseases in plant, including cultivated plants, resulting in economic losses. For example, grapevine powdery mildew (*Erysiphe necator*), rusts and smuts. Although the Oomycota are not true fungi and do not belong to the fungal kingdom, several species cause irreversible damage to crops, for example potato blight (*Phythophthora infestans*) and grapevine downy mildew (*Plasmopara viticola*), and their life cycle is similar to that of fungi. Furthermore, they have been responsible for spectacular and catastrophic epidemics in crops in the past, such as the Irish potato famine caused by *P. infestans* from 1845 to 1852. Several of them are still difficult to control even today, despite the use of fungicides. Pathogenic obligate biotrophs grow asymptomatically during the early stages of infection (RoyChowdhury et al. 2022). Symptoms sometimes appear during inter- or intracellular growth, but more often during reproduction. These pathogens have adapted so that they can reproduce asexually and/or sexually on the same or different host plants, as is the case with certain rusts such as *Puccinia striiformis* (Zhao et al. 2016).

During the infection stage, biotrophs develop specialised structures (appressoria, haustoria) which enable them to derive their nutrients from living plant cells. The hyphae or appressoria attach themselves firmly to the cuticle and, through a combination of mechanical and enzymatic action, penetrate the plant's epidermal cells. Several species additionally secrete effectors to circumvent the plant's defence mechanisms. In turn, the fungus can respond by activating virulence genes which enable it to counter the plant's immune response.

3.2.4   Current Classification of Fungi

3.2.4.1   Evolution of the Concept of Classification

The classification of fungi began at the time when only visible mushrooms were studied. The origins of classification can be found in the writings of Theophraste in 300 BCE (*De Historia Plantarum*). He considered that fungi were imperfect plants entirely devoid of roots. Nonetheless, he proposed what is thought to be the first attempt at classification, grouping them in four divisions: round mushrooms growing underground, those growing at the surface and attached to a stalk, those with neither stalk nor cap, and round ones resembling a human head.

Following the works of l'Ecluse, Colonna, Bauhin and Hooke, the Flemmish mycologist van Sterbeeck (1675), wrote *Theatrum fungorum* in which he classified fungi in 15 different groups according to morphological criteria. Various publications by the English naturalist John Ray (1686–1688) enabled criteria to be defined on the basis of the growing environment, resulting in the creation of three divisions (terrestrial, arboreal and subterranean fungi) and subdivisions based on their morphology and edibility. De Tournefort (1694) revolutionised the approach to the classification of plants, dedicating the 17th class to plants considered to be fungi. He created seven distinct groups according to objective criteria associated with their structure and morphology. According to his definition, the Fungi are a genus of plants with a stem and cap, with or without gills, the Fungoides are hollow and cup- or funnel-shaped, the Boletus have a lattice-like structure or are filled with cavities, the Agaricus grow on tree trunks, the Lycoperdon are hard and fleshy initially before turning to dust, the Coralloides have a coral-like structure, and finally the Tubers form a genus of mainly rounded, subterranean plants. Micheli (1729) revolutionised the classification once again, including microscopic criteria in his taxonomic system. The work carried out by Micheli was continued by Gleditch (1753), among others, who created four sections according to the location of the spores, paving the way for a more modern and experimental form of mycology; and the most eminent mycologists of our times are still refining the morphological classification of fungi and lichens.

3.2.4.2 Taxonomy and Current Classification

The taxonomy (rules of classification) of the fungi kingdom is governed by the International Code of Nomenclature for algae, fungi, and plants, Shenzhen Code, China 2017 (Turland et al. 2018). Fungi are defined by a specific Latin nomenclature according to the following principles:

- fungi: kingdom (regnum)
- -mycota: division or phylum
- mycotina: subdivision (subphylum)
- mycetes: class
- mycetidae: subclass
- ales: order
- ineae: suborder
- aceae: family (sometimes divided in subfamily, tribe, subtribe)

The family is subdivided into genus (*genus*), subgenus, sections, subsections, series and subseries, then the name of the species (*species*), which may in turn be divided into subspecies, variety, subvariety, form and subform. Table 3.1 defines as example the taxonomy of *Botrytis cinerea* f. *cinerea*, *Armillaria mellea*, *Neophysella ampelopsidis* and *Diaporthe ampelina*.

The classification that existed until the 1990s—prior to advances in molecular analysis—classified the fungal kingdom in four main groups:

- The Gymnomycota: including the Dictyostellidae, the Labyrinthulomycota, the Acrsiomycota, the Plasmodiophorida and the Myxomycota. Most members of this group are now no longer considered part of the fungi kingdom.
- The Deuteromycota: fungi deemed 'imperfect', reproducing by asexual means alone. This notion is now obsolete and considered false.

Table 3.1 Examples showing the taxonomy of four fungal pathogens of grapevine; bunch rot (*Botrytis cinerea* f. *cinerea*), root rot (*Armillaria mellea)*, rust (*Neophysopella ampelopsidis*) and *Phomopsis* dieback (*Diaporthe ampelina*)

Kingdom	Fungi			
Phylum	Ascomycota	Basidiomycota	Basidiomycota	Ascomycota
Subphylum	Pezizomycota	Agaricomycotina	Pucciniomycotina	Pezizomycotina
Class	Leotiomycetes	Agaricomycetes	Pucciniomycetes	Sordariomycetes
Subclass	Leotiomycetidae	Agaricomycetidae		Sordariomycetidae
Order	Helotiales	Agaricales	Pucciniales	Diaporthales
Suborder		Marasmiineae		
Family	Sclerotiniaceae	Physalacriaceae	Neophysopellaceae	Diaporthaceae
Genus	*Botrytis*	*Armillaria*	*Neophysopella*	*Diaporthe*
Species	*Botrytis cinerea*	*Armillaria mellea*	*Neophysopella ampelopsidis*	*Diaporthe ampelina*
Form	*Botrytis cinerea* f. *cinerea*			

- The Mastigomycota: including the Chytridiomycota and the Chromista (including the Oomycetes). Currently, the Chromista are no longer considered part of the fungi kingdom.
- The Eumycota including Zygomycota, Ascomycota and Basidiomycota.

Advances in DNA amplification and sequencing techniques as well as more recent democratisation of genome sequencing have revolutionised the notion of taxonomic groups and profoundly altered classifications based on the cellular organisation, trophic modes and habitat. These classifications have been replaced by more rigorous definitions at phylogenetic level, but the names, criteria for definition and number of clades are yet to be decided (Richards et al. 2017). The current classification system is constantly being revised due to the time-consuming and laborious process of phylogenetic analysis, which requires expert knowledge of both mycology and bioinformatics. By the end of the 1990s, of the four clades previously defined, only the Chytridiomycota, Zygomycota, Ascomycota and Basidiomycota remained. With advances in analytical techniques and based on current knowledge, most recent classifications of the fungi kingdom define nine phylum (Naranjo-Ortiz and Gabaldón 2019), described in Table 3.2. Together, these nine phyla form a monophyletic clade, the true fungi. To these can be added several groups of organisms known as the *Fungi incertae sedis*—fungal organisms whose taxonomic affiliation remains elusive.

Table 3.2 Currently, the classification of fungi defines the nine *phylum-level* clades (major lineages) listed below

Major lineages	Secondary lineages	Lifestyle
Zoosporic fungi		
1. Opisthosporidia	Apelidea Rozellidea Microsporidia	Parasites and parasitoids of Eucaryotes
2. Chytridiomycota	Chytridiomycetes Monoblepharidomycetes Hyaloraphidiomycetes	Free-living saprobes; animal parasites and plant pathogens
3. Neocallimastigomycota	Neocallimastigaceae	Obligate anaerobic, non-parasitic fungi. Found in gut of herbivorous mammals, sea urchins and iguanas
4. Blastocladiomycota	Blastocladiomyceta	Saprobes; animal parasites, plant pathogens
Zygomycetous fungi		
5. Zoopagomycota	Zoopagomycotina Entomophthoromycotina Kickxellomycotina	Saprobes, invertebrate parasites, mycoparasites, amoebophagus
6. Mucoromycota	Mortierellomycotina Mucoromycotina	Saprobes, ectomycorrhizal, mycoparasites, plant pathogens
7. Glomeromycota	Paraglomerales Archaesporales Diversisporales Glomerales	Plant symbionts of land plants

(continued)

Table 3.2 (continued)

Major lineages	Secondary lineages	Lifestyle
Dikarya		
8. Basidiomycota	Pucciniomycotina	Biotrophic plant pathogens, insect parasites, saprobes, endophytes and mycorrhizal
	Ustilagomycotina	Saprobes, biotrophic plant pathogens
	Agaricomycotina	Saprobes, plant parasites, ectomycorrhizal, endophytes, mycoparasites, amoebophagous, symbionts and lichens
9. Ascomycota	Taphrinomycotina	Plant pathogens, saprotrophs, endophytes, animal pathogens
	Saccharomycotina	Saprobes, commensals, extremotolerants and parasites
	Pezizomycotina	Saprobes, lichens, plant necrotrophic and biotrophic pathogens, animal parasites, mycorrhizal, endophytes, amoebophagous, extremophiles

Four major lineages belong to the zoosporic fungi (one flagella at some point of their life cycle), three to the Zygomycetous fungi (loss of flagella and formation of zygospores, with the exception of the Glomeromycota) and two to the Dicarya (sexual cycle characterised by the formation of dikaryotic hyphae). The secondary lineages as well as their different lifestyles are listed for each of the *major lineages*. This table was created using data published in review by Naranjo-Ortiz and Gabaldón (2019). The groups classed in *FUNGI INCERTAE SEDIS* are not shown

A detailed description of the nine phyla is beyond the remit of this book. The main phyla are described in simple terms to give the reader a broad understanding of the structure and organisation of the main taxonomic groups addressed in this work. Outside the fungi kingdom, the Oomycota, classed in the Stramenopiles, are also discussed because this phylum contains a vast number of plant pathogens, including grapevine downy mildew, *Plasmopara viticola*.

3.2.4.3 Zygomycetous Fungi: Mucoromycota and Glomeromycota

The Zygomycetous fungi include more than 1800 species divided into two major lineages: one that is composed mostly of parasites, the Zoopagomycota, which are not addressed in this book, and a second composed mostly of saprophytes and plant symbionts; the Mucoromycota and Glomeromycota (Naranjo-Ortiz and Gabaldón 2019).

Mucoromycota

The Mucoromycota are a major phylum within the Zygomycetes with more than 740 described species, mostly saprophytes. They also include plant and animal pathogens (including species in the order Entomophthorales which colonise insects), and even human allergens. Some species are known to cause food spoilage, as in the case of grapes, which at the end of the season may be covered in the fruiting bodies of some species belonging to the genera *Rhizopus* or *Mucor*. The hyphae are generally coenocytic. They reproduce asexually by producing sporangiospores contained within sporangia (Fig. 3.29).

Sexual reproduction takes place by gametangial fusion resulting in the formation of a zygospore, whose name is derived from the Zygomycota, a former phylum now obsolete. Some species are homothallic, with zygospores formed from the development of a single sporangiospore. However, most species are heterothallic, requiring the presence of two compatible individuals to produce a zygospore. The order of the Mucorales, comprising around 300 species, is the most widely studied and most widely used in research and industrial and medical fields: as producers of secondary metabolites and other value-added products (Mohamed et al. 2021). When two strains within the Mucorales are compatible, two hyphae differentiate to produce an apical swelling which fuse with one another to produce a zygospore (Fig. 3.30). This spore can germinate directly to form a sporangiospore, or form a filamentous thallus.

Glomeromycota

This phylum consists almost entirely of fungi forming obligate symbiotic relationships with terrestrial plants (herbaceous and woody). These symbioses take place in the roots and are called mycorrhizas. Coenocytic fungal hyphae grow intercellularly within the root tissue. Nutrient exchange between the fungus and its host takes place in special vesicles and/or arbuscules. Depending on the type of structures produced, mycorrhizas may be arbuscular (AM fungi), vesicular (VM fungi) or both vesicular and arbuscular (VAM fungi). In general, the fungus supplies minerals and water to the plant and in return the plant supplies the fungus with the products of photosynthesis. More than 230 species of mycorrhiza have been described. The asexual spores are often very large and multinucleate, produced at the tip of a hyphal segment. These spores are chlamydospores, having a secondary internal wall often

Fig. 3.29 Mature (black) and immature (white) sporangia of a Mucorale. Each pinhead contains hundreds of sporangiospores

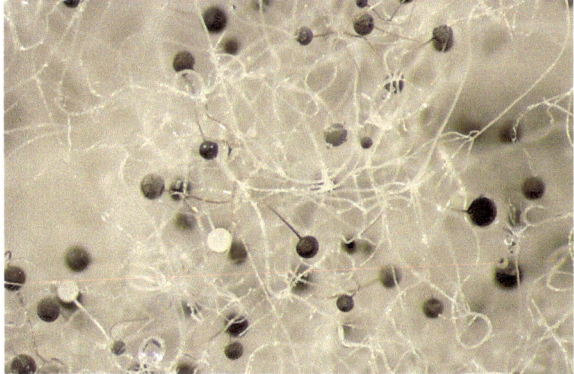

Fig. 3.30 Formation of a zygospore (sexual) in the Mucorales. (**a**) From left to right, differentiation of two hyphae which fuse to form a zygospore. (**b**) Zygospore of *Mucor hiemalis*. Scale bar represents 250 µm

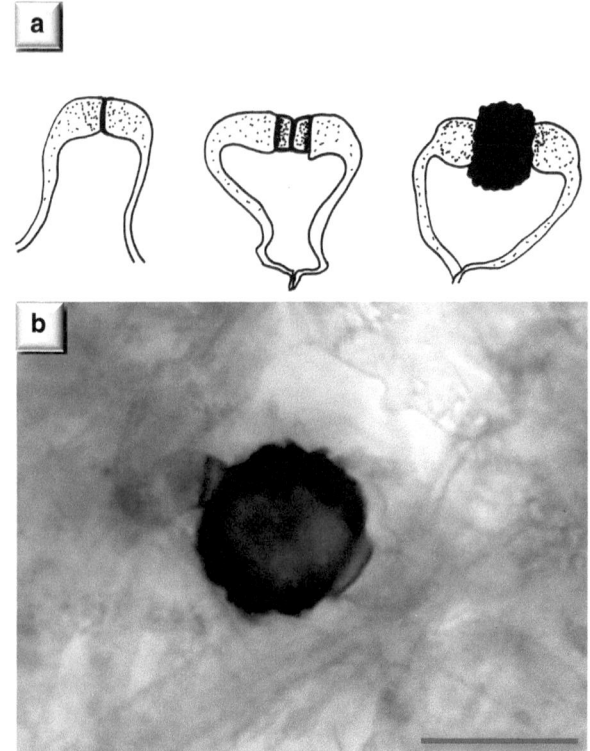

impregnated with hydrophobic compounds (Griffiths 1974). *Rhizoglomus intraradices* (formerly *Glomus intraradices*) produces terminal chlamydospores with lengths ranging from 147–383 µm depending on the strain (Fig. 3.31) (Walker et al. 2021).

3.2.4.4 Basidiomycota

The phylum of the Basidiomycota, or Basidiomycetes, reproduce by means of basidia. The dikaryotic hyphae produced by anastomosis and plasmogamy can survive for long periods and contain two stable but genetically different populations of nuclei. Following this period of latency, the nuclei fuse (karyogamy) and undergo meiosis to produce two to four external basidiospores (exospores) by a process of exogenisation, borne by specialised cells called basidia. The basidiospores are often borne on fine denticles called sterigmata (Fig. 3.32).

Basidia are generally borne on or encased in a carpophore, in this case called a basidiocarp or basidioma. The hyphae of the Basidiomycetes are generally septate, with cell membranes composed mainly of ergosterol. The Basidiomycetes contain more than 32,000 species, making it the second most abundant phylum described to date. The Basidiomycota include the main clades responsible for breaking down

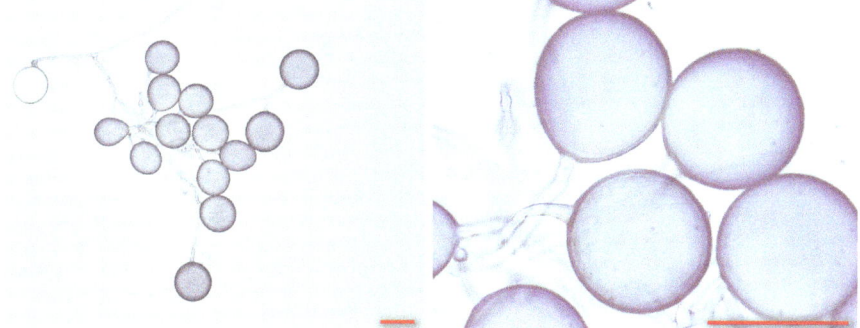

Fig. 3.31 Terminal chlamydospores of *Rhizoglomus intraradices*. Scale bar represents 200 μm

Fig. 3.32 *Suillellus luridus* basidia and basidiospores on fine sterigmata (arrow). *b* basidium. Scale bar represents 5 μm. (Photograph © Heinz Clémençon, Switzerland)

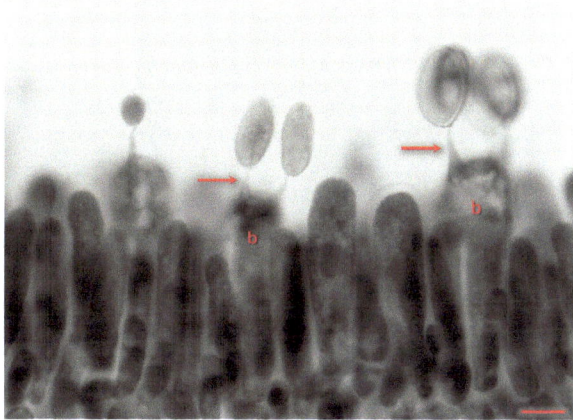

wood and plant litter—the key players in the carbon cycle (Oberwinkler 2012). Like the Ascomycetes, the Basidiomycetes form a single phylum divided into three subphyla:

1. **The Pucciniomycotina** (including rusts that cause major damage to plants), with septate basidia: this abundant group contains more than 8400 described species, including the fungi with the most complex life cycle (for example, rusts in the genus *Puccinia*), alternating between several host plants and producing structures specific to each one (Fig. 3.33).

2. **The Ustilaginomycotina** have septate basidia. This subphylum contains more than 1700 species, mostly plant pathogens (smuts) (Fig. 3.33), but also animal saprophytes and pathogens. Many species develop asexually in the form of yeasts, others are dimorphic. The phytopathogenic species often alternate between asexual states in the form of yeasts and dikaryotic filamentous states.

3. **The Agaricomycotina**, most species of which have non-septate basidia. This subphylum is the largest in the Basidiomycota, comprising more than two thirds of the described species. Most Agaricomycotina have a saprophytic, phytopatho-

Fig. 3.33 Puccionomycotina and Ustilaginomycotina. A left: *Puccinia recondita*, wheat leaf rust.
A right: maize smut, *Ustilago maydis*

genic and/or ectomycorrhizal lifestyle. Most white and brown wood rots are
caused by the growth of species in this group. The Agaricomycotina include all
fungi with true or false gills (such as *Armillaria mellea*, honey fungus, and
Schizophyllum commune respectively), pores (like the polypores), tubes (like the
bolets) as well as species with an internal hymenium (fertile part) (*Lycoperdon*
sp., *Phallus impudicus*) and species which form crusts on dead wood, such as
Terana coerulea (Fig. 3.34). Some representatives, notably the orders Tremellales
(Tremells) and Auriculariales (Judas ears), can be distinguished from other
Agaricomycotina by their jelly-like texture and septate basidia, called phragmo-
basidia. This group also contains fungi that trap nematodes, such as species in
the genera *Corpinus* or *Pleurotus*, fungi which have created stable associations
with diverse insects (such as *Termitomyces* sp. associated with *Macrotermes bellicosus*) and lichens, endophytes, hyperparasites and mycoparasites.

3.2.4.5 Ascomycota

The Ascomycota (or Ascomycetes) are the largest and most diverse phylum, containing more than 90,000 described species, which amounts to more than half the
described species in the fungi kingdom. Unlike the Basidiomycota, their dikaryotic

Fig. 3.34 Example of
Agaricomycotina in the
form of crusts on dead
wood (*Terana coerulea*).
(Photograph © Carole
Parodi, Switzerland)

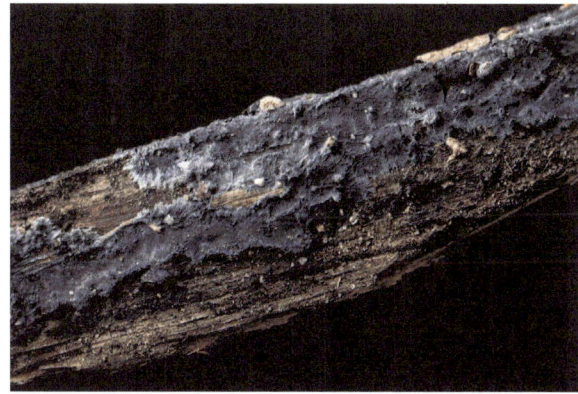

Fig. 3.35 Asci (red arrow)
and ascospores (black
arrow) of *Peziza* sp.

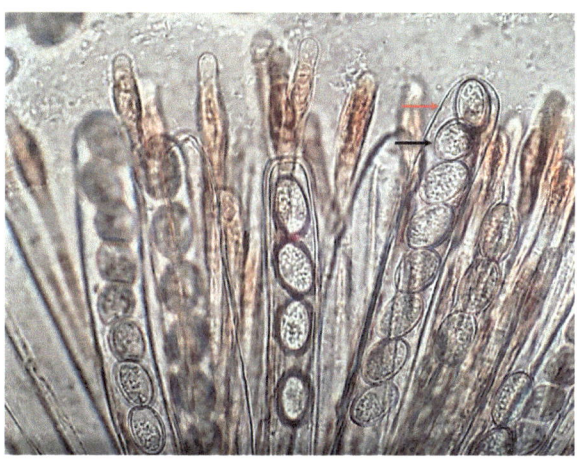

phase is short-lived, rapidly leading to karyogamy and the formation of asci; cells shaped like a sac or glove finger in which (usually) eight ascospores develop following meiosis (Fig. 3.35). The asci are produced inside an ascocarp. Asexual production is the dominant form of propagation in this vast group and the sexual phase is unknown in many species in this phylum. As with the Basidiomycota, the Ascomycota can have very different morphologies ranging from yeasts to highly complex macroscopic structures.

This phylum is divided into three main subphyla:

1. **The Saccharomycotina** (formerly known as the Hemiascomycota) contain more than 1000 species, growing almost exclusively in the form of yeasts (Shen et al. 2018). In the dimorphic species, the filamentous form can switch to the yeast form. Asci, where present, are not encased in a fruiting body and are thus described as 'naked'. Several species are of major importance, either for fermentation processes and in biotechnology (*Saccharomyces cerevisiae*, the yeast used to make bread, beer and wine), or in medicine, (e.g. some species in the genus *Candida* are human pathogens).

2. **The Taphrinomycotina** contain around 140 species. These species grow in the form of fission yeasts, for example *Schizosaccharomyces pombe*. Despite the small number of described species, the Taphrinomycotina comprise biotrophic plant pathogens such as *Taphrina deformans* (peach leaf curl) and human pathogens (e.g. *Pneumocystis carinii*, which causes pneumonia in people infected with HIV), saprophytic yeasts and another clade (including the genus *Neolecta*) forming carpophores which, in morphological terms, resemble the ascocarps found in certain Pezizomycota, but whose histological structure is very different (Nguyen et al. 2017).

3. **The Pezizomycotina** (formerly Euascomycota) contain more than 82,000 species, most of which are filamentous. This is the most diverse subphylum in the Ascomycota. It contains several plant pathogens (*Botrytis, Fusarium Penicillium, Aspergillus, Alternaria, Erysiphe, Phyllosticta*), human and animal pathogens (including the genus *Aspergillus* with *Aspergillus fumigatus*) and entomopathogens (such as fungi in the genera *Cordyceps* and *Beauveria*), lichens (98% of lichens are Pezizomycotina), mycorrhizal species, endophytes, symbionts, and mycoparasites. Most species are microscopic. However, some have ascocarps that are visible to the naked eye which may be brightly coloured, either simple (apothecia) or complex (stromatic apothecial fruiting bodies, such as the morel *Morchella esculenta*). Three types of asci can be found: prototunicate (ascus wall formed by two fused layers), unitunicate (ascus wall formed by a single layer) and bitunicate (ascus wall formed by two separate layers). Finally, the Pezizomycota have four main types of fruiting body: protothecia, cleistothecia, perithecia and apothecia (Fig. 3.36).

   (a) **Protothecium**: asci (usually 10) are embedded in a loosely woven mass of protective mycelial hyphae. This type of organisation is only found in the prototunicates. A greater number of protective hyphae have tended to evolve over time to create a denser structure.

   (b) **Cleistothecium**: asci are encased in a globose fruiting body with no opening to the outside. This structure is normally less than 1 mm in size. Deliquescence or total desiccation of the entire structure is required to expel the spores. This structure is characteristic of powdery mildews, notably *Erysiphe necator*, grapevine powdery mildew.

   (c) **Perithecium:** pear- or flask-shaped fruiting body which opens by a pore (ostiole). The wall of the perithecium is formed by sterile cells derived from hyphae which surround the ascogonium during its development. Asci in this group are unitunicate. The fruiting body is often covered in melanin, causing it to resemble a tiny piece of charcoal. This structure is characteristic of *Phyllosticta ampelicida*, the cause of grapevine black rot.

   (d) **Apothecium**: cup- or saucer-shaped fruiting body, bearing asci. At maturity, the tips of the asci are exposed to the air. Apothecia are often brightly coloured structures, either sessile (without foot) or pedunculate (with foot). The genus *Botrytis*, including *B. cinerea*, produces small brownish apothecia during its sexual cycle. Similarly, the morels (*Morchella* sp.) are each a mass of pedunculate apothecia fused together (called an apothecial stromata) (Fig. 3.36).

Fig. 3.36 Different types of apothecia in the Pezizomycotina. (**a–d**) Sessile apothecia: (**a**) *Caloscypha fulgens*. (**b**) *Lasiobolus macrotrichus*. (**c**) *Elaiopezia polaripapulata*. (**d**) *Scutellinia legaliae*. (**e**) Pedunculate apothecium *Rutstroemia sydowiana*. (**f**) Pedunculate apothecial stromatocarp: *Morchella costata*. (Photographs © René Dougoud, Switzerland)

3.2.4.6 Non-Fungal Microorganisms Causing Plant Diseases: The Oomycota

Since Whittaker (1959), the eukaryote Tree of Life has evolved enormously and undergone constant revisions and major regroupings. For 15 years, mainly due to the emergence of phylogenomics, the eukaryote tree has been divided into five to eight supergroups. The current tree is derived purely from phylogenetic molecular data, marking a major departure from earlier versions which were based on biological data (Burki et al. 2020). The fungi kingdom is part of the Amorphea supergroup, where it shares the Opisthokonta group with animals and some unicellular organisms. The Oomycetes belong to the TSAR supergroup (Telonemids, Stramenopiles, Alveolates and Rhizaria) and the Stramenopiles group within it, which is characterised by the presence of at least one hairy flagellum. The Oomycetes are filamentous organisms performing very similar ecological roles to fungi, which is why they were long considered to be fungal organisms. In fact, the Oomycetes are far removed from fungi in phylogenetic terms. They are divided into several orders (McCarthy and Fitzpatrick 2017), including the Peronosporales, which contains the genera *Phythphtora* and *Plasmopara*. They have various structural and chemical characteristics which distinguish them from the fungal kingdom. At cell level, the hyphae are coenocytic, lacking the *Spitzenkörper,* and the mitochondrial cristae are tubular, though flattened like those of fungi. Their chemical storage component is mycolaminarin (glycogen in fungi), fucosterol is the dominant sterol (egosterol in fungi), and the cell wall is composed of cellulose, with chitin very rarely present.

The Oomycetes are a group of 800–1000 species of aquatic and terrestrial saprophytes and pathogens of plants and animals. Several species produce a large hyphal network. The Oomycetes can reproduce sexually by oogamy and asexually by forming sporangia. These develop from an oospore derived from sexual reproduction (winter egg, encysted spore), which produces spores with two flagella called zoospores. The zoospores of the Oomycetes are characterised by having two morphologically distinct flagella, one of which has lateral filaments (mastigonema) which enable them 'swim'. The smooth flagellum acts as a rudder. The phytopathogenic species infect plants with zoospores initially, then sporangia appear on the sporangiophores during vegetative growth. Sporangia and sporangiophores form the characteristic blanket of downy mildew. Depending on the species, these sporangia may immediately produce a germ tube to infect a new plant (for example in *Phytophthora infestans*, potato blight) or produce infectious zoospores with two flagella, actively released by the sporangia (as with *Plasmopara viticola*, grapevine downy mildew) (Fig. 3.37).

Fig. 3.37 Zoospores with two flagella (arrows) of *Plasmopara viticola*, characteristic of the Oomycetes

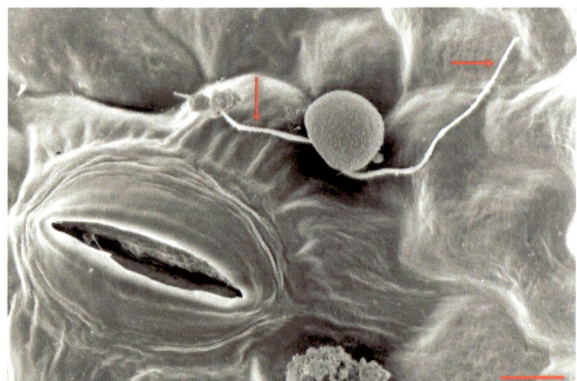

3.2.4.7 Why Do Fungal Names Change?

Fungal Species Identification

There are estimated to be between 2.2 and 3.8 million different species of fungus (Hawksworth and Lücking 2017), of which approximatively 150,000 are described. The Ascomycota make up the largest and most diverse phylum, containing more than 90,000 described species. The vast majority of species involved in grapevine diseases belong to the Ascomycota. Most species in this fungal lineage are pleomorphic, meaning that they can occur in both the asexual (anamorphic) and sexual (teleomorphic) form (Manawasinghe et al. 2021). This characteristic has often led experts to give two different names to the sexual and asexual forms of the same species, and thus to erroneously consider that these two forms correspond to two different species (Fig. 3.38). Moreover, the teleomorph in around 10% of Ascomycota has never been observed, suggesting that these species only reproduce asexually (Senanayake et al. 2022) and explaining their earlier classification as *Fungi imperfecti* or Deuteromycota.

Three principal species concepts have been defined in biology: morphological, biological and phylogenetic. Historically, species were first identified based on their morphology (macro- and microscopic characters) and most of the morphological characteristics allowing reliable identification were essentially provided by the sexual form of the fungus. In plant pathology, scientists isolate fungi from small pieces of plant tissue and grow them on Petri dishes. Unfortunately, most Ascomycota fungi produce only anamorphs when growing *in vitro* because they require specific, often unknown, conditions for sexual reproduction (Sun and Heitman 2011). Also, anamorphic (asexual) forms can exhibit different morphologies, depending on the

Fig. 3.38 *Botrytis cinerea*. (**a**) The asexual form has been more often described because it has been more frequently isolated in plants: the asexual stage of this species was named *Botrytis cinerea* by Micheli (1729). (**b**) Sexual stage of *Botrytis cinerea*. The connection between the two forms was first discovered by de Bary (1866) then validated by Fuckel (1869) and the sexual form was named *Sclerotinia fuckeliana* in his honour. Whetzel (1945) placed *Sclerotinia fuckeliana* in the genus *Botryotinia* and renamed the sexual form *Botryotinia fuckeliana*. Today, both the sexual and asexual form are called *Botrytis cinerea*

culture medium, or even between subcultures of the same fungal isolate when grown on the same medium (Senanayake et al. 2017). Therefore, the teleomorphic (sexual) characteristics necessary for species identification are absent. The definition of the biological concept of species was formulated by Dobzhansky (1937). This definition states that if two individuals are capable of procreating and generating fertile offspring from their union, they belong to the same species. However, as described above, it is very rare to obtain the sexual form of fungi, when it exists, in vitro (Sun and Guo 2012). Therefore, the biological concept of species, like that of morphology, is not applicable to most fungi. Finally, with advances in molecular biology over the past few decades, two molecular approaches to species identification have emerged: barcode DNA sequences and phylogeny. The idea of barcode sequences is to associate each species with a DNA sequence, allowing it to be identified by sequence similarity search in public sequence databases. Ideally, this implies that a reliable sequence coming from a representative collection (type sequence) for each described species has been deposited in these databases. For fungi, the chosen barcode sequence is the internal transcribed spacers (ITS) of nuclear ribosomal DNA (Schoch et al. 2012). However, this identification method fails in many cases because the names of fungal species have been wrongly attributed to sequences for several reasons (Hofstetter et al. 2019). Among these reasons is the description of cryptic species, i.e., closely related species that are indistinguishable from each other at macro- and microscopic levels and have identical ITS sequences. This is the case for several important pathogen genera. The ITS sequences of a series of different cryptic species will be 100% similar (Beker et al. 2016; Shivas and Cai 2012) so other genes, evolving faster than the ITS, are used to describe these species. Another main reason for ITS-based misidentification is related to sequence deposition. The database provides an accession number for each deposited sequence. Authors, if not careful enough when depositing sequences, may attribute a sequence to the wrong fungal collection at the time of deposit and/or in the corresponding publication. Anyone who obtains a sequence identical to one of these misidentified sequences will often adopt the associated fungal name and perpetuate the problem in the database. Also, not all described fungal species have been sequenced and databases still have poor taxon coverage, which is the main limitation of the sequence-based identification of fungi. Phylogeny is another approach to molecular identification, using DNA sequences to reconstruct the evolutionary history of species. The phylogenetic approach to identifying fungal species has led to dramatic changes in the taxonomy of fungal species.

Changes in Fungal Nomenclature and Taxonomy

Fungal phylogeny has profoundly changed species concepts in two ways. First, it has made it possible to link anamorphs to their corresponding teleomorphs. For example, a particular anamorphic species involved in grapevine esca disease was, until recently, named *Phaeoacremonium aleophilum* W. Gams, Crous, M.J. Wingf, and Mugnai 1996. However, its teleomorph, *Togninia minima* (Tul. & C. Tul.) Berl 1900, was described over 100 years ago. One century later, Mostert et al. (2003) ran phylogenetic analyses and established that the genus *Togninia* was the teleomorphic form of the anamorphic genus *Phaeoacremonium*. Keeping two different names for the same species, once the link between anamorph and teleomorph has been established, is confusing. Therefore, the scientific community decided to keep the first published genus name in such cases, unless there are practical arguments to maintain one genus name over the other. Due to the widely accepted use of the name *Phaeoacremonium*, the more recent genus name *Phaeoacremonium* was retained, and the rarely used *Togninia minima* was abandoned in favour of *Phaeoacremonium minimum* (Tul. & C. Tul.) Gramaje, L. Mostert and Crous (2015). This new species name combines the newer anamorphic genus name with the older teleomorph species epithet. Depending on the historical usage of names, the choice may also favour one of the older genus names, as with the anamorphic *Phomopsis viticola* (Sacc.) Sacc. 1915, a well-known grapevine wood pathogen name. It is now named *Diaporthe ampelina* after its teleomorph (which typically has priority). The species epithet *'viticola'* had to be abandoned in the process because an older anamorph epithet for this *Diaporthe*, namely *Phoma ampelina* Berk. and M.A. Curtis 1873, takes precedence as the oldest anamorph name for this very same species. This complex problem of finding the correct name for species that were known under different names depending on their asexual or sexual forms is handled by specific scientists who are experts in this field.

Apart from purely nomenclatural issues, phylogenetic analysis of DNA sequences allows us now to re-examine the true evolutionary relationships of species that were previously identified based on morphological similarity. This too very often leads to name changes. Indeed, traditional morphological characteristics appeared to misleadingly infer real species evolutionary relationships because of parallelisms or convergences. Evolutionary parallelism refers to the fact that two or more lineages have evolved in a similar way, with all descendants looking very much as their ancestors, while convergence corresponds to the appearance of similar traits in totally unrelated organisms because they adapted to a similar environment, for example (whales, which are mammals like humans, look like fish because they live in the water, and bats, equally mammals, have wings like birds because they fly). An example of such evolutionary convergence among grapevine diseases is the downy mildew, *Plasmopara viticola*. This species was long considered to be a fungal species because it produces sporangiophores highly similar to those produced by many Ascomycota. However, it has flagellate spores like Chytridiomycota, another important division of fungi, and phylogeny showed that this species is not a fungus but belongs to a totally different lineage, the Stramenopiles, a group that includes brown algae which also have flagellate spores (Dick 2013). Another example is the anamorphic genus *Botryosphaeria* that includes many species, several of which are involved in grapevine trunk diseases. Based on molecular systematics and anamorph-teleomorph correspondence, morphologically similar *Botryosphaeria* anamorphs are now renamed/transferred to different genera as diverse as *Peyronellaea* Goid. ex Togliani 1952, *Lasiodiplodia* Ellis & Everhart 1896, *Dothiorella* Sacc. 1880, or to new genera such as *Neofusicoccum* and *Pseudofusicoccum* (Crous et al. 2006), or *Spencermartinsia* (Phillips et al. 2008).

Increase in the Number of Fungal Species Associated with Grapevine Diseases

Studies conducted in different geographical regions are one of the main factors that have contributed to the increase in the number of fungal species associated with grapevine diseases, especially grapevine trunk diseases. Fungal communities associated with grapevine, as with other plants, depend on the location and different environmental conditions (González and Tello 2011; Lade et al. 2022). Most species that can associate with grapevine are not specific to *Vitis vinifera* (Hofstetter et al. 2012) and can be hosted by many other plants in the vicinity of vineyards. The direct environment thus represents a reservoir of grapevine-associated fungal endophytes and pathogens (Granett et al. 2001; Terral et al. 2010). Since phylloxera destroyed almost all Eurasian vineyards, the grafting process using Eurasian cultivars on resistant American rootstocks resulted in the worldwide introduction of many non-native endophytic fungal species, including several latent pathogens. Molecular phylogenies, based on many genes, have revealed differences leading to the description of numerous new 'cryptic' species, i.e. very closely related species that cannot be distinguished from each other based on morphology (Dugan and Everhart 2016).

3.3   Holobiome, Microbiome and Mycobiome

(Pélissier 2022)

3.3.1   Plant-Microbe Interactions

Microbes play a critical role early in life. All living organisms seem to be embedded in a network of microbial interactions that contribute to many of the functions (nutrition, development, immunity, behaviour) on which their lives depend (Rodriguez and Redman 2008). Plants, animals and humans are 'never alone' because of their close relationship with microorganisms whose diversity is remarkable (Selosse 2017). Whether in animals or plants, these microbial communities are the result of more than 500 million years of coevolution (Heckman et al. 2001). The idea that plants are not autonomous entities but are associated with a multitude of microbes is not new, but technical advances in high-throughput sequencing and omics technologies have changed this area of research, allowing it to be studied from a new angle. These approaches complement functional studies at the individual scale, facilitating investigations of the entire plant-microbe association, at both genetic and molecular levels (Müller et al. 2016). Understanding the relationship between plants and microbes is an area of research that has attracted increasing interest given their involvement in ecology, chemistry, biology, health and agriculture (Martin et al. 2017).

It should be noted that these interactions are becoming better documented but most of the mechanisms and laws that govern them are often still assumptions and concepts at the ecological level.

3.3.2   The Holobiont and Holobiome Concepts

Like humans and animals, plants are now considered to be part of a holobiont, i.e. the entire host organism and its associated microbiota. First proposed by Margulis in 1991, this term is defined as a simple biological entity involving a host and its single inherited endosymbiont (Margulis and Fester 1991). It was further extended to encompass all microorganisms associated with the host, including viruses, protists and archaea (Rohwer et al. 2002). The holobiont represents a dynamic functional entity in which coevolutionary selection probably occurs between the host and its linked microorganisms and among the microorganisms, maintaining the overall stability of the system over ecological and evolutionary timescales (Trivedi et al. 2020). Mechanisms in the holobiont likely allow a fine-tuning of host and symbiont behaviours. As a result, the measure of plant fitness should correspond to the measure of the holobiont fitness itself (Vandenkoornhuyse et al. 2015). The fact that the huge gene pool of the associated microorganisms extends the host genome and contributes to its phenotype led to the hologenome concept. This term was introduced and defined as the sum of genetic information of the host and that of its associated microbes; that is to say, the collective genome of a holobiont (Rosenberg et al. 2007; Zilber-Rosenberg and Rosenberg 2008). As there are about 20,000 genes in the human genome but 33 million genes in its hologenome, the genetic information provided by the microbiota is far greater than that of the host. Notably, the establishment of a microbe from soil or atmosphere in the host will add new genetic material to the holobiont (Rosenberg and Zilber-Rosenberg 2016).

By taking a closer look at what constitutes the plant holobiont, the plant being the host itself, a definition of the plant microbiota can be established. The plant microbiota comprises a variety of microorganisms including bacteria and fungi, which are the dominant forms, and additionally other groups such as archaea, algae, oomycetes, protists, nematodes and viruses, which are beneficial, neutral or pathogenic microorganisms (Rosenberg and Zilber-Rosenberg 2016) (Fig. 3.39).

The sum of the genetic information of the microbiota is defined as the microbiome and encompasses the functional attributes of the microbiota (Mishra et al. 2021). In the same way that the human microbiome is an 'organ' in itself, the plant microbiome can be considered as an extension and forms a second genome or pangenome (Christian et al. 2015; Turner et al. 2013). Genetic variations in the microbiome or in the host genome can therefore lead to genetic variations of the hologenome. However, as the microbiome is able to adjust more quickly and efficiently to environmental processes, it is more likely to play a main role in adaptation and evolution of the holobiont (Rosenberg and Zilber-Rosenberg 2016). The concept of 'core microbiome' suggests that a fraction of the microbial community is

Fig. 3.39 Microbiome concept. The plant (right) and the human (left) microbiome can be considered as an extension and form a second genome or pan-genome. It comprises a variety of microorganisms including bacteria and fungi, which are the dominant forms, and additionally other groups such as archaea, algae, oomycetes, protists, nematodes and viruses, which are beneficial, neutral or pathogenic microorganisms. (Illustration © Léonie Pélissier, Switzerland)

selectively recruited to be ubiquitous, steady and maintained in high numbers in most of the communities associated with a specific host, and has functions essential to the fitness of the host taxon (Vandenkoornhuyse et al. 2015). It has also been postulated that keystone species described as 'hub microbes' could directly and indirectly function as mediators between the host and the microbiome and affect its organisation as a network (Agler et al. 2016).

Microbiome assembly is determined by selective forces such as dispersal, species interactions, host plant metabolites and adaptation to a niche or a compartment ecosystem. Early colonisers may be transmitted vertically and might have seed traits selected by the host plant. After germination, microbiome assembly is more likely to be driven by horizontal transfer and be highly dynamic in the vegetative phase, stabilising during the reproductive phase (Müller et al. 2016). Communities seem to become highly plant-specific and less diverse as the plant grows and develops, suggesting a robust habitat selection over relatively short timescales for the plant (Morella et al. 2020).

3.3.3   Distribution of the Microbiota in the Host Plant

The microbial communities associated with plants are microbial cells with different functions, structure and composition distributed in distinct plant compartments from the soil to the rhizosphere, phyllosphere and endosphere (Anal et al. 2020). The soil is regarded as a 'seed bank' for the root microbiota, functioning as a reservoir for both below-ground and above-ground communities (Vandenkoornhuyse et al. 2015). The rhizosphere is a 'growth chamber', a rich, highly dynamic ecosystem, in which root exudates (depending on plant species and growth stage) determine the microbiome structure (Turner et al. 2013). In contrast to the rhizosphere, the phyllosphere is relatively poor in nutrients and subject to a harsher and rapidly fluctuating environment with variations in temperature, radiation and moisture levels (Lindow and Brandl 2003). The inoculum source is more variable and only a microbiota able to adapt to such an environment is able to sustain itself. Selection of specialised microorganisms able to efficiently colonise the leaf surface can also lead to competitive exclusion of other microbes, including pathogens (Vandenkoornhuyse et al. 2015).

The endosphere is seen as a 'restricted area' comprising microorganisms called endophytes which are able to penetrate and colonise all internal plant tissues (roots, stems, leaves and xylem vessels). It is constituted of highly specific microbial communities as the endospheric microbiota is limited by the plant's innate immune defence system (Mishra et al. 2021). Endophyte microbes spend at least part of their life inside living plant tissues and are considered as asymptomatic (no visible symptoms) (Compant et al. 2021). In contrast to rhizosphere fungi, which help to promote plant growth directly through their accessi to nutrients and water, endophytes provide more indirect benefits for protection against pathogens through defence triggering and metabolite production (Vandenkoornhuyse et al. 2015). Colonisation of the endosphere is said to originate partly from the roots and shoots ('endosphere continuity'), and be actively selected by the plant, and partly from horizontal transfer and the external environment through atmospheric deposition and colonisation, and growth on the leaf surface prior to penetration (Uroz et al. 2016) (Fig. 3.40).

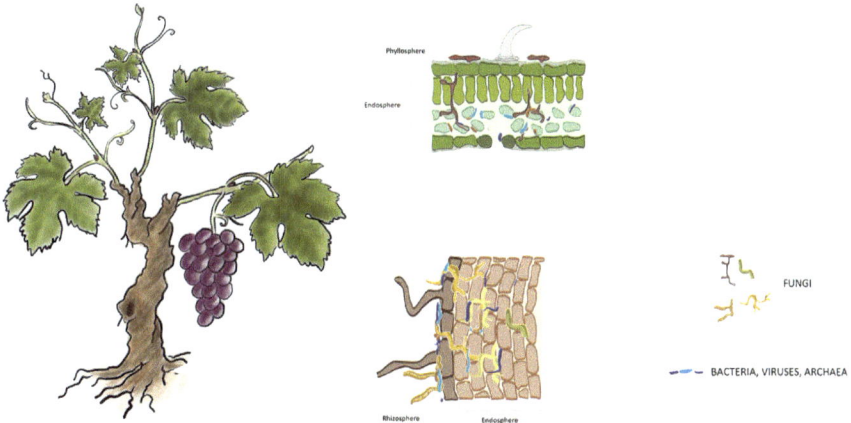

Fig. 3.40 Distribution of the microbial communities (bacteria, viruses, archaea, fungi) associated with grapevine plants in distinct plant compartments from the soil to the rhizosphere (roots), phyllosphere (surface of green organs and fruits) and endosphere (internal cell tissues). (Illustration © Léonie Pélissier, Switzerland)

3.3.4 Roles of the Microbiota in the Plant Host

The association of plants and their microbiomes is known to be ancient and its evolution over millions of years has led to the emergence of highly specialised ecosystems, supporting primordial ecological functions (fitness, structure, growth, health). By providing additional genes to the host, the microbiota is involved in its dynamic adaptation to biotic and abiotic environmental conditions and extends the ability of plants to adapt to many types of stress (Perez-Alonso et al. 2020). This is of primary relevance in view of the lack of locomotion and sessile lifestyle of plants. Several studies have reported a wide range of benefits brought to the plant by the microbiota to help it minimise the impact of stress (Rodriguez et al. 2009).

Those benefits include increasing nutrient acquisition and bioavailability through the transformation and translocation of soil nutrients that are not readily available for the plant. The differentiation between fast and slow growing plants might even be due to the influence of the microbiota (Richardson and Simpson 2011). The plant will also select a microbiome able to promote stress resistance under biotic or abiotic stress conditions. Some communities, for example, can be involved in the secretion of surfactants to increase water permeability or wettability of the plant, or to fortify cell walls (Burch et al. 2014; Cha et al. 2016). The microbiota is also involved in plant health and disease resistance, by directly affecting plant pathogens through competition or antimicrobial action and selection of members able to produce enzymes and specialised metabolites against a pathogen (Hassani et al. 2018; Carrión et al. 2019). More indirectly, the microbiota can act as an 'acquired immune system' for the host in priming and modulating plant defences (via jasmonic acid signalling for example) and induced systemic resistance (Pieterse et al. 2012). Some of those selected stress resistance traits accessed through the microbiome can be passed from mother plants to their offspring, and contribute to the fitness of individual plant genotypes (Trivedi et al. 2020). Plants can therefore be seen as 'genetic mosaics', because the microbiome confers to each organ a unique combination of genes and functions (Vega and Blackwell 2005). The plant-microbiota relationship is one of give-and-take since, in exchange, the host plant provides the microbes with an essential niche to adapt to a nutrient-poor environment and a stable availability of metabolites to compensate for metabolite deficiencies (Choi et al. 2021). This reduces the selective pressure on the microbiome and helps maintain its biosynthetic capabilities (Vandenkoornhuyse et al. 2015). In this way, members of the microbial communities can also adapt interactions with their host and other microbes within the community to maximise their own fitness (Hassani et al. 2018).

The highly complex interactions of the mycobiomes in the different compartments of a plant and its environment are widely unknown. In agroecosystems, hypothetically balanced mycobiomes contribute to the active defence mechanisms of the plants against pathogens or pests but are generally not sufficient to prevent any use of plant protection products.

3.3.5 Insights into the Plant Mycobiome

While the bacterial part of the plant microbiota has been the focus of research in recent years (Fitzpatrick et al. 2020), the plant-associated fungal communities, i.e. the plant mycobiota, have been less studied, in part due to technical hurdles arising from the latter's phenotypic and genotypic complexity (Pagano et al. 2017). As part of the microbiome, the mycobiome is also involved in the management of biotic and abiotic stress, and more specifically acts as a protective shield against phytopathogens (Weller et al. 2002). Fungal microbes also play a role in structuring the microbiota assemblages within a single plant host, notably through rhizosphere signalling, for example with fungal volatiles (Hassani et al. 2018).

The mycobiota is composed of five main fungal functional groups: saprotrophic, pathogenic, epiphytic, endophytic and mycorrhizal (ecto/arbuscular/ericoid mycorrhizal fungi) (Porras-Alfaro and Bayman 2011). Mycobiota assemblages are also allocated to the different plant compartments and controlled by different factors (host genotype, environment, nutrients, other members of the microbiota) (Pozo et al. 2021). The huge diversity of fungi colonising plants below and above ground mostly belong to the phyla Ascomycota and Basidiomycota, although arbuscular and/or vesicular mycorrhizal (endomycorrhizal) fungi from the Glomeromycota phylum are the most abundant in the rhizosphere (Martin et al. 2017). The rhizosphere is mostly colonised by mycorrhizal fungi coexisting with dark septate endophytes. Working in synergy with soil microbes, they are particularly involved in nutrient use efficiency, plant productivity and antioxidant activity (Ehrmann and Ritz 2014).

Fungal microbial communities in the phyllosphere include epiphytic fungi living on leaf surfaces and endophytic fungi inhabiting the leaf cells (Andrews and Harris 2000). They play major roles in ecosystem functions but also in contributing to decomposition of leaf litter and recycling of carbon and nutrients (Voříšková and Baldrian 2013). Some analyses have shown that endophytic communities are more specialised and more resistant than epiphytic communities. Endophytic diversity is more influenced by plant host, likely because of the plant selection process (Yao et al. 2019).

The endosphere is colonised by endophytic fungal communities. Fungal endophytes are an important component of fungal biodiversity and plant microbiome, residing symbiotically and asymptomatically inside the plant tissue and interacting with the other fungal groups (Porras-Alfaro and Bayman 2011). They have significant functional roles for plant fitness, growth and development. Specifically, they are considered as plant 'bodyguards' through their antipathogen capacities as producers of antimicrobial metabolites and plant-like phytohormones (Anal et al. 2020).

Endophytic fungi are seen as a potential treasure of hidden biodiversity, but the diversity, composition and ecological relevance of phyllosphere and endosphere fungi have been much less studied than those of below-ground and pathogenic fungi (Hardoim et al. 2015).

3.4 Fungal Endophytes

3.4.1 Definition of the Term 'Endophyte'

Higher plants host very diverse and rich assemblages of microorganisms since they provide complex, multilayered, spatially and temporally diverse habitats. Fungal microbes are dominant in these assemblages, colonising various compartments and, among them, increasing attention has been paid to fungi living in the internal tissues of healthy plants, i.e. fungal endophytes.

German botanist Heinrich Fredrich Link (1809) was the first to observe and describe what appeared to be endophytes, as a distinct group of partly parasitic fungi living inside plants, calling them 'Entophytae'. During the nineteenth century, however, it was assumed that healthy plants were sterile, and thus devoid of microorganisms (Compant et al. 2012). Nevertheless, more than 130 years ago, Galippe was the first scientist to investigate the presence of bacteria and fungi living inside different plant tissues and postulated on their migration from the soil into the plant and fought for this idea (Galippe 1887). Further studies supported the occurrence of microorganisms hidden within plants, but conflicting opinions still prevailed at the time (Laurent 1889).

More than a century ago, the presence of 'endotrophic' mycorrizha dispersed from the roots to the aerial part of the *Calluna* plant was already evidenced by Rayner et al. (1916, 1925a). Later on, endophytic fungi in *Festuca rubra* and *Lilium* spp. among others were described (Rayner 1925b; Sampson 1935), but it was only in 1977 that this relationship between fungi and plants was clarified by showing that the needles of *Pseudotsuga menziesii* (Mirb.) Franco can host endophytes that do not seem to provoke any symptoms in the host (Carroll et al. 1977). Although described in the twentieth century, endophytic microorganisms did not receive significant attention until the first 20 years of the twenty-first century, when their capacity to protect their hosts against insect pests, pathogens and even domestic herbivores such as sheep and cattle was recognised.

Finding an ideal definition of the term endophyte has posed difficulties for the scientific community and the term has therefore continued to evolve and be reshaped over the years. De Bary, a famous nineteenth century German pathologist, rescued and re-defined the term 'endophytes' in 1866 for fungi 'living in plant tissues' (De Bary 1866). The organisms usually associated are bacteria and fungi. At first, endophyte simply referred to the location of the organism: 'in the plant' (Greek endon = within, phyton = plant), in contrast to epiphyte, which refers to organisms living on the surface of plant organs. As this definition is considered broad and vague, various authors have proposed a wide range of definitions over the last decades.

De Bary defined endophytes as all organisms living in plant tissues, but Carroll (1986) restricted this term to those that cause asymptomatic infections, excluding pathogenic, saprotrophic and mutualistic fungi. In 1991, Orlando Petrini, a Swiss mycologist, proposed extending it to all organisms that, at some stage in of their life, can colonise internal plant tissues without causing apparent harm to the host (Petrini 1991) (Fig. 3.41). This included endophytic organisms which have a certain epiphytic phase and latent pathogens which may cause symptoms at some point in their life. As defined by Wilson (1995), the term 'endophyte' describes more

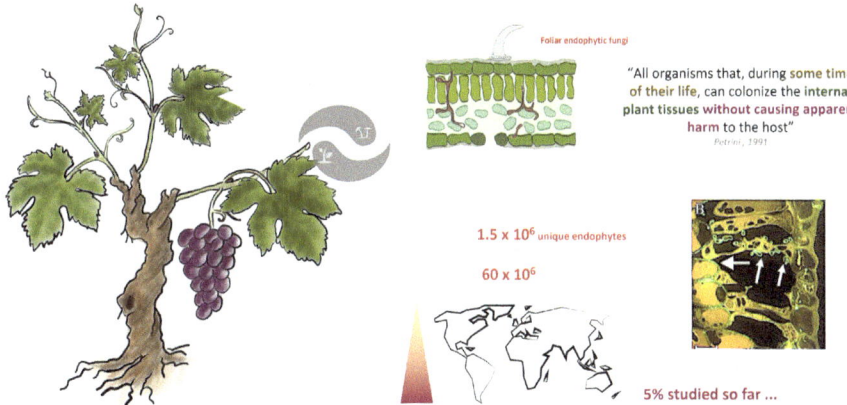

Fig. 3.41 Concept of fungal endophytes (Petrini 1991). Considering that every plant species hosts between 2 and 5 specific endophytes, with an average of 5 endophytes per plant among the 300,000 estimated plant species on Earth, there could be up to 1.5 million unique cultivable endophytes. This number could potentially escalate to 60 million if uncultivable endophytes are included. (Illustration © Léonie Pélissier, Switzerland)

importantly the nature of the interaction strategy of a particular fungus or bacteria, which is that the organisms found inside the plant do not elicit symptoms of disease.

Endophytic microorganisms were more recently and precisely defined as a group of organisms colonising diverse living internal tissues and organs of terrestrial and marine plants, whose colonisation and development is imperceptible and at least temporarily asymptomatic for the host tissues (Stone et al. 2000; Schulz and Boyle 2006). This implies that the internal localisation of the microbial colonisation can be evidenced, via either histological means or isolation from sterilised tissues, or by amplification by nuclear DNA. This definition notably encompasses bacteria, fungi, archaea and viruses (Ryan et al. 2008; Bao and Roossinck 2013; Müller et al. 2015). Fundamentally, this definition describes a momentary status, thus theoretically includes all microorganisms with different life history strategies and development such as endophytes that became saprophytes, mutualistic fungi, latent pathogens and strains with altered virulence (Freeman and Rodriguez 1993; Stone 2011). Hence the importance of stressing the absence of macroscopically visible symptoms; fungal endophytism is not a stably trophic state, but rather a transient trophic mode. It is important to keep in mind that the term endophyte is useful for communication purposes but is not yet biologically well delineated.

3.4.1.1 Types of Endophytic Fungi

Historically, endophytic fungi have been classified in two major groups based on phylogeny, biodiversity, host range and life history trends (mode of transmission, colonisation, ecology): Clavicipitaceous (C) and Non-Clavicipitaceous (NC) endophytes. C-endophytes represent class 1 endophytes, systemically colonising warm and cool season grasses (Saikkonen et al. 2006). NC-endophytes are found in the tissues of a broad range of non-vascular plants, ferns and allies, conifers and angiosperms. They are grouped into 3 classes representing fungi systemically colonising host tissues (class 2), fungi growing only in above-ground plant tissues (class 3) and fungi exclusively colonising plant roots (class 4).

3.4.1.2 Possible Evolution of Endophytic Fungi

The ubiquity of endophytic fungi (particularly NC) among plant tissues, and their obvious association since the dawn of terrestrial life, implies that they share a long and intimate coevolutionary history. However, the evolution of endophytic fungi appears to be more complex than the common belief that they are plant-defending mutualists. Indeed, endophyte-plant interactions extend to a more dynamic ecological landscape, involving multi-species interactions in diverse abiotic and biotic environmental contexts. Consequently, a specialised symbiosis will need well-defined structural, physiological and life-strategy parameters in the host plant and the fungi to evolve and persist (Saikkonen et al. 2004).

Mutualism through defence against herbivores has rapidly been considered as the primary selective force driving the evolution of endophytes. However, both kinds of endophytes are believed to have evolved directly from pathogenic fungi. Woody plant endophytes are likely pre-adapted to mutualism by extension of their latency period in the host and reduction of their virulence. The similarities and taxonomic proximity between endophytes (NCs) and pathogens suggest that this transition may have occurred several times in the fungal kingdom, depending on a/biotic factors.

Endophytism would consequently be rather unstable from an evolutionary point of view and fungi seem to switch repeatedly from one mode of life to another and vice versa. This phenomenon is known as the symbiotic continuum (Rodriguez and Redman 2008).

3.4.2 Biodiversity and Repartition of Endophytic Fungi

Because of the cryptic and often concealed lifestyles of fungi, it is difficult to assess the exact extent of their diversity. According to Purvis and Hector (2000), fungi are the second most species-rich after the insects. Estimates of fungal diversity at a global scale are still highly debated, but recent reports estimate that there might be between 2.2 and 3.8 million species on Earth, which is 6–10 times more than the estimated number of plant species (Hawksworth and Lücking 2017). Approximately 150,000 species have been identified so far (mostly belonging to Ascomycota and Basidiomycota) but, according to scientists, more than 90% of the species remain unknown to science (Antonelli et al. 2020). Indeed, the unveiling of an ever-increasing number of non-cultivated fungi has shown that this diversity is generally underestimated. The increasing use of molecular techniques such as high-throughput sequencing has significantly accelerated the discovery and description of previously unknown diversity. Wu et al. (2019) estimated that, taking into account culture-independent methods, the diversity could be set at about 7.8–8.8 times the current figure, that is about 12 million species. These head-spinning estimates are galvanising scientists to understand and fill the gaps in fungal and in particular, endophyte diversity.

Endophytic fungi consequently clearly represent a key component of fungal and global biodiversity. More than 30 years ago, the postulate of endophytic species on Earth proposed that endophytic fungi alone might represent up to 1.3 million fungal species. This estimate considered only cultivable fungi and was made on the assumption that each plant hosted 2–5 unique endophytes (Dreyfuss and Chapela 1994). More recent studies attempted to estimate global endophytic diversity by considering culture-independent analyses such as amplicon sequencing or OTU-based surveys (Aghdam and Brown 2021). Considering that there is at least one non-cultivable host-specific fungal endophyte for every cultivable endophyte, there may be at least 60 million species of fungal endophytes in the estimated 300,000 plant species on Earth (Hawksworth and Lücking 2017).

Moreover, endophytes have been evidenced within all plant species studied so far, including diverse ecosystems such as lichens, mosses and ferns, and in numerous angiosperms and gymnosperms such as grasses, palms, shrubs, coniferous and deciduous trees (Sun and Guo 2012). They are found in any kind of tissues within the plant, from roots to foliage, bark, xylem and needles. Leaves and twigs generally harbour a diverse endophytic assemblage, and the distribution may even vary within the crown itself, and by tissue age (Eberl et al. 2019). This diversity exists even on a very small scale and within species (genotype level): even small volumes of tissue, such as a single conifer needle, can host several dozen species and even genotypes (Müller et al. 2001). This is not surprising given that their distribution and presence is likely favoured by the organisation of the plant as a complex molecular, genetic and structural construct, as well as by climatic and ecosystem conditions (Eberl et al. 2019).

However, characterising endophytic diversity still remains extremely challenging, notably because of the efforts required for adequate sampling, culturing and sequencing. The extent of diversity is difficult to assess without the complementarity of culture-free methods to fully understand the diversity and composition of endophyte communities. Indeed, the accounted endophytes might be those that grow the most rapidly or easily in culture, for example. Consequently, there is still a lack of knowledge concerning endophyte distribution and biodiversity at global, inter-biome and host plant levels alike (Harrison and Griffin 2020).

From a taxonomical point of view, the NC-endophytes include species mostly belonging to Dikarya (Ascomycota and Basidiomycota phyla) (Roy and Banerjee 2018). Fungal endophytes associated with foliage (NC class 3) are dominated by members of the Ascomycota phylum. The majority belongs to the Dothideomycetes and Sordariomycetes classes, which account for more than 75% of endophytes in sites ranging from the arctic to the tropics. Studies have shown that the phylogenetic composition of endophytic communities varies with latitude: Sordariomycetes appear to be prevalent in tropical communities, and Dothidomycetes in boreal communities (103,105). Some of the most common genera include *Acremonium*, *Alternaria*, *Cladosporium*, *Coniothyrium*, *Epicoccum*, *Fusarium*, *Phoma* and *Pleospora* and correspond to ubiquitous genera. In tropical sites, endophytes seem to be consistently represented by a subset of common genera including *Colletotrichum*, *Xylaria*, *Fusarium*, *Botryosphaeria*, *Phomopsis*, *Phyllosticta* and *Pestialotopsis* (Roy and Banerjee 2018).

3.4.3   Interaction of Endophytic Fungi with the Host Plant

3.4.3.1   Transmission

Endophytes have also been classified by their preferred transmission mode in the host, as vertically-transmitted or horizontally-transmitted endophytes. The transmission mode depends on the ecological and evolutionary link between the host and the microbes. In vertical transmission, endophytes are transmitted directly from the host plant (maternal plant) to the offspring through host tissues like seeds or vegetative propagules (Bamisile et al. 2018). Under appropriate conditions, the seed then germinates and the endophytes that were inside enter into the newly formed plant progeny (Hodgson et al. 2014). Hence, contrary to horizontal transmission, seedlings raised under sterile conditions will contain culturable endophytes. This type of transmission is found mostly in C-endophytes, which are non-sporulating and cause systemic infection. This type of transmission implies that reproduction and fitness of fungi and host plants are tightly related, leading to very specific mutualistic interactions. The communities of vertically transmitted grass endophyte fungi are thus very dependent on host genotype and tend to have a long generation time (Saikkonen et al. 2010).

   In horizontal transmission, fungi travel via their fungal propagules, including sexual or asexual spores as well as hyphal fragments, between different individuals of a given population (Verma et al. 2017). Spores and dry propagules are dispersed from plant to plant via biotic factors like herbivores or insects and abiotic factors like air currents, wind, rain splashes or precipitation (Feldman et al. 2008; Swamy and Sandhu 2021). The germinating spores then penetrate through the plant cuticle to enter the plant tissues. To allow the rain dispersal of the spores, many endophytes of woody plants produce slimy spore masses that helps them adhere to host surfaces and surrounding offspring during wet periods (Slippers and Wingfield 2007). This is notably the case for Botryosphaeriaceae.

   This type of transmission is the preferred mode for NC-endophytes, especially class 3 endophytes colonising above-ground tissues, notably in woody plants and leaves of tropical trees. They form local latent infections and are rather non-specific and highly diverse (Verma et al. 2017). Due to the presence of the inoculum in the air and the relatively high humidity of the leaf surface, colonisation of leaves by class 3 endophytes can happen rapidly, independently of leaf firmness or chemistry

(Arnold and Herre 2003). In woody plants, newly emergent leaves are endophyte-free but are quickly invaded by fungal spores. It has been shown for example that, in a tropical forest site, more than 80% of the leaves of *Theobroma cacao* L. emerging from endophyte-free seedlings were colonised by endophytes after 2 weeks (Wilson and Carroll 1994; Faeth and Hammon 1997). Although this type of transmission is less specific than the vertical one, host genotype specificity has been reported, indicating a tight relation between endophyte and host physiology (Ahlholm et al. 2002).

Vertically transmitted endophytes may infect evergreen woody plants and perennial grasses all year round and accumulate in older leaves, whereas horizontally transmitted endophytes re-infect the foliage of woody trees yearly and accumulate seasonally.

3.4.3.2 Colonisation

Endophytic colonisation depends on many parameters inherent to the host plant (genotype, tissue type, plant physiology), the microbe (taxon, strain type, availability of spores, hydrolytic enzymes), and biotic or abiotic environmental factors (surrounding vegetation, plant architecture, weather, light). Endophytes can colonise plants in various ways. In shoots, for example, colonisation can be intracellular and limited to single cells, or intercellular and localised, whereas it might be more systemic and both intra- and intercellular in roots (Schulz and Boyle 2005). It can also be confined to the leaves or the needles or adapted to the growth in the bark (Wang and Guo 2007). Probably due to selective pressure, colonisation is organ- and tissue-specific and will trigger a tissue-specific defence in the host (Mengistu 2020). C-endophytes are virtually exclusively detected in grasses and colonise plant shoots to induce a systemic infection. Within NC-endophytes, class 2 colonise both above-ground and below-ground tissues, roots, stems and leaves, and can induce extensive infections in plants. Class 3 are mainly restricted to aerial tissues and induce highly localised infections (Rodriguez et al. 2009). In woody plants newly flushed leaves are considered free of endophytes but are rapidly invaded by fungal spores dispersed via air, rain and animal vectors, and from senescent elder leaves.

3.4.3.3 Construction of Associations Between Endophytic Fungi and Their Host Plants

In order to establish such remarkable symbiotic relationships, the microbe and its plant host must coordinate their biology. The interaction generally involves the following steps (Sieber 2007; Plett and Martin 2018):

- Recognition: pre-symbiotic exchange of diffusible and volatile signals
- Adhesion: first adhesion of the microbe to the plant cell (Toti et al. 1992)
- Germination: aggregation or and inter- or intracellular growth

- Penetration
- Exchange

To then distinguish from pathogens and assume a quiescent state within the plant, the endophyte must be able to overcome the plant's innate and induced defence response.

At the beginning, germinating endophytic spores will attach to the host surface (Viret et al. 1994). Recognition of a specific host is thought to be often mediated by lectin-like molecules that can bind chitin or ß-glucan (Chapela et al. 1993; Wawra et al. 2019). Once attached, the spores penetrate directly through the leaf cuticle thanks to the secretion of hydrolytic exoenzymes (lipase, cutinase, cellulase) that weaken the plant cell walls (Gindro and Pezet 1999), the possible formation of appressoria or infection cushions (flattened hyphae that make bulbous structures) and/or haustoria (root-like hyphae) (Choi et al. 2005; Viret and Petrini 1994) or by direct penetration of the germinating tubes through natural openings, for example through stomata or wounds (made for instance by herbivores). After surviving the super oxidative environment within the plant, notably thanks to detoxifying enzymes that deactivate reactive oxygen and nitrogen species, attached fungi can then migrate to colonise the internal tissues of the host (Nogueira-Lopez et al. 2018).

In some cases, this process is done gently and microscopic observations show that the cell integrity was not disrupted during early colonisation, contrary to a pathogenic colonisation, as exemplified by the tomato roots endophyte *Trichoderma* sp., or *Fusarium* sp. in bean plants (Boyle et al. 2001).

3.4.3.4 Distinction of Different Fungal Lifestyles (Endophytic/Pathogenic) by the Host Plant

Plants possess a dynamic immune system, a physiological response and a labile transcriptional tool set that differ in their activation/repression depending on the lifestyle or identity of the microbe approaching their tissues. Consequently, interactions with endophytes are characterised by dynamic change in real time, where controlled by signalling and perception pathways emitted by the microbes to the plants, leading to the appropriate plant immune response (Zipfel and Oldroyd 2017). Among those signals, microbe-associated molecular patterns (MAMPs), and their equivalents in pathogens (pathogen-associated molecular patterns, PAMPs) constitute carbohydrate and protein-based signals essential for microbial survival, and not present in the plant (Boller and Felix 2009). They include peptidoglycans, lipopolysaccharides, β-glycans and chitin (Newman et al. 2013). These patterns are recognised via pattern recognition receptors (PRRs) located on the surface of plant cells. This leads to MAMP-triggered immunity (MTI). Molecules (called effectors) produced by microbes are recognised by intracellular receptors and activate effector-triggered immunity (ETI). In pathogenic interaction, the recognition of PAMPs and DAMPs (Damage-Associated Molecular Patterns produced by the disruption of cell integrity) initiates an immune response. In symbiotic interactions, DAMPs receptors are thought to be inactivated, leading to repression of the immune response, along with MAMPs receptor signalling to lower the cellular defences. In this way, plants might be able to differentiate between a friend or a foe (Zamioudis and Pieterse 2012; Yan et al. 2019).

Studies have found that the plant defence signal is further moderated by plant microRNAs (miRNAs), small non-coding RNAs folded into a stem-loop structure and recognised by multiple defence pathways (Woloshen et al. 2011). It appears that miRNAs are differentially expressed and regulate different targets depending on the type of plant-microbe interaction (pathogenic or symbiotic). In symbiotic interactions, the majority of miRNAs are thought to target hormone-response pathway, protein methylation and innate immune functions to turn off defence pathways. Conversely, they seem to turn on defence and detoxification functions in pathogenic interactions (Li et al. 2010; Formey et al. 2014).

The plant also responds selectively to different microbial lifestyles through hormonal pathways. Typically, the plant hormones jasmonic acid (JA), ethylene (ET) and salicylic acid (SA) play a major role in plant defence strategy, notably against necrotrophic and biotrophic pathogens, but also in certain mutualistic interactions (Fesel and Zuccaro 2016). For example, it was shown that colonisation of A. thaliana by root endophytic fungi led to the induction of genes involved in SA and JA biosynthesis, followed by JA accumulation, but then to the induction of SA catabolism. When plants were colonised by biotrophic parasitic fungi, the opposite was observed, with the activation of SA-mediated defence and the repression of JA-mediated defence (Lahrmann et al. 2015). Similarly, Chen et al. showed that distinctly expressed genes linked to JA signalling pathways were constantly activated by beneficial endophytes in rice, whereas the SA pathway was activated only when the roots were colonised by pathogens. This indicated that JA was involved in controlling endophyte density in the roots (Chen et al. 2020).

These studies suggest that, in pathogenic interactions, these pathways are triggered relatively early and serve to defend against any microbial colonisation, while they play a later role in symbiotic interactions, where they serve to maintain a low level of defence in the host and keep the microbes at bay, preventing them from overwhelming the plant.

3.4.3.5 Evasion of Host Immune System by Endophytic Fungi

Moreover, endophytes also have the ability to go unnoticed by evading and manipulating plant immune system responses and can adapt the way they interact with the plant to selfishly maximise their fitness. They seem able to short-circuit plant defence responses like ETI via removing or mutating some effectors (Zamioudis and Pieterse 2012). For example, β-glucan is a fungal cell wall MAMP that can trigger the plant immune system. The endophyte *Piriformospora indica*, a root endomycorrhizal fungus belonging to the Glomales (root endophyte), has a gene encoding a specific lectin that binds to fungal β1,6-glucan to alter the fungal cell wall composition and hence protects β-glucan polymers from recognition in plants and induces the resulting plant immunity. As shown in *Pestalotiopsis* sp., endo-phytes seem able to hide their chitin from the host immune system by modifying it with a chitin deacetylase that prevents the plant specific receptors from recognising the chitosan motifs. Some endophytes have also developed resistance to cytotoxic compounds produced by the host. For instance, *Fusarium solani* has developed resistance to camptothecin, produced by the plant to disrupt fungal DNA topoisom-erase, by altering its camptothecin-binding domain along with the catalytic domain of the topoisomerase I (Kusari et al. 2011). Another way for the endophyte to keep a low profile would be to hijack the plant hormone signalling system by, for instance, short-circuiting the JA pathway or synthesising auxin and gibberellin analogs that could attenuate SA signalling (Jacobs et al. 2011; Bastias et al. 2017).

In the end, colonisation by endophytic fungi, even if initially identified as patho-genic invasion, results in an asymptomatic residence in the host through the devel-opment of sophisticated ways to mutually maintain control of the plant immunity for a friendly and balanced interaction. This leads to the idea of 'balanced antagonism'.

3.4.3.6   Endophytes: Friends or Foe?

The result of endophytic colonisation shows that what lies behind the apparent asymptomatic colonisation of the plant by an endophyte is actually the result of a finely tuned balanced interaction between endophytes and their hosts. This idea was originally postulated by Schulz et al. (2002) and Schulz and Boyle (2005) as the 'balanced antagonism' hypothesis. These interactions are characterised by a 'momentary status' during which there is a fragile balance of antagonisms between the fungal virulence for colonisation, shelter and nutrient access and the host defence that tries to limit this colonisation (Schulz et al. 2015). When the balance is maintained, the plant-fungus type of association is mutualistic (beneficial to both fungus and host) or commensal (neutral existence and nutrient provision without affecting the host), establishing the fungus as an 'endophyte'. When the balance is disrupted, either the endophyte overcomes plant immunity and becomes pathogenic, resulting in plant disease, or it is blocked by the plant defence system resulting in plant resistance (Kogel et al. 2006). In this context, the disease repre-sents an unbalanced state of the symbiosis.

Due to this ambivalence, endophytes appear to have great potential for creativity in evolutionary development, as well as great phenotypic plasticity as they repre-sent, as such and as a community, a 'continuum' of associations from latent to active pathogenicity to mutualism. Indeed, it seems that there is no defined life history strategy for endophytic fungi in their host, and that endophytism is unstable and is sometimes only a part of their life cycle. Hence, after entering the plant internal system to associate as endophytes, fungi might go through various dynamic life-styles among the endophyte-pathogen-saprotroph patterns to adapt to the changing environment, and the host genotype and physiology (Redman et al. 2001). This phenomenon is known as the symbiotic continuum and has been exemplified notably by phylogenetic studies reviewed in Delaye et al. 2013. Hence, even different strains from the same species can have different life patterns (Ma et al. 2010).

Several studies supported this fact by conducting pathogenicity tests showing that fungi isolated as endophytes (from healthy leaves) caused disease symptoms when inoculated in a stressed host (Guske et al. 2004). Some endophytes are also considered latent pathogens whose virulence is triggered in a different host or when the host weakens (weather, wounds, etc.). Pathogenic fungi will sporulate and secrete toxin to kill the host cells, while endophytes do not sporulate and live in mutualism. Between sporulation and germination, there can be a latency period in which there is no growth and the strain remains inactive (Pillai 2017). This behaviour has been observed in species having a wide host range, such as Botrysosphaeriaceae or *Diaporthe* isolates (Sessa et al. 2018). Some will switch to saprophytic mode once the leaf falls and becomes decaying litter, with the advantage of having a ready nutritional source in living leaves compared to non-endophytic saprophytic fungi (Szink et al. 2016).

The change of life strategy is also influenced by various factors linked to environment and natural ecosystem, stress conditions, and imbalance in nutrient exchange and communication. Álvarez-Loayza et al. (2011) demonstrated for instance that light was a factor in the transition from endophyte to pathogen of *Diplodia mutila* colonising the palm *Iriartea deltoidea*. Under high light, the endophyte became pathogenic, while low light favoured endosymbiosis, assuring better survival. High light makes the endophyte pathogenic by secreting hydrogen peroxide inducing cell death response in plants, while low light favours endosymbiosis. Due to this effect, the colonised palm will survive better protected by the shade of the canopy. The switch is often the result of a slight change in fungal gene expression and sometimes involves a single mutation. This is the case for instance for *Colletotrichum magna*, in which the mutation of a single gene may trigger the transition from pathogenesis to mutualism (Redman et al. 1999).

The delineation between endophyte and pathogen is therefore very fine, as they also both produce many similar virulence factors such as exoenzymes or phytotoxic metabolites, and the host can counter-attack with the same defence system such as induced defence metabolites. Comparative genomic analysis of endophytes and pathogens has indeed revealed the existence of similarities in their virulence factors. However, certain endophytes lack the key virulence factors which may act as the distinguishing feature (Mattoo and Nonzom 2021). The modulation of this fine balance depends on the status of both partners (virulence and defence), which is influenced by physiological and genetic factors. Taking into consideration that, in addition to the host defences, the endophytes also have to tolerate microbial competitors (other beneficial microbes) or invaders (pathogens), Schulz et al. (2015) recently modified the concept as 'multiple balanced antagonism', represented by the compatible and multipartite symbiosis in a healthy plant. The plant-endophyte complex is thus nested in intra- and interkingdom interactions involving crosstalk with the plant and with the other microbes, notably through chemical communication strategies. In asymptomatic conditions, communication occurs for example through the production by endophytes of secondary metabolites that mimic plant metabolites to maintain beneficial microbes in the plant, or that can be antimicrobial to keep competitors at bay (Kusari et al. 2012).

Quorum sensing (QS) and quorum quenching (QQ) communication systems are also used by endophytes to ensure the stability of their colonisation and combat potential pathogens. Probably through this coevolution, plants are also able to produce QS inhibitors and molecules mimicking the microbial QS autoinducers (Teplitski et al. 2011). This leads to a crosstalk effect between plant and endophyte communities, allowing the endosphere to be a healthy habitat. Any disruption of these communication systems can destabilise the cordial symbiosis. Due to its frequency over the past millions of years, it seems that the 'endophytic continuum' is also an 'evolutionary continuum' characterised by dynamic evolutions and developmental changes. The symbiotic continuum is a good ecological strategy for fungi to capitalise on nutrients. Their establishment in tissues allows them direct access to available nutrients during plant senescence. This could represent an evolutionary transition or be a consequence of ecological flexibility to ensure prosperity in various host ranges, where plants either participate in or potentially disrupt the balance causing the change. However, the omnipresence of endophytes and their long-term association with plants evidences that endophytism must be a successful life strategy for fungi.

3.4.3.7 Role of the Endophytic Fungi in the Protection of the Plant

Some endophytic fungi play an important role in protecting the plant against biotic and abiotic stresses and may participate in plant fitness either by modulating host physiology and/or host interaction with external attacks. Endophytes have demonstrated abilities to help their host plant resist biotic stresses such as pathogens, herbivores and nematodes. This implies direct mechanisms through the secretion of antimicrobial agents and lytic enzymes, as well as indirect mechanisms via the induction of systemic resistance and plant physiology enhancement.

Endophytes have been shown to produce a wide range of specialised metabolites with antimicrobial properties including alkaloids, polyketides, terpenoids, phenols and flavonoids (Mousa and Raizada 2013). For example, the alkaloid altersetin produced by the fungal endophyte *Alternaria* spp. was reported to strongly inhibit gram positive pathogenic bacteria (Hellwig et al. 2002). Endophytic fungi from finger millet (*Eleusine coracana*) have demonstrated activity against *Fusarium* fungal species, contributing to the resistance of the plant to pathogens (Mousa et al. 2015). Some of these compounds are also produced as diverse mixtures of easily diffusible volatile organic compounds (VOCs) and have been shown to be important in the defence against numerous microbes and plant pathogens (Kaddes et al. 2019).

Some compounds (alkaloids, terpenoids) are toxic to herbivores and protect plants against grazing. For example, the cooperation between an endophytic and a mycorrhizal fungus led to enhanced monoterpene and sesquiterpene levels in tomato plants, resulting in stronger defence against beet armyworm (*Spodoptera exigua* Hübner) (Shrivastava et al. 2015).

Endophytes have played a crucial role in the survival of plants since their first appearance on earth because they also help plants to cope with the different abiotic stresses they face in their environment (e.g. drought, salinity, metals). These stresses lead to several dysfunctions and to accumulation of reactive oxygen species, membrane impairments and phytohormone imbalance, damaging the morphology and physiology of plants (Egamberdieva et al. 2017).

Endophytes are able to regulate the plant hormonal balance through the production of phytohormones such as indole acetic acid (IAA), abscisic acid (ABA), gibberellins (GA) or salicylic acid (SA). For example, Kahn et al. showed that the fungal endophyte *Chaetomium globosum* Kunze improved the drought tolerance of its host plant *Capsicum annuum* L. through the production of IAA and GA (Khan et al. 2012). Similarly, *Aspergillus fumigatus* Fresen. synthesises GA which increases plant growth in soybean plants under salt and drought stresses (Khan et al. 2011). Several endophytic fungi (e.g. *Phoma glomerata* (Corda) Qian Chen & L. Cai, *Penicillium roqueforti* Thom.) have shown production of IAA, notably involved in tolerance to metal toxicity and increased nutrient uptake (Ikram et al. 2018). Some fungi might also act through antioxidant activity, for example through the stimulation of plant antioxidant enzymatic system (Sun et al. 2010). Endophytes may also play a role in the lipidic composition of the cell membranes to prevent electrolyte leakage due to stress conditions (Khan et al. 2015).

3.4.3.8 The Mycobiota of Grapevine

Vines, like all woody plants, coexist with a microbiota which varies depending on the plant parts and phenological development stage:

The microbiota of grapevine is significantly influenced by the climate where (temperature and humidity), the *terroir*, the growing system, the use of plant protection products, the variety and/or rootstock, and finally the landscape (limitrophe to a wooded area, to other crops, residential areas, bodies of fresh or salt water) (Bettenfeld et al. 2022). The microbial assemblages associated with the vine (bacteria and fungi) are influenced by a myriad of factors which determine their structure and composition. Most taxa associated with plant organs are soil-borne and their distribution reflects the influence of highly localised biogeographic factors and the management of the vineyard (Zarraonaindia et al. 2015). Grafting of *V. vinifera* on different types of rootstocks can be important in initially establishing the vine microbiome. Studies by Gramaje et al. (2018) show a continuum of the vine mycobiota on grafted plants during the plant's development. The diversity of fungal communities varies depending on the sampling period. At the same time, fungal richness and diversity within the vascular system diminishes during the process of propagation. These authors showed the existence of a core mycobiome comprising the genera *Cadophora*, *Cladosporium*, *Penicillium* and *Alternaria*, and identified the pathogenic genus *Neofusicoccum* as a persistent taxon throughout the propagation process. Knapp et al. (2021) showed that *V. vinifera* cv. Furmint harbours a common core group of fungal species (core mycobiota), regardless of the organ, season or sampling site. A core fungal community is always present, dominated by *Aureobasidium pullulans*, *Cladosporium* spp. and a species complex of *Alternaria alternata*. The concept of a 'core microbiome' is predicated on the existence of a species continuum extending from the rhizosphere to the aerial parts. The core microbiome can be defined as a set of communal species found within different compartments of a single individual, but also as a core set of species found in a given compartment in several individuals, whatever their age, location or edaphic properties (Bettenfeld et al. 2022). In the case of the vine, communal microbial communities (bacteria and fungi) were found to exist that are independent of the region or climate but dependent on the sampled organ. Regardless of the organ studied, the microorganisms, and specifically the fungi, interact with themselves and with the vine. Depending on various endogenous and exogenous factors, some species of fungus can pass from a latent to an active state, whether pathogenic, beneficial or neutral, and whatever the organ studied. The ecological function of an endophyte (for example, neutrality, pathogenicity or antagonism) can be modulated by environmental factors such as soil characteristics, nutrient status and climate conditions or by host-related factors such as physiological state (Romeralo et al. 2022). Furthermore, abiotic conditions can indirectly influence the relationship between host and endophytes. This issue is key to understanding vine wood diseases, especially esca. The incidence of esca varies from one vineyard to another and is influenced by climate, soil type, cultivar and cultivation method (Gramaje et al. 2018; Claverie et al. 2020). Esca is a disease involving multiple fungal pathogens (repeatedly isolated from symptomatic plant parts). The multidimensional interactions between them and the vine are still not fully understood (Graniti et al. 2000). A recent study (Monod 2024) has helped to narrow down the factors determining the increased incidence of esca and wood diseases in general in a single variety susceptible to this disease (*Vitis vinifera* cv. Gamaret) planted in 21 vineyards spread over a wide geographic area. However, it did not find any difference in the fungal communities sampled in asymptomatic and symptomatic plants. The presence and/or absence of specific species traditionally associated with esca is not enough to explain the presence of symptoms. Analysis of the mycobiome of asymptomatic and symptomatic plants shows that the fungal composition is highly diverse but does not differ significantly as a function of the health of the sampled plants, with weak variation recorded at the level of terroir or region. Generally, it can be thought of as a region-specific mycobiota whose pathogenic evolution is determined by crucial abiotic factors, notably the water retention capacity of the soil (Monod et al. 2024).

3.5   Fungi in Plant Pathology

For millennia, hunter-gatherers have relied on nature's bounty for sustenance. As humans became farmers, over the centuries they learnt how to cultivate the soil to provide their needs, eventually producing more abundant harvests that enabled exchange and commerce. Today, our global food supply is ensured by a tiny fraction of the population in industrialised countries, equipped with modern infrastructures and tools. This evolution has been achieved by continuously improving farming methods, selecting high-yielding cultivated plants, and controlling the innumerable pests and diseases that ravage the crops.

Despite this, it is currently estimated that around one quarter of agricultural production each year is destroyed by crop pests and diseases, representing a net loss that would be enough to feed half a billion people (Oerke 2006). Without pest control measures, losses would have catastrophic consequences inevitably resulting in famine (Oerke and Dehne 2004). The growing impact of crop pests can be explained by the following factors:

- The practice of monoculture, which consists of growing a single species of plant on the same vast areas year after year.
- The use of varieties specifically bred to increase yields and quality without considering their susceptibility to pathogens and pests that wild species may possess.
- The cultivation of some vegetables outside their original geographic areas in colder or more humid conditions.
- The opening up of trade borders on a vast scale worldwide, leading to the global spread or pests and diseases.

Plant diseases have a significant impact on all crops, causing major losses in quality and yield. The presence of toxic substances produced by the secondary metabolism of some fungi can compromise food and feed security. On a global scale, the control of fungal diseases incurs considerable costs each year, and in the event of crop failure, the financial losses can amount to several hundred billion (Gula 2023).

Plant pathology, or phytopathology, is a branch of biology involving the study of plant diseases, including their causes, development, spread and control. This interdisciplinary science draws on microbiological, genetic, agronomic and molecular biological knowledge to understand the complex interactions between plants, pathogens and their environment. It encompasses all the phenomena associated with physical and physiological changes in the plants. Plant pathology is the study of plant diseases caused by living organisms (biotic factors) and environmental or physiological factors (abiotic factors). It is difficult to give a precise estimate of the number of pathologies that plants can be subject to, as the number of diseases is influenced by several factors such as the emergence of new diseases, changes in cultivation practice, genetic evolution of causative agents, and the environment. There are currently estimated to be more than 50,000 diseases of wild and cultivated plants worldwide. This number is continuing to grow as new diseases are identified and characterised, each one designated by a common name and the scientific name of the genus and species of the organism involved, for example Saint Anthony's fire or ergot caused by the fungus *Claviceps purpurea* (Fig. 3.42).

The pathogens responsible for diseases comprise different organisms such as fungi, oomycetes, bacteria, viruses, viroids, phytoplasms, protozoa, nematodes and plant parasites (Nazarov et al. 2020). Ectoparasitic pests such as insects, mites and vertebrates are excluded from plant pathology unless they are vectors of disease. Phytopathology encompasses a wide range of activities, notably the identification of pathogens, the aetiology of diseases, their infection cycle, their economic impact,

Fig. 3.42 Rye ergot.
Sclerotium of *Claviceps
purpurea*. (Photograph ©
Carole Parodi, Agroscope)

their epidemiology, interactions between plant and pathogen, resistance mechanisms and control measures.

While fungal diseases of crops are a major problem in global agriculture, the study of fungal plant pathogens is a relative recent discipline linked to the development of mycological sciences. Aetiology—the study of the causative agents of disease—emerged only with the invention of the microscope due to the microscopic size of fungal pathogens. Hooke (1665) was the first to use a microscope to observe and identify organisms previously invisible to the naked eye, yet it would be another two centuries for the discoveries and concepts of plant pathology to evolve, notably through the efforts of Berkeley and De Bary in the mid-nineteenth century. In 1884, Koch proposed a series of principles, including the postulates that bear his names, still applied to this day, to confirm the aetiological role of a microorganism in the development of a disease (Gradmann 2014).

Fungi can infect all plant organs, from the roots to the aerial parts, resulting in considerable yield and economic losses (Fig. 3.43). Some of the most significant fungal diseases of cultivated plants are reported below:

- **Downy mildews**: downy mildews are caused by Oomycetes belonging mostly to the genera *Phytophthora* (Fig. 3.43), *Peronospora*, *Pseuperonospora*, *Bremia* and *Plasmopara*, which can infect a wide range of plants, including those in families of Solanaceae (tomatoes, potatoes, aubergines, etc....), Cucurbitaceae (cucumbers, courgettes, melons, etc....), Fabaceae (peas, beans, etc....), Vitaceae (cultivated vines), Lamiaceae (basil...) and Asteraceae (leaf vegetables, sunflowers). Mildews appear as brown and/or necrotic spots on aerial plant organs and powdery white coatings on the under and upper side of the leaves under humid conditions.
- **Powdery mildews**: generic term for fungal diseases caused by species of the Erysiphaceae family, including fungi belonging mainly to the genera *Erysiphe*, *Blumeria*, *Podosphaera*, *Phyllactinia*, *Microsphaera* and *Oïdium*. Powdery mildews affect a very wide range of woody and herbaceous plants, both wild and

cultivated. They generally grow in climates with mild temperatures and moderate humidity and can spread rapidly in crops when conditions are favourable. Typical symptoms of powdery mildew present as white or greyish powdery spots on leaves, often accompanied by deformation and reduced growth, and can severely impact yields. *Erysiphe necator* is the causative agent of grapevine powdery mildew. This disease can lead to leaf deformation, poor photosynthesis, delayed ripening of fruits and reduced yield and quality of the grape harvest. *Blumeria graminis* is responsible for powdery mildew in cereal crops, especially wheat, oats and barley. It can cause severe yield losses in temperate regions when environmental conditions favour its growth.

- **Rusts**: caused by various fungi belonging to the Basidiomycetes, for example in the genera *Puccinia* (Fig. 3.43), *Magnaporthe, Gymnosporangium* and *Phakopsora*. They infect cereals (wheat, barley, maize, rice, etc...) as well as other crops such as soyabean, asparagus, cotton, pear and coffee. Typical symptoms of rust include brownish-orange spots on leaves, stems, and the ears of cereals. Rice rust (pyricularia), caused by *Magnaporthe oryzae*, mainly affects aerial parts, causing lesions on the leaves, stems, panicles and grains. In some cases it results in severe yield losses (Dean et al. 2005). Rusts generally have complex life cycles involving several different types of spore (uredospores, teliospores, aecidiospores, basidiospores), which form successively on different host plants that are essential for the pathogen to complete its life cycle.

- **Smuts**: grain smuts caused by Basidiomycetes in the genus *Ustilago* produce powdery masses of sooty black spores which spread the disease and result in partial or total destruction of the grains. Smuts fall into two categories: covered smuts, where the fungus transforms the entire grain leaving only the awns intact, and loose smuts, which do not destroy the ear's integrity. Smuts infect wheat (*U. tritici*), barley (covered smut, *U. hordei*; loose smut, *U. nuda*), oats (covered smut, *U. kolleri*; loose smut, *U. avenae*) and maize (*U. maydis*). Formerly of concern, the destructive nature of these diseases has diminished though improved selection and generalised seed treatments. *U. maydis*, for example, produces large galls on stems, leaves and ears which make maize unfit for consumption and ensiling.

- **Bunts**: wheat bunts are caused by Basidiomycetes in the genus *Tilletia*. Systematic treatment of seeds has diminished the impact of these once very harmful diseases. Common bunt of wheat (*T. caries* and *T. foetida*) and dwarf bunt (*T. controversa*) cause total devastation of the ears, leaving only the wrinkled, dull yellow external seed membrane. Grains infected with bunt give off a noxious smell reminiscent of rotten fish (hence its common name 'stinking bunt of wheat') and are unfit for consumption.

- **Fruit rots and moulds**: several groups of fungi are responsible for rots affecting practically all fruits before (Fig. 3.43) or during ripening and during storage. Some of the most common fungi belong to the genera *Botrytis* (e.g. *Botrytis cinerea* causing grey rot in fruits including grapes), *Rhizopus* (e.g. *Rhizopus stolonifer* causing soft rot in bananas, mangos, papayas and citrus fruit), *Penicillium* spp. (blue and green rots in citrus fruit and various pome fruit), *Aspergillus* spp.

Fig. 3.43 Some significant fungal diseases of cultivated plants. (**a**) *Phytophthora infestans* on potatoes. (**b**) *Venturia inaequalis* on apples. (**c**) *Fusarium* spp. on corn. (**d**) *Sclerotinia sclerotiorum* on sunflowers. (**e**) *Monilia laxa* on apples. (**f**) *Puccinia striiformis* f.sp. *tritici* on wheat

(moulds in dried and stored fruit), *Monilinia* (Fig. 3.43) and *Colletotrichum* (anthracnose).

- **Fusariosis**: fusariosis is the generic term used to describe a very large number of plant diseases caused by fungi in the genus *Fusarium*. These diseases affect a wide range of cultivated plants such as cereals (Fig. 3.43), legumes, fruit trees and market garden crops. Symptoms include wilting, yellowing of the leaves and sometimes root rot. The most frequently involved species are:

    - *Fusarium oxysporum*, the cause of tomato vascular wilt (*Fusarium oxysporum* f. sp. *lycopersici*), basal rot of onion (*Fusarium oxysporum* f. sp. *cepae*) and lettuce fusarium wilt (*Fusarium oxysporum* f. sp. *lactucae*),
    - *Fusarium graminearum*, which causes diseases such as head blight of wheat and maize (also called scab) (Fig. 3.43), resulting in yield losses and quality problems associated with mycotoxins produced by the pathogen.

There are estimated to be more than 10,000 diseases caused by fungi and oomycetes which affect a huge variety of cultivated, wild, and ornamental plants worldwide (Fisher et al. 2020; Nazarov et al. 2020). These organisms can grow on dead or living plant tissue and remain dormant until conditions are favourable for growth. The fungi may penetrate the vegetative tissue (endoparasites) or grow only on the surface of plant organs, using special mycelial structures to anchor themselves in the epidermis (ectoparasites).

Fungal spores, the vectors of disease, are transported mainly by wind, water and soil microfauna. In most cases, only the asexual form of the fungus can be isolated from diseased plants in order to identify the species concerned. Several characteristics must be observed to permit identification, including the presence or absence of asexual or sexual fruiting bodies, their mode of development, shape, pigmentation and the presence or absence of conidiophores. Fungal diseases have long been diagnosed on the basis of visible symptoms such as the presence of fungal propagules (sporulation, sclerotia, mycelium, asexual or sexual fruiting bodies) as well as the typicality of the symptoms (organs affected, discolouration, tissue necroses, deformation and wilting). The fungal pathogen is then isolated, cultivated (unless it is an obligate biotrophic) and examined under a microscope to determine hypha structure and colour, spore size and shape, growth rate, etc. It is not unusual to isolate several species of fungus from one infected organ. If the disease has not yet been recorded, new plants must be artificially infected with the isolated fungal species to confirm the identity of the causative agent of the disease according to the postulates developed by Koch in the 1890s (Kaufmann and Winau 2005) which define four rules:

- The fungus is present in the infected plant but absent in healthy plants.
- The fungus can be isolated and grown in pure culture or isolated from the host (spores) in the case of obligate biotrophs.
- The fungus induces the same disease when inoculated into a susceptible healthy plant of the same species.
- The fungus isolated from the newly infected plant must be identical to the one inoculated.

This diagnostic approach is not suitable for more complex diseases caused by an assembly of fungi forming a dynamic fungal community. For example, esca, a vine wood disease caused by a convergence of biotic (complex fungal community) (Fig. 3.44) and abiotic factors (available water in soil, climate, terroir). In these cases, traditional diagnostic techniques are replaced by a molecular method of identification which uses target DNA amplification, thereby avoiding the need for isolation and microscopic examination of the fungi. This broad approach highlights the host plant mycobiome and the causative fungal pathogens. Typical molecular techniques currently used are based on DNA analysis, namely polymerase chain reaction (PCR) and associated techniques (for example nested PCR, real-time PCR, loop-mediated isothermal amplification LAMP, nucleic acid sequence-based amplification NASBA). These highly sensitive techniques enable fungal pathogens to be diagnosed at the early stages of infection, before the emergence of plant symptoms. The development of molecular techniques has also enabled the systematic diagnosis of diseases caused by bacteria, phytoplasms, viruses and viroids that would otherwise necessitate recourse to a transmission electron microscope or an immunologist.

Epidemiology, a discipline of phytopathology, is dedicated to the study of the distribution, frequency and determinants of fungal diseases affecting cultivated, wild, and ornamental plants. It aims to understand the factors influencing the emergence, dissemination, dynamics and persistence of fungal diseases in conjunction with the host plant and the climate and environmental conditions. Epidemiology also explores interactions between fungal pathogens, host plant responses at different stages of phenological development, genetic resistance and cultivation practices. These different aspects allow direct preventive control strategies to be defined involving the application of specific fungicides which are designed to minimise their impact on the crop. To control fungal diseases indirectly, farmers generally use a range of integrated measures, including crop rotations, resistant varieties and seeds coated with plant protection products, targeted application of fungicides and management of the canopy microclimate to mitigate conditions favourable to fungal infections.

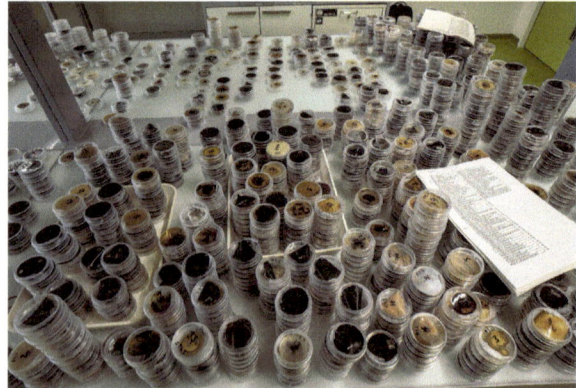

Fig. 3.44 Complex fungal community isolated from a vine presenting symptoms of esca and wood diseases. More than 100 fungal species were isolated and cultivated on an agar medium from a single vine plant

The study of the epidemiology of fungal diseases entails the analysis of factors influencing their appearance, spread and persistence within the host plant, the main steps of which are:

- Monitoring: it is essential to have a monitoring system in place to detect the presence and spread of the disease through the routine collection of data on symptoms, weather conditions and cultivation practices.
- Identifying the risk factors: identifying the factors favouring the development and spread of the disease, for example, favourable weather conditions, poor cultivation practices or the presence of reserves of the fungal inoculum makes it possible to predict the probability of infection.
- Modelling the epidemiology: using mathematical models whose algorithms help to understand and predict disease dynamics in time and space and incorporating significant meteorological and cultural parameters to define a control strategy.
- Analysing the mode of transmission: studying the spread of the disease from one plant to another, either through fungal spores, biological vectors (such as insects) or agricultural practices, is another important aspect of preventive disease control.
- Evaluating the economic impact of the disease on yields and harvest quality makes it possible to calculate the financial risk in relation to the control costs.
- Defining direct preventive control strategies using all the information gathered helps to ensure the success of control measures while respecting the environment. It is estimated that more than 150 fungicidal compounds with diverse modes of action are used globally in agriculture, resulting in an increased risk of resistance and unwanted environmental contamination (Nazarov et al. 2020).

References

Aghdam SA, Brown AMV (2021) Deep learning approaches for natural product discovery from plant endophytic microbiomes. Environ Microbiome 16(1):6. https://doi.org/10.1186/s40793-021-00375-0

Agler MT, Ruhe J, Kroll S, Morhenn C, Kim S-T, Weigel D et al (2016) Microbial hub taxa Link host and abiotic factors to plant microbiome variation. PLoS Biol 14(1):e1002352. https://doi.org/10.1371/journal.pbio.1002352

Ahlholm JU, Helander M, Henriksson J, Metzler M, Saikkonen K (2002) Environmental conditions and host genotype direct genetic diversity of *Venturia ditricha*, a fungal endophyte of birch trees. Evolution 56(8):1566–1573. https://doi.org/10.1111/j.0014-3820.2002.tb01468.x

Álvarez-Loayza P, White JF Jr, Torres MS, Balslev H, Kristiansen T, Svenning JC, Gil N (2011) Light converts endosymbiotic fungus to pathogen, influencing seedling survival and niche-space filling of a common tropical tree, *Iriartea deltoidea*. PLoS One 6(1):e16386. https://doi.org/10.1371/journal.pone.0016386

Anal AKD, Rai S, Singh M, Solanki MK (2020) Plant mycobiome: current research and applications. In: Solanki M, Kashyap P, Kumari B (eds) Phytobiomes: current insights and future vistas. Springer, Singapore. https://doi.org/10.1007/978-981-15-3151-4_4

Andrews JH, Harris RF (2000) The ecology and biogeography of microorganisms on plant surfaces. Annu Rev Phytopathol 38:145–180. https://doi.org/10.1146/annurev.phyto.38.1.145

Antonelli A, Fry C, Smith RJ, Simmonds MSJ, Kersey PJ, Pritchard HW et al (2020) State of the world's plants and fungi. Royal Botanic Gardens, Kew. https://doi.org/10.34885/172

Arastehfar A, Carvalho A, Houbraken J, Lombardi L, Garcia-Rubio R, Jenks JD, Rivero-Menendez O, Aljohani R, Jacobsen ID, Berman J, Osherov N, Hedayati MT, Ilkit M, Armstrong-James D, Gabaldón T, Meletiadis J, Kostrzewa M, Pan W, Lass-Flörl C, Perlin DS, Hoenigl M (2021) *Aspergillus fumigatus* and aspergillosis: from basics to clinics. Stud Mycol 100:100115. https://doi.org/10.1016/j.simyco

Arnold AE, Herre EA (2003) Canopy cover and leaf age affect colonization by tropical fungal endophytes: ecological pattern and process in Theobroma cacao (Malvaceae). Mycologia 95(3):388–398. https://doi.org/10.1080/15572536.2004.11833083

Balasundaram SV, Hess J, Durling MB, Moody SC, Thorbek L, Progida C, LaButti K, Aerts A, Barry K, Grigoriev IV, Boddy L, Högberg N, Kauserud H, Eastwood DC, Skrede I (2018) The fungus that came in from the cold: dry rot's pre-adapted ability to invade buildings. ISME J 12(3):791–801. https://doi.org/10.1038/s41396-017-0006-8

Bamisile BS, Dash CK, Akutse KS, Keppanan R, Wang L (2018) Fungal endophytes: beyond herbivore management. Front Microbiol 9:544. https://doi.org/10.3389/fmicb.2018.00544

Bao X, Roossinck MJ (2013) Multiplexed interactions: viruses of endophytic fungi. Adv Virus Res 86:37–58. https://doi.org/10.1016/B978-0-12-394315-6.00002-7

Bastias DA, Martínez-Ghersa MA, Ballaré CL, Gundel PE (2017) *Epichloë* fungal endophytes and plant defenses: not just alkaloids. Trends Plant Sci 22(11):939–948. https://doi.org/10.1016/j.tplants.2017.08.005

Batsch AJGC (1783) Elenchus fungorum: accedunt icones 57 fungorum nonnullorum agri jenensis, secundum naturam ab autore depictae. Halae Magdeburgicae, Gebauer JJ

Beker HJ, Eberhardt U, Vesterholt J (2016) Hebeloma (Fr.) P. Kumm. Fungi Europaei 14:1–1218

Bertrand S, Schumpp O, Bohni N, Monod M, Gindro K, Wolfender JL (2013) De novo production of metabolites by fungal co-culture of Trichophyton rubrum and *Bionectria ochroleuca*. J Nat Prod 76(6):1157–1165

Bettenfeld P, Cadena I, Canals J, Jacquens L, Fernandez O, Fontaine F, van Schaik E, Courty PE, Trouvelot S (2022) The microbiota of the grapevine holobiont: a key component of plant health. J Adv Res 40:1–15. https://doi.org/10.1016/j.jare.2021.12.008

Boller T, Felix G (2009) A renaissance of elicitors: perception of microbe-associated molecular patterns and danger signals by pattern-recognition receptors. Annu Rev Plant Biol 60(1):379–406. https://doi.org/10.1146/annurev.arplant.57.032905.105346

Boyle C, Gotz M, Dammann-Tugend U, Schulz B (2001) Endophyte-host interactions III. Local vs. systemic colonization. Symbiosis 31:259–281

Bruisson S, Maillot P, Schellenbaum P, Walter B, Gindro K, Deglène-Benbrahim L (2016) Arbuscular mycorrhizal symbiosis stimulates key genes of the phenylpropanoid biosynthesis and stilbenoid production in grapevine leaves in response to downy mildew and grey mould infection. Phytochemistry 131:92–99. https://doi.org/10.1016/j.phytochem.2016.09.002

Brundrett MC, Tedersoo L (2018) Evolutionary history of mycorrhizal symbioses and global host plant diversity. New Phytol 220(4):1108–1115. https://doi.org/10.1111/nph.14976

Brunswik H (1924) Untersuchungen über die Geschlechts- und Kernverhältnisse bei der hymenomycetengattung *Coprinus*. Bot Abhandl Heft 5:1–152

Buller AHR (1933) Researches in fungi, vol 5. University of Toronto Press, Toronto, ON

Burch AY, Zeisler V, Yokota K, Schreiber L, Lindow SE (2014) The hygroscopic biosurfactant syringafactin produced by *Pseudomonas syringae* enhances fitness on leaf surfaces during fluctuating humidity. Environ Microbiol 16(7):2086–2098. https://doi.org/10.1111/1462-2920.12437

Burki F, Roger AJ, Brown MW, Simpson AGB (2020) The new tree of eukaryotes. Trends Ecol Evol 35(1):43–55. https://doi.org/10.1016/j.tree.2019.08.008

Carrión VJ, Perez-Jaramillo J, Cordovez V, Tracanna V, de Hollander M, Ruiz-Buck D et al (2019) Pathogen-induced activation of disease-suppressive functions in the endophytic root microbiome. Science 366(6465):606–612. https://doi.org/10.1126/science.aaw9285

Carroll GC (1986) The biology of endophytism in plants with particular reference to woody perennials. In: Fokkenna NJ, Van den Heuvel J (eds) Microbiology of the Phylosphere. Cambridge University Press, Cambridge, pp 205–222

Carroll FE, Müller E, Sutton BC (1977) Preliminary studies on the incidence of needle endophytes in some European conifers. Sydowia 29:87–103

Cha JY, Han S, Hong H-J, Cho H, Kim D, Kwon Y et al (2016) Microbial and biochemical basis of a fusarium wilt-suppressive soil. ISME J 10(1):119–129. https://doi.org/10.1038/ismej.2015.95

Chapela IH, Petrini O, Bielser G (1993) The physiology of ascospore eclosion in *Hypoxylon fragiforme*: mechanisms in the early recognition and establishment of an endophytic symbiosis. Mycol Res 97(2):157–162. https://doi.org/10.1016/S0953-7562(09)80237-2

Chen X, Marszałkowska M, Reinhold-Hurek B (2020) Jasmonic acid, not salicyclic acid restricts endophytic root colonization of rice. Front Plant Sci 10:1758. https://doi.org/10.3389/fpls.2019.01758

Choi YW, Hodgkiss IJ, Hyde KD (2005) Enzyme production by endophytes of *Brucea javanica*. J Agric Technol:55–66

Choi K, Khan R, Lee S-W (2021) Dissection of plant microbiota and plant-microbiome interactions. J Microbiol 59(3):281–291. https://doi.org/10.1007/s12275-021-0619-5

Christian N, Whitaker BK, Clay K (2015) Microbiomes: unifying animal and plant systems through the lens of community ecology theory. Front Microbiol 6:869. https://doi.org/10.3389/fmicb.2015.00869

Claverie M, Notaro M, Fontaine F, Wery J (2020) Current knowledge on grapevine trunk diseases with complex etiology: a systemic approach. Phytopathol Mediterr 59(1):29–53

Cole GT, Samson RA (1979) Patterns of development in conidia fungi. Pitman Publishing, London, p 190. ISBN: 0-273-08407-0

Compant S, Sessitsch A, Mathieu F (2012) The 125th anniversary of the first postulation of the soil origin of endophytic bacteria—a tribute to M.L.V. Galippe. Plant Soil 356:299–301. https://doi.org/10.1007/s11104-012-1204-9

Compant S, Cambon MC, Vacher C, Mitter B, Samad A, Sessitsch A (2021) The plant endosphere world—bacterial life within plants. Environ Microbiol 23(4):1812–1829. https://doi.org/10.1111/1462-2920.15240

Costa JH, Wassano CI, Angolini CFF, Scherlach K, Hertweck C, Pacheco Fill T (2019) Antifungal potential of secondary metabolites involved in the interaction between citrus pathogens. Sci Rep 9(1):18647. https://doi.org/10.1038/s41598-019-55204-9

Crous PW, Slippers B, Wingfield MJ, Rheeder J, Marasas WF, Philips AJ, Alves A, Burgess T, Barber P, Groenewald JZ (2006) Phylogenetic lineages in the Botryosphaeriaceae. Stud Mycol 55:235–253. https://doi.org/10.3114/sim.55.1.235

De Bary A (1866) Morphologie und Physiologie der Pilze, Flechten und Myxomyceten [Internet]. W Engelmann, Leipzig, p 338. (Handbuch der physiologischen Botanik). https://www.biodiversitylibrary.org/bibliography/120970

De Tournefort JP (1694) Eléments de botanique ou méthode pour connître les plantes. Imprimerie Royale, Paris

Dean RA, Talbot NJ, Ebbole DJ, Farman ML, Mitchell TK, Orbach MJ, Thon M, Kulkarni R, Xu JR, Pan H, Read ND, Lee YH, Carbone I, Brown D, Oh YY, Donofrio N, Jeong JS, Soanes DM, Djonovic S, Kolomiets E, Rehmeyer C, Li W, Harding M, Kim S, Lebrun MH, Bohnert H, Coughlan S, Butler J, Calvo S, Ma LJ, Nicol R, Purcell S, Nusbaum C, Galagan JE, Birren BW (2005) The genome sequence of the rice blast fungus *Magnaporthe grisea*. Nature 434(7036):980–986. https://doi.org/10.1038/nature03449

Delaye L, García-Guzmán G, Heil M (2013) Endophytes versus biotrophic and necrotrophic pathogens—are fungal lifestyles evolutionarily stable traits? Fungal Divers 60(1):125–135. https://doi.org/10.1007/s13225-013-0240-y

Dick MW (2013) Straminipilous fungi: systematics of the Peronosporomycetes including accounts of the marine straminipilous protists, the plasmodiophorids and similar organisms. Springer, Dordrecht. https://doi.org/10.1007/978-94-015-9733-3

Dobzhansky T (1937) Genetics and the origin of species. Columbia University Press, New York
Dreyfuss MM, Chapela IH (1994) Potential of fungi in the discovery of novel, low-molecular weight pharmaceuticals. Biotechnology 26:49–80. https://doi.org/10.1016/b978-0-7506-9003-4.50009-5
Dugan FM, Everhart SE (2016) Cryptic species: a leitmotif of contemporary mycology has challenges and benefits for plant pathologists. Plant Health Prog 17:250–253. https://doi.org/10.1094/PHP-RV-16-0046
Eberl F, Uhe C, Unsicker SB (2019) Friend or foe? The role of leaf-inhabiting fungal pathogens and endophytes in tree-insect interactions. Fungal Ecol 38:104–112. https://doi.org/10.1016/j.funeco.2018.04.003
Egamberdieva D, Wirth SJ, Alqarawi AA, Abd Allah EF, Hashem A (2017) Phytohormones and beneficial microbes: essential components for plants to balance stress and fitness. Front Microbiol 8:2104. https://doi.org/10.3389/fmicb.2017.02104
Ehrmann J, Ritz K (2014) Plant: soil interactions in temperate multi-cropping production systems. Plant Soil 376(1):1–29. https://doi.org/10.1007/s11104-013-1921-8
Faeth SH, Hammon KE (1997) Fungal endophytes in oak trees: long-term patterns of abundance and associations with leafminers. Ecology 78(3):810–819. https://doi.org/10.1890/0012-9658(1997)078[0810:FEIOTL]2.0.CO;2
Favre-Godal Q, Gourguillon L, Lordel-Madeleine S, Gindro K, Choisy P (2020) Orchids and their mycorrhizal fungi: an insufficiently explored relationship. Mycorrhiza 30(1):5–22. https://doi.org/10.1007/s00572-020-00934-2
Feldman TS, O'Brien HE, Arnold AE (2008) Moths that vector a plant pathogen also transport endophytic fungi and mycoparasitic antagonists. Microb Ecol 56(4):742–750. https://doi.org/10.1007/s00248-008-9393-8
Fesel PH, Zuccaro A (2016) Dissecting endophytic lifestyle along the parasitism/mutualism continuum in Arabidopsis. Curr Opin Microbiol 32:103–112. https://doi.org/10.1016/j.mib.2016.05.008
Fisher MC, Gurr SJ, Cuomo CA, Blehert DS, Jin H, Stukenbrock EH, Stajich JE, Kahmann R, Boone C, Denning DW, Gow NAR, Klein BS, Kronstad JW, Sheppard DC, Taylor JW, Wright GD, Heitman J, Casadevall A, Cowen LE (2020) Threats posed by the fungal kingdom to humans, wildlife, and agriculture. mBio 11:10. https://doi.org/10.1128/mbio.00449-20
Fitzpatrick CR, Salas-González I, Conway JM, Finkel OM, Gilbert S, Russ D et al (2020) The plant microbiome: from ecology to reductionism and beyond. Ann Rev Microbiol 74(1):81–100. https://doi.org/10.1146/annurev-micro-022620-014327
Formey D, Sallet E, Lelandais-Brière C, Ben C, Bustos-Sanmamed P, Niebel A, Frugier F, Combier JP, Debellé F, Hartmann C, Poulain J, Gavory F, Wincker P, Roux C, Gentzbittel L, Gouzy J, Crespi M (2014) The small RNA diversity from *Medicago truncatula* roots under biotic interactions evidences the environmental plasticity of the miRNAome. Genome Biol 15(9):457. https://doi.org/10.1186/s13059-014-0457-4
Freeman S, Rodriguez RJ (1993) Genetic conversion of a fungal plant pathogen to a nonpathogenic, endophytic mutualist. Science 260(5104):75–78. https://doi.org/10.1126/science.260.5104.75
Fuckel L (1869) Symbolae mycologicae. Beiträge zur Kenntniss der rheinischen Pilzen. Wiesbaden Jehr. Nassau Ver Naturk 23:330. https://doi.org/10.5962/bhl.title.47117
Galippe V (1887) Note sur la présence de micro-organismes dans les tissus végétaux. CR Séances Soc Biol Fil 39:410–416
Gindro K, Pezet R (1999) Purification and characterization of a 40.8-kDa cutinase in ungerminated conidia of *Botrytis cinerea* Pers.: Fr. FEMS Microbiol Lett 171(2):239–243. https://doi.org/10.1111/j.1574-6968.1999.tb13438.x
Gleditch JG (1753) Methodus Fungorum: exhibens genera, species et varietates cum charactere, differentia specifica, synonomis, solo, loco et observationibus. Berolini sumtibus scholae realis
González V, Tello ML (2011) The endophytic mycota associated with *Vitis vinifera* in Central Spain. Fungal Divers 47:29–42. https://doi.org/10.1007/s13225-010-0073-x

Gradmann C (2014) A spirit of scientific rigour: Koch's postulates in twentieth-century medicine. Microbes Infect 16(11):885–892. https://doi.org/10.1016/j.micinf.2014.08.012

Graf T, Franziska S, Pascal NA, Grabenweger G (2023) From lab to field: biological control of the Japanese beetle with entomopathogenic fungi. Front Insect Sci 3:1138427. https://doi.org/10.3389/finsc.2023.1138427

Gramaje D, Úrbez-Torres JR, Sosnowski MR (2018) Managing grapevine trunk diseases with respect to etiology and epidemiology: current strategies and future prospects. Plant Dis 102(1):12–39. https://doi.org/10.1094/PDIS-04-17-0512-FE

Granett J, Walker MA, Kocsis L, Omer AD (2001) Biology and management of grape phylloxera. Annu Rev Entomol 46:387–412. https://doi.org/10.1146/annurev.ento.46.1.387

Graniti A, Surico G, Mugnai L (2000) Esca of grapevine : a disease complex or a complex of diseases. Phytopathol Mediterr 39(1):16–20

Griffiths DA (1974) The origin, structure and function of chlamydospores in fungi. Nova Hedwigia 25(3/4):503–549

Grimm M, Grube M, Schiefelbein U, Zühlke D, Bernhardt J, Riedel K (2021) The Lichens' microbiota, still a mystery? Front Microbiol 12:623839. https://doi.org/10.3389/fmicb.2021.623839

Gula LT (2023) Researchers helping protect crops from pests. Web publication of USDA National Institute of Food and Agriculture. www.nifa.usda.gov/about-nifa/blogs/researchers-helping-protect-crops-pests

Guskc S, Schulz B, Boyle C (2004) Biocontrol options for Cirsium arvense with indigenous fungal pathogens. Weed Res 44(2):107–116. https://doi.org/10.1111/j.1365-3180.2003.00378.x

Hardoim PR, van Overbeek LS, Berg G, Pirttilä AM, Compant S, Campisano A et al (2015) The hidden world within plants: ecological and evolutionary considerations for defining functioning of microbial endophytes. Microbiol Mol Biol Rev 79(3):293–320. https://doi.org/10.1128/MMBR.00050-14

Harrison JG, Griffin EA (2020) The diversity and distribution of endophytes across biomes, plant phylogeny and host tissues: how far have we come and where do we go from here? Environ Microbiol 22(6):2107–2123. https://doi.org/10.1111/1462-2920.14968

Hassani MA, Durán P, Hacquard S (2018) Microbial interactions within the plant holobiont. Microbiome 6(1):58. https://doi.org/10.1186/s40168-018-0445-0

Hawksworth DL, Lücking R (2017) Fungal diversity revisited: 2.2 to 3.8 million species. Microbiol Spectr 5(4). https://doi.org/10.1128/microbiolspec

Heckman DS, Geiser DM, Eidell BR, Stauffer RL, Kardos NL, Hedges SB (2001) Molecular evidence for the early colonization of land by fungi and plants. Science 293:1129. https://doi.org/10.1126/science.1061457

Hellwig V, Grothe T, Mayer-Bartschmid A, Endermann R, Geschke FU, Henkel T, Stadler M (2002) Altersetin, a new antibiotic from cultures of endophytic Alternaria spp. Taxonomy, fermentation, isolation, structure elucidation and biological activities. J Antibiot (Tokyo) 55(10):881–892. https://doi.org/10.7164/antibiotics.55.881

Hodgson S, de Cates C, Hodgson J, Morley NJ, Sutton BC, Gange AC (2014) Vertical transmission of fungal endophytes is widespread in forbs. Ecol Evol 4(8):1199–1208. https://doi.org/10.1002/ece3.953

Hofstetter V, Buyck B, Croll D, Viret O, Couloux A, Gindro K (2012) What if esca disease of grapevine were not a fungal disease? Fungal Divers 54:51–67. https://doi.org/10.1007/s13225-012-0171-z

Hofstetter V, Buyck B, Eyssartier G, Schnee S, Gindro K (2019) The unbearable lightness of sequenced-based identification. Fungal Divers 96:243–284. https://doi.org/10.1007/s13225-019-00428-3

Hooke R (1665) Micrographia: or some physiological descriptions of minute bodies made by magnifying glasse: with observations and inquiries thereupon. Jo Martyn and Ja Allestry, London

Ikram M, Ali N, Jan G, Jan FG, Rahman IU, Iqbal A, Hamayun M (2018) IAA producing fungal endophyte Penicillium roqueforti Thom., enhances stress tolerance and nutrients uptake in

wheat plants grown on heavy metal contaminated soils. PLoS One 13(11):e0208150. https://
    doi.org/10.1371/journal.pone.0208150
Jacobs S, Zechmann B, Molitor A, Trujillo M, Petutschnig E, Lipka V, Kogel KH, Schäfer P (2011)
    Broad-spectrum suppression of innate immunity is required for colonization of Arabidopsis
    roots by the fungus *Piriformospora indica*. Plant Physiol 156(2):726–740. https://doi.
    org/10.1104/pp.111.176446
Jones SK Jr, Bennett RJ (2011) Fungal mating pheromones: choreographing the dating game.
    Fungal Genet Biol 48(7):668–676. https://doi.org/10.1016/j.fgb.2011.04.001
Kaddes A, Fauconnier M-L, Sassi K, Nasraoui B, Jijakli M-H (2019) Endophytic fungal vola-
    tile compounds as solution for sustainable agriculture. Molecules 24(6):1065. https://doi.
    org/10.3390/molecules24061065
Kaufmann S, Winau F (2005) From bacteriology to immunology: the dualism of specificity. Nat
    Immunol 6:1063–1066. https://doi.org/10.1038/ni1105-1063
Keller M, Viret O, Cole FM (2003) *Botrytis cinerea* infection in grape flowers: defense reac-
    tion, latency, and disease expression. Phytopathology 93(3):316–322. https://doi.org/10.1094/
    PHYTO.2003.93.3.316
Khan AL, Hamayun M, Kim Y-H, Kang S-M, Lee J-H, Lee I-J (2011) Gibberellins producing
    endophytic *Aspergillus fumigatus* sp. LH02 influenced endogenous phytohormonal levels, iso-
    flavonoids production and plant growth in salinity stress. Process Biochem 46(2):440–447.
    https://doi.org/10.1016/j.procbio.2010.09.013
Khan A, Shinwari Z, Kim Y-H, Waqas M, Hamayun M, Kamran M, Lee IJ (2012) Role of endo-
    phyte *Chaetomium globosum* lk4 in growth of capsicum annuum by production of gibberellins
    and indole acetic acid. Pak J Bot 44(5):1601–1607
Khan AL, Hussain J, Al-Harrasi A, Al-Rawahi A, Lee IJ (2015) Endophytic fungi: resource for
    gibberellins and crop abiotic stress resistance. Crit Rev Biotechnol 35(1):62–74. https://doi.
    org/10.3109/07388551.2013.800018
Kim H, Metzenberg RL, Nelson MA (2002) Multiple functions of mfa-1, a putative pheromone
    precursor gene of *Neurospora crassa*. Eukaryot Cell 1(6):987–999. https://doi.org/10.1128/
    EC.1.6.987-999.2002
Knapp DG, Lázár A, Molnár A, Vajna B, Karácsony Z, Váczy KZ, Kovács GM (2021) Above-
    ground parts of white grapevine *Vitis vinifera* cv. Furmint share core members of the fungal
    microbiome. Environ Microbiol Rep 13(4):509–520. https://doi.org/10.1111/1758-2229
Kogel KH, Franken P, Hückelhoven R (2006) Endophyte or parasite—what decides? Curr Opin
    Plant Biol 9(4):358–363. https://doi.org/10.1016/j.pbi.2006.05.001
König JG (1779) Naturgeschichte der sogennanten weiben Ameise. Beschaftigungen der
    Berlinischen Gesellschaft Naturforschender Freunde, vierter Band
Kusari S, Košuth J, Čellárová E, Spiteller M (2011) Survival-strategies of endophytic *Fusarium
    solani* against indigenous camptothecin biosynthesis. Fungal Ecol 4(3):219–223. https://doi.
    org/10.1016/j.funeco.2010.11.002
Kusari S, Hertweck C, Spiteller M (2012) Chemical ecology of endophytic fungi: origins of second-
    ary metabolites. Chem Biol 19(7):792–798. https://doi.org/10.1016/j.chembiol.2012.06.004
L'Ecluse C (1601) Plantarum Historia: fungorum in Pannoniis observatorum brevia historia. Ex
    Officina Plantiniana, apud Joannem Moretum, Antwerpiae
Lade SB, Štraus D, Oliva J (2022) Variation in fungal community in grapevine (*Vitis vinifera*) nurs-
    ery stock depends on nursery, variety and rootstock. J Fungi 8(1):47. https://doi.org/10.3390/
    jof8010047
Lahrmann U, Strehmel N, Langen G, Frerigmann H, Leson L, Ding Y, Scheel D, Herklotz S,
    Hilbert M, Zuccaro A (2015) Mutualistic root endophytism is not associated with the reduc-
    tion of saprotrophic traits and requires a noncompromised plant innate immunity. New Phytol
    207(3):841–857. https://doi.org/10.1111/nph.13411
Laurent E (1889) Sur l'existence de microbes dans les tisus des plantes supérieures. Bulletin de la
    Société Royale de Botanique de Belgique 28:233–244

Li Y, Zhang Q, Zhang J, Wu L, Qi Y, Zhou J-M (2010) Identification of micrornas involved in pathogen-associated molecular pattern-triggered plant innate immunity. Plant Physiol 152(4):2222–2231. https://doi.org/10.1104/pp.109.151803

Lindow SE, Brandl MT (2003) Microbiology of the phyllosphere. Appl Environ Microbiol 69(4):1875–1883. https://doi.org/10.1128/AEM.69.4.1875-1883.2003

Link HF (1809) Observationes in ordines plantarum naturales: dissertation I.ma complectens Anandrarum ordines Epiphytas, Mucedines Gastromycos et Fungos. Aut Henr Frid Link, professore Rostochiensi. Der Gesellscchaft Naturforschender Freunde zu Berlin 3:3–42

Ma LJ, van der Does HC, Borkovich KA, Coleman JJ, Daboussi M-J, Di Pietro A et al (2010) Comparative genomics reveals mobile pathogenicity chromosomes in *Fusarium*. Nature 464(7287):367–373. https://doi.org/10.1038/nature08850

Manawasinghe IS, Phillips A, Xu J, Balasuriya A, Hyde K, Stepien L, Harishchandra D, Karunarathna A, Yan J, Weerasinghe J, Luo M, Dong Z, Cheewangkoon R (2021) Defining a species in fungal plant pathology: beyond the species level. Fungal Divers 109(10):267–282. https://doi.org/10.1007/s13225-021-00481-x

Margulis L, Fester R (eds) (1991) Symbiosis as a source of evolutionary innovation: speciation and morphogenesis. MIT Press, Cambridge, MA, p 470

Markham P, Collinge AJ (1987) Woronin bodies of filamentous fungi. FEMS Microbiol Rev 3(1):1–11. https://doi.org/10.1111/j.1574-6968.1987.tb02448.x

Martin FM, Uroz S, Barker DG (2017) Ancestral alliances: plant mutualistic symbioses with fungi and bacteria. Science 356(6340):eaad4501. https://doi.org/10.1126/science.aad4501

Matthioli PA (1560) Medici Senensis medici, Commentarii in sex Libros Pedacii Dioscoridis Anazarbei de Materia Medica, Adjectis quam plurimis plantarum & animalium imaginibus. In officina Valgrisiana, Venetiis

Mattoo AJ, Nonzom S (2021) Endophytic fungi: understanding complex cross-talks. Symbiosis 83(3):237–264. https://doi.org/10.1007/s13199-020-00744-2

McCarthy CGP, Fitzpatrick DA (2017) Phylogenomic reconstruction of the oomycete phylogeny derived from 37 genomes. mSphere 2:e00095–e00017. https://doi.org/10.1128/mSphere.00095-17

Mengistu AA (2020) Endophytes: colonization, behaviour, and their role in defense mechanism. Int J Microbiol 2020:6927219. https://doi.org/10.1155/2020/6927219

Micheli PA (1729) Nova plantarum genera: iuxta Tournefortii methodum disposita. Typis Bernardi Paperinii, Florentiae

Mishra S, Bhattacharjee A, Sharma S (2021) An ecological insight into the multifaceted world of plant-endophyte association. Crit Rev Plant Sci 40(2):127–146. https://doi.org/10.108 0/07352689.2021.1901044

Mohamed H, Naz T, Yang J, Shah AM, Nazir Y, Song Y (2021) Recent molecular tools for the genetic manipulation of highly industrially important Mucoromycota fungi. J Fungi 7(12):1061. https://doi.org/10.3390/jof7121061

Monod V (2024) Deciphering the multifaceted determinants of esca incidence across a vineyards network. PhD Thesis, University of Neuchâtel, Faculty of Science, Switzerland

Monod V, Zufferey V, Wilhelm M, Viret O, Gindro K, Croll D, Hofstetter V (2024) A systemic approach allows to identify the pedoclimatic conditions most critical in the susceptibility of a grapevine cultivar to esca/Botryosphaeria dieback. bioRxiv 2023.05.23.541976. https://doi.org/10.1101/2023.05.23.541976

Morella NM, Weng FC-H, Joubert PM, Metcalf CJE, Lindow S, Koskella B (2020) Successive passaging of a plant-associated microbiome reveals robust habitat and host genotype-dependent selection. PNAS 117(2):1148–1159. https://doi.org/10.1073/pnas.1908600116

Mostert L, Crous PW, Ewald Groenewald JZ, Gams W, Summerbell RC (2003) Togninia (Calosphaeriales) is confirmed as teleomorph of Phaeoacremonium by means of morphology, sexual compatibility and DNA phylogeny. Mycologia 95(4):646–659. https://doi.org/10.108 0/15572536.2004.11833069

Mousa WK, Raizada MN (2013) The diversity of anti-microbial secondary metabolites produced by fungal endophytes: an interdisciplinary perspective. Front Microbiol 4:65. https://doi.org/10.3389/fmicb.2013.00065

Mousa WK, Schwan A, Davidson J, Strange P, Liu H, Zhou T, Auzanneau FI, Raizada MN (2015) An endophytic fungus isolated from finger millet (*Eleusine coracana*) produces anti-fungal natural products. Front Microbiol 6:1157. https://doi.org/10.3389/fmicb.2015.01157

Müller MM, Valjakka R, Suokko A, Hantula J (2001) Diversity of endophytic fungi of single Norway spruce needles and their role as pioneer decomposers. Mol Ecol 10(7):1801–1810. https://doi.org/10.1046/j.1365-294X.2001.01304.x

Müller H, Berg C, Landa BB, Auerbach A, Moissl-Eichinger C, Berg G (2015) Plant genotype-specific archaeal and bacterial endophytes but similar *Bacillus* antagonists colonize Mediterranean olive trees. Front Microbiol 6:138. https://doi.org/10.3389/fmicb.2015.00138

Müller DB, Vogel C, Bai Y, Vorholt JA (2016) The plant microbiota: systems-level insights and perspectives. Annu Rev Genet 50(1):211–234. https://doi.org/10.1146/annurev-genet-120215-034952

Naranjo-Ortiz MA, Gabaldón T (2019) Fungal evolution: diversity, taxonomy and phylogeny of the fungi. Biol Rev Camb Philos Soc 94(6):2101–2137. https://doi.org/10.1111/brv.12550

Nazarov PA, Baleev DN, Ivanova MI, Sokolova LM, Karakozova MV (2020) Infectious plant diseases: etiology, current status, problems and prospects in plant protection. Acta Naturae 12(3):46–59. https://doi.org/10.32607/actanaturae

Newman MA, Sundelin T, Nielsen J, Erbs G (2013) MAMP (microbe-associated molecular pattern) triggered immunity in plants. Front Plant Sci 4:139. https://doi.org/10.3389/fpls.2013.00139

Nguyen TA, Cissé OH, Yun Wong J, Zheng P, Hewitt D, Nowrousian M, Stajich JE, Jedd G (2017) Innovation and constraint leading to complex multicellularity in the Ascomycota. Nat Commun 8:14444. https://doi.org/10.1038/ncomms14444

Ni M, Feretzaki M, Sun S, Wang X, Heitman J (2011) Sex in fungi. Annu Rev Genet 45:405–430. https://doi.org/10.1146/annurev-genet-110410-132536

Nogueira-Lopez G, Greenwood DR, Middleditch M, Winefield C, Eaton C, Steyaert JM, Mendoza-Mendoza A (2018) The apoplastic secretome of *Trichoderma virens* during interaction with maize roots shows an inhibition of plant defence and scavenging oxidative stress secreted proteins. Front Plant Sci 9:409. https://doi.org/10.3389/fpls.2018.00409

Oberwinkler F (2012) Evolutionary trends in Basidiomycota. Stapfia 96:45–104

Oerke EC (2006) Crop losses to pests. J Agric Sci 144(1):31–43. https://doi.org/10.1017/S0021859605005708

Oerke EC, Dehne HW (2004) Safeguarding production—losses in major crops and the role of crop protection. Crop Prot 23:275–285. https://doi.org/10.1016/j.cropro.2003.10.001

Pagano MC, Correa EJA, Duarte NF, Yelikbayev B, O'Donovan A, Gupta VK (2017) Advances in eco-efficient agriculture: the plant-soil mycobiome. Agriculture 7(2):14. https://doi.org/10.3390/agriculture7020014

Pélissier L (2022) Comprehensive study of the fungal endophyte community of the long-lived Amazonian palm *Astrocaryon sciophilum*: a model for deciphering plant-microbe interactions at chemical and bioactive levels. Doctorat ès Sciences en Sciences de la Vie des Facultés de Médecine et des Sciences, Mention Sciences Pharmaceutiques, thèse no. 150

Perez-Alonso M-M, Guerrero-Galan C, Scholz SS, Kiba T, Sakakibara H, Ludwig-Mueller J et al (2020) Harnessing symbiotic plant-fungus interactions to unleash hidden forces from extreme plant ecosystems. J Exp Bot 71(13):3865–3877. https://doi.org/10.1093/jxb/eraa040

Petrini O (1991) Fungal endophytes of tree leaves. In: Andrews JH, Hirano SS (eds) Microbial ecology of leaves, Brock/Springer series in contemporary bioscience, 1st edn. Springer, New York, pp 179–197. https://doi.org/10.1007/978-1-4612-3168-4_9

Phillips AJ, Alves A, Pennycook SR, Johnston PR, Ramaley A, Akulov A, Crous PW (2008) Resolving the phylogenetic and taxonomic status of dark-spored teleomorph genera in the Botryosphaeriaceae. Persoonia 21:29–55. https://doi.org/10.3767/003158508X340742

Pieterse CMJ, Van der Does D, Zamioudis C, Leon-Reyes A, Van Wees SCM (2012) Hormonal modulation of plant immunity. Annu Rev Cell Dev Biol 28(1):489–521. https://doi.org/10.1146/annurev-cellbio-092910-154055

Pillai TG (2017) Pathogen to endophytic transmission in fungi—a proteomics approach. SOJ Microbiol Infect Dis 5(3):1–5. https://doi.org/10.15226/sojmid/5/3/00173

Plett JM, Martin FM (2018) Know your enemy, embrace your friend: using omics to understand how plants respond differently to pathogenic and mutualistic microorganisms. Plant J 93(4):729–746. https://doi.org/10.1111/tpj.13802

Pontecorvo G (1956) The parasexual cycle in fungi. Ann Rev Microbiol 10:393–400. https://doi.org/10.1146/annurev.mi.10.100156.002141

Porras-Alfaro A, Bayman P (2011) Hidden fungi, emergent properties: endophytes and microbiomes. Annu Rev Phytopathol 49(1):291–315. https://doi.org/10.1146/annurev-phyto-080508-081831

Pozo MJ, Zabalgogeazcoa I, Vazquez de Aldana BR, Martinez-Medina A (2021) Untapping the potential of plant mycobiomes for applications in agriculture. Curr Opin Plant Biol 60:102034. https://doi.org/10.1016/j.pbi.2021.102034

Purvis A, Hector A (2000) Getting the measure of biodiversity. Nature 405(6783):212–219. https://doi.org/10.1038/35012221

Rajarammohan S (2021) Redefining plant-ecrotronph interactions: the thin line between hemibiotrophs and necrotrophs. Front Microbiol 12:673518. https://doi.org/10.3389/fmicb.2021.673518

Ray J (1688) Historia plantarum, vol 2. Clark, London

Rayner MA (1916) Recent developments in the study of endotrophic mycorrhiza. New Phytol 15(8):161–175

Rayner MA (1925a) The nutrition of mycorrhiza plants: *Calluna vulgaris*. J Exp Biol 2(2):265–292

Rayner MC (1925b) An endotrophic fungus in the Coniferæ. Nature 115:14–15. https://doi.org/10.1038/115014b0

Redman RS, Ranson JC, Rodriguez RJ (1999) Conversion of the pathogenic fungus *Colletotrichum magna* to a nonpathogenic, endophytic mutualist by gene disruption. MPMI 12(11):969–975. https://doi.org/10.1094/MPMI.1999.12.11.969

Redman RS, Dunigan DD, Rodriguez RJ (2001) Fungal symbiosis from mutualism to parasitism: who controls the outcome, host or invader? New Phytol 151(3):705–716. https://doi.org/10.1046/j.0028-646x.2001.00210.x

Richards TA, Leonard G, Wideman JG (2017) What defines the "Kingdom" fungi? Microbiol Spectr 5(3). https://doi.org/10.1128/microbiolspec.FUNK-0044-2017

Richardson AE, Simpson RJ (2011) Soil microorganisms mediating phosphorus availability update on microbial phosphorus. Plant Physiol 156(3):989–996. https://doi.org/10.1104/pp.111.175448

Rigotti S, Gindro K, Richter H, Viret O (2002) Characterization of molecular markers for specific and sensitive detection of *Botrytis cinerea* Pers.: Fr. in strawberry (*Fragaria×ananassa* Duch.) using PCR. FEMS Microbiol Lett 209(2):169–174. https://doi.org/10.1111/j.1574-6968.2002.tb11127.x

Rodriguez R, Redman R (2008) More than 400 million years of evolution and some plants still can't make it on their own: plant stress tolerance via fungal symbiosis. J Exp Bot 59(5):1109–1114. https://doi.org/10.1093/jxb/erm342

Rodriguez R, White J, Arnold A, Redman R (2009) Fungal endophytes: diversity and functional roles. New Phytol 182:314–330. https://doi.org/10.1111/j.1469-8137.2009.02773.x

Rohwer F, Seguritan V, Azam F, Knowlton N (2002) Diversity and distribution of coral-associated bacteria. Mar Ecol Prog Ser 243:1–10. https://doi.org/10.3354/meps243001

Romeralo C, Martín-García J, Martínez-Álvarez P, Muñoz-Adalia EJ, Gonçalves DR, Torres E, Witzell J, Diez JJ (2022) Pine species determine fungal microbiome composition in a common garden experiment. Fungal Ecol 56:101137. https://doi.org/10.1016/j.funeco.2021.101137

Rosenberg E, Zilber-Rosenberg I (2016) Microbes drive evolution of animals and plants: the Hologenome concept. mBio 7(2):e01395. https://doi.org/10.1128/mBio.01395-15

Rosenberg E, Koren O, Reshef L, Efrony R, Zilber-Rosenberg I (2007) The role of microorganisms in coral health, disease and evolution. Nat Rev Microbiol 5(5):355–362. https://doi.org/10.1038/nrmicro1635

Roy S, Banerjee D (2018) Diversity of endophytes in tropical forests. In: Pirttilä A, Frank A (eds) Endophytes of forest trees. Forestry sciences, vol 86. Springer, Cham. https://doi.org/10.1007/978-3-319-89833-9_3

RoyChowdhury M, Sternhagen J, Xin Y, Lou B, Li X, Li C (2022) Evolution of pathogenicity in obligate fungal pathogens and allied genera. PeerJ 10:e13794. https://doi.org/10.7717/peerj.13794

Ryan RP, Germaine K, Franks A, Ryan DJ, Dowling DN (2008) Bacterial endophytes: recent developments and applications. FEMS Microbiol Lett 278(1):1–9. https://doi.org/10.1111/j.1574-6968.2007.00918.x

Saikkonen K, Hel M, Faeth SH (2004) Evolution of endophyte—plant symbioses. Trends Plant Sci 9:275–280

Saikkonen K, Lehtonen P, Helander M, Koricheva J, Faeth SH (2006) Model systems in ecology: dissecting the endophyte–grass literature. Trends Plant Sci 11(9):428–433. https://doi.org/10.1016/j.tplants.2006.07.001

Saikkonen K, Wäli PR, Helander M (2010) Genetic compatibility determines endophyte-grass combinations. PLoS One 5(6):e11395. https://doi.org/10.1371/journal.pone.0011395

Sampson K (1935) The presence and absence of an endophytic fungus in Lolium temulentum and L. perenne. Trans Br Mycol Soc 19(4):337–343. https://doi.org/10.1016/S0007-1536(35)80031-4

Schoch CL, Seifert KA, Huhndorf S, Robert V, Spouge JL, Levesque CA, Chen W (2012) Nuclear ribosomal internal transcribed spacer (ITS) region as a universal DNA barcode marker for fungi. Proc Natl Acad Sci USA 109(16):6241–6246. https://doi.org/10.1073/pnas.1117018109

Schulz B, Boyle C (2005) The endophytic continuum. Mycol Res 109(6):661–686. https://doi.org/10.1017/S095375620500273X

Schulz B, Boyle C (2006) What are endophytes? In: Schulz BJE, Boyle CJC, Sieber T (eds) Microbial root endophytes, Soil biology, vol 9. Springer, Berlin/Heidelberg, pp 1–13. https://doi.org/10.1007/3-540-33526-9_1

Schulz B, Boyle C, Draeger S, Römmert A-K, Krohn K (2002) Endophytic fungi: a source of novel biologically active secondary metabolites. Mycol Res 106(9):996–1004. https://doi.org/10.1017/S0953756202006342

Schulz B, Haas S, Junker C, Andrée N, Schobert M (2015) Fungal endophytes are involved in multiple balanced antagonisms. Curr Sci 109(1):7

Schürch S, Gindro K, Schnee S, Dubuis PH, Codina JM, Wilhelm M, Riat A, Lamoth F, Sanglard D (2023) Occurrence of Aspergillus fumigatus azole resistance in soils from Switzerland. Med Mycol 61(11):myad110. https://doi.org/10.1093/mmy/myad110

Selosse MA (2017) Jamais seul. Actes Sud, Paris, p 370

Senanayake IC, Pem D, Rathnayaka AR, Wijesinghe SN, Tibpromma S, Wanasinghe DN, Phookamsak R et al (2022) Predicting global numbers of Teleomorphic ascomycetes. Fungal Divers 114(19):237–278. https://doi.org/10.1007/S13225-022-00498-W

Senanayake IC, Crous PW, Groenewald JZ, Maharachchikumbura SSN, Jeewon R, Phillips AJL, Bhat JD, Perera RH, Li QR, Li WJ, Tangthirasunun N, Norphanphoun C, Karunarathna SC, Camporesi E, Manawasighe IS, Al-Sadi AM, Hyde KD (2017) Families of Diaporthales based on morphological and phylogenetic evidence. Stud Mycol 86:217–296. https://doi.org/10.1016/j.simyco.2017.07.003

Sessa L, Abreo E, Lupo S (2018) Diversity of fungal latent pathogens and true endophytes associated with fruit trees in Uruguay. J Phytopathol 166(9):633–647. https://doi.org/10.1111/jph.12726

Shen XX, Opulente DA, Kominek J, Zhou X, Steenwyk JL, Buh KV, Haase MAB, Wisecaver JH, Wang M, Doering DT, Boudouris JT, Schneider RM, Langdon QK, Ohkuma M, Endoh R, Takashima M, Manabe R, Čadež N, Libkind D, Rosa CA, DeVirgilio J, Hulfachor AB,

Groenewald M, Kurtzman CP, Hittinger CT, Rokas A (2018) Tempo and mode of genome evolution in the Budding Yeast Subphylum. Cell 175(6):1533–1545.e20

Shi J, Wang X, Wang E (2023) Mycorrhizal symbiosis in plant growth and stress adaptation: from genes to ecosystems. Annu Rev Plant Biol 74(1):569–607. https://doi.org/10.1146/annurev-arplant-061722-090342

Shivas R, Cai L (2012) Cryptic fungal species unmasked. Microbiol Aust 33:35–36

Shrivastava G, Ownley BH, Augé RM, Toler H, Dee M, Vu A, Köllner TG, Chen F (2015) Colonization by arbuscular mycorrhizal and endophytic fungi enhanced terpene production in tomato plants and their defense against a herbivorous insect. Symbiosis 65(2):65–74. https://doi.org/10.1007/s13199-015-0319-1

Sieber TN (2007) Endophytic fungi in forest trees: are they mutualists? Fungal Biol Rev 21(2–3):75–89. https://doi.org/10.1016/j.fbr.2007.05.004

Slippers B, Wingfield MJ (2007) Botryosphaeriaceae as endophytes and latent pathogens of woody plants: diversity, ecology and impact. Fungal Biol Rev 21(2–3):90–106. https://doi.org/10.1016/j.fbr.2007.06.002

Stone J (2011) Initiation and development of latent infections by *Rhabdocline parkeri* on Douglas-fir. Can J Bot 65:2614–2621. https://doi.org/10.1139/b87-352

Stone JK, Bacon CW, White JF (2000) An overview of endophytic microbes: endophytism defined. In: Bacon CW, White JF (eds) Microbial endophytes. CRC, Boca Raton, FL, pp 3–29. https://doi.org/10.1201/9781482277302

Sun X, Guo LD (2012) Endophytic fungal diversity: review of traditional and molecular techniques. Mycology 3(1):65–76. https://doi.org/10.1080/21501203.2012.656724

Sun S, Heitman J (2011) Is sex necessary? BMC Biol 9:56. https://doi.org/10.1186/1741-7007-9-56

Sun C, Johnson JM, Cai D, Sherameti I, Oelmüller R, Lou B (2010) *Piriformospora indica* confers drought tolerance in Chinese cabbage leaves by stimulating antioxidant enzymes, the expression of drought-related genes and the plastid-localized CAS protein. J Plant Physiol 167(12):1009–1017. https://doi.org/10.1016/j.jplph.2010.02.013

Swamy N, Sandhu SS (2021) Fungal endophytes: entry, establishment, diversity, future prospects in agriculture. In: Sharma VK, Shah MP, Parmar S, Kumar A (eds) Fungi bio-prospects in sustainable agriculture, environment and nano-technology. Academic, Cambridge, MA. https://doi.org/10.1016/B978-0-12-821394-0.00004-4

Szink I, Davis EL, Ricks KD, Koide RT (2016) New evidence for broad trophic status of leaf endophytic fungi of *Quercus gambelii*. Fungal Ecol 22:2–9

Teplitski M, Mathesius U, Rumbaugh KP (2011) Perception and degradation of n-acyl homoserine lactone quorum sensing signals by mammalian and plant cells. Chem Rev 111(1):100–116. https://doi.org/10.1021/cr100045m

Terral JF, Tabard E, Bouby L, Ivorra S, Pastor T, Figueiral I, Picq S, Chevance JB, Jung C, Fabre L, Tardy C, Compan M, Bacilieri R, Lacombe T, This P (2010) Evolution and history of grapevine (Vitis vinifera) under domestication: new morphometric perspectives to understand seed domestication syndrome and reveal origins of ancient European cultivars. Ann Bot 105(3):443–455

Toti L, Viret O, Chapela IH, Petrini O (1992) Differential attachment by the conidia of the endophyte *Discula umbrinella* (Berk. & Br.) Morelet to host and non-host surfaces. New Phytol 121:469–475. https://doi.org/10.1111/j.1469-8137.1992.tb02947.x

Trivedi P, Leach JE, Tringe SG, Sa T, Singh BK (2020) Plant–microbiome interactions: from community assembly to plant health. Nat Rev Microbiol 18(11):607–621. http://www.nature.com/articles/s41579-020-0412-1

Trouvelot S, Bonneau L, Redecker D, van Tuinen D, Adrian M, Wipf D (2015) Arbuscular mycorrhiza symbiosis in viticulture: a review. Agron Sustain Dev 35(4):1449–1467. https://doi.org/10.1007/s13593-015-0329-7

Turland NJ, Wiersema JH, Barrie FR, Greuter W, Hawksworth DL, Herendeen PS, Knapp S, Kusber WH, Li DZ, Marhold K, May TW, McNeill J, Monro AM, Prado J, Price MJ, Smith GF (2018) International Code of Nomenclature for algae, fungi, and plants (Shenzhen Code) adopted by the Nineteenth International Botanical Congress Shenzhen, China, July 2017. Regnum Vegetabile 159. Koeltz Botanical Books, Glashütten. https://doi.org/10.12705/Code.2018

Turner TR, James EK, Poole PS (2013) The plant microbiome. Genome Biol 14(6):209. https://doi.org/10.1186/gb-2013-14-6-209

Uroz S, Buée M, Deveau A, Mieszkin S, Martin F (2016) Ecology of the forest microbiome: highlights of temperate and boreal ecosystems. Soil Biol Biochem 103:471–488. https://doi.org/10.1016/j.soilbio.2016.09.006

Van Kan JA, Shaw MW, Grant-Downton RT (2014) *Botrytis* species: relentless necrotrophic thugs or endophytes gone rogue? Mol Plant Pathol 15(9):957–961. https://doi.org/10.1111/mpp.12148

van Sterbeeck F (1675) Theatrum fungorum oft het Tooneel der Campernoelien. T'Antwerpen By I. Iacobs

Vandenkoornhuyse P, Quaiser A, Duhamel M, Van AL, Dufresne A (2015) The importance of the microbiome of the plant holobiont. New Phytol 206(4):1196–1206. https://doi.org/10.1111/nph.13312

Vega FE, Blackwell M (2005) Insect-fungal associations: ecology and evolution. Oxford University Press, Oxford, p 352

Verma SK, Gond SK, Mishra A, Sharma VK, Kumar J, Singh DK, Kumar A, Kharwar RN (2017) Fungal endophytes representing diverse habitats and their role in plant protection. In: Satyanarayana T, Deshmukh SK, Johri BN (eds) Developments in fungal biology and applied mycology. Springer, Singapore. https://doi.org/10.1007/978-981-10-4768-8_9

Viret O, Petrini O (1994) Colonisation of beech leaves (*Fagus sylvatica*) by the endophyte *Discula umbrinella* (teleomorph: *Apiognomonia errabunda*). Mycol Res 98(4):423–432

Viret O, Toti L, Chapela IH, Petrini O (1994) The role of the extracellular sheath in recognition and attachment of conidia of *Discula umbrinella* (Berk. & Br.) Morelet to the host surface. New Phytol 127:123–131. https://doi.org/10.1111/j.1469-8137.1994.tb04266.x

Von Linné (1753) Species plantarum: exhibentes plantas rite cognitas ad genera relatas. Holmiae Impensis Laurentii Salvii

Von Kaub J (1491) Hortus Sanitatis : de Herbis et Plantis – de Animalibus et Reptilibus – de Avibus et volatilibus- de Piscibus et natitilibus- de Lapidibus et in Terre Venis Nascentibus – de Urinis et Earum Speciebus. Frankfurt, ed. Jacob Meydenbach, Mainz, Germany

Von Linné (1735) Systema naturae. Ex Typographia, de Groot JW

Voříšková J, Baldrian P (2013) Fungal community on decomposing leaf litter undergoes rapid successional changes. ISME J 7(3):477–486. https://doi.org/10.1038/ismej.2012.116

Walker C, Schüßler A, Vincent B, Cranenbrouck S, Declerck S (2021) Anchoring the species *Rhizophagus intraradices* (formerly *Glomus intraradices*). Fungal Syst Evol 8:179–201. https://doi.org/10.3114/fuse.2021.08.14

Wang Y, Guo L (2007) A comparative study of endophytic fungi in needles, bark, and xylem of *Pinus tabulaeformis*. Can J Bot 85(10):911–917. https://doi.org/10.1139/B07-084

Wawra S, Fesel P, Widmer H, Neumann U, Lahrmann U, Becker S, Hehemann JH, Langen G, Zuccaro A (2019) FGB1 and WSC3 are in planta-induced β-glucan-binding fungal lectins with different functions. New Phytol 222(3):1493–1506. https://doi.org/10.1111/nph.15711

Weller DM, Raaijmakers JM, Gardener BBM, Thomashow LS (2002) Microbial populations responsible for specific soil suppressiveness to plant pathogens. Annu Rev Phytopathol 40:309–348. https://doi.org/10.1146/annurev.phyto.40.030402.110010

Whetzel HH (1945) A synopsis of the genera and species of the Sclerotiniaceae, a family of stromatic inoperculate Disomycetes. Mycologia 37:648-714. https://doi.org/10.1080/00275514.1945.12024025

Whittaker RH (1959) On the broad classification of organisms. Q Rev Biol 34(3):210–226

Willdenow KL (1810) Anleitung zum Selbstudium der Botanik. Ein Handbuch zu öffentlichen Vorlesungen. Zweite Aufl, Berlin

Wilson D (1995) Endophyte: the evolution of a term, and clarification of its use and definition. Oikos 73(2):274–276. https://doi.org/10.2307/3545919

Wilson D, Carroll GC (1994) Infection studies of *Discula quercina*, an endophyte of *Quercus garryana*. Mycologia 86(5):635–647. https://doi.org/10.1080/00275514.1994.12026463

Woloshen V, Huang S, Li X (2011) RNA-binding proteins in plant immunity. J Pathog 2011:e278697. https://doi.org/10.4061/2011/278697

Woronin M (1864) Zur Entwicklungsgeschichte des *Ascobolus pulcherrimus* Cr. und einiger Pezizen. Abh Senkenb Naturforsch 5:333–344

Wu B, Hussain M, Zhang W, Stadler M, Liu X, Xiang M (2019) Current insights into fungal species diversity and perspective on naming the environmental DNA sequences of fungi. Mycology 10(3):127–140. https://doi.org/10.1080/21501203.2019.1614106

Yan L, Zhu J, Zhao X, Shi J, Jiang C, Shao D (2019) Beneficial effects of endophytic fungi colonization on plants. Appl Microbiol Biotechnol 103(8):3327–3340. https://doi.org/10.1007/s00253-019-09713-2

Yao H, Sun X, He C, Maitra P, Li X-C, Guo L-D (2019) Phyllosphere epiphytic and endophytic fungal community and network structures differ in a tropical mangrove ecosystem. Microbiome 7(1):57. https://doi.org/10.1186/s40168-019-0671-0

Zamioudis C, Pieterse CMJ (2012) Modulation of host immunity by beneficial microbes. MPMI 25(2):139–150. https://doi.org/10.1094/MPMI-06-11-0179

Zarraonaindia I, Owens SM, Weisenhorn P, West K, Hampton-Marcell J, Lax S, Bokulich NA, Mills DA, Martin G, Taghavi S, van der Lelie D, Gilbert JA (2015) The soil microbiome influences grapevine-associated microbiota. mBio 6(2):e02527–e02514. https://doi.org/10.1128/mBio.02527-14

Zhao J, Wang M, Chen X, Kang Z (2016) Role of alternate hosts in epidemiology and pathogen variation of cereal rusts. Annu Rev Phytopathol 54:207–228. https://doi.org/10.1146/annurev-phyto-080615-095851

Zilber-Rosenberg I, Rosenberg E (2008) Role of microorganisms in the evolution of animals and plants: the hologenome theory of evolution. FEMS Microbiol Rev 32(5):723–735. https://doi.org/10.1111/j.1574-6976.2008.00123.x

Zipfel C, Oldroyd GED (2017) Plant signalling in symbiosis and immunity. Nature 543(7645):328–336. https://doi.org/10.1038/nature22009

Chapter 4
Fungal Diseases of Green Organs

4.1 Downy Mildew

Current name:

- *Plasmopara viticola* (Berk. & M.A. Curtis) Berl. & De Toni 1888, Oomycota

The most important disease for winegrowers in economic terms, grapevine downy mildew is present in all the world's vineyards, although climates with dry springs (Mediterranean, California, western Australia, northern Chile) are less conducive to the disease. In Europe, it was imported into the south of France in 1878 from the United States on phylloxera-resistant American rootstocks used to reestablish European vineyards. The disease was identified in France by Jules Emile Planchon in 1868. In 1878 Planchon noted that grapevine downy mildew, the causal agent of which had first been identified in the United States by Farlow (1876) under the name of *Peronospora viticola*, was now widespread in France. The latter had written (Farlow 1882): "I made the statement that practically no harm was done to the grape crop in our Northern State by this fungus. Should the fungus be introduced into central Europe, the case might be different." In less than 5 years, the disease had spread to all winegrowing regions of Europe, notably the French, Italian, German and Swiss vineyards (Koledenkova et al. 2022). By the end of the nineteenth century, downy mildew, along with powdery mildew and phylloxera, was the cause of major declines in vineyard acreage. Recent population genetics analysis has revealed that downy mildew populations exhibit widespread traces of recombination, indicating frequent episodes of sexual reproduction. Invasive European populations show low genetic diversity and a weak population structure, albeit significant at continental scale. All invasive populations in Europe appear to originate from just one of the five native North American lineages; the one parasitising *Vitis aestivalis* (Fontaine et al. 2021). By accumulating phylogenetic data and studying populations on more than 2000 downy mildew samples collected from all winegrowing regions of the world, the same authors proposed an invasion scenario for *P. viticola;* after an initial introduction into Europe, invasive European populations served as the secondary source

O. Viret, K. Gindro, *Science of Fungi in Grapevine*,
https://doi.org/10.1007/978-3-031-68663-4_4

of introduction of the disease into vineyards throughout the world, including China, South Africa and, in two separate events, Australia. The invasion of Argentina is probably the only tertiary introduction, in this case from Australia. Downy mildew can cause serious economic damage—both quantitatively and qualitatively—in regions with a temperate, humid climate. Infected bunches can substantially reduce yields and lead to early defoliation, which alters the ripening process. All European varieties of grapevine belonging to the species *Vitis vinifera* are susceptible to downy mildew, hence the development of breeding programmes to select resistant varieties by backcrossing *V. vinifera* with highly resistant American and/or Asiatic vines.

4.1.1 Causal Organism

Plasmopara viticola is an obligate biotrophic microorganism belonging to the class Oomycetes, the order Peronosporales and the family Peronosporaceae, which is linked to the family Vitaceae. *P. viticola* was first described by Lewis David von Schweinitz (1834), who classified it as *Botrytis cana*. It was then reclassified *Botrytis viticola* by Berkeley and Curtis in 1848 before being renamed *Peronospora viticola* by Heinrich Anton de Bary (1863). In 1886, Joseph Schröter described the genus *Plasmopara* and modified the name *Peronospora viticola* to *Plasmopara viticola*. Currently, *P. viticola* is classed in the Stramenopiles (or heterokonts), a subclass of the chromalveolata, characterised by having, at some stage in their life cycles, a biflagellate cell bearing two flagella of different length and structure. All species of this genus are plant pathogens and most are obligate biotrophes. The asexual phase of *P. viticola* is characterised by the formation of arbuscular sporangiophores, simple or fasciculate, emerging from the stomata of the host plant (Fig. 4.1).

The sporangiophores branch monopodially, more or less at right angles, and produce elliptical sporangia (14–25 × 10–15 μm), hyaline to grey, each of which can release 1–10 motile, biflagellate zoospores (Fig. 4.2) (6–8 × 4–5 μm). In rare cases, the sporangia can germinate directly. The sexual phase of *P. viticola* starts after the fertilisation of an oogonium by an antheridium. The oospores (or winter eggs, 20–120 μm in diameter) are globular and hyaline, with a ridged, sometimes reticulate external wall. The mycelium is branched. The oospores germinate in water, producing a hypha at the end of which is a primary, piriform sporangium of varying dimensions (28–50 μm) capable of producing between 30 and 60 infectious zoospores.

Population Genetics of Plasmopara viticola
(Rouxel et al. 2013, 2014; Fontaine et al. 2021)
   Five cryptic species (also known as *special forms*) have been identified in the *P. viticola* species complex with genetic differentiation and a range of hosts encompassing various species of *Vitis* (Rouxel et al. 2013, 2014; Fontaine et al. 2021). The five haplotypes of special forms of *P. viticola* (f.

sp.) are present in wild *Vitis* species throughout North America: namely *P. viticola* f.sp. *vulpina*, *P. viticola* f. sp. *aestivalis*, *P. viticola* f. sp. *quinquefolia*, *P. viticola* f. sp. *vinifera* and *P. viticola* f. sp. *riparia*. Fontaine et al. (2021) analysed a dataset of 2000 downy mildew samples collected from the most important winegrowing regions of the world using nuclear and mitochondrial gene sequences and microsatellite markers. Population genetic analysis revealed very low genetic diversity in invasive downy mildew populations worldwide and little evidence of admixture. All invasive populations appear to originate from just one of the five native North American lineages; the one parasitising *Vitis aestivalis*.

The authors confirmed that *P. viticola* f. sp. *aestivalis* was first introduced into Western Europe from where it spread to Central and Eastern Europe. Invasive populations in Europe then served as the source for secondary introductions into other winegrowing regions of the world, such as Northeast China, South Africa and Australia.

Fig. 4.1 (**a**) Sporangiophores of *Plasmopara viticola* on the abaxial surface of a vine leaf. (**b**) Detail of sporangiophores and sporangia emerging from the stomata

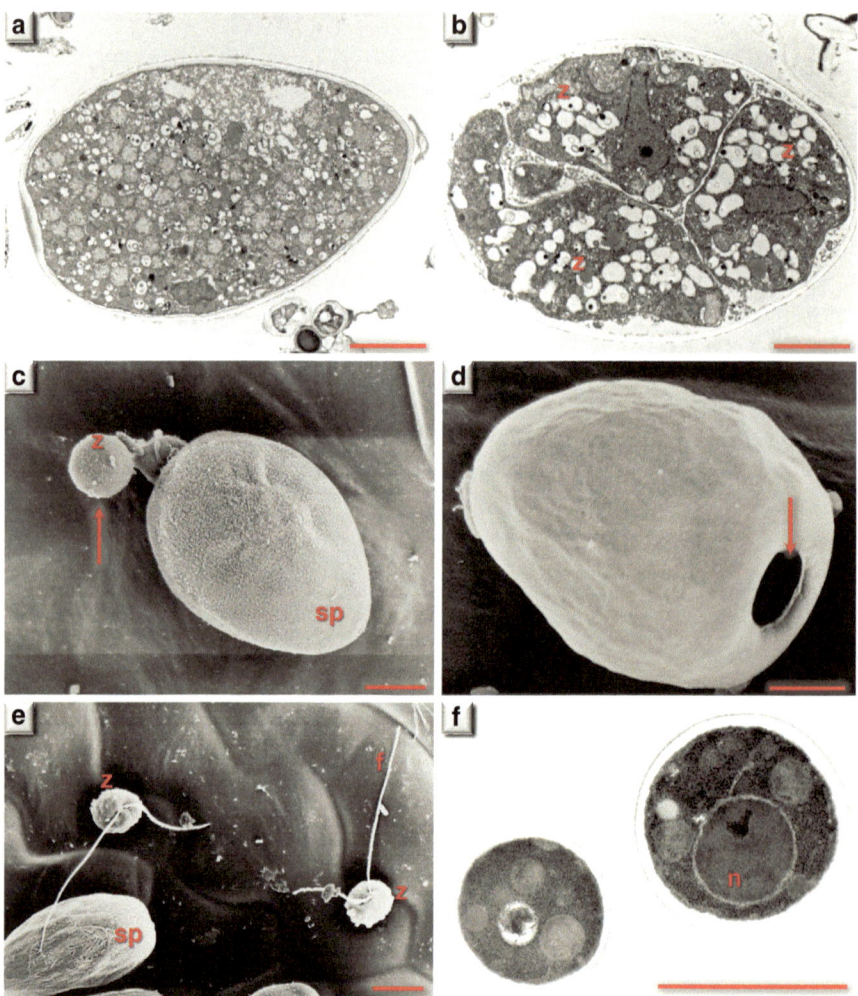

Fig. 4.2 Different stages of maturity of *Plasmopara viticola* sporangia. (**a**) Thin section of an undifferentiated sporangium viewed under transmission electron microscope (TEM). (**b**) Thin section of a sporangium with three differentiated zoospores. (**c**) Sporangium releasing a zoospore (viewed under scanning electron microscope, SEM). (**d**) Empty sporangium after release of zoospores via the apical 'lid' or operculum (arrow), viewed under SEM. (**e**) Biflagellate zoospores viewed under SEM. (**f**) Transverse section of zoospores observed under TEM. Scale bar represents 5 µm. *f* flagellum, *n* nucleus, *sp* sporangium, *z* zoospore

4.1.2   Symptoms

All green organs of the vine can be infected by the pathogen. The first leaf symptoms are characterised by circular, semi-translucent yellow spots, known as 'oil spots' due to the olive-coloured halo surrounding them (Fig. 4.3), which eventually become necrotic. In hot, humid weather, the underside of these spots turns white due to asexual sporulation involving numerous sporangiophores bearing downy mildew sporangia (Fig. 4.3). Severely infected leaves turn brown and fall prematurely. Inflorescences are particularly susceptible. When colonised before or during flowering, they turn yellow and curl into a 'shepherd's crook' (Fig. 4.4) before turning brown and withering partially or entirely. Colonised shoots develop superficial, longitudinal, purplish-brown lesions, bend downward, and may have sporulations of varying abundance (Fig. 4.5).

Sporulations on the surface of young berries at fruit set stem from infected inflorescences (Fig. 4.6). When young berries are covered by sporangia, the bunches are infected with grey rot (Fig. 4.7). As the disease progresses, the bunch peduncles develop extensive brown patches and berries present with brown rot: turning purple, then brown and withering (Fig. 4.8). At the end of the season, when conditions are favourable for mildew, later attacks of the foliage appear as a tapestry-like (mosaic) pattern of spots of varying severity (Fig. 4.9). Young leaves from lateral shoots are particularly susceptible to downy mildew, which can sporulate abundantly in this tissue (Fig. 4.10).

Fig. 4.3 Leaf symptoms of downy mildew. (**a**) 'Oil spot': yellow blotch on upper surface of the leaf. (**b**) Sporulation on the underside of the same spot. (**c**). Extensive infection of foliage

Fig. 4.4 Inflorescence infected by *Plasmopara viticola* showing peduncle bent like a shepherd's crook, which eventually withers

Fig. 4.5 Shoot infected by *Plasmopara viticola*, bent, purplish-brown and showing abundant sporulation (arrows)

4.1.3   Biology and Epidemiology

4.1.3.1   Disease Cycle

The epidemiological development of downy mildew is summarised in Fig. 4.11. Like many other Oomycetes, downy mildew generally develops both asexually and sexually. During the asexual phase, the formation of multinucleate sporangia (sporangiogenesis) is followed by the individualisation of biflagellate zoospores

Fig. 4.6 Downy mildew sporulation on the berries after fruit set (**a**) and detail showing major sporulation on the surface of young berries (**b**)

Fig. 4.7 Grey rot: abundant sporulation of downy mildew on the surface of berries, which gradually shrivel

(zoosporogenesis), which are the propagules of infection. The zoospores released by the sporangia (Fig. 4.11 no. 1) via an apical bud swim towards a stomate (Fig. 4.11 no. 2), encyste (flagella detach) (Fig. 4.11 no. 3) and extend a germ tube into the substomatal cavity (Fig. 4.11 no. 4). The intercellular mycelium develops in the parenchyma and produces several intracellular suckers called haustoria (Fig. 4.11 no. 5). The mycelium gradually colonises the green tissue and on reaching a substomatal cavity, produces a thick cushion from which a sporangiophore emerges and forms sporangia (sporulation) (Fig. 4.11 no. 6). Sporulation takes place in the dark and requires a minimum of 4 h with minimum 98% relative humidity and a temperature above 19 °C (Bläser and Weltzien 1979). The period of parenchymatous growth

Fig. 4.8 Symptoms of brown rot. (**a**) Berries turn purplish, then brown before shrivelling. (**b**) Advanced symptoms on a white grape variety showing shrivelled berries. (**c**) Advanced symptoms on a red grape variety showing shrivelled berries

Fig. 4.9 Mosaic spots caused by late infections of foliage with downy mildew. Spots on the upper side (**a**) are covered in downy mildew sporangia on the underside (**b**)

varies and may be asymptomatic until an oil spot appears on the adaxial surface of the leaf. During the sexual phase, compatible heterothallic individuals (mating-types P1 and P2 for *P. viticola*) (Wong et al. 2001) recognise one another through hormones, as seen in certain oomycetes (Lee et al. 2012; Koledenkova et al. 2022). Gametangia then develop in which meiosis occurs. Fertilisation of the oogonium by an antheridium leads to the formation of a diploid oospore, the wall of which thickens as the oospore reaches maturity (Fig. 4.11 no. 7). The oospore then enters a period of winter dormancy (Fig. 4.11 no. 8). These oospores (Fig. 4.12) develop in late summer and autumn in infected leaves which then fall to the ground. They can survive in the soil for several years (Fig. 4.11 no. 8)—constituting a source of

Fig. 4.10 Abundant downy mildew sporulation on the underside of a vine leaf

infection that is invariably sufficient to trigger an epidemic when climatic conditions favour the pathogen. Their germination depends on several climatic factors such as light, temperature and humidity.

Disease pressure is determined not by the previous year's epidemic, but the prevailing conditions in the current year. In spring, prediction of oospore maturity in dead leaves is based on the sum of temperatures above the threshold of 8 °C (Gehmann 1987). This method involves adding up the positive values of mean daily temperatures above 8 °C since the first of January (in the Northern Hemisphere). When this sum exceeds 140 °C, the oospores germinate and produce macrosporangia (Fig. 4.11 no. 9) (primary sporangia) (Fig. 4.13). The macrosporangia are deposited on the organs of the vine in water droplets splashed up from the ground by wind and heavy rainfall (Fig. 4.11 no. 10). After zoosporogenesis, the infectious zoospores are released and 'swim' through the water to the stomata of the vine leaves. Infections caused by oospores and primary sporangia on the ground can occur

Fig. 4.11 Disease cycle of *Plasmopara viticola*. Carried by wind or rain, the sporangia release biflagellate zoospores (1), which swim toward a stomate on the underside of the leaf, or any other available green organ (2), lose their flagella (3) and encyst before extending a germ tube into the substomatal cavity (4). Hyphae colonise the parenchyma intercellularly and produce intracellular suckers (haustoria) (5). Depending on environmental conditions, sporangiophores bearing numerous sporangia emerge from the stomata (6), and release zoospores which then reinfect the green vine tissue (11). Oospores start to develop at the end of summer after sexual reproduction (7). They overwinter in plant debris (8) and germinate to produce macrosporangia when spring arrives (9). In spring the primary sporangia release zoospores which are deposited on the green organs of the vine by rain splash (10). (Illustration © Virginie Duquette, Gravir un Monde d'Illustration, Switzerland)

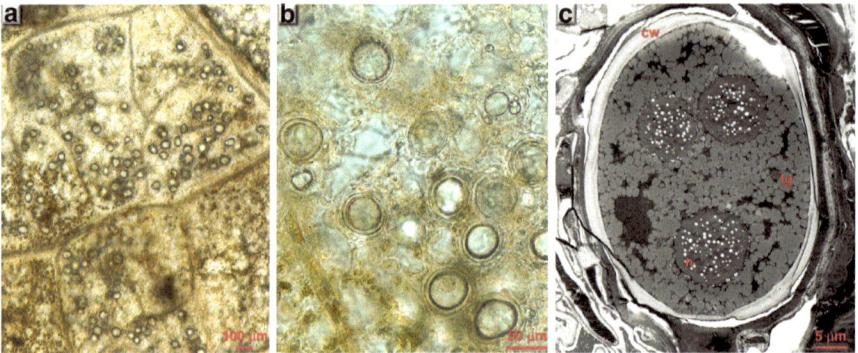

Fig. 4.12 (**a** and **b**) Oospores of *Plasmopara viticola* observed transparently through a fragment of dead leaf. (**c**) Transverse section of an oospore viewed under a transmission electron microscope, comprising a thick cell wall (CW), several lipid globules (lg) and the nucleus (n)

Fig. 4.13 Primary
sporangia resulting from
oospore germination in
spring

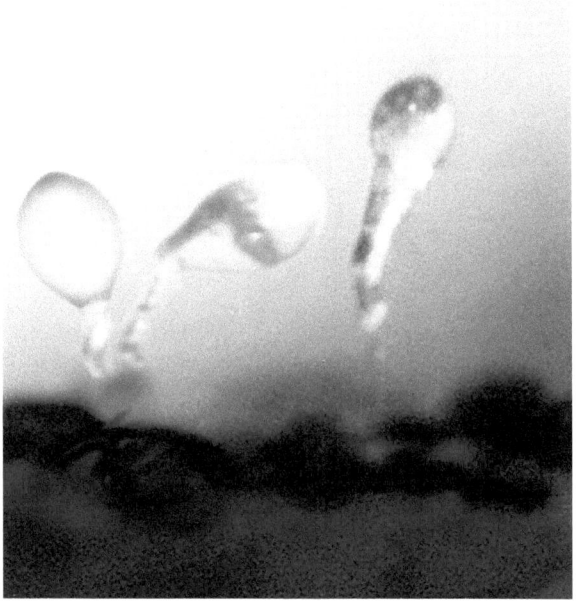

throughout the growing season and combine with infections transmitted between
aerial organs of the vine following the cycle of transmission and downy mildew
sporulation during the asexual phase. When conditions are particularly favourable
for downy mildew, the combination of these two types of infection is instrumental
in explaining the very rapid exponential development of the disease. The yearly
number of infectious cycles varies depending on the weather conditions. In years
when disease pressure is high, dozens of infectious cycles may occur throughout the
growing season of the vine. However, downy mildew is most virulent and/or the
plant most receptive during the exponential growth phase around the time of flowering (May to June in the Northern Hemisphere, November to December in the
Southern Hemisphere).

Having entered the organ through the stomata, the fungus gradually colonises the
infected plant tissue. Incubation corresponds to the time elapsing between the start of
infection and the emergence of new sporangia via stomata on the underside of the
leaves. According to the curve of Müller et al. (1923), incubation lasts between 4 and
12 days based solely on temperature. At the end of this period, sporangia appear on
the underside of the leaves if the following conditions of sporulation are met: dark-
ness, leaves damp or relative air humidity above 92%, temperature two metres above
the ground at least 12 °C at the start of the period of leaf wetness and maintained for
at least 4 h. Sporangia produced on the underside of the leaf, or on other organs, are
dispersed by hydrochory (rain) and anemochory (wind). When a sporangium or zoospores arrive on a healthy, damp leaf, the zoospores encyst and produce a germ tube
which penetrates the tissues via the stomata (Fig. 4.14). The factors determining the
development of infection are duration of leaf wetness and temperature. According to

Fig. 4.14 Zoospores of
Plasmopara viticola. (**a**)
Biflagellate zoospore
beside a stomate. (**b**)
Encysted zoospore (flagella
detached) produces a germ
tube (arrow) which
penetrates the opening of a
stomate (st) as far as the
substomatal cavity. Scale
bar represents 10 µm

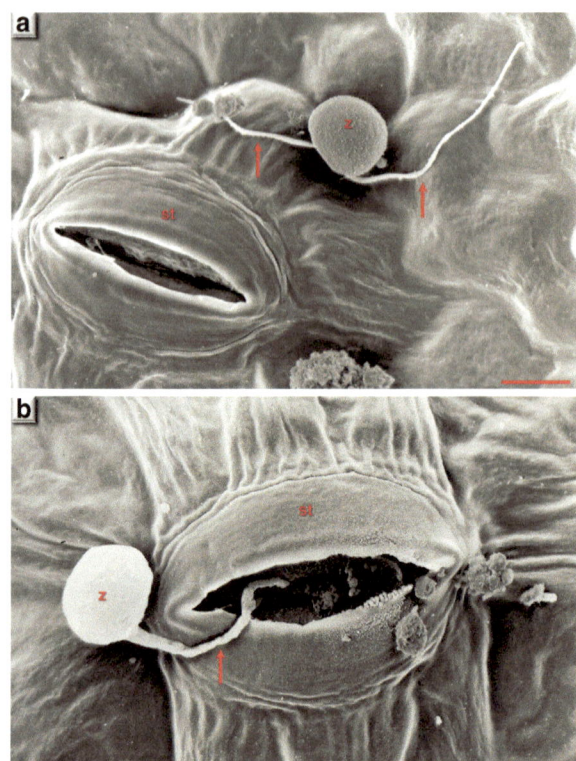

Bläser (1978), infections can occur only when the product of the mean daily temperature multiplied by the duration of leaf wetness reaches a value of 50. This means that
at 10 °C, leaves must be damp for at least 5 h for the fungus to penetrate the tissue via
the stomata. If the leaf dries before this time, the zoospores die. In the climate conditions of northern vineyards (frequent rain, dew), duration of leaf wetness is rarely a
limiting factor for infection. In high summer temperatures (T > 25 °C), the presence
of downy mildew on leaves may permit sporulation without rainfall, but only on the
basis of transpiration on the underside of the leaf.

According to Bläser (1978), the sporangia remain infectious for 8 days at 15 °C
and 70–90% relative humidity. At 30 °C and low relative humidity, the sporangia die
within a few days. Inside the tissues however, the fungus remains active and can
produce new sporangia on the same lesion when conditions become favourable
again. The work of Hill (1989) shows that oil spots produce a maximum number of
sporangia between 15 and 20 °C, indicating that the risk of infection is greatest at
these temperatures when oil spots are present. Sporangia preserved at −80 °C
remain viable for several years, like the conidia of *Botrytis cinerea* Pers. (Gindro
and Pezet 2001; Gindro et al. 2003). This observation suggests that airborne sporangia could reach the upper layers of the atmosphere on rising air currents, where they
could freeze and be transported over long distances.

4.1.3.2 Infection of Bunches and Systemic Development

Pre-bloom Infection (BBCH 55)

Infection of inflorescences and bunches by *Plasmopara viticola* can occur from the pre-bloom stage (BBCH 55: flower buds tightly clustered) until the onset of veraison (ripening of the grapes). The severity of symptoms diminishes with the phenological development of the bunches. When infections occur at the BBCH 55 stage, the entire inflorescence can be completely colonised. Symptoms of this type of infection cause the tip of the inflorescence to become hook-shaped, evolving into the characteristic shepherd's crook appearance (Fig. 4.4) and eventually leading to complete desiccation of the bunch (Fig. 4.15).

At this stage, *P. viticola* infects via the stomata, which are abundant on the calyptra, receptacle, pedicel and rachis (Bessis 1972; Bernard 1977; Nakagawa et al. 1980). With the exception of the pollen sac, all flower parts can be extensively colonised by downy mildew (Fig. 4.16), as well as the pedicels and rachis, though not their vascular elements (Fröbel and Zyprian 2019). If the infection starts at stage BBCH 69, symptoms are generally limited to certain sections of

Fig. 4.15 An infection caused by *Plasmopara viticola* at BBCH55 can lead to complete desiccation of the bunches

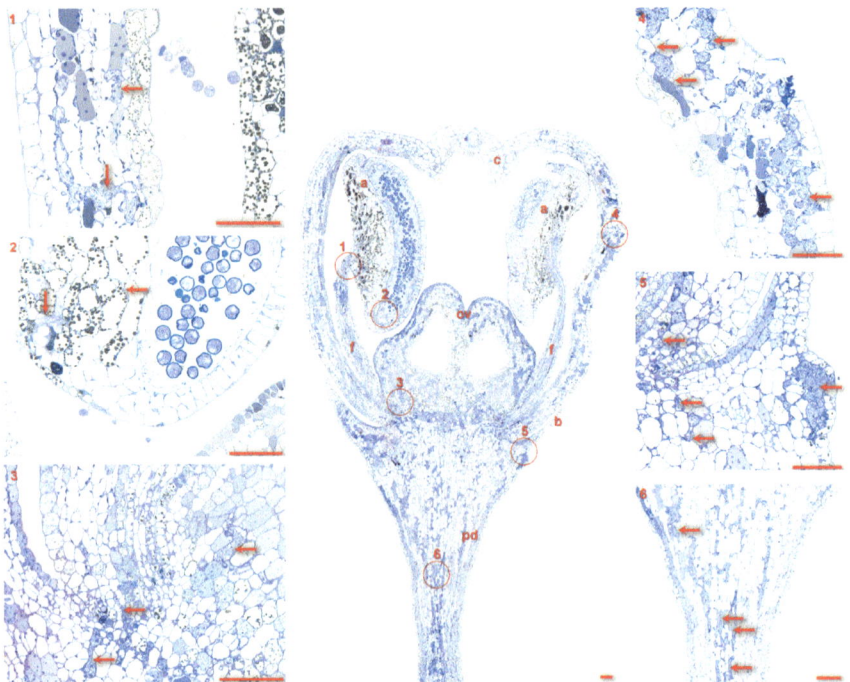

Fig. 4.16 Colonisation of a vine flower by *Plasmopara viticola* at BBCH 55, viewed under an optical microscope on semi-thin sections dyed with toluidine blue. The arrows indicate the presence of intercellular downy mildew in different tissues of the flower. *A* anther, *b* bract, *c* calyptra, *f* stamen filament, *pd* pedicel. Scale bar represents 10 μm

the bunch, resulting in desiccation of parts of the stems and certain pedicels (brown rot) (Fig. 4.8), while at the same time allowing uninfected berries to develop normally. Colonisation of the infected parts of the bunch spreads to all tissues except those of the endosperm and vascular bundle. At this stage, functional stomata are present on the surface of the ovary, pedicels and rachis, paving the way for infection. Observation of the limitrophic regions between symptomatic and asymptomatic parts of the rachis showed the presence of sparse hyphae and rachis in healthy tissue adjacent to necrotic tissue; however, no structural barrier or visible response by the plant was observed (Gindro et al. 2022). The reason for the cessation of the pathogen's development is not clear. It could be due to plant defensive responses such as localised stilbene production (Alonso-Villaverde et al. 2011), early change of primary metabolism and lipid compounds (Chitarrini et al. 2017), or the allocation and distribution of mineral elements (Cesco et al. 2020). The arrested development of *P. viticola* does not appear to be linked to structural modifications such as lignification (Vance et al. 1980; Nicholson and Hammerschmidt 1992) or localised accumulation of callose (Chen and Kim 2009).

Post-bloom Infection (BBCH 75)

If bunches become infected at stage BBCH 75 (pea-sized), only some berries present typical symptoms of brown rot. From a functional perspective, it is important to remember that infections occur primarily through the functional stomata, probably by chemotaxis (Kortekamp et al. 1997), enabling downy mildew to invade plant tissue via intercellular spaces by producing intracellular haustoria (Gessler et al. 2011) (Fig. 4.17).

During berry development, functional stomata are transformed into lenticels (Nakagawa et al. 1980; Kennelly et al. 2005; Gindro et al. 2012) on the berry surface. However, functional stomata remain on the pedicel, leading to the colonisation of the entire berry.

Late Infection (BBCH 81)

Infections that occur from the onset of veraison (BBCH 81) do not produce any symptoms. Maturation and ripening of the grape berries are senescence processes. The onset of grape berry maturation (veraison) is a complex process characterised by

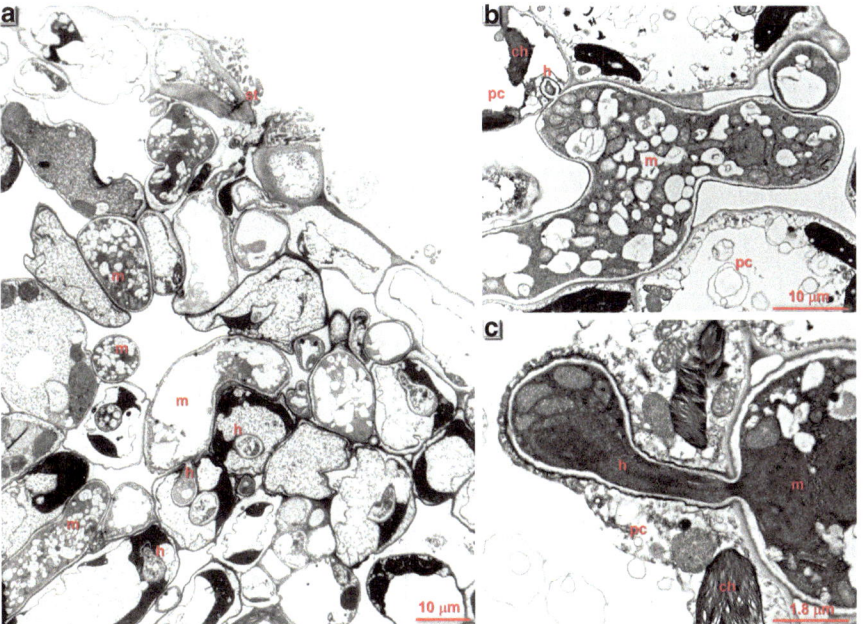

Fig. 4.17 Intercellular growth of *Plasmopara viticola* hyphae in mesophyll cells. Numerous haustoria (h) are produced intracellularly. (**a**) General view of the intercellular growth of grapevine downy mildew hyphae (m) after infection via stomata (st). (**b**) An intracellular haustorium starting to form in a plant cell (pc). (**c**) Detail of a haustorium at a more advanced stage

initiation of sugar accumulation, diminution of organic acids, rapid pigmentation of berries by anthocyanins (in red grape varieties), and both structural and hormonal changes to the berry (Ollat et al. 2002; Pezet et al. 2003). In fact, several hormones may be involved in controlling the ripening of grape berries, which are non-climacteric fruit whose ethylene production is limited by the fact that they do not have a typical peak of respiration (Coombe and Hale 1973). Certain studies have highlighted the role of abscisic acid as a trigger for berry ripening (Wheeler et al. 2009; Sun et al. 2010) and steroids such as endogenous brassinosteroids, levels of which increase at the onset of ripening (Ziliotto et al. 2012). Furthermore, jasmonic acid, which is principally associated with defence against pathogens, seems to promote the production of antho- cyanins during ripening (Belhadj et al. 2008). The synergistic effect of these metabo- lites could contribute to reducing the susceptibility—referred to as ontogenic resistance (Kennelly et al. 2005)—of susceptible varieties.

The development of ontogenic resistance has a major impact on control strate- gies. Protecting the inflorescence and bunches until veraison is a key objective, especially during the pre-bloom and bloom phases when infections lead to complete destruction of the harvest.

4.1.4 Disease Control

The extent of damage observed from 1890 encouraged scientists, already galvan- ised by the impact of phylloxera and powdery mildew, to find effective means of combatting downy mildew. Alexis Millardet, professor of botany at the University of Bordeaux, discovered by chance in 1882 in Saint-Julien du Médoc that copper inhibits the germination of downy mildew spores thanks to a winegrower who had sprayed verdigris (essentially copper) on to vines planted along the path to prevent passers-by stealing his grapes. While all the surrounding vines were attacked by downy mildew, this row was completely unharmed.

After a great deal of trial and error, Millardet and Ulysse Gayon, chemistry pro- fessor at Bordeaux, arrived at the definitive formula for the 'Bordeaux mixture' (three parts copper sulphate in 2% water solution to one part lime). From 1886 (Millardet 1887), it was common practice to apply the mixture with a heather brush prior to the appearance of the first sprayers, notably the Vermorel. From 1887, lime was gradually replaced with soda ash. Copper sulphate was the active ingredient in both versions and proved equally effective in controlling black rot, which appeared soon after 1885.

The evolution of application techniques, development of regulated crop protec- tion products and widening knowledge of the aetiology and epidemiology of downy mildew enabled adapted control strategies to be developed that are less harmful to the user and the environment. These strategies are based on decision support sys- tems (risk prevention models) that enabled fungicidal treatments to be applied at the right time and dosage to be adjusted to the leaf surface or the phenological develop- ment stage of the vine.

Copper in Grapevine

Prior to the fortuitous discovery of copper, downy mildew was impossible to control, despite using sulphur solution that effectively controlled powdery mildew, iron sulphate, sulphuric acid, tannins, acetic acid, sodium borate, chromic acid, potassium thiocarbonate or carbolic acid (Viala 1885).

Copper was the first active ingredient in the history of plant pathology to effectively combat oomycetes and bacteria and is without doubt one of the most widely used active ingredients in global agriculture today. Copper is virtually insoluble in water, so in order to apply it as a fungicide in viticulture, it was combined with quicklime (calcium oxide, CaO) in a mixture developed by Millardet (1887). Extensively used over the centuries, copper remains indispensable in viticulture and for many other crops. This active ingredient, which is controversial due to its stability and accumulation in soil, has been applied in the form of copper hydroxide, oxychloride, sulphate and oxide, among others, for more than 130 years. Copper ions are the only components of the Bordeaux mixture to have a biocidal effect on the downy-mildew particles; quicklime is simply added to enable the copper to form a solution that adheres to the plant organs.

Extracts from Millardet's original document (1887) (translated from French).

Citation:

> The formula for the mixture generally used thus far is as follows: 100 litres water, 8 kilos copper sulphate. Mix with 30 litres of water and 15 kilos of hydrated limestone. 30 litres of water. All these mixtures must be made in the following way: pour 100 litres of water into a container made of wood, an old barrel for example. Place the copper sulphate in a basket or a small cloth bag and keep it immersed in the upper layers of the liquid. After a dozen or so hours, the copper sulphate will have completely dissolved. Place the lime in a container of some kind and pour over 2 to 4 litres of water, starting with small quantities for three minutes at a time, then adding larger amounts at increasingly speed. In this way, in half an hour, a milk of lime is obtained that is thick enough to make one think that less water or a better quality lime had been used. Having carefully crushed the lumps and removed any impurities (small stones, etc.), gradually pour the milk of lime into the copper sulphate solution while stirring. Stir the mixture for two minutes, using a bundle of sticks, for example. The mixture is now ready. It should be a beautiful shade of blue (if it were grey, it would be worthless). Resting the mixture leads to the formation of an equally blue deposit and the surface liquid clarifies and becomes perfectly transparent. This liquid must be colourless.

> Small-scale winegrowers are able to spray the foliage with a heather brush of the type used to apply the first treatments in Dauzac and Beaucaillou (Médoc) in 1885. With a bucket or watering can in one hand and the brush in the other, a labourer can do a third of a hectare a day. Large-scale winegrowers with larger areas to treat must work faster and would be well advised to procure a sprayer. All things being equal, they should choose instruments which spray from top to bottom and deliver the finest spray with the greatest pressure.

Several comparative experiments conducted in 1886, as well as all the treatments that have been applied as a whole, whether in France or abroad, have shown that, to produce their maximum effect, all treatment methods must be applied in a completely preventive manner, and before downy mildew has made its appearance. Indeed, it has been proven that the fateful fungus can exist within the leaves for eight to ten days, especially in cold weather, without revealing any indication of its existence. In these conditions, the simplest and most certain approach appears to be to treat downy mildew in the same way as powdery mildew, against which one applies the first sulphur treatment at a fixed period, without worrying whether or not the disease has been detected.

One may well wonder whether the copper treatment introduced metal into the harvest in sufficiently high quantities to render it harmful. This question has been resolved in the most favourable sense by M. Gayon, my colleague and collaborator. He was the first to show that, whether free-run wines, press wines, second wines or sour *piquettes*, they contain no copper whatsoever, or only negligible traces (one tenth to one hundredth of a milligram per litre).

4.1.4.1 Disease Forecasting

The control of downy mildew is essentially based on the preventive application of fungicides. In reality, no product can completely eradicate downy mildew. Preventive or curative active ingredients are all that is available. It is virtually impossible to totally eradicate the pathogen in a severely infected plot. Thus the ability to predict the development of downy mildew is an important objective of epidemiological research, enabling pathogen development to be correlated with environmental parameters and host physiology (Gessler et al. 2011). By deciphering and analysing these factors, it is possible to develop mathematical relationships that can be integrated into models to simulate and predict the development of *Plasmopara viticola* on the basis of biological and environmental data. Use of these predictive models enables the targeted application of fungicides based on the development of the pathogen. Predictive models underpinned by climatic or microclimatic (plot-level) data are designed to indicate favourable conditions for infection and continuously update the incubation period, enabling more targeted preventive intervention before downy mildew sporulates and multiplies exponentially.

The prediction of downy mildew infections is generally based on the works of Bläser (1978), Gehmann (1987), Hill (1989), the incubation curve of Müller et al. (1923), optionally the vine growth model of Schultz (1992) and more recently the models of Rossi et al. (2008) or Bleyer et al. (2011). Models generally take into account the conditions determining the maturation of oospores as well as the potential transmissibility of downy mildew. All determining parameters are integrated into various computer systems which can access centralised data thanks to a network of monitoring stations.

4.1.4.2   Control Strategies

After more than 150 years' experience of battling downy mildew, this disease remains a scourge which can lead to severe economic losses. Copper is still an indispensable active substance in organic and biodynamic viticulture, despite the pollution it causes due to its toxic effects on soil fertility, soil-dwelling organisms and microorganisms, and human health (Anant et al. 2018; Taylor et al. 2020; Karimi et al. 2021). Apart from the copper-based treatments available to practitioners (Gessler et al. 2011), several multi-site fungicides such as dithiocarbamates, phthalimides, chloronitriles and quinones are used to control downy mildew infections. These preventive treatments form a chemical barrier on the surface of the green organs. Single-site fungicides with both a preventive (pre-infection) and curative (post-infection) action are also available, such as phenylamides (acylalanines), strobilurins or quinone outside inhibitors (QoI fungicides), quinone inside inhibitors (QiIs), oxysterol-binding protein homologue inhibitors (OSBPIs), dinitroanilines, or carboxylic acid amides (CAA). The recommended strategy consists of waiting until the incubation period for the primary infection calculated by the predictive models reaches 80% of the incubation period—generally just before the appearance of the first oil spots—before applying a preventive active ingredient ahead of the next rainfall or heavy dew that triggers sporulation and the first secondary infections. If the first treatment cannot be applied before it rains, it is essential to use a curative penetrant active substance 1–3 days after it has rained at the latest. The maximum curative effect of penetrant or systemic fungicides is obtained 2–3 days after infection. They work only at the start of the incubation phase of downy mildew.

A second, riskier approach involves waiting until the incubation period reaches 100% and the model indicates the first secondary infection. If the presences of oil spots from the primary infection is confirmed, a curative penetrant active substance must be applied in the following days. This treatment protects the foliage from the first secondary infections. Although this approach carries certain risks, it enables the first anti-mildew treatment to be delayed as long as possible. In plots with recurrent downy mildew infections or organic vineyards, preventive control is recommended just before the primary infection to prevent symptoms appearing. The first treatment protects the leaves for 8–12 days, depending on the rate of vine growth and the type of active substance. To extend protection, the same strategies can be applied, taking into account the effective life of the active substances used at the time of the first treatment (8–10 days for contact products, 10–12 for penetrant and systemic products) and the phytosanitary status of the plot. When the risk of infection is high (oil spots present, combination of soil-borne infection and infection between aerial organs), it is advisable to intervene before the next rains. If there is a low risk of infection, the second treatment can be postponed until conditions for sporulation and infection are met. A penetrant active substance should then be applied 2–3 days later. The setting up and monitoring of an untreated control area of around 20 m^2 in a susceptible zone of the vineyard provides valuable indications of the disease development. Incorporating this information into a predictive model enables the control of downy mildew to be optimised. Anti-mildew measures must be part of an integrated approach to tackling other diseases, especially powdery mildew (*Erysiphe necator*), black rot (*Guignardia bidwellii*) and Brenner disease (*Pseudopezicula tracheiphila*), enabling treatments to be combined.

In addition to chemical controls, several studies have focused on discovering alternative treatments based on natural extracts, herbal decoctions or microorganisms (organic control) (Koledenkova et al. 2022). In terms of alternative controls, the principle of elicitation has also been widely studied and various extracts have been developed which aim to stimulate the vine's defence mechanisms, such as lamarin extracted from the brown kelp *Laminaria digitata* (Aziz et al. 2003). Unfortunately however, even when promising results are obtained in the laboratory or greenhouse, no treatment is sufficiently effective in the field to control downy mildew when disease pressure is high.

4.2 Powdery Mildew

Current name:

- *Erysiphe necator* Schwein 1832, Ascomycota, syn. *Uncinula necator* (Schwein.) Burrill 1892

Erysiphe necator, the causative agent of grapevine powdery mildew or '*maladie du blanc*' in Canada, is a fungus originating in the eastern United States, where it developed in vines of the genus *Vitis* as well the genera *Parthenocissus*, *Cissus* and *Ampelopsis*. Powdery mildew was the first economically important fungal disease of the vine to appear in Europe. The first case was recorded in Great Britain in 1845 on vines cultivated under glass. In 1847, Berkeley described the asexual form of the fungus for the first time in Europe, naming it *Oïdium tuckeri*. Having already been observed in the Paris region that same year, in less than 4 years the disease had spread from France to all European and Mediterranean vineyards, as well as to California in 1859 and Australia in 1866. Powdery mildew is currently present in all vineyards throughout the world and, together with downy mildew and grey mould, regarded as one of the major fungal diseases of the vine, capable of completely destroying the harvest if adequate control measures are not taken. Wild American vines are generally resistant or less susceptible to this pathogen. Qualitatively speaking, wines made from severely infected grapes are impaired, tainted by a persistent, musty smell.

> **Volatile Compounds Produced by Erysiphe necator**
> (Darriet et al. 2002; Lopez Pinar et al. 2017)
>   The development of *Erysiphe necator* on the surface of grape berries is accompanied by a characteristic and often intense musty odour which significantly impairs the quality of the grapes. More than 22 volatile compounds

linked to the development of powdery mildew on grapes have been detected by gas chromatography and olfactometry analysis. Of these, three strong-smelling main components were isolated from the musts; 1-octen-3-ol (almost metallic, mushroom-like odour), (Z)-1.5-octadien-3-one (*Pelargonium* leaf odour) and one unidentified component with a persistent, nauseous fishy odour. The monitoring of these molecules during the alcoholic fermentation process showed that the first two were enzymatically transformed by *Saccharomyces cerevisiae* into much less odorous compounds, namely 3-octanone and (Z)-5-octen-3-one (Fig. 4.18). These results show that alcoholic fermentation can to some extent reduce odours associated with powdery mildew. The fishy-smelling compound has yet to be reported in the literature. The most important effects of the colonisation of berries by powdery mildew are an increase in concentrations of phenylacetic acid, acetic acid and γ-decalactone.

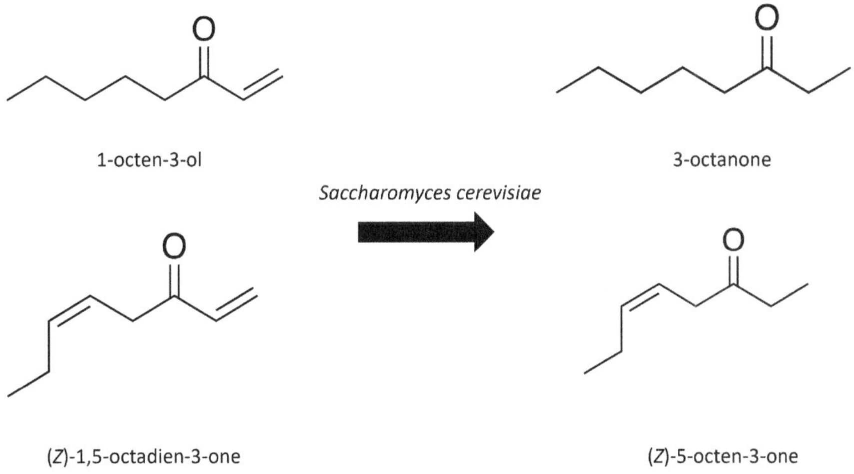

1-octen-3-ol

3-octanone

Saccharomyces cerevisiae

(Z)-1,5-octadien-3-one

(Z)-5-octen-3-one

Fig. 4.18 Transformation of two strong smelling compounds (1-octen-3-ol and (Z)-1.5-octadien-3-one) produced by *Erysiphe necator* in infected grapes and musts to less odorous compounds by the enzymes of *Saccharomyces cerevisiae* during the process of alcoholic fermentation. (Illustration © Robin Huber, University of Geneva, Switzerland)

4.2.1 Causal Organism

Erysiphe necator is a fungus belonging to the Ascomycetes, the order Erysiphales and the family Erysiphaceae (Bolay 2005). Lewis David von Schweinitz (1834) described the sexual structures of the fungus, which he named *Erysiphe necator* [*Erysiphe necator*, L. v. S., *multo rarius in Uvis Vitis labruscae varietatibus cultis in vineis nostris...tenuissimum albidum, floccis valde tenuibus, orbiculatum, non constringens. Sporangiolis minutissimis raris fusco-nigris, globosis. Ubi omnino evoluta, etiam haec species destruit uva.* [Translation: *Erysiphe necator*, L.v.S. occurs more rarely in bunches of the wild vine than in the cultivated varieties. Very fine, white, round, it has numerous small, thin hairs, not binding, sparse, very small brownish black sporangia. Having colonised the entire plant, this species destroys the bunches.] In Europe, Berkeley reclassified the asexual form of the species and named it *Oïdium tuckeri*. Over a period of several years, different names were given to the sexual and/or asexual form of the fungus. The name *Uncinula necator* persisted until the works of Braun and Takamatsu (2000), who retransferred the species to the genus *Erysiphe*, as initially proposed by Pier Andrea Saccardo (1882), and reattributed the species to Schweintiz, according to the rule of anteriority. *E. necator* is an obligate ectoparasite of the genera *Ampelopsis*, *Cissus*, *Parthenocissus* and *Vitis* in the Vitaceae family. The hyphae are superficial, septate and hyaline (4–5 µm in diameter); they form appressoria, thick-walled swellings which penetrate the host tissue. Inside the tissue, haustoria (suckers) absorb the nutrients required for fungal development (Fig. 4.19). Multiseptate conidiophores (40–400 µm long) on the mycelium are responsible for asexual reproduction. Hyaline conidia (asexual spores) (23–48 × 12–22 µm), cylindro-ovoid to doliiform, are produced at the tips of the conidiophores. Contrary to the numerous descriptions of *E. necator*, the conidiophores do not produce multiple conidia at once in the form of a chain; they are the *Pseudoidium*-type, bearing a single conidium. Nevertheless, the conidia may be attached to one another by a mucilage to give the appearance of a chain (Fig. 4.20).

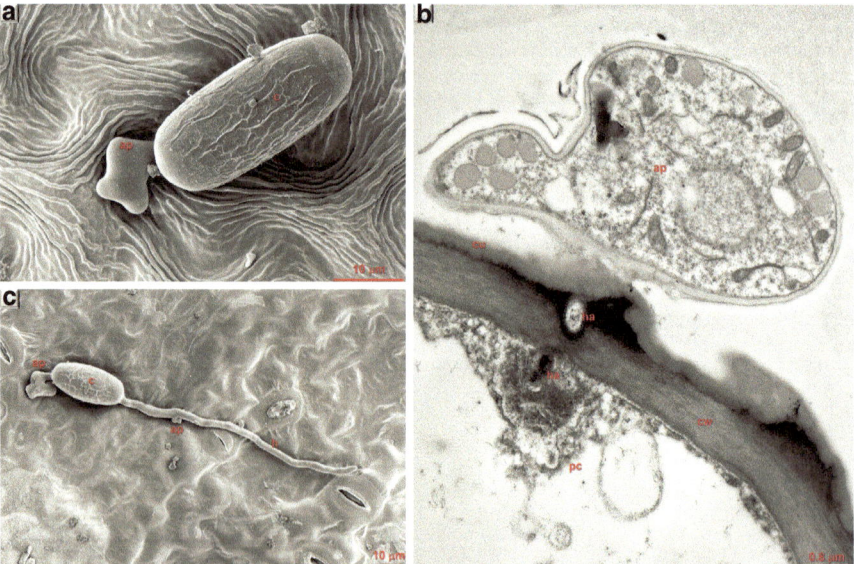

Fig. 4.19 Conidia of *Erysiphe necator* germinate to form an appressorium (**a**), then a germ tube, which develops appressoria at more or less regular intervals (**b**). Beneath each appressorium, an intracellular haustorium forms to absorb nutrients (**c**). *ap* appressorium, *c* conidium, *cu* cuticle, *cw* cell wall, *h* hypha, *ha* haustorium, *pc* plant cell

Fig. 4.20 Formation of the conidia of *Erysiphe necator*. A single conidium forms at the tip of each conidiophore (red arrows); however, they may be attached to one another by a mucilage to give the appearance of a chain (called a pseudochain) (black arrow)

The fruiting bodies or cleistothecia (80–130 μm in diameter) are formed by the fusion of two different types of hyphae (mating types). They are globose, yellowish-white at first, turning blackish-brown at maturity. The outer surface of the cleistothecia has 10–30 septate, bristle-like appendages (length: 1–6 times the diameter of the cleistothecia) called fulcra, which are hooked at the tip (Fig. 4.21). Each cleistothecium contains four to eight asci, ovoid to piriform, (40–70 × 25–45 μm) enclosing four to six hyaline ascospores, ovoid to ellipsoid, measuring 15–25 × 9–15 μm (Fig. 4.22). Like conidia, the ascospores produce short germ hyphae which rapidly differentiate at their tips to form appressoria.

Population Biology of Erysiphe necator: Genetic Groups A and B

(Délye et al. 1997; Amrani and Corio-Costet 2006; Gadoury et al. 2012; Csikós et al. 2020)

In Europe and Australia, two genetically distinct groups, A and B, have been identified based on their marked differentiation using genetic markers (random amplification of polymorphic DNAs (RAPDs) and amplified fragment length polymorphisms (AFLPs)). More recently, new markers have been identified on the basis of sequence-characterised amplified region (SCAR) and single nucleotide polymorphisms (SNPs) on several regions of the DNA which correlate perfectly with the previously identified groups. At biological level, one hypothesis is that populations in group A overwinter as fragments of mycelium in dormant buds, causing 'flag' symptoms in spring, and are more susceptible to fungicides than B populations, which overwinter in the form of ascospores and are dispersed later in the season. The recent results of Csikós et al. (2020) do not confirm this hypothesis. The two genotypes appear to be randomly distributed within the vineyard, with A populations also found later in the season. Nonetheless, these two genotypes are genuinely differentiated; genotype studies covering the whole grape-growing season should make it possible to characterise the temporal and/or spatial distribution of A and B genotypes and their role in the epidemiological development of powdery mildew.

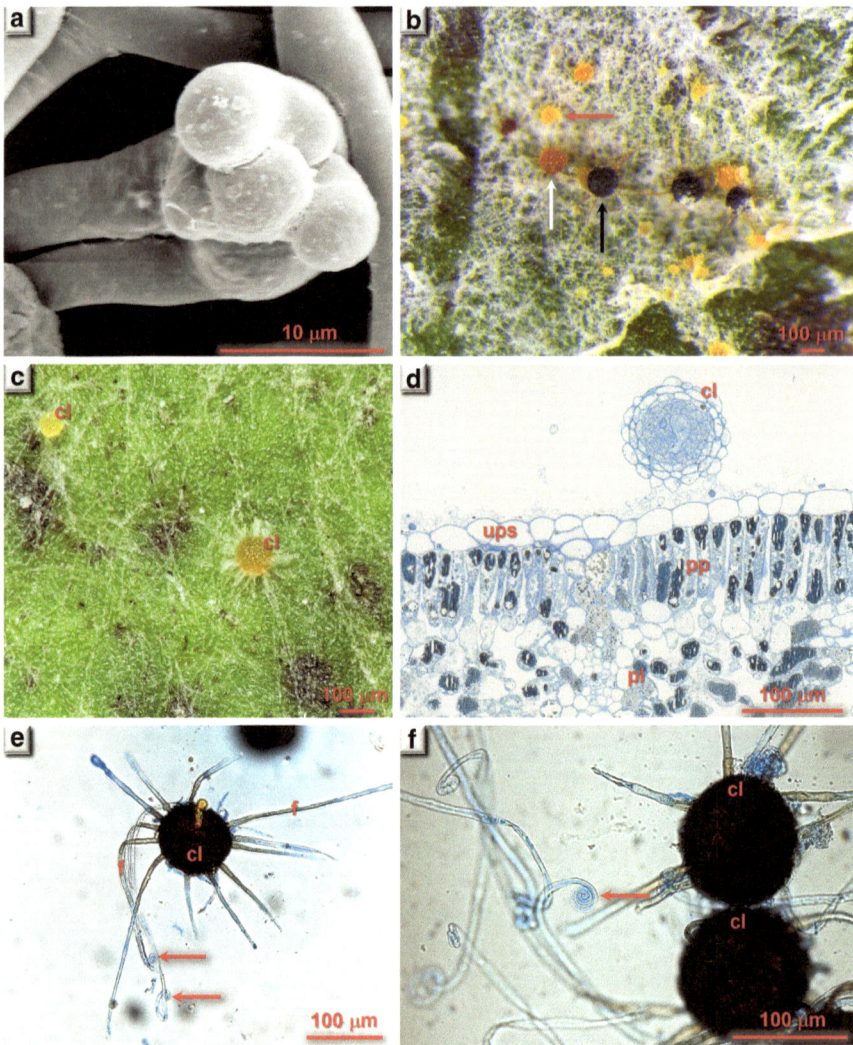

Fig. 4.21 The cleistothecia of *Erysiphe necator* are formed from the fusion of two different mating types of hyphae (**a**) They are globose, yellow at first (red arrow), turning yellowy orange to rust-coloured (white arrow), then blackish-brown (black arrow) at maturity (**b**) The young cleistothecia are firmly attached to the upper surface of the vine leaf (**c**); the asci and ascospores are not yet differentiated (**d**) The outer surface of the cleistothecia is covered in bristle-like appendages, the fulcra, which are hooked to a greater or lesser degree at the tip (**e** and **f**) (arrows). *Cl* cleistothecium, *f* fulcrum, *pl* spongy parenchyma, *pp* palisade parenchyma, *ups* upper epidermis

Fig. 4.22 Asci and ascospores of *Erysiphe necator*. Each cleistothecium contains four to eight asci, ovoid to piriform (red arrows), enclosing four to six hyaline ascospores (black arrows)

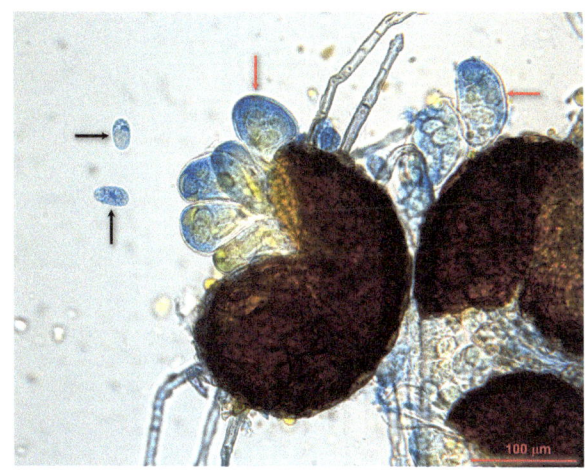

4.2.2   Symptoms

All green organs of the vine can be infected by the pathogen. *Erysiphe necator* is an ectoparasite which primarily develops on the surface of green tissue, producing haustoria which penetrate the cells of the epidermis. On leaves, the fungus initially manifests as curling of the leaf blade, particularly affecting young leaves, which become distorted. The first symptoms are often very difficult to detect: slight yellowing appears on the upper surface of the leaves which can be confused with the early signs of downy mildew. On the underside of these spots, infected cells turn brownish-grey beneath the grey fungal mycelium (Fig. 4.23). Depending on the inherent pilosity of the grapevine variety, the mycelium can only be observed with a hand-held magnifying glass or binocular magnifier. At a later stage, leaf necrosis becomes clearly visible and is characterised by a greyish-white powdery coating on the under and upper side of the leaves, which produces multiple conidia (Fig. 4.24). Infected sites have a characteristic musty odour.

Infected green shoots have small greyish-white blotches which expand and merge as the mycelium grows, forming large brown net-like patches. In winter, reddish-brown blotches appear on woody shoots (Fig. 4.25), indicating the severity of the disease in the previous year.

At the end of summer, small globose bodies, barely visible to the naked eye, yellowish at first, then blackish-brown, intermittently appear on all infected organs. These are the cleistothecia, the organs of the sexual phase (fruiting bodies) which enable the fungus to survive the winter and can cause primary infections the following year (Fig. 4.26).

Fig. 4.23 Initial symptoms of *Erysiphe necator* on vine leaves. Upper surface (**a**), slight localised yellowing (arrow) corresponding to pale brown blotches indicating the infected zone (arrow) on the underside of the leaf blade (**b**). Insert: magnification of the greyish-brown mycelium network of powdery mildew

Fig. 4.24 Abundant colonisation of whitish-grey mycelium and sporulation (arrows) of *Erysiphe necator* on vine leaf (**a**) and detail of sporulation (conidiophores and conidia) observed under binocular magnifying glasses (**b**). Scale bar represents 100 μm

These primary infections can also be caused by fragments of mycelium that have overwintered in the buds. In this case, the shoots emerging from contaminated buds are entirely covered in a powdery coating and have the appearance of flags at

Fig. 4.25 Symptoms caused by *Erysiphe necator* on green shoots appear as merged brown net-like patches (**a**) and on woody shoots as reddish-brown patches (**b**)

Fig. 4.26 The cleistothecia of *Erysiphe necator* appear on all organs of the vine such as woody shoots (**a**), berries (**b**) and leaves (**c**) (arrows). These small, globose bodies are yellow at first (**d**), gradually darkening to black as they mature (**e**). Scale bar represents 500 μm

Fig. 4.27 Primary infections from infected buds. Infected shoots are stunted and covered in powdery mildew (**a**). The base of the shoot and the petiole of the first leaves are infected (arrows) (**b**)

half-mast, so-called "flag shoots". They can be distinguished from healthy shoots by their distorted, stunted growth (Fig. 4.27). These relatively rare 'flag shoots' can be observed from the 4–6 leaf stage (BBCH 14–16).

Powdery mildew is visible at all stages of berry development (Fig. 4.28). Inflorescences can be infected before or shortly after flowering. Young green tissues are particularly susceptible to powdery mildew. The fungus gradually destroys the epidermal cells of infected berries. As a result, when the volume of pulp increases, the infected epidermis loses its elasticity, causing the berries to burst and eventually wither (Fig. 4.29). Severely infected grapes emit a persistent musty odour and are unfit for winemaking.

4.2.3   Biology and Epidemiology

4.2.3.1   Disease Cycle

The epidemiological development of powdery mildew is summarised in Fig. 4.30. Two overwintering strategies have been observed in *E. necator*. In regions with relatively mild winters, notably in the south, the fungus generally overwinters in the form of mycelium on the leaf primordia inside dormant buds. The following spring, mycelium growth resumes, giving rise to severely infected, deformed shoots known as 'flag shoots'. The fungus sporulates on these shoots, producing a vast number of conidia which are carried on the wind to healthy plant tissue. This form of infection is relatively rare in the climate conditions of northern vineyards and can be considered secondary. In the second case, the fungus overwinters in the form of cleistothecia (syn. chasmothecia) in the bark, stems, fallen fruit and leaf litter on the ground.

Fig. 4.28 Development of powdery mildew on berries at different phenological stages

Fig. 4.29 Berries covered in powdery mildew: the infected epidermis loses its elasticity, causing the berries to burst

These structures do not appear to be viable in the long term on fallen leaves as they are often destroyed by antagonistic microorganisms or hyperparasites such as the fungus *Ampelomyces quisqualis*, which has been widely studied for its potential to degrade the cleistothecia of powdery mildews, including *E. necator* (Falk et al. 1995). Cleistothecia form on the surface of severely diseased tissue from midsummer through to autumn. With the onset of spring rains, the cleistothecia open and release ascospores (Fig. 4.30 no. 1). Ascospore release is always associated with rainfall, leaf wetness lasting more than 2.5 h and temperatures exceeding 11 °C (Jailloux et al. 1999). Released ascospores land on buds and young leaves, where they germinate and cause primary infections (Fig. 4.30 no. 2). Continuing wetness is not necessary for ascospore germination and infection of the host tissue. An appressorium forms at each new site of infection. From its underside, a germ tube emerges, pierces the cuticle and penetrates an epidermal cell, where a haustorium

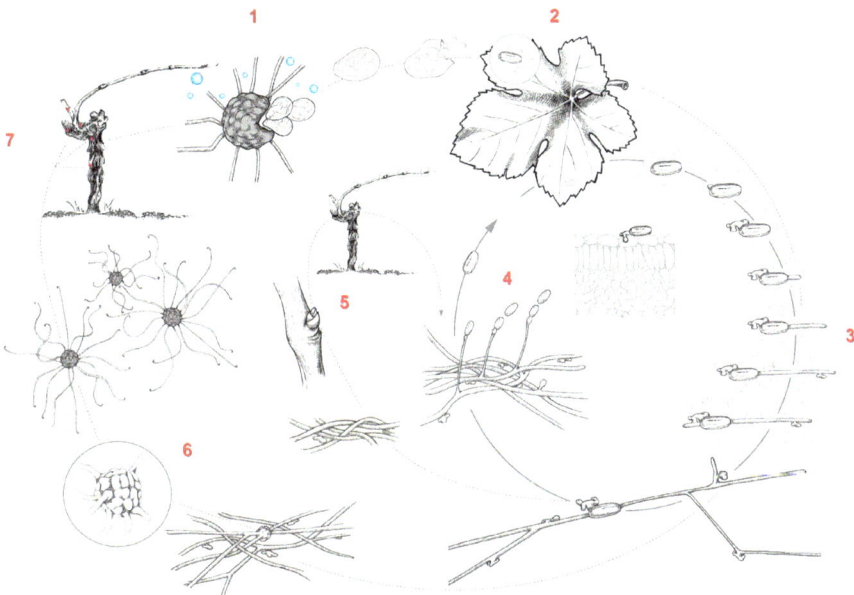

Fig. 4.30 Disease cycle of *Erysiphe necator*. Cleistothecia (syn. chasmothecia) release asci and ascospores (1), initiating an infection on the young green organs of the vine (2). The spores germinate, forming an ectophytic mycelium with numerous appressoria. Haustoria beneath the appressoria penetrate the epidermal cells (3). Mycelium growth rapidly leads to sporulation (4) and the production of numerous conidia within a few days, thereby completing the cycle. Fragments of hyphae can invade dormant buds (5), serving as an inoculum the following spring. Sexual reproduction from summer leads to the formation of very large numbers of cleistothecia (6) which overwinter mainly in vine bark (7) and trigger infections the following spring. (Illustration © Virginie Duquette, Gravir un Monde d'Illustration, Switzerland)

forms. The mycelium develops on the surface of the vine epidermis (Fig. 4.30 no. 3) (ectophytic growth on green organs), leading to sporulation and the production of numerous conidia within a few days, thereby completing the cycle (Fig. 4.30 no. 4). Conidiogenesis occurs at temperatures between 20 and 25 °C and is favoured by high relative humidity. This stage marks the start of the outbreak. The cycle repeats itself throughout the vine's growing season, leading to a rapid increase in the incidence of the disease. In parallel, fungal hyphae can invade dormant buds, where they remain latent from autumn until the following spring (Fig. 4.30 no. 5). The period between germination and sporulation can last from 7 to 13 days, depending on temperatures. For conidia, temperature appears to be the determining factor for sporulation and infection (optimum: 20–27 °C), whereas the conditions required for ascospore development are still largely unknown. In contrast to downy mildew, the presence of a film of water on the surface of the leaves inhibits germination of powdery mildew conidia. Sexual reproduction in summer and autumn produces numerous cleistothecia (Fig. 4.30 no. 6) which overwinter mainly in the vine bark, providing a source of inoculum the following spring (Fig. 4.30 no. 7).

Practical experience shows that under natural conditions primary infections are generally precocious and intervene before flowering (Rossi et al. 2010). When the first symptoms are visible on the bunches, it is generally very difficult to halt the disease progression. Bunches exhibit strong ontogenic resistance, with their susceptibility to powdery mildew diminishing as the berries develop. Inflorescences are most susceptible at the start of flowering (BBCH 61), until just after fruit set, when berries are groat-sized (BBCH 73). During this period, young cells on both stems and berries provide a particularly favourable substrate for powdery mildew. Artificial infection of bunches after stage BBCH 73 indicates very limited development of powdery mildew, regardless of the variety's susceptibility (Stark-Urnau and Kast 1999; Gadoury et al. 2003). At the end of the season, second crop fruits from the second flowering of the vine may again be very severely infected, thereby confirming the low ontogenic resistance of young tissue.

Ontogenic Resistance
(Stark-Urnau and Kast 1999; Gadoury et al. 2003, 2012)

Interaction between the powdery mildew and the vine is dynamic, with tissue susceptibility evolving over time as a function of growth stage and age. This reduction in susceptibility linked to the age of the tissue, which can culminate in almost total resistance, is referred to as ontogenic resistance. The inflorescences and bunches of two varieties (Trollinger and Lemberger) were artificially inoculated at different stages of growth to evaluate their susceptibility. Stark-Urnau and Kast (1999) observed a reduction in susceptibility only a few days after the end of flowering, from stage BBCH 73 (groat-sized), as well as the development of a relatively high level of resistance from stage BBCH 75 (pea-sized). Similar results were obtained with Chardonnay, Riesling, Gewürztraminer and Pinot Noir. Bunches are highly susceptible to infection during the 2 weeks after flowering, after which susceptibility rapidly declines. Four weeks after flowering, the bunches are virtually immune. Integrating these findings into strategies to control powdery mildew and fully protecting bunches during the critical period enables more targeted use of fungicides.

4.2.4 Disease Control

The control of powdery mildew should be primarily preventive. The earliness of leaf outbreaks determines the frequency and severity of bunch infections, with an important correlation between the incidence of affected leaves at flowering and the severity of berry symptoms at bunch closure (Calonnec et al. 2006, 2008).

The main objective of controlling powdery mildew is to prevent or delay the onset of the disease on the leaves and so reduce the amount of inoculum present from flowering to fruit set, when bunches are most vulnerable. Effective spraying is essential to achieve this, especially when bunches are most susceptible. Only a perfectly adjusted and calibrated sprayer enables uniform, targeted application and good penetration of plant protection products within the bunch zone. On this subject, preventive canopy management such as aeration of the bunch zone (trellising, shoot thinning, leaf removal) and vigour control (fertilisation, cover cropping, choice of rootstock) are of prime importance, as they enable plant protection products to penetrate the bunch zone while at the same time preventing the creation of microclimates that favour the disease.

Sulphur in Grapevine

Sulphur is the oldest active ingredient used in agriculture to repel insects and fungi affecting crops. In viticulture, its efficacy was discovered when powdery mildew first arrived in Europe, where its application in the form of a powder rapidly became widespread from the mid-nineteenth century. It works as a fungicide through the process of sublimation (sublimed sulphur or flower of sulphur), a physical phenomenon whereby sulphur changes directly from a solid to a gaseous form, producing biocidal vapours. This phenomenon is accelerated by high temperatures (>20 °C) and intense light. Sulphur sublimation also depends on the size of the particles. The larger the particles in crushed sulphur (>25 μm), the slower the rate of sublimation. Conversely, the finer the particles in micronised sulphur (<1 μm), the greater their tendency to penetrate the plant and induce phytotoxicity. Like copper, sulphur is one of the most widely used fungicides in the world. Due to its poor solubility in water, it was initially used in the form of powder prior to the development of wettable micronised sulphur (sulphur particles of 1–5 μm). The formulation most commonly used nowadays is micronised sulphur granules, which are fully soluble in water.

The best documented findings about the mode of action of sulphur and its application in viticulture comes from Marès (1857).

Citations:

Flower of sulphur indeed possesses all the qualities required to constitute an excellent curative agent. Firstly, it destroys powdery mildew on coming into contact with it; secondly, in the form of a very fine powder, simply by virtue of dusting it is possible to coat all vegetative parts of the vine, and its volatility at the daily temperatures produced on the ground and green surfaces exposed to the sun in summer ensure its effectiveness against the harmful germs. Moreover, it possesses the property, no less remarkable than unexpected, of stimulating vine growth, thereby conferring to it the necessary vigour to vanquish the attacks of the parasite.

The fundamental principle of sulphuring diseased vines can be summed up as follows: spread the sulphur dust on all their green parts as soon as the first symptoms of the disease appear and repeat the application each time it reappears on the vines.

The vine disease was first discovered in the year 1845 on cultivated vines in hot houses and on trellises in Margate, a port on the south-east coast of England… from the hot houses of England, the disease gradually invaded the entire continent of

Europe wherever the vine is cultivated, as well as the Mediterranean basin and the islands of the Ocean.

The idea of applying sulphur to cure grapevine powdery mildew is ancient; it dates back almost to the emergence of this disease, having been proposed by an English gardener from Leyton, Mr. Kyle, in 1846 and by Mr. Tucker, who first discovered powdery mildew and combined sulphur with quicklime; but little attention was paid at the time.

Thus the problem of finding the means of combatting grapevine powdery mildew is also one of destroying the disease or its germs, at all stages of their development, and on whatever parts of the vine they may reside. To this end, I have conducted a large number of experiments of all kinds over several years and have come to the conclusion that it is almost impossible to destroy the disease by attacking the germs that reproduce it during the dormant period. The means employed for this purpose can achieve it, and yet the disease does not disappear.

Since one cannot definitively identify the disease except by the presence of powdery mildew, and this is found only on the green parts of the vine, the parasite must be attacked and destroyed as soon as it makes an appearance on the leaves.

4.2.4.1   Disease Forecasting

Available models must incorporate two parameters that are essential for calculating the risk of powdery mildew: the susceptibility of the vine at different phenological stage linked to ontogenic resistance (Stark-Urnau and Kast 1999; Gadoury et al. 2003, 2012) and the relative favourability of weather conditions on disease development (Kast 1997; Caffi et al. 2011, 2012). The model strategy is to fully protect the vine when it is very susceptible and weather conditions favour the disease. In concrete terms, models must offer recommendations for the timing of the first treatment at the start of the season. Treatment intervals must then be determined according to the indices of calculated risk infection and effective life of the plant protection products applied. If full protection has been provided until the point of fruit set and powdery mildew is absent, the risk of new contaminations declines as a function of the ontogenic resistance of the bunches.

4.2.4.2   Control Strategies

Prevention is the best method of controlling powdery mildew. Several synthetic active ingredients can be used against this disease. However, since powdery mildew is capable of adapting to single-site active substances, it is important to alternate active substances belonging to different groups to prevent the development of resistant strains. Marès (1857) devised a technique for applying dry sulphur in the mid-nineteenth century and it remains the most widely used active ingredient to this day, whether in wettable form or as a dusting powder. Sulphur's multi-site mode of action has prevented the selection of resistant strains of powdery mildew.

Due to its formulation, and the difficulties associated with application and protection of users and wildlife, dusting should be reserved exclusively for eradicating visible outbreaks. The efficacy of sulphur powder is linked to its vapour phase, of which temperature is the limiting factor (optimum: 25–30 °C). Sulphur does not enter its vapour phase below 18 °C and at high relative humidity, which limits its effectiveness. Above 30 °C, sulphur can become phytotoxic and cause leaf burn. The powder acts directly on the fungus due to its vapourising effect, whereas wettable sulphur acts solely on contact. Wettable sulphur must be applied immediately, before or even at the same time as the first interventions against downy mildew. Copper-based products used to control downy mildew after flowering slow the development of powdery mildew. It was not until 1978 that triazoles (inhibitors of the synthesis of IBS sterols) became popular as a means of replacing or complementing sulphur. Currently, penetrant fungicides such as the strobilurins (quinone outside inhibitors, QoIs), demethylation inhibitors (triazols, DMIs), phenyl-acetamides, dinitroanilines, azanaphtalenes (AZNs), succinate dehydrogenase inhibitors (SDHIs), amines-piperidines, benzophenones families are widely used on an international scale.

4.3   Grey Mould

Current name:

- *Botrytis cinerea* Pers. 1794, Ascomycota

Botrytis cinerea is one of the most widespread pathogens in the world, and the cause of rots in a vast number of wild and cultivated plants. In viticulture it causes grey mould (grey rot), also known as bunch rot. This disease was described in vineyards in Roman times when the legions settled in Gallic territories (Gouvernet 1978). Thus *Botrytis cinerea* has been causing serious damage to grapes since ancient times, regularly leading to economic losses and impairing the quality of the fruit.

The symptoms of this insidious disease generally appear a few weeks before harvesting at a time when intervention is no longer possible. With particularly susceptible varieties, grey mould may dictate the date of harvest, even if this means sacrificing the sought-after level of ripeness. *B. cinerea* is a polyphagous fungus which lives as a necrogenic saprophyte on a multitude of host plants. Its rapid development during fruit ripening leads to quantitative losses and depreciation of the harvest. As a result, the grape harvest from affected plots requires painstaking and labour-intensive triage. The development of grey mould has negative repercussions for the winemaking process which are difficult to overcome. For example, the production of laccases by *B. cinerea* is responsible for undesirable oxidative reactions. The successive evolution of resistance to different specific fungicides (benzimidazoles, dicarboximides, anilinopyrimidines, phenylpyrazoles…) combined with the exceptional adaptation capacity of this fungus makes control increasingly complex and uncertain under favourable weather conditions.

However, grape berry infection by *B. cinerea* can be desirable under specific conditions that favour 'noble rot'. In the past, fortuitous circumstances forced wine-growers to delay harvesting, leading to the chance discovery of the benefits of noble rot on overripe grapes. This occurred in Sauternes in 1815, when the grape harvest was delayed because of incursions into French territories by Germano-Russian armies during the Napoleonic Wars. Noble rot is the result of fungal growth that is restricted to the epidermis of the berries. Enhanced permeability leads to water loss, causing sugar concentrations in the berry to increase, while acidity remains stable. Noble rot leads to the production of highly prized sweet wines of great quality (Ribéreau-Gayon et al. 1980).

4.3.1 Causal Organism

The genus *Botrytis* belongs to the Ascomycetes, the order Leotiales and the family Sclerotiniaceae. *Botrytis* was first described by P.A. Micheli in 1729 in his work *Nova Plantarum Genera* (Fig. 4.31). The authenticity of the genus was confirmed by Persoon (1801), who defined the species *B. cinerea* according to Linnaeus' binomial system of nomenclature. The link between the asexual form *Botrytis* and the sexual form *Botryotinia* was demonstrated experimentally by Istvanffi in 1903 and validated definitively in 1953. Species of the genus *Botrytis* are ubiquitous and opportunistic saprophytic fungi which are also major pathogens of economically important crops.

Several species of *Botrytis* have been described), most of which specialise in a limited number of hosts, such as *B. aclada* (Allium spp.), *B. fabae* (*Vicia faba*) or *B. tulipae* (*Tulipa* spp.). In this respect, *B. cinerea* appears to be an exception, with an impressive range of host plants, including herbaceous and woody crops, ornamentals, mosses and ferns (Williamson et al. 2007; Abbey et al. 2019; Esterio et al. 2020). In addition to its great host diversity, *B. cinerea* can also colonise a wide range of environments: it has been found from the coldest regions of Alaska and Canada to subtropical regions such as Egypt. Its lifestyle is generally described as saprophytic and necrogenic (killing its host and consuming the dead tissue), although endophytic behaviour can be observed on rare occasions (Van Kan et al. 2014). It can also behave as a saprophyte (living on decaying or dead matter). Due to its versatility and necrotrophic lifestyle, *B. cinerea* is one of agriculture's most problematic fungi. Its capacity to develop at low temperatures means that it can also pose problems in food storage facilities. *Botrytis cinerea* is characterised by septate hyphae, branched, multinucleate and heterokaryotic, with frequent anastomoses (Brooks 1908), which differentiate into branched conidiophores bearing hyaline conidia, ovoid to ellipsoid, brown to grey coloured, thought to be the primary mechanism of dispersal of the fungus (Fig. 4.32). These multinucleate conidia (6–18 × 4–11 μm) are anemochorous (adapted for wind dispersal) with hydrophobic walls. The conidia are produced synchronously by holoblastic ontogeny from a bulbous swelling on the tips of the conidiophores (Fig. 4.33). The conidia are held in place by fine denticles which remain on the swelling after release of the conidia.

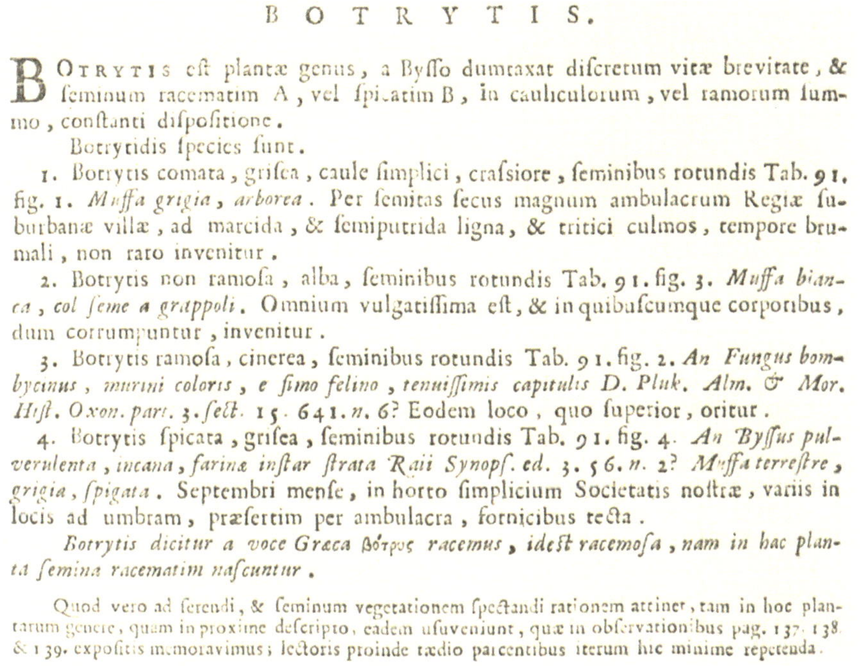

B O T R Y T I S.

B Otrytis eſt plantæ genus, a Byſſo dumtaxat diſcretum vitæ brevitate, &
  feminum racematim A, vel ſpicatim B, in cauliculorum, vel ramorum ſum-
mo, conſtanti diſpoſitione.

  Botrytidis ſpecies ſunt.

  1. Botrytis comata, griſea, caule ſimplici, craſsiore, feminibus rotundis Tab. 91,
fig. 1. *Muffa grigia, arborea*. Per ſemitas ſecus magnum ambulacrum Regiæ ſu-
burbanæ villæ, ad marcida, & ſemiputrida ligna, & tritici culmos, tempore bru-
mali, non raro invenitur.

  2. Botrytis non ramoſa, alba, feminibus rotundis Tab. 91. fig. 3. *Muffa bianca, col ſeme a grappoli*. Omnium vulgatiſſima eſt, & in quibuſcumque corporibus,
dum corrumpuntur, invenitur.

  3. Botrytis ramoſa, cinerea, feminibus rotundis Tab. 91. fig. 2. *An Fungus bom-
bycinus, murini coloris, e ſimo felino, tenuiſſimis capitulis D. Pluk. Alm. & Mor.
Hiſt. Oxon. part. 3. ſect. 15. 641. n. 6?* Eodem loco, quo ſuperior, oritur.

  4. Botrytis ſpicata, griſea, feminibus rotundis Tab. 91. fig. 4. *An Byſſus pul-
verulenta, incana, farinæ inſtar ſtrata Raii Synopſ. ed. 3. 56. n. 2? Muffa terreſtre,
grigia, ſpigata*. Septembri menſe, in horto ſimplicium Societatis noſtræ, variis in
locis ad umbram, præſertim per ambulacra, fornicibus tecta.

  Botrytis dicitur a voce Græca βότρυς racemus, ideſt racemoſa, nam in hac planta femina racematim naſcuntur.

  Quod vero ad ſerendi, & feminum vegetationem ſpectandi rationem attinet, tam in hoc plantarum genere, quam in proxime deſcripto, eadem uſuveniunt, quæ in obſervationibus pag. 137. 138.
& 139. expoſitis memoravimus; lectoris proinde tædio parcentibus iterum hic minime repetenda.

Fig. 4.31 First description of the genus *Botrytis* by Pietro Antonio Micheli in 1729 in his work
Nova Plantarum Genera. (Photograph © reproduced by kind permission of the *Bibliothèque des
Conservatoire et Jardin botaniques de Genève*, Switzerland)

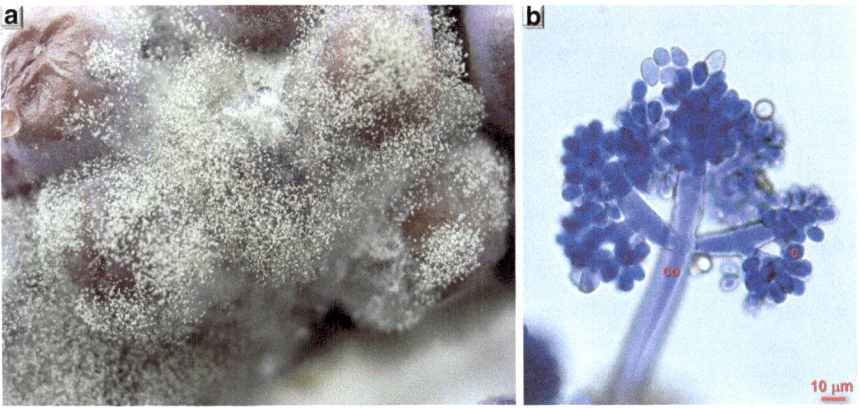

Fig. 4.32 Sporulation and conidia of *Botrytis cinerea* (**a**) and microscopic view (**b**) of a conidiophore (co) bearing numerous conidia (c)

  Microconidia are produced on phialides emerging from certain hyphae, sclerotia or
conidia. Brierly (1918) showed that microconidia were capable of germinating in
nutrient-rich environments, producing a micromycelium. More recently, Drayton

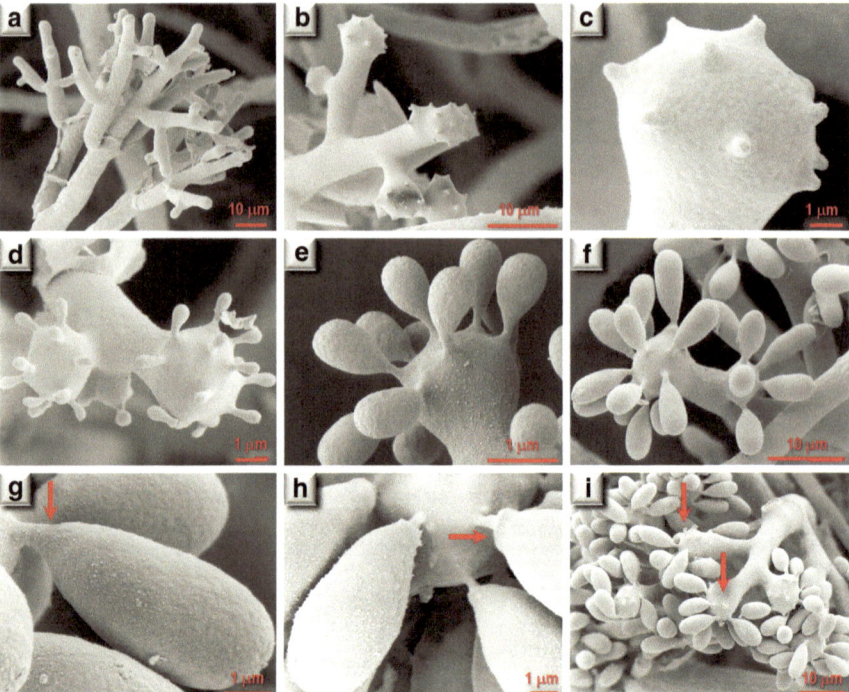

Fig. 4.33 Conidiogenesis of *Botrytis cinerea* viewed under scanning electron microscope. (**a**) Non-differentiated conidiophore with swellings appearing at the tips of the fertile branches. (**b**) Synchronous holoblastic development of early conidia structures on each swelling. (**c**) Detail of (**b**). (**d–h**) Continued development of conidia with progressive refinement of the denticles and emergence of a dehiscent plate (arrow). (**i**) Bulbous swellings bearing mature conidia: some denticles are no longer attached to their conidia after release (arrows)

(1932) showed that the microconidia were sterile and involved only in spermatisation of the sclerotia. These sclerotia—survival structures which form when environmental conditions are unfavourable—can produce conidiophores asexually as well as apothecia sexually (De Bary 1866; De Istvanffi 1903; Groves and Loveland 1953). The apothecia are cupulate, brownish and stipitate with a maximum size of 8–10 mm (Fig. 4.34). These apothecia form asci, unitunicate, inoperculate, generally clavate, containing eight haploid, multinucleate ascospores with an apical pore which produces a blue positive reaction to Melzer's reagent (Melzer 1924) (Fig. 4.35). The ascospores germinate and produce an abundant mycelium, conidiophores, conidia and sclerotia. *B. cinerea* also produces thick-walled chlamydospores and globose oidia (flat-ended asexual spores formed by the breaking up of a hypha into cells).

During the last 20 years, new experimental evidence has shown that populations of *B. cinerea* often coexist with another cryptic species (indistinguishable at morphological level): *Botrytis pseudocinerea* (Walker et al. 2011). In previous studies,

Fig. 4.34 Apothecia (a),
stipitate (st), and
conidiophores (co) of
Botrytis cinerea growing
on sclerotia (s)

Fig. 4.35 Asci (as) and
ascospores (sp) of *Botrytis
cinerea*. Scale bar
represents 10 μm

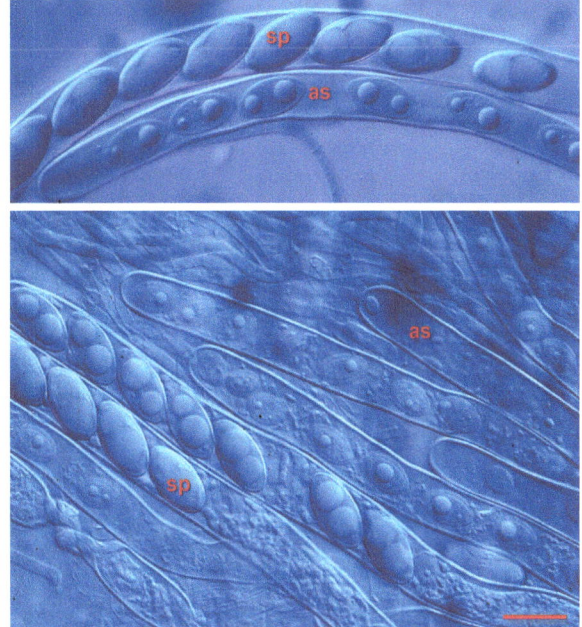

this species had been incorrectly described as *B. cinerea*. Surprisingly, phylogenetic studies have shown that *B. cinerea* and *B. pseudocinerea* are not sister species, *B. fabae* comes in between them (Walker et al. 2011). These two species form a complex of two cryptic species living in sympatry on several hosts. Identifying the species is made difficult by the fact that different strains of *B. cinerea* exhibit significant morphological differences (mycelial growth, conidiogenesis, sclerotia formation). These two species can only be distinguished at genetic level (Plesken et al. 2021).

Genetic Variability of Botrytis cinerea
(Plesken et al. 2021)

Botrytis cinerea strains show considerable morphological variability and a high degree of genetic polymorphism. Studies by Giraud et al. (1997, 1999) detected transposable elements within the genome of *B. cinerea* indicating the existence of two distinct genetic groups based on the presence or absence of these transposable elements (TE): *B. cinerea* var. *vacuma* (without TE) and *B. cinerea* var. *transposa* (with TE). The authors also showed that these two groups frequently occurred in sympatry but appeared to differ in terms of their host range, genetic diversity, temporal succession and size of conidia. Martinez et al. (2005) showed that the presence of transposable elements correlated with higher virulence. More recently, thanks to the development of mass sequencing, Plesken et al. (2021) analysed over 100 strains from diverse geographic origins and harvest years, revealing an as-yet-unknown population of strains—referred to as B strains—in which the gene cluster (*boa*) for the biosynthesis of the phytotoxin botcinic acid was missing. However, the absence of this one gene cluster does not appear to affect the virulence of the fungus. B strains of *B. cinerea* were found on numerous hosts plants and in different countries, which shows that this population is present in Europe, and presumably worldwide. The overall results obtained by Plesken et al. (2021) tend to reveal new insights into the genetic diversity of *B. cinerea* and provide evidence for intraspecific differentiation and different degrees of adaptation of this polyphagous necrotrophic fungus.

4.3.2   Symptoms

4.3.2.1   Grey Mould

Grey mould is omnipresent and can cause dieback in different aerial parts of the plant. Broadly speaking, it produces a grey mould on the surface of affected organs. Three types of damage can occur, depending on the affected area: rotting of green parts, rachis rot and bunch rot.

Infection of Green Parts

The fungus can attack shoots during humid weather in spring, giving rise to reddish-brown necrotic spots on the surface of the leaves (Fig. 4.36). In severe attacks, elongated patches with black borders form on shoots, causing them to wilt and shrivel.

Rachis Infection

Before or after flowering, grey mould can attack the peduncles (rachis) of young inflorescences and bunches, which may be completely or partially infected, resulting in browning and withering (Fig. 4.37).

Fig. 4.36 Initial leaf
necroses caused by
Botrytis cinerea (top),
converging and causing
parts of the leaf blade to
wither (bottom)

Fig. 4.37 Peduncular necroses (necrotic rachis) on young inflorescences caused by *Botrytis cinerea* (**a**), which progressively spread along the stem (**b**), causing the entire inflorescence to wither (**c**)

Bunch and Berry Infections

Appearing at veraison, bunch rot is responsible for economic losses and quality impairment. Some visible grey mould on berries stems from floral infections which have remained dormant until veraison, while other berry infections arise via the epidermis or through wounds. Berry susceptibility increases with ripening, especially in wet, temperate climate conditions. Infected berries turn brown then become covered in a greyish mycelium consisting of the fruiting bodies of the pathogen (Fig. 4.38).

During winter pruning, characteristic black elongated patches around 2–5 mm long can be seen on poorly lignified vine shoots and dead leaves on the ground (Fig. 4.39). These are the sclerotia of *B. cinerea;* survival structures capable of resisting the most extreme conditions.

4.3.2.2   Noble Rot

Unlike grey mould, noble rot (Fig. 4.40) causes berries to turn brown and shrivel generally without visible sporulation. The fungus contained within the berry's skin modifies epidermal permeability, causing water to evaporate. This increases the sugar concentration while maintaining acidity levels. Noble rot occurs only under certain climate conditions which prevent the exponential development of the

Fig. 4.38 Grey mould on bunches (bunch rot). (**a**) Infected berries covered in the asexual fruiting bodies (conidiophores and conidia) of *Botrytis cinerea* (arrows). (**b**) Largely sporulated *B. cinerea* on a colonised berry

Fig. 4.39 Sclerotia (survival structures) of *Botrytis cinerea* on woody stems (**a**) and on dead leaves (**b**) (arrow). These sclerotia, variable in form, produce an abundance of conidiophores and conidia (**c**) (arrow)

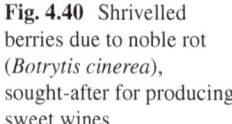
Fig. 4.40 Shrivelled berries due to noble rot (*Botrytis cinerea*), sought-after for producing sweet wines

pathogen. In the climate conditions of northern vineyards, the characteristic shrivelled 'roasted' grapes resulting from noble rot are not produced each year. Overripening may be accompanied by noble rot and/or grey rot, depending on the climate conditions, producing wines with different aromatic profiles. Certain regions and varieties particularly favourable to noble rot provide some of the most sought-after sweet wines in the world, such as the Sauternes in France, the *Trockenbeerenauslese* in Germany, the *Aszu* in Hungary and several other botrytised wines in different regions of the world (Vannini and Chilosi 2013). The most adapted varieties according to their aromatic profile are Sauvignon blanc, Sémillon blanc, Marsanne, Roussanne, Pinot gris, Arvine, Amigne, Riesling or Gewürztraminer.

4.3.3 Infection Process

Botrytis cinerea is a very well-studied fungus at epidemiological, metabolomic and proteomic level (Pezet et al. 2004). Analysis of the stages of penetration through plant tissue has identified different enzymes involved in the degradation of cellular components such as the layer of epicuticular wax, the cuticle and cutin, and the cell wall. Particular attention has also been paid to studying the enzymes involved in the detoxification of certain complex polymers as well as plant defence mechanisms. *B. cinerea* invades plant tissue primarily via dead and dying tissue. Consequently, vine flowers are often contaminated with the asexual spores of the fungus (conidia), which can remain at the base of the young berries. This enables the conidia to obtain sufficient nutrients from dead or dying floral parts (such as dehiscent stamens or remnants of flower caps) to germinate, secrete enzymes and penetrate healthy tissue. The first stage of successful infection is the adherence of conidia or ascospores to the surface of the plant tissue, namely the cuticle. This adherence is the result of hydrophobic interactions between the fungal cell and components of the cuticle and epicuticular wax. During germination, the conidia excrete an extracellular matrix which anchors them firmly to the cuticle. Around 50% of this matrix is composed of carbohydrates, lipids and proteins, including numerous enzymes such as esterases and lipases. The remaining 50% appears to be composed of melanin, a virtually insoluble black pigment which provides the fungus with rigidity and protection. Apart from its anchoring function, the matrix contains enzymes such as esterases, cutinases (Arya and Cohen 2022) and lipases which bring about the gradual erosion of the cuticle (Fig. 4.41) and enable the hyphae to penetrate the epidermal cells (Doss et al. 1993, 1995; Gindro and Pezet 1997)—the prerequisite for successful infection and invasion of plant tissue. The development of *B. cinerea* within the plant tissue is then facilitated by the activity of a wide range of enzymes controlling the breakdown of plant tissue, its degradation and the detoxification of released metabolites. These enzyme panels include for example polygalacturonases, pectin methyl esterases, glucosaminidases, laccases, catalases, pectinases and xylanases.

4.3.4   Disease Cycle

The latest findings about the disease cycle of grey mould are summarised in Fig. 4.42.

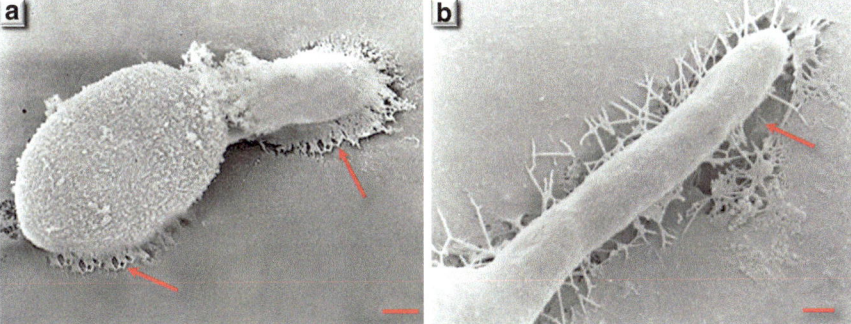

Fig. 4.41 Development of *Botrytis cinerea* on the cutin. Both the conidium (**a**) and germ tubes (**b**) produce an extracellular matrix containing the range of enzymes required for the gradual erosion of the cuticle and the penetration of hyphae into healthy tissue (arrows). Scale bar represents 2 μm

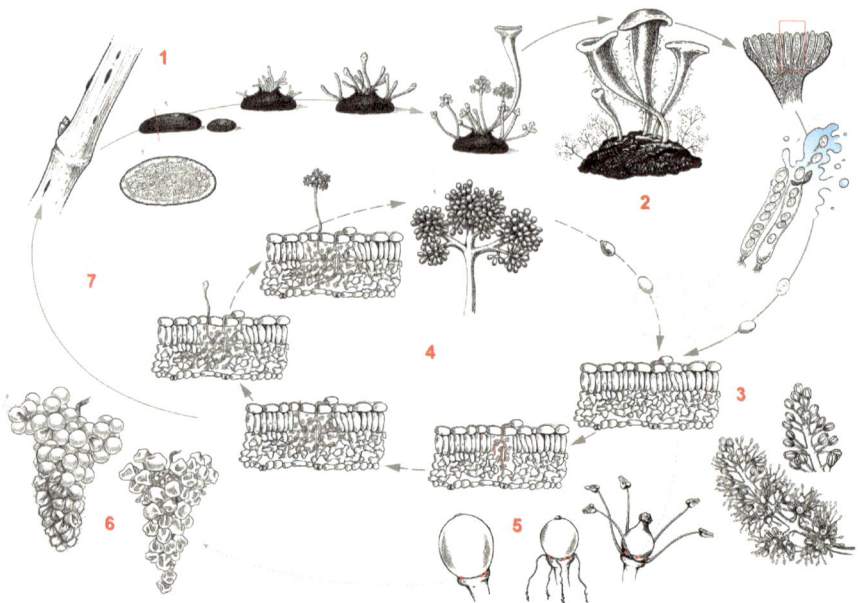

Fig. 4.42 Disease cycle of *Botrytis cinerea*. The sclerotia enable the fungus to survive the winter (1). The sclerotia resume their metabolic activity in early spring, triggering the development of mycelium, conidiophores, conidia and apothecia (2). The sexual and asexual spores produced are the primary inoculum of the fungus, enabling infection of the green tissues of the vine (3). Depending on weather conditions, the fungus can produce numerous infection cycles accompanied by the production of vast numbers of conidia, thereby increasing potential infection pressure (4). At the flowering stage, *B. cinerea* can infect young flower tissue and remain dormant in berry tissue (5) until the disease expresses itself near the time of harvesting (6), when the necessary levels of humidity are reached. Wet conditions also favour berry infection by airborne spores (7). (Illustration © Virginie Duquette, Gravir un Monde d'Illustration, Switzerland)

Botrytis cinerea overwinters as a saprophyte in the vegetative organs of the vine (leaves, petioles, stems, berries) as well as on plant debris, where they form sclerotia (Fig. 4.42 no. 1). It can persist in the bark and buds in the form of mycelial fragments (Gindro et al. 2014). Furthermore, conidia detached from their conidiophore are capable of surviving even harsh winters or prolonged periods of drought at low temperatures and maintaining their infectious potential until the following season (Walker 1926; Maas 1969; Shiraishi et al. 1970; Gindro and Pezet 2001). When temperature and humidity increase in spring, the sclerotia produce conidiophores bearing conidia. This process is subject to specific environmental parameters such as light intensity (Tan and Epton 1973; Epton and Richmond 1980), temperature (De Istvanffi 1903; Epton and Richmond 1980), humidity (Hopkins 1921; Jarvis 1977) or availability of carbohydrates and minerals in the substrate (Leach and Moore 1966). The conidia are generally responsible for primary infections in young shoots. Each outbreak of infection produces fresh conidia, enabling the infection to be propagated by wind and rain. The flowers have maximum susceptibility to the conidia of *B. cinerea* throughout the entire flowering season. The conidia can germinate at temperatures between 1 and 30 °C, the optimum being 18 °C (Blakeman 1980), in the presence of a film of water or 90% relative humidity, in a wide pH range (Webb 1921). The conidia and hyphae of *B. cinerea* are heterokaryotic and multinucleate. These characteristics confer considerable genetic variability and thus a high degree of adaptability to environmental pressures, which can vary depending on the host plant and environmental conditions. To control the disease effectively, it is essential to understand its epidemiology under real conditions. In the vine, the disease generally appears in an unpredictable manner on ripe berries in the run-up to or during harvest, whereas in other fruit, the fungus causes damage before the harvest (Sommer 1982), or mainly infects the vegetative parts of the plant. As a necrogenic saprophyte, *B. cinerea* can use the remains of flowers (Bristow et al. 1986), or more broadly, senescent vegetative organs (Koike 1997) to infect healthy tissue. There are five different phases of infection in the vine: primary inoculum, infection of the vegetative parts of the plant, infection of the inflorescences, period of latency and expression of the disease.

4.3.4.1 Primary Inoculum

The main condition for successful infection is the presence of a primary inoculum and susceptible plant tissue. However, this prerequisite is not a limiting factor for *Botrytis cinerea*. In fact, fungal particles, including sclerotia, can form at a wide range of temperatures and survive for several years under diverse conditions, including freezing and drought. Similarly, propagules of *B. cinerea*, such as mycelial fragments, can survive in and on bud tissue (Bisiach et al. 1980; Gindro et al. 2014). As a result, vineyards are under constant pressure from the primary inoculum of *B. cinerea*, a fact which is supported by field observations. In other words, this fungus is an integral part of the vineyard microflora. The practice of leaving shredded woody prunings on the ground in winter, where they mix with plant debris containing sclerotia, constitutes a major source of primary inoculum (Nair and Nadtotchei 1987).

The sclerotia produce mycelium, conidiophores and conidia or apothecia (Fig. 4.42 no. 2). However, conidia are the most widespread main source of primary inoculum and infection. These dry conidia are generally dispersed by the wind. Jarvis (1962) observed that conidia are released by a hygroscopic mechanism, with airborne conidia reaching a peak when relative humidity fluctuates between 65% and 85%. These data confirm that bunch infections can occur in the absence of rain since sclerotia present on the ground can form conidiophores over a long period of time at a wide range of temperatures (3–27 °C) and produce conidia repeatedly (Coley-Smith 1980). According to De Istvanffi (1903), they remain viable for more than 20 years. Dry conidia can remain viable for several months or even years when stored without pretreatment at −80 °C without losing their germination potential (Gindro and Pezet 2001). This suggests that spores carried to the upper layers of the atmosphere—where they freeze—could be transported over long distances before landing on plant tissue, thereby playing a major role in long-distance dispersal. The role played by apothecia and ascospores under field conditions is poorly documented, although the presence of *B. cinerea* apothecia is regularly reported on leaves and goods stored in controlled conditions (De Istvanffi 1903; De Bary 1886; Faretra and Antonnaci 1987). The molecular findings of Giraud et al. (1999) show that sexual recombinations can be a common feature in complex populations of *Botrytis cinerea*.

Preservation of the Conidia of Botrytis cinerea as a Function of Temperature
(Gindro and Pezet 2001)

The conidia of *B. cinerea* are the most important structures involved in the infection process. Biochemical and molecular studies linked to their metabolism during dormancy, adhesion to vegetative organs, and germination and penetration of the host tissue provide a better understanding of the pathogen's epidemiological development. Different hypotheses have been proposed to explain the resistance of conidia to cold and desiccation. With no prior treatment, conidia were harvested in sporulating cultures, then sealed in tubes and stored at 21 °C, 4 °C, −20 °C and −80 °C for more than 30 months. Their viability was evaluated based on metabolic parameters such as cellular respiration, adenylate concentration, energy charge, enzyme activity, aggressiveness and ultrastructure of the conidia. The results showed that conidia stored at 21 °C were no longer viable after less than 1 month, while 0.2%, 8% and 80% still germinated after 30 months when stored at 4, −20 and −80 °C respectively. A general decrease in cellular energy level, oxygen consumption and aggressiveness was observed as the storage temperature increased. Ultrastructural analyses showed that conidia stored for 1 month at 21 °C no longer possessed recognisable cellular structures as the cytoplasmic content was totally disrupted. The same results were obtained with conidia stored for more than 30 months at 4 °C. After 30 months at −80 °C, the ultrastructure of the conidia showed more pronounced vacuolisation, the mitochondria were denser and the cytoplasm was retracted slightly, resulting in only minor loss of viability. The cellular water content—considered to be a biological solvent—plays a crucial role in this respect: the hydrogen bonds maintain cohesion between the proteins, carbohydrates, nucleic acids and hydration layers. Loss of water completely disrupts this equilibrium, leading to an accumulation of non-aqueous compounds such as membrane lipids. Anemochorous conidia, such as those of *Botrytis*, have a very low water content. In such biological systems, vitrification of water at −80 °C does not produce ice crystals that are harmful to the cell. This special state of water may play a role in protecting cell membranes during dehydration and freezing. However, even at these very low temperatures, long-term storage at normal atmospheric pressure inevitably results in gradual sublimation of water. This slow but progressive loss of water is responsible for the ultrastructural modifications observed in conidia and a loss of viability in the region of 20%.

4.3.4.2   Infection of Green Parts

In spring, infection of the vegetative organs of the vine by *Botrytis cinerea* can be considered the first stage of its epidemiological development (Fig. 4.42 no. 3), corresponding to the formation or maintenance of the primary inoculum. Such infections do not generally have an economic impact. For the vine, however, high relative humidity and heavy rainfall can in rare situations lead to severe infections of the green organs before flowering, resulting in critical epidemic outbreaks on buds, shoots, leaves and rachis. On the leaves, the first symptoms are often visible on the veins or petiole, forming large necrotic patches. On the young shoots and rachis, infections are responsible for the total breakdown of apical tissue, sometimes causing major damage to young plants in vine nurseries. Once the infection has become established in plant tissue, suitable humidity and temperature conditions enable the fungus to sporulate on affected organs, thereby creating significant potential for inoculum through successive cycles of infection and sporulation (Fig. 4.42 no. 4).

4.3.4.3   Bloom Infection

The crucial stage in the epidemiological development of *B. cinerea* on the vine is the infection of flowers, culminating in a latent period lasting until veraison (Fig. 4.42 no 5). Grey mould of grape has always been associated with infection of mature berries after rainfall or prolonged periods of high relative humidity close to harvest. In 1973, McClellan and Hewitt added a new element to the epidemiological cycle of *B. cinerea*, which they called 'early rot'. It occurs during flowering through infection of the stigma and style and remains dormant until berry ripening. Before flowering, the flowers are covered by a cap (calyptra), which normally detaches from the base of the ovary and falls off. However, in humid conditions during flowering and in certain varieties, the calyptra and stamens can remain attached to the berry's pedicel. These dehiscent tissues are particularly conducive to the development of *B. cinerea*, although the germination rate does not appear to be higher on calyptra detached from the base of the ovary than on other floral organs (Keller et al. 2003). Nair and Allen (1993) studied the conditions for flower and berry infection and showed that the duration of green tissue wetness was the key difference. Bloom infections require 1.3 h of wetness, while berry infections need at least 14 h. These results highlight the great vulnerability of flowers to grey mould infections. Nair et al. (1995) established a direct quantitative relationship between flower infection and the incidence of *Botrytis* at harvest and developed a model to measure the potential risk of expression of the disease at harvest. Flower infections, especially at the most susceptible full-bloom stage, are potentially important for disease development and the severity of its impact on harvests (Keller et al. 2003; Pezet et al. 2003).

Pezet and Pont (1986) showed that, after inoculation with conidia at flowering, the fungus can infect the base of the stamens, colonise the receptacle area, then grow within the pedicel. They also showed that infections initiated before flowering led to abscission of the rachis, whereas infections at the time of flowering led to

B. cinerea latency. The receptacle area of grape flowers has an open interstitial zone (channel-like gap) at the base of the calyptra where the stamen filaments are attached, corresponding to the point of attachment of the future berry on the pedicel (Fig. 4.43). The conidia of *B. cinerea* can become established in this zone, and no doubt germinate and remain dormant there (Keller et al. 2003; Viret et al. 2004; Gindro et al. 2005).

4.3.4.4   Latency

The term 'latent infection' was first defined by Gäumann (1951) as a dormant parasitic relationship which can become active after a certain time. From an epidemiological point of view, latency is the period from infection until the infected tissue becomes infectious, the 'spore-to-spore' principle defined by Vanderplank (1963). Latency is a common feature in plant-fungus interactions and has been reported for *B. cinerea* in the main small-berry crops such as the strawberry. Infection occurs during flowering and the disease generally expresses itself in ripe fruit. In the grapevine, the latent period runs from the end of flowering (BBCH 69) until the start of veraison (BBCH

Fig. 4.43 Semi-thin section of a vine flower showing an interstitial zone (it) in the receptacle area in which the conidia of *Botrytis cinerea* (arrows) can remain dormant until the disease expresses itself. The presence of *B. cinerea* in this interstice has been validated by PCR (Gindro et al. 2005). *ca* calyptra, *nt* necrotic tissue

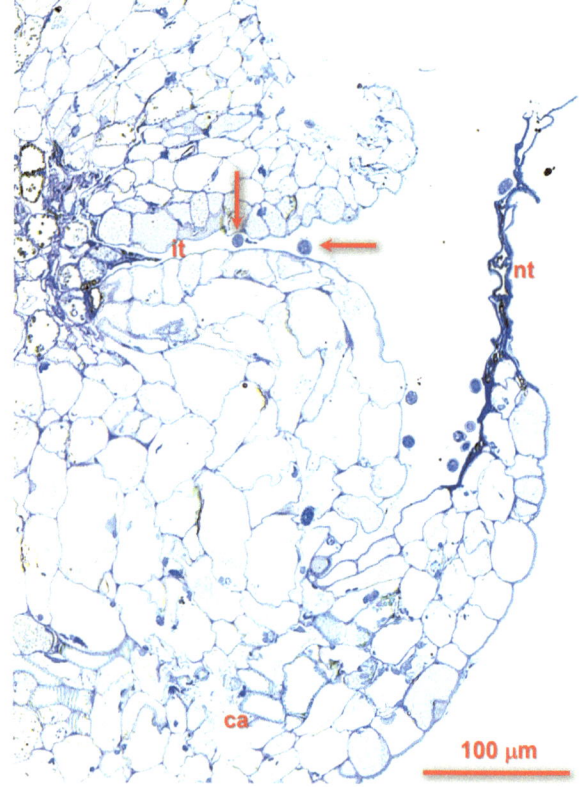

81). This period lasts around 2 months, depending on the grape variety and the weather conditions. During this relatively long period, immature berries are highly resistant to *B. cinerea*, even though the fungus can be readily isolated from the tissues after surface disinfection (Keller et al. 2003; Pezet et al. 2003). The vine's natural biochemical defence mechanisms (biochemical factors, either constitutive or induced by the fungus) are at play here, keeping the pathogen dormant.

Infection Pathway in Inflorescences and Latent Period
(Pezet and Pont 1986; Keller et al. 2003; Viret et al. 2004; Gindro et al. 2005)

The pathway for *Botrytis cinerea* infection of grape flowers has been long disputed. Artificial infections have shown that full bloom is the most susceptible period. Depending on the weather conditions and variety, this contamination increases visible mould by 20–30% at harvest. For a long time, the infection pathway for *B. cinerea* in flowers was thought to be located in the stigma, in parallel with fertilisation. However, *B. cinerea* was detected in the receptacle area with the aid autoradiography (conidia marked with ^{14}C); but never in the ovary tissue (Pezet and Pont 1986). To pinpoint the pathway for the pathogen, different floral organs of the grapevine, including the calyptra, stigma and receptacle area, were artificially inoculated with an aqueous suspension of conidia. The initial stages involved in the colonisation and infection of host tissue were studied under a microscope for several days. The results showed that conidia germinated on all floral organs examined and remained attached to the surface of the host for 48 h after inoculation (Keller et al. 2003; Viret et al. 2004). In all cases, most of the conidia accumulated in the channel-like gap between the ovary and the receptacle (Fig. 4.43), which tapers inwards and corresponds to the point where the future berry is attached to the tip of the pedicel (Viret et al. 2004). Conidial germ tubes were rarely detected on the style following inoculation of the stigma, and no evidence of their growth towards the ovaries was found. In contrast, hyphae were observed in abundance in the receptacle area, regardless of the site of inoculation. The tips of the necrotic receptacle and the mycelium formed in the gap between the ovary and the receptacle are key indications of colonisation and infection. This explains why the first outbreaks of rot in susceptible varieties generally start inside the bunches and often remain invisible. *B. cinerea* also colonises dehiscent calyptras, which provide a potential source of inoculum if they remain stuck within the cluster. The receptacle area and calyptras are thus the main infection pathways for *B. cinerea*. To consolidate the results, a molecular assay was developed which enables a single propagule of *B. cinerea* to be detected in the vegetative tissues without extracting DNA in advance (Gindro et al. 2005). Using this technique, the latency of *B. cinerea* in artificially inoculated flowers was monitored in the field, clearly establishing that it was restricted to the receptacle area of asymptomatic berries at the pea-sized stage, other parts of the berry being exempt from the latent pathogen.

4.3.4.5 Development of Visible Berry Rot

The next and most destructive stage in the epidemiological development of grey mould is the expression of the disease, culminating in the growth of latent fungal biomass (Fig. 4.42 no. 6) combined with infections from external sources at veraison (Fig. 4.42 no. 7). When relative humidity is high, infected berries become covered in fungal mycelium, conidiophores and conidia. The process generally starts inside the bunches, where adequate moisture levels are maintained by the microclimate, especially on clones and varieties with compact bunches. In vineyards in the Northern Hemisphere, this economically important phase depends mainly on prevailing weather conditions from veraison to harvest, varietal characteristics and cultivation methods. All these factors correlate with the berry ripening process, especially the structural and biochemical modifications of the berry surface (Padgett and Morrison 1990), the berries themselves (Hardie et al. 1996) and tissue senescence during the ripening phase (Pezet et al. 2003). It is therefore not surprising that *B. cinerea* becomes active when the berries' natural defence mechanisms begin to weaken.

Hot, humid conditions during berry ripening favour the expression of grey mould. In the Northern Hemisphere, *B. cinerea* can totally destroy the harvest or significantly impair grape quality if wet weather occurs in late August and September. At this stage, both healthy berries and those that have been mechanically damaged can be infected externally. In some varieties, heavy rainfall before harvest can cause the berry skin to split, enabling *B. cinerea* to become established and sporulate (Fig. 4.44). Wet conditions seem to accelerate the senescence and maturation process in plant tissue (De Luca d'Oro and Trippi 1987), further increasing the risk of infection.

Fig. 4.44 In some varieties, heavy rainfall close to harvest can cause the berries to split, thereby favouring the establishment of *Botrytis cinerea*

4.3.5   Disease Control

The susceptibility of grapes to grey mould varies greatly depending on the variety and the climate conditions. Varieties with compact bunches or thin-skinned berries are generally more susceptible to grey mould. Any preventive control measures aimed at encouraging a dry microclimate within the bunch zone should be adopted before using plant protection products. Choice of variety, clone and rootstock, row orientation, planting density and pruning system are factors to consider before planting a vineyard. Cultivation practices such as restricting vigour through reduced nitrogen fertilisation or choice of rootstock, suppressing axillary shoots and leaves in the bunch zone, controlling larvae (grapevine moths: *Eupoecilia ambiguella*; *Lobesia botrana*), or cover cropping can significantly reduce the impact of grey mould.

Chemical control can only be preventive, and it is essential to consider the resistance of the fungus to products that have previously been used when choosing an active ingredient. Without performing a costly resistance analysis, the observation that control measures are becoming less effective indicates the need to modify the control strategy by switching to active substances in different chemical groups to those previously applied. Fungicides applied to control downy mildew at flowering have a secondary effect against grey mould that is sufficient to control floral infections. Chemical control is more effective at bunch closure and the beginning of veraison. The main active chemical groups against grey mould are, dinitroanilines, hydroxyanilides, phenylpyrroles (PP fungicides), dicarboximides, methyl benzimidazoles carbamates (MBCs), anilinopyrimidines (AP fungicides), succinate dehydrogenase inhibitors (SDHI) and keto-reductase inhibitors (KRIs). The critical moment in terms of efficacy is just before bunch closure, with the second treatment at veraison reserved for situations and varieties that are particularly susceptible to grey mould. In susceptible varieties, splitting the bunches in two after veraison can significantly reduce sour rot and grey rot, as can the application of gibberellic acid during flowering, which modifies the stem structure and rate of fruit set (Spring and Viret 2009, 2011).

4.4   Black Rot

Current name:

• *Phyllosticta ampelicida* (Engleman) Aa 1973, Ascomycota, syn. *Guignardia bidwellii* (Ellis) Viala & Ravaz 1892

Black rot, also called dry rot, is a specific disease of the grapevine and Virginia creepers (genera *Parthenocissus*, *Ampelopsis* and *Cissus*) originating from North America. All varieties of the European grapevine (*V. vinifera*) are susceptible (Hausmann et al. 2017) and harvest losses are inevitable without suitable phytosanitary treatments. This fungus was first described in North America in 1853, then in France in 1885 by Viala and Ravaz (1886). Its Europe-wide introduction from the end of the nineteenth century in the wake of downy mildew and powdery mildew can be attributed to the importation of phylloxera-resistant rootstocks. Today, black rot is found in all the world's major winegrowing regions. This pathogen, which has been known of for some time, does not generally pose major problems except in the north-eastern United States, Canada and certain regions characterised by a temperate climate and heavy spring and summer rainfall. In Europe, climate change and restrictions on the use of plant protection products have led to the re-emergence of this somewhat overlooked pathogen. Black rot has regularly been observed in the Mosel (Germany) since 2002 and is making a comeback in all the vineyards of Germany, Luxembourg, Austria, Switzerland, Romania, and more recently, in the warmer Mediterranean climate of Portugal and Italy (Jermini and Gessler 1996; Molitor et al. 2011; Rinaldi et al. 2013). Its presence is generally associated with abandoned or neglected vineyard plots and inadequate control of downy mildew and powdery mildew through the use of less effective active ingredients. The disease's resurgence can also be explained by the planting of interspecific varieties resistant to downy mildew and powdery mildew but susceptible to black rot, and use of plant protection products to only a very limited extent, if at all (Pertot et al. 2017). The scale of infection can be controlled with plant protection measures and the systematic application of effective fungicides, as well as the removal of abandoned vines.

4.4.1 Causal Organism

Phyllosticta ampelicida belongs to the Ascomycetes, the order Botryosphaeriales and the family Phyllostictaceae. The causal agent of black rot was first described under the name of *Sphaeria bidwellii* by Job Bicknel Ellis (1880), an American mycologist, in honour of Dr. E.C. Bidwell, who discovered the fungus on mummified berries in May 1880. *S. bidwellii* was subsequently renamed *Guignardi bidwellii* by Viala and Ravaz (1892). The name of the asexual form *Phyllosticta ampelicida* is prioritised over the more common name *Guignardia bidwellii* (sexual form) because *Phyllosticta* references more species of plant pathogen on a broad range of host plants and due to the name's seniority (Zhou et al. 2015). However, in the literature grapevine black rot is more often cited under the name *Guignardia bidwellii* (Szabó et al. 2023).

Three specialised forms of *G. bidwellii* (Luttrell 1948; Wicht et al. 2012; Rinaldi et al. 2017) with different host plants within the Vitaceae family have been described:

- *G. bidwellii* f. sp. *euvitis*, pathogen of the American *Vitis* and *Vitis vinifera*
- *G. bidwellii* f. sp. *muscadinii* on *Vitis rotundifolia* syn. *Muscadinia rotundifolia* and *V. vinifera*
- *G. bidwellii* f. sp. *parthenocissi* on *Parthenocissus*

Under the name *Phyllosticta*, grapevine black rot is regarded as a complex of four species (Zhang et al. 2013; Zhou et al. 2015): *P. ampelicida*, *P. parthenocissi*, *P. partricuspidatae* and *P. vitis-rotundifolia*.

P. ampelicida, generally described as *G. bidwellii*, is the principal pathogen of *Vitis vinifera* and the interspecific hybrids. *P. ampelicida* is a hemibiotrophic fungus which infects living tissue and completes its developmental cycle on dead tissue. In spring, the perithecia (Fig. 4.45)—also called pseudothecia, (61–199 µm in diameter)—embedded in the skin of mummified berries and tipped with an ostiole (Janex-Favre et al. 1996), ripen and release unicellular ascospores, hyaline, oblong, rounded at the ends, (10.6–18.4 × 4.8–9 µm) contained in cylindrical, bitunicate asci (45–65 × 9–14 µm). The black pycnidia (59–196 µm in diameter), arranged in concentric circles on the lesions, have an ostiole at the tip (Janex-Favre et al. 1993) (Fig. 4.46). Conidiogenesis inside the pycnidia starts before the conidia are released (Janex-Favre et al. 1993). It generates a multitude of unicellular pycnidiospores (conidia), ovoid to ellipsoidal, hyaline, binucleate (7.1–14.6 × 5.3–9.3 µm) (Fig. 4.46).

Fig. 4.45 Cross-section of a perithecia (pe) of *Phyllosticta ampelicida* on a mummified berry skin (bs) containing several asci (a) in which ascospores (as) are produced; ascospores and conidia are responsible for the primary inoculum after overwintering on infected plant parts. Scale bar represents 50 µm

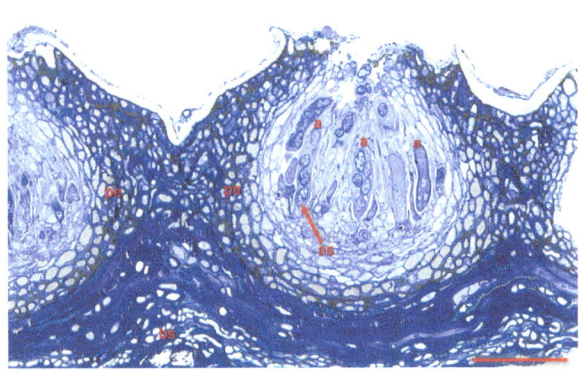

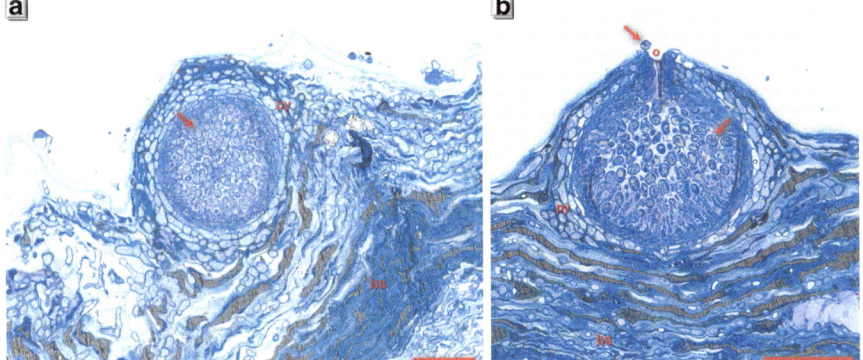

Fig. 4.46 (**a** and **b**) Pycnidia (py) of *Phyllosticta ampelicida* on mummified berry skin (bs) containing pycnidiospores (arrows) released through the ostiole (o). Scale bar represents 50 µm

4.4.2 Symptoms

All green organs of the vine can be infected by black rot. Young leaves are particularly susceptible, as well as inflorescences and berries immediately after fruit set (BBCH 71). Bunches can be infected by the pathogen until veraison (BBCH 83–85). Leaf symptoms are characterised by orbicular to polygonal spots, pale greyish-brown initially, turning dark brown, from 2 to over 10 mm in diameter (Fig. 4.47). Numerous isolated or confluent spots can form on the surface of the leaves. The edge of these necrotic lesions is darker and when they appear, they can easily be confused with damage caused by contact herbicide drift. A few days after they appear, small black dots become visible inside the lesions, confirming the presence of black rot. These pustules correspond to the pycnidia (asexual fruiting bodies of the fungus) which contain masses of conidia that infect healthy plant organs (Figs. 4.48 and 4.49). Pycnidia also develop on elongated lesions appearing on shoots, stems, petioles and tendrils (Fig. 4.49), causing confusion with the symptoms of anthracnose (*Elsinoë ampelina*). Individual affected berries turn brown within the bunch, gradually shrivel and eventually become mummified (Fig. 4.50). Entire bunches can be affected by black rot. The skin of the berries is covered in tiny black pustules containing a mixture of pycnidia and perithecia. These asexual and sexual fruiting bodies are distinguishable only under a microscope. In summer, the pycnidia alone serve as a source of infection by releasing conidia through their ostiole. The perithecia develop in late summer, differentiating to form asci and ascospores. These structures, as well as pycnidia present on different organs of the vine, fall to the ground, where they provide the primary inoculum for the following year's infections.

Fig. 4.47 Leaf symptoms of black rot initially appear as brown, isolated or confluent necroses (arrow) (**a**), initially without fruiting bodies (insert). Later, several black pycnidia containing conidia appear in a more and less circular configuration (**b–d**) (arrows)

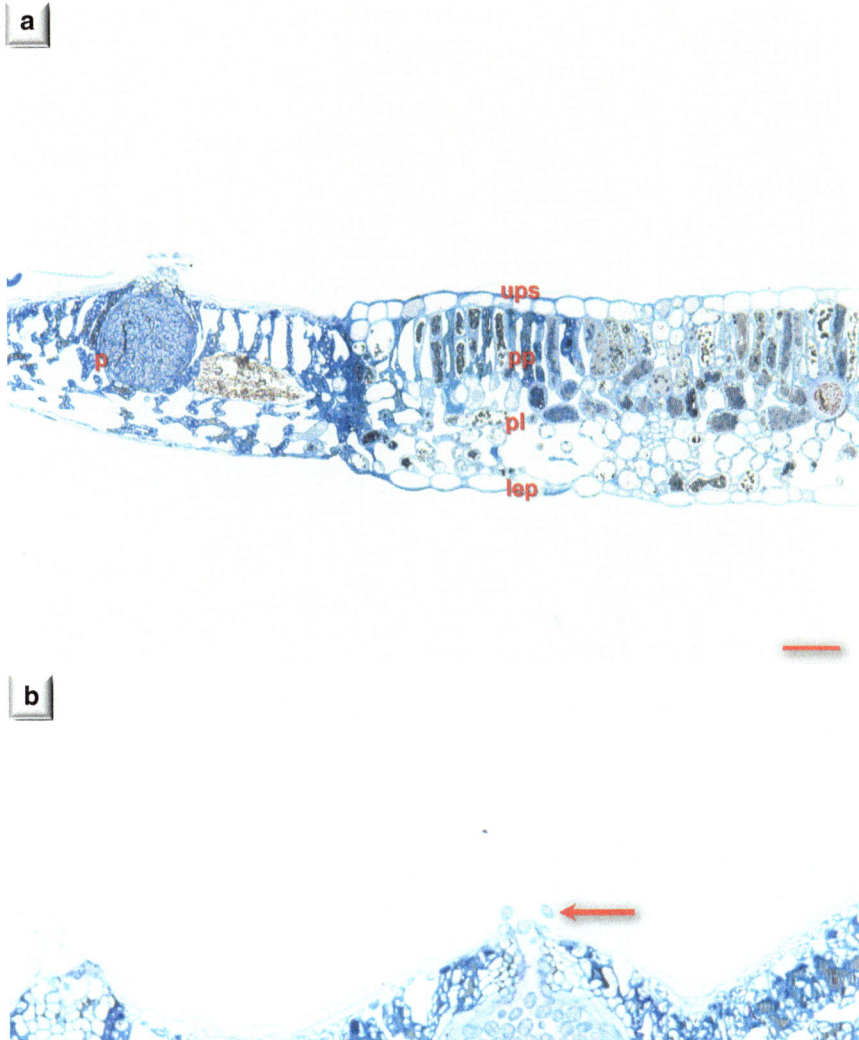

Fig. 4.48 Cross-section of a grapevine leaf showing a mature pycnidia of *Phyllosticta ampelicida* (**a**) containing conidia (**b**) (arrow) able to infect all green parts of the vine throughout the summer. *lep* lower epidermis, *pl* lacunar parenchyma, *pp*: palisade parenchyma; ups: upper epidermis. Scale bars represent 50 μm

Fig. 4.49 First lesions of black rot on shoots (**a**). Under high disease pressure, shoots can be severely infected, developing long dark brown necrotic areas which produce several pycnidia (**b**)

4.4.3   Biology and Epidemiology

The disease cycle of black rot is summarised in Fig. 4.51. Black rot is a polycyclic disease which begins its cycle with the primary inoculum (perithecia and pycnidia) present on parts of the vine infected the previous year (Fig. 4.51 no. 1). This inoculum can survive for several years, leading to an accumulation of infectious particles and strengthening the disease over time. The fungus overwinters on the ground in mummified berries or necrotic shoots, and even in grape pomace from contaminated plots which has been used as an organic mulch. Primary infections can be initiated by either ascospores (Fig. 4.51 no. 2′) or conidia (Fig. 4.51 no. 2). Conidia produced in vast numbers are the main source of infection in the summer (Molitor and Beyer 2014). In humid weather, ascospores and conidia can be released from bud break until the end of July. They are deposited on different organs of the vine by the wind and by water droplets splashed up from the ground, where they germinate and penetrate the plant tissue (Fig. 4.51 no. 3). Pycnidia developing on necrotic lesions during the growing season release masses of conidia which go on to infect healthy organs (Fig. 4.51 no. 4). One to three hours of rainfall is sufficient for their dispersal, although more prolonged rainfall washes the particles from the plant and is less conducive to the development of the disease. Ascospores can be released after 3 mm of rainfall, whereas at least 10 mm of rain is required for pycnidia to ripen. Conidia continue to be produced for a period of almost 3 months, even when humidity levels are low (Onesti et al. 2017). Leaves can be infected at any age. The hyphae of

Fig. 4.50 Black rot infections appear as pale creamy-brown berries, similar to sour rot, surrounded by healthy berries (**a** and **b**). Later, the infected berries dry completely, developing a mummified appearance (**c**). Perithecia and pycnidia appear on the surface of diseased berries, ensuring the primary inoculum for the next year (**d**) (arrows)

P. ampelicida penetrate directly by producing appressoria which digest the cell walls. The subcuticular mycelium gradually colonises the mesophyll, causing necrosis, degeneration and desiccation of the vine tissue (Fig. 4.51 no. 5).

The severity of black rot infections is linked to leaf wetness duration, relative humidity and phenological development of the vine (Spotts 1977; Molitor et al. 2016). The longer and hotter the humid periods, the more severe the infections. The incubation period depends on temperature. Regardless of the age of the leaf, the first symptoms on leaves appear after a cumulative temperature of 175 °C is reached (sum of average daily temperatures between minimum 6 °C and maximum 24 °C). This temperature threshold also applies to bunch infections until bunch closure (BBCH 77–79) (Molitor et al. 2012). From this stage onwards, their ontogenic resistance increases. The emergence of leaf symptoms starts with localised bleaching of the leaf blade, followed by tissue necrosis the following day. The first pycnidia are usually visible 1 day after the appearance of the necrotic lesions. The incubation period for black rot is particularly long (period between penetration of the host tissue and emergence of symptoms); 3 weeks for leaf and young berry infections and up to 5 weeks for bunch infections (Hoffman et al. 2002, 2004).

Fig. 4.51 Disease cycle of *Phyllosticta ampelicida*. The fungus overwinters in infected tissue (mummified berries, woody shoots, leaf matter) during the previous season in the form of pycnidia or perithecia, which produce pycnidiospores and ascospores respectively (1). The conidia (2) and the ascospores (2′) constitute the primary inoculum for the infection of green tissue. The conidia germinate on the surface of grapevine organs and their hyphae form an appressorium which penetrates the tissue by enzymatic digestion of the cellular components (3). After colonising the mesophyll, new pycnidia are produced throughout the winegrowing season (4) on all infected tissue, producing masses of conidia ready to infect. Mummified tissue can contain hundreds of perithecia and pycnidia, enabling the fungus to overwinter until the next season (5). (Illustration © Virginie Duquette, Gravir un Monde d'Illustration, Switzerland)

Black rot is a disease which develops gradually, especially in varieties resistant to downy mildew and powdery mildew which lack the resistance genes for *P. ampelicida* and require only a limited number of plant protection treatments. Although the first mummified berries or leaf lesions may initially appear insignificant, they provide the primary inoculum for the following years. If the disease is not brought under control before becoming established, economic losses will be difficult to avoid after a few years.

4.4.3.1 Ontogenic Bunch Resistance

Berries are at their most susceptible between flowering (BBCH 61–69) and the start of bunch closure (BBCH 77). Their ontogenic resistance increases until 450 cumulative degree-days above 10 °C (Molitor and Berkelmann-Löhnertz 2011). Infections can no longer occur after the onset of veraison (BBCH 81–83). On bunches, the duration of the incubation period increases continuously with the berries ongoing development. Consequently, the incubation period of an infection at a late stage of berry development can be twice as long as that of an infection occurring at flowering (Hoffman et al. 2002; Molitor et al. 2016).

4.4.4   Disease Control

4.4.4.1   Disease Forecasting

To simulate the epidemiological development of black rot, data on wetness duration obtained by Spotts (1977) were integrated into algorithms for the first predictor model (Ellis et al. 1986). Subsequent models incorporated a 3-day weather forecast, allowing a strategy for fungicide applications to be defined. However, these early models did not consider the severity of infection, the incubation period or the evolution of vine susceptibility as a function of its phenological development (Spotts 1980; Hoffman et al. 2002; Molitor et al. 2012).

The mechanistic dynamic model developed by Rossi et al. (2014) incorporates knowledge about the biology and conditions required for the epidemiological development of black rot on leaves and bunches. This model defines the ripening of ascospores and conidia on mummified berries and predicts primary infections when rainfall is equal to or greater than 0.5 mm in 1 h. The severity of infection is then calculated by integrating the temperature and wetness duration as well as the emergence of symptoms after an incubation period of 175–305 degree-days between 6 °C and 24 °C (cardinal temperature corresponding to the limit of development of black rot). Pycnidia appear on lesions between 262 and 392 degree-days at the same thresholds of cardinal temperatures and the quantity of conidia released is calculated as a function of temperature.

Similarly, the model developed by Molitor et al. (2016) incorporates climatic data (temperature, rainfall, leaf wetness duration and relative humidity), 5-day weather forecasts, the grapevine growth model (Schultz 1992) and bunch susceptibility (Molitor and Berkelmann-Löhnertz 2011) to calculate the infection index on leaves and bunches; zero (<85), low (85–150), medium (150–300) or high (>300). Like the model developed by Rossi et al. (2014), infections are followed by an incubation period (Molitor et al. 2012) which ends after 175 cumulative degree-days (using 6 and 24 °C as cardinal temperatures) when symptoms appear, but Molitor applies a correction factor for bunch infection.

When incorporated into decision support systems, these models enable the targeted use of fungicide treatments based on the predicted incidence of black rot infections. By adopting this targeted approach, it is possible to reduce the number of applications compared with a fixed treatment regime (Onesti et al. 2016; Molitor et al. 2016).

Host Sensitivity and Resistance Genes to Black Rot
The host plants of black rot belong to the genera *Vitis, Cissus, Parthenocissus* and *Ampelopsis*. The species *Vitis vinifera* is the most susceptible to the disease, while *V. riparia, V. rupestris, V. mustangensis, V. vulpina* and *V. sylvestris* are virtually resistant. While most North American and French hybrids that are resistant to downy mildew and powdery mildew vary in terms of their susceptibility to black rot, they are generally less susceptible than *V. vinifera* (Hausmann et al. 2017). To date, QTLs (Quantitative Trait Loci) associated with black rot (which is probably polygenic) have been identified; Rgb1 located on chromosome 14 (Dalbò et al. 2000; Rex et al. 2014) and Rgb2 on chromosome 16 (Rex et al. 2014) identified on the genotype of the 'Börner' rootstock (*V. riparia* Gm183 × *V. cinerea* Arnold). This resistance, originating in *V. cinerea*, is defined as partial. More recently, Bettinelli et al. (2023) have identified, on Merzling, a major new QTL associated with black rot bunch resistance, designated *Rgb3*, and, like *Rgb1*, located on chromosome 14. The authors showed that the physical region encompassing the two QTLs does not comprise annotated resistance (R) genes. The *Rgb1* locus is linked to phloem dynamics and mitochondrial proton transfer, while *Rgb3* presents a cluster of pathogenesis-related genes, promoters of programmed cell death (Bettinelli et al. 2023). Breeding programmes for selecting resistance to downy mildew and powdery mildew which do not take account of susceptibility to black rot are creating a new plant protection problem with this previously insignificant disease. It is therefore important to include black rot resistance in selective breeding programmes, to continue to explore diverse sources of resistance to this fungus to prevent the emergence of resistant strains and ensure long-term resistance to black rot, and to avoid using fungicides even when disease pressure is high.

4.4.4.2 Control Strategies

Primary infections of black rot are generally linked to abandoned or neglected neighbouring vineyards or to the minimal application of plant protection products on varieties which are resistant to downy mildew and powdery mildew but susceptible to *P. ampelicida*. In plots infected with black rot, the primary inoculum can be reduced using preventive measures designed to eliminate necrotic organs, such as removing mummified berries before harvesting and removing shoots and leaves from contaminated plots.

When preventive measures are insufficient, plant protection products must be applied. Active substances from dithiocarbamate, strobilurine (QoIs inhibitors) or demethylation inhibitors—triazol (DMIs) families used to control downy mildew and powdery mildew are equally effective against black rot (Molitor et al. 2011). Copper and sulphur are the only option in organic vineyards, and their effectiveness is limited. Control measures should be implemented from bud break in correlation with rainfall events and in combination with the control of downy mildew. The period around flowering is the most susceptible. Warm, dry springs are particularly unfavourable to the development of black rot.

4.5 Anthracnose, Black Spot, Bird's Eye Rot

Current name:

* *Elsinoë ampelina* (De Bary) Shear 1929, Ascomycota, syn. *Sphaceloma ampelinum* De Bary 1874

The fungus *Elsinoë ampelina* is the causal agent of a vine disease which affects all winegrowing regions of the world, especially Argentina, Australia, Brazil, Canada, China, India, Europe, Japan, Korea, New Zealand, South Africa, Thailand, the United States and Uruguay. This disease, commonly known as anthracnose, can cause extremely serious damage to both *Vitis vinifera* and American vines by attacking the foliage, bunches and woody shoots. Anthracnose originated in Europe. According to Viala (1893), the disease described by Théophraste and Pline appears to be anthracnose of the vine. In Europe, this fungal disease was described at the start of the seventeenth century and caused severe economic impacts prior to the arrival of downy mildew and powdery mildew towards the end of the nineteenth century. Thomas J. Burril, one of the pioneers of plant pathology, was the first to observe the disease in the United States, in Illinois, in 1881 (Burril 1886). De Bary studied the disease from 1873, giving precise descriptions of the fungus, which at the time was called *Sphaceloma ampelinum*. Since then, anthracnose has been very widely studied in all winegrowing regions of the world. As a result, the name of the genus and species frequently changed before the name of the ascogenic form of the fungus described by Shear (1929), namely *Elsinoë ampelina,* was finally adopted on the basis of nomenclatural priority. Preventive measures to control downy mildew and anthracnose of the vine, initially using Bordeaux mixture based on copper, then other active substances, has practically eradicated anthracnose. However, the disease is still a concern in hot and humid regions, especially in tropical zones, in mother vine fields for American vines and in hybrids resistant to downy mildew and powdery mildew.

4.5.1 Causal Organism

Elsinoë ampelina belongs to the Ascomycota, the order Myriangiales, the family Elsinoaceae and the genus *Elsinoë*. Anthracnose was mentioned for the first time at the start of the seventeenth century under the name 'vine smut'. This term was subsequently abandoned to avoid confusion with 'grain smut' (*Ustilago nuda, U. tritici*), also discovered at that time. The fungus forms isolated or confluent acervuli on the surface of lesions (<60 μm diameter) containing condiogenous cells, mono- or polyphialidic, hyaline to light brown (3.5–6 × 3–5.5 μm) which produce unicellular conidia, hyaline, cylindrical to oblong, at their rounded tips (3.4–7.5 × 2–3.5 μm, Li et al. 2018) throughout the summer (Fig. 4.52). In autumn, sclerotia appear at the edges of the necrotic lesions, ensuring the pathogen's survival through the winter. In spring, conidia produced directly by the sclerotia or developing within the acervuli initiate primary infections. Asci (12.9–34.3 × 12.6–29.6 mm), globose or elliptical, bitunicate, develop in the pyriform cavities of localised stroma inside the lesions. The asci contain eight hyaline ascospores comprising two to four cells (one to three-septate), 15–16 × 4–5 μm. Braga et al. (2020) showed that the sexual and asexual stages of *E. ampelina* are present from the start of the epidemic, which may help to explain the explosive nature of the disease when the first stages of vine growth coincide with the rainy season. *E. ampelina* produces a phytotoxin similar to cercosporin, the red/orange elsinochromes which appear to be vital to the infection process and are probably virulence factors of the pathogen (Jiao et al. 2019).

4.5.2 Symptoms

Anthracnose can infect all green organs of the vine (leaves, petioles, bunches, stems, berries, tendrils, shoots), especially in the juvenile stage when tissues are particularly vulnerable (Magarey et al. 1993). Circular or angular necrotic spots several centimetres in diameter appear on the leaves. The spots are brownish-purple to begin with, then the border blackens and the centre turns greyish-white and shrivels (Fig. 4.53). These necrotic centres subsequently drop out, giving the impression that the leaves have been peppered with lead shot. Young leaves curl up and further growth is restricted. The foliar symptoms can be mistaken for those of excoriose. Affected shoots also have brownish-purple necrotic lesions with greyish-white centres, but

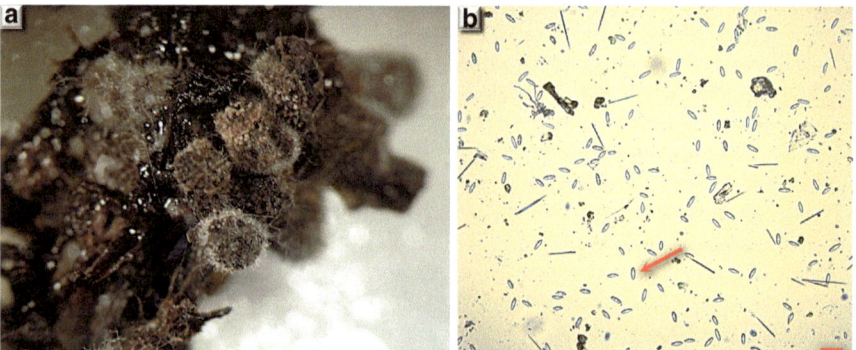

Fig. 4.52 *Elsinoë ampelina* on a canker (mass of confluent acervuli) (**a**) and (**b**) cylindrical to oblong conidia (arrow) observed under optical microscope. Scale bar represents 5 μm. (Photos Odile Carisse from Agriculture and Agri-Food Canada, © His Majesty the King in Right of Canada, represented by Agriculture and Agri-Food Canada 2024)

Fig. 4.53 Symptoms of anthracnose on a leaf. The leaf blade is speckled with necrotic patches, circular or angular, purplish-brown initially (black arrow), then blackening: the centre of the patches turns greyish-white, shrivels and drops out (red arrow). (Photos Odile Carisse from Agriculture and Agri-Food Canada, © His Majesty the King in Right of Canada, represented by Agriculture and Agri-Food Canada 2024)

Fig. 4.54 Symptoms of anthracnose on a green shoot showing purplish-brown necrotic spots (black arrow), also with a greyish-white centre (red arrow). (Photos Odile Carisse from Agriculture and Agri-Food Canada, © His Majesty the King in Right of Canada, represented by Agriculture and Agri-Food Canada 2024)

more elongated (Fig. 4.54). These lesions, sometimes extending as far as the medullary rays of the shoot, can be mistaken for the symptoms of black rot or excoriose (Fig. 4.55). On shoots, hail can cause similar necroses. Severe infections of *E. ampelina* can cause the shoots to break, leading to total loss of the crop and threatening the vine's survival. These symptoms also affect the tendrils, stems and petioles of the leaves. On immature berries, anthracnose manifests itself as sunken purplish spots with a greyish centre resembling a bird's eye, hence its common name in English (Fig. 4.56). In severe infections, affected berries split at the lesions. Vines infected by anthracnose typically have low yields and delayed berry ripening. In severe cases, shoots defoliate prematurely, berries shrivel and sap flow is interrupted. The leaves of susceptible varieties of the genus *V. vinifera* infected by *E. ampelina* have reduced levels of chlorophyll, carotenoids and ascorbic acid compared with healthy leaves, while their malondialdehyde levels are significantly elevated (Murria et al. 2018).

Fig. 4.55 Severe symptoms of anthracnose on shoots (**a**) show large confluent brownish-black necrotic zones, light grey in the centre, where acervuli and conidia are formed. They can be confused with symptoms of *Phomopsis* dieback (*Diaporthe ampelina*) (**b**) or black rot (*Phyllosticta ampelicida*) (**c**)

Fig. 4.56 On immature berries, anthracnose manifests itself as purplish spots (left), becoming sunken necrotic lesions with a greyish centre resembling a bird's eye, hence its common name in English. (Right-hand photo: Odile Carisse from Agriculture and Agri-Food Canada, © His Majesty the King in Right of Canada, represented by Agriculture and Agri-Food Canada 2024)

4.5.3   Biology and Epidemiology

Elsinoë ampelina is specific to the genus *Vitis* and also infects interspecific hybrids resulting from crossings between *V. vinifera* and *Vitis* sp. The disease cycle of *E. ampelina* is summarised in Fig. 4.57. Anthracnose is a polycyclic disease whose conidia are infectious particles that can form at relatively low temperatures (>2 °C). The pathogen overwinters in the form of sclerotia in the cankers of infected organs

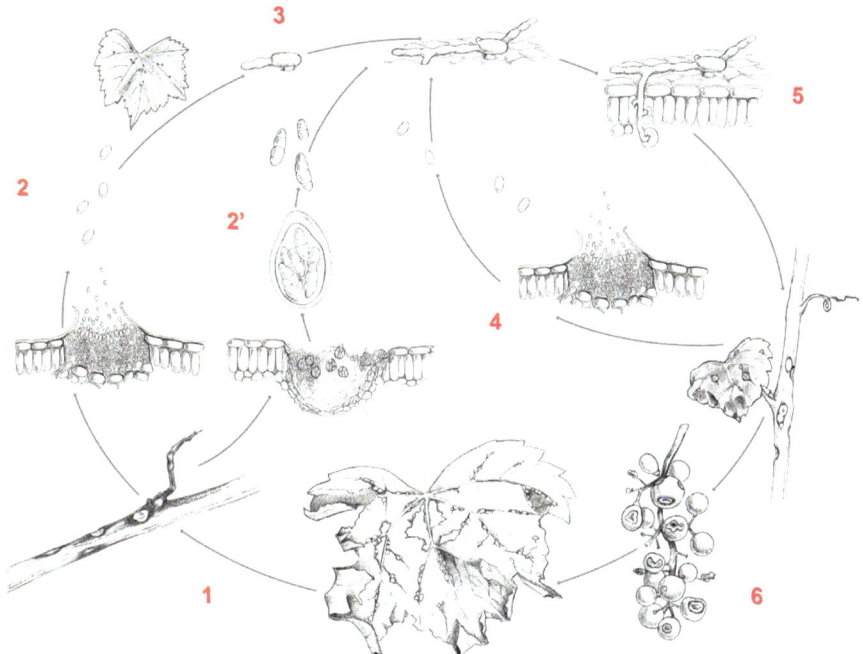

Fig. 4.57 Disease cycle of *Elsinoë ampelina*. The pathogen overwinters in the form of sclerotia in the cankers of infected organs (berries, leaves, shoots) (1). In spring, the sclerotia germinate directly, discharging conidia after at least 24 h of leaf wetness and temperatures above 2 °C (2–3). Primary infections can also be caused by ascospores, although their epidemiological role remains hypothetical (2′). The pathogen then produces acervuli containing the conidia responsible for secondary infections, which amplify the disease and cause the most damage to the vine (4). Once the first lesions have formed, conidia are produced continuously. On germinating, they produce one to five hyphae which break down the cuticle and penetrate the vegetative organs with or without forming appressoria (5). The fungus then colonises the tissues both intra- and intercellularly, causing severe necroses which correspond to a breakdown of the epidermis or parenchyma at the centre of the lesions (6). (Illustration © Virginie Duquette, Gravir un Monde d'Illustration, Switzerland)

(berries, leaves, shoots). These survival structures can remain viable in the soil for more than 5 years. Primary infections can also be caused by ascospores, although their epidemiological role remains hypothetical. In spring, the sclerotia germinate directly, discharging conidia after at least 24 h of leaf wetness and temperatures above 2 °C. The conidia, encased in a gelatinous mass, are projected a short distance from the ground onto green parts of the vine by splashing raindrops. This mode of transport explains the spread of the disease in localised outbreaks. The pathogen then produces acervuli containing the conidia responsible for secondary infections, which amplify the disease and cause the most damage to the vine. Rainfall and relatively high humidity promote the development of anthracnose. In contrast, temperature is not a limiting factor as the conidia can germinate between 2 and 32 °C in a film of water. Sporulation reaches a peak at around 20 °C, with the disease developing optimally between 24 and 26 °C (Carisse et al. 2020). The sclerotia or acervuli must be wet for conidia to be released and to germinate. The higher the temperature, the shorter the duration of leaf wetness needed for the conidia to germinate (Carisse et al. 2020). Once the first lesions have formed, conidia are produced continuously. On germinating, they produce one to five hyphae which break down the cuticle and penetrate the vegetative organs with or without forming appressoria. The fungus then colonises the tissues both intra- and intercellularly, causing severe necroses which correspond to a breakdown of the epidermis or parenchyma at the centre of the lesions (Li et al. 2019; Braga et al. 2020).

Vine organs are most susceptible during the pre-bloom and bloom stage, after which susceptibility diminishes until veraison. Inflorescences of the hybrid varieties Vandal-Cliche, Marquette and Vidal are most susceptible at flowering, with susceptibility diminishing until veraison, after which the berries are practically resistant to *E. ampelina* (Carisse et al. 2020).

Host Resistance

All species of the genus *Vitis* are susceptible to anthracnose, especially *Vitis vinifera*. In contrast, *V. labrusca* and certain hybrids are considered resistant to moderately susceptible, while *Muscadinia rotundifolia* (syn. *Vitis rotundifolia*) is considered immune, although its resistance level varies depending on the genotype. Observation of 54 different cultivars under controlled vineyard conditions shows that the susceptibility of *V. rotundifolia* ranges from highly resistant to highly susceptible and that the expression of genes for chalcone synthesis, stilbene synthesis and inhibitor proteins for polygalacturonase, chitinase and lipid-transferase are not present in the genotypes resistant to *Elsinoe ampelina* (Clifford et al. 2011). According to the references, 26 species of *Vitis* are described as resistant to *E. ampelina*, although for some, such as *V. labrusca*, the level of resistance depends on the disease pressure coupled with the moisture conditions (Santos et al. 2018). In China, the reservoir of resistance genes in the wild vine gene pool reveals a vast potential for selection in 13 species of *Vitis* all resistant to anthracnose (*V. amurensis*, *V. quinquangularis*, *V. romanetii*, *Vitis adstricta*, *V. pseudoreticulata*, *V. piazezkii*, *V. davidii*, *V. d. var. cyanocarpa*, *V. liubanensis*, *V. quinlingensis*, *B. bashanica*, *V. yeshanensis*, and *V. hancockii*) (Li et al. 2008).

Understanding the resistance genes is key to obtaining varieties that are less susceptible to anthracnose. For marker-assisted selection, researching the QTLs for resistance or other molecular markers at transcriptome level is the next hurdle to overcome in the search for a sustainable means of controlling anthracnose in winegrowing areas where control is essential.

4.5.4   Disease Control

Anthracnose requires systematic control in the north-eastern United States, Canada, China, Brazil, India and all regions with a hot and humid climate (Li et al. 2021). Like black rot, anthracnose can also reappear in varieties resistant to downy mildew and powdery mildew requiring few, if any, fungicidal treatments.

Predictive models have been developed to better target the control of anthracnose by firstly establishing multiple regressions and correlation coefficients between disease incidence and air temperature, relative humidity and rainfall. The primary inoculum has been shown to overlap with secondary infections during the exponential development phase of the vine, enabling a non-linear sigmoid model to be developed (Carisse and Lefebvre 2011). A recent mechanistic model based on all the published findings on *E. ampelina* provides reliable prediction of infections and more targeted control based on the epidemiological development of the disease (Ji et al. 2021a, b, c).

Spreading of contaminated conidia and long-term establishment of the disease can generally be avoided with a limited number of fungicidal treatments, starting at bud break.

4.6   White Rot

Current name:

- *Coniella diplodiella (Speg.)* Petrak & Sydow 1927, Ascomycota, syn. *Coniothyrium diplodiella* (Speg.) Sacc 1884, *Pilidiella diplodiella* (Speg.) Crous & Van Nieker 2004, syn. *Phoma diplodiella* Speg. 1878
- *Coniella vitis* Chethana, J.Y. Yan, X.H. Li & K.D. Hyde 2017, Ascomycota
- *Coniella fragariae* (Oudem) B. Sutton 1977, Acomycota
- *Coniella castaneicola* (Ellis & Everh.) B. Sutton 1980, Ascomycota

The fungus responsible for grapevine white rot was first described by Spegazzini (1878) under the name *Phoma diplodiella,* following its discovery in Conegliano in northern Italy. During the course of further studies, the fungus was renamed *Coniothyrium diplodiella* by Saccardo (1884) and finally *Coniella diplodiella* by Petrak and Sydow (1927). It was not until the late nineteenth century that the disease was observed in France, Italy, Spain and the United States. Grapevine white rot is commonly referred to as 'hail disease' because of its sudden appearance after hailstorms; however, it can also develop in the absence of hailstorms. The fungus can colonise grape berries through any wound and produce the symptoms of white rot. The disease is currently found in all winegrowing regions of the world, and can cause major economic losses, especially in some regions of China where the monsoon season in Southeast Asia coincides with the development of the grape berries (He et al. 2017). In these vineyards, white rot is mainly due to *Coniella vitis*, which can cause harvest losses of up to 16% (Li et al. 2008). In the other major regions of the world, *Coniella* sp. is a wound pathogen which is difficult to detect. Its phytosanitary importance is relative, because active substances used to control downy mildew, powdery mildew and grey mould are also effective against *Coniella* sp.

4.6.1 Causal Organism

Coniella diplodiella is an Ascomycota, of the order Diaporthales and the family Schizoparmaceae. The fungus produces yellowish-brown pycnidia embedded in the surface of infected berries, but clearly prominent (85–130 μm diameter). Arising from the four layers of cells lining the interior of the pycnidia are the conidiophores, bearing unicellular conidia, hyaline then brownish at maturity, elliptical to obvoid, rounded at the tip and truncated at the base (8–16 × 5.5–7.5 μm).

4.6.2 Symptoms

White rot infects all green tissues of the vine, but the most severe damage occurs on ripening clusters, generally a few days after a hailstorm. Wounded berries turn pale yellow, then gradually brown with alternating concentric circles of a darker shade, before eventually blackening. This metamorphosis is accompanied by shrivelling and ultimately, complete desiccation of the berries (Fig. 4.58). Shrivelled berries are covered in small purplish-brown pustules. These are the pycnidia or fruiting bodies of the fungus (Fig. 4.58), containing masses of conidia which can transmit the infection to other organs. These symptoms on berries can be confused with those of black rot or excoriose (Fig. 4.59), while partial desiccation of the bunches can be confused with brown rot caused by downy mildew, which does not produce pycnidia. The name 'white rot' is linked to the greyish-white colour of mature pycnidia on the berries and stems. The disease spreads rapidly from one berry to another via the stems, eventually infecting the entire bunch. In favourable conditions, a few initially infected berries are enough for white rot to colonise the whole bunch. Woody shoots, especially in mother vine fields, can also be infected following hail damage. Elongated patches appear, greyish in the centre with a black border. Cankers surrounded by swollen scar tissue can cause the bark to split and peel off in strips (Fig. 4.60). On the leaves, circular brown lesions appear on the leaf blade, evolving into concentric circles, ultimately leading to desiccation of the leaf. In humid weather, greyish-white pycnidia are generally visible on the necrotic tissues. On green shoots, similar, more elongated necrotic areas can develop and evolve into cankers.

Fig. 4.58 Symptoms of white rot on bunches. Affected berries turn pale yellowish-beige, then gradually become browner, giving rise to alternating concentric circles (**a**) of a darker shade (arrow). When fully brown, they become covered in pycnidia (**b**) before eventually blackening completely. This metamorphosis is accompanied by shrivelling (**c**) (arrow) and ultimately, complete desiccation of the berries (**d**)

Fig. 4.59 White rot symptoms on berries (**a**) can be confused with black rot (**b**) or *Phomopsis* dieback (**c**)

Fig. 4.60 Symptoms of white rot on woody shoots damaged by hail. (**a**) Elongated necrotic areas develop on the shoots and form cankers with black borders (arrows). (**b**) Cankers surrounded by swollen scar tissue (arrow) can cause the bark to split and peel off in strips

4.6.3   Biology and Epidemiology

C. diplodiella overwinters as mycelium or pycnidia on parts infected the previous year (mummified berries, stems, shoots) that fall to the ground. Pycnidia can remain viable for around 15 years and can repeatedly discharge conidia when conditions are sufficiently moist. During hailstorms or heavy rainfall, conidia are splashed up from the ground onto berries, where they germinate. This fouling is clearly visible on bunches and leaves after violent storms, with or without hail. The conidia infect green tissue through stomata, microfissures, wounds, or direct penetration by digesting the epidermis. Wounds favour infections, with the severity of symptoms proportional to the inoculum of conidia. The incubation period is longer on uninjured berries. Artificial inoculation of berries produces symptoms regardless of whether berries are injured or intact (Ji et al. 2021a, b, c).

Pruning systems that bring the fruit closer to the ground, such as goblet- low Guyot cane- spur- or cordon-pruning, increase the likelihood of infection. The progressive mechanisation of vineyards has changed the parameters of vine cultivation. Terrace cultivation, wire cordon training techniques and inter-row cover cropping in vineyards on steep slopes significantly reduce the risk of infection by *Coniella* by eliminating the primary inoculum.

Conidia germinate at optimum temperatures of 24–27 °C. The thresholds for germination are below 12 °C above 33 °C. Tissues must be wet for at least 2 h for infection to occur (Chen et al. 1980). Recent studies of conditions for infection showed that 1 h of wetness was sufficient to cause infection of the berries at all temperatures tested (10–35 °C), with 23.8 °C being the optimum temperature (Ji et al. 2021a, b, c).

The sugar concentration in the berries does not have a positive impact on white rot. Berries wounded before or after veraison incur the same risk of infection. The growth of *C. diplodiella* mycelium does not depend on the presence of glucose, starch, tartaric or malic acid.

4.6.4   Host Resistance

All varieties of the species *Vitis vinifera* are susceptible to *Coniella* sp. In hot and humid regions of South East Asia affected by this disease, the creation of resistant hybrids provides opportunities to improve the management of white rot without using fungicides. More than 35 of the 70 known species of *Vitis* in the world originate in China, some of which are resistant to *Coniella* (*V. amurensis, V. piazezkii, V. davidii, V. davidii var. cyanocarpa, V. liubanensis, V. quinlingensis, B. bashanica, V. yeshanensis* and *V. hancockii* (Li et al. 2008). The QTLs (quantitative trait loci) for resistance to *Coniella diplodiella*, identified on *V. davidii*, offer opportunities to selectively breed varieties resistant to white rot (Zhang et al. 2020). The findings of Li et al. (2023) on an interspecific crossing (*Vitis vinifera* L. × *Vitis davidii* Foex.) have revealed a new, stable resistance QTL located on chromosome 3, which explains 17.9% of phenotypic variation.

4.6.5   Disease Control

Control is required only in situations where there is a known risk of white rot due to the frequency and intensity of hail events, the susceptibility of varieties, the use of low pruning systems or the absence of inter-row crops. In these situations, the closer the bunches are to bare ground, the greater the risk of infection. Good results can be achieved with a single application of active substances in the phthalimide or dithiocarbamate family 12–18 h after hail. However, beyond 24 h after a hail event, its effectiveness is uncertain. This treatment also protects wounds from other fungal pathogens present in the atmosphere.

In hot and humid regions of China where white rot is a common pathogen (Chethana et al. 2017), when temperatures ranging from 24 to 27 °C and high humidity coincide with berry ripening, preventive control requires several fungicidal applications either of the active substances mentioned previously, or of pyraclostrobin (Li et al. 2020) or fluazinam, an active substance which inhibits mycelial growth by interrupting adenosine triphosphate (ATP) synthesis (Wang et al. 2018). The application of the antagonistic bacterium *Bacillus velezensis* isolated in the vineyard, which promotes plant growth, presents an opportunity to control white rot organically (Yin et al. 2013). The relation between temperature and wetness duration at different stages of berry infection by *Coniella diplodiella* has led to the development of algorithms which could be used to calculate the risk of infection and thus schedule fungicide applications with greater precision based on the epidemiological development of the disease (Ji et al. 2021a, b, c).

4.7 Sour Rot (Acid Rot)

Current name:

- Fungal genera:

  - *Candida* spp. Berkhout 1923, Ascomycota
  - *Hanseniaspora* spp. Zikes ex Klöcker 1912, Acomycota
  - *Metschnikowia* spp. Kamienski 1899, Acomycota
  - *Pichia* spp. E.C. Hansen 1904, Acomycota
  - *Saccharomyces* spp. Meyen ex E.C. Hansen 1883, Ascomycota
  - *Torulaspora* spp. Lindner 1904, Ascomycota
  - *Williopsis* spp. Zender 1925, Ascomycota
  - *Zygoascus* spp. M.T. Sm. 1986, Ascomycota
  - *Zygosaccharomyces* spp. B.T.P. Barker 1901, *Incertae sedis*

- Bacterial genera:

  - *Acetobacter spp. Beijerinck 1898*
  - *Gluconobacter spp. Asai 1935*

- Insects:

  - *Drosophila melanogaster Meigen 1830, syn. Drosophila fasciata Meigen 1830*
  - *Drosophila suzukii Matsumura 1931*

Sour rot is a vine disease that is particularly harmful to the quality of the grapes. From 20% to 30% of affected grapes, the harvest becomes unfit for vinification because of the very high volatile acidity that is released from rotten grapes. Present in all vineyards around the world, sour rot results from the co-infection of berries by yeasts and bacteria (Hall et al. 2019; Haviland et al. 2017; Barata et al. 2012), a process mediated by *Drosophila* species (Hall et al. 2018; Barata et al. 2012), or any kind of mechanical damages on the berries. Some grape varieties are particularly susceptible to sour rot because of their compact clusters and/or thin berry epidermis. Pinot (black, grey, white), Gamay, Chasselas, Gewürztraminer, Muscat of Alexandria, Muscat à petits grains, Riesling, Müller-Thurgau, Sauvignon Blanc, Mourvèdre, Chardonnay, Bondola, Barbera, Dolcetto, Grignolino, Teroldego, Marzemino and Vernatsch are known to be susceptible to sour rot. The causes of this disease include biotic and abiotic factors inherent to vine cultivation that must be integrated in the prevention measures.

4.7.1   Causal Organisms

A complex variety of yeasts and bacteria are involved in the symptomatology of sour rot. Several genera of yeasts, belonging to several taxa, have been identified and isolated from berries infected with sour rot. Among the main ones, the genera *Candida, Hanseniaspora, Metschnikowia, Saccharomyces, Zygosaccharomyces, Zygoascus* and *Pichia* (three species of which formerly belonged to the genius *Issatchenkia*) are frequently identified as phylloplane microorganisms forming part of the mycobiome on the surface of the berries (Hall et al. 2019). Yeasts occur in all environments. A large number of these unicellular fungi have been assessed as biocontrol agents for use in crop protection due to their antagonistic ability (competition, enzyme secretion, toxin production, volatiles, mycoparasitism, induction of resistance). Paradoxically, the genera associated with sour rot described above include the species currently used in biotechnology and biocontrol applications (Freimoser et al. 2019). The actual involvement of yeasts in the development of sour rot is closely connected to the presence of specific bacteria and the action of vectors such as *Drosophila* spp.

The genus *Candida* is in the phylum Ascomycota, the order Serinales and the family Debaryomycetaceae. This genus comprises a very large number of species (over 1000), including some human pathogens—such as *Candida parapsilosis*. The most studied species is *Candida albicans*, part of the human microbiota (gastrointestinal, cutaneous and vaginal) which can become pathogenic in certain circumstances. The species identified in connection with grapevine sour rot include *C. amapae, C. diversa, C. ethanolica, C. tropicalis* and *C. oleophila* (Barata et al. 2008a).

The genus *Hanseniaspora* comprises a smaller number of around 30 species. This genus is in the phylum Ascomycota, the order Saccharomycodales and the family Saccharomycodaceae. *Hanseniaspora uvarum* is one of the species associated with sour rot. The oval cells of *H. uvarum* have a pointed tip. Occurring individually or in pairs, they multiply by bipolar budding.

The genus *Metschnikowia* is also an Ascomycota, of the order Serinales and the family Metschnikowiaceae. This family contains around 100 species, some of which are thought to be responsible for off-flavours in the wine and incomplete fermentation. One species in particular, *M. pulcherrima*, forming part of the berry mycobiome and also isolated in berries infected with sour rot, produces beneficial enzymes involved in the formation of metabolites that can enrich the wine's aromas. *M. pulcherrima* is a commercially available yeast that is already used for this purpose (Giménez et al. 2023; Lebleux et al. 2023; Windholtz et al. 2023). *M.*

chrysoperlae is another species isolated from bunches infected with sour rot (Hall et al. 2019).

The genus *Saccharomyces* is a member of the Ascomycota, the order Saccharomycodales and the family Saccharomycodaceae. This genus includes more than 600 species, the best known of which is undoubtedly *Saccharomyces cerevisiae* (brewer's yeast), responsible for alcoholic fermentation and the production of fermented products such as wine, bread and beer. This yeast is an integral part of the grapevine mycobiome and has frequently been isolated from bunches infected with sour rot, although its presence is not necessarily implicated in the emergence of the disease.

The genus *Issatchenkia* has only five species. These yeasts also belong to the Saccharomycetaceae family. The works of Barata et al. (2008a) demonstrated the presence of certain species associated with the symptoms of sour rot, including *I. occidentalis*, renamed *Pichia occidentalis*; *I. orientalis*, renamed *Pichia kudria-vzevii* and *I. terricola*, renamed *Pichia terricola*. The species cited by Barata et al. (2008a) in connection with sour rot, formerly generically affiliated to the genus *Issatchenkia*, are now affiliated to the genus *Pichia*. Other species belonging to the genus *Pichia*, such as *P. kluyveri*, *P. guilliermondii*, renamed *Meyerozyma guillier-mondii* and *P. galeiformis*, renamed *Pichia mandshurica*, have been isolated on bunches infected with sour rot. The genus *Pichia*, comprising over 280 species, is in the phylum Ascomycota, the order Pichiales and the family Pichiaceae. Most of these species are found on plant debris and decomposing plant material, some live in symbiosis with various insects, and others are potential sources of defects during the vinification process.

The genus *Zygosaccharomyces* (*incertae sedis*), and specifically *Zygosaccharomyces bailii* (Barata et al. 2008b) is often isolated from bunches con-taminated with sour rot. *Z. bailii* is known to be one of the most aggressive food spoilage microorganisms—contaminating wine during fermentation, for example. Aggressive food spoilage microorganisms, for example, contaminating wine during fermentation. It is also present in many acidic, high-sugar canned foods (Kuanyshev et al. 2017). Its spoilage ability relies on the yeast's unique feature of tolerating the most common preservatives such as sulphite, dimethyl dicarbonate, acetic acid and sorbic acid.

Comprising ten species, the genus *Zygoascus* belongs to the Ascomycota of the order Dipodascales and the family Trichomonascaceae and includes *Z. hellenicus*, the species most frequently isolated from berries infected with sour rot (Barata et al. 2008a).

4.7.2   Symptoms

The disease is characterised by rapid development from veraison. The berries take on a brown hue, of varying intensity, initially around the base of the pedicel. In white grape varieties, they become light brown, in red grape varieties rather pinkish, and undergo strong oxidation (Fig. 4.61). Berries generally remain turgid but can

Fig. 4.61 Sour rot on a white variety with berries turning brown (left) and on a red variety showing pinkish-brown berries (right)

also shrivel and lose their juice if their epidermis breaks. A strong odour of acetic acid is released from affected clusters, attracting numerous fruit flies (*Drosophila melanogaster* and *D. suzukii*) (Fig. 4.62). Inside the infected and partially empty berries, a slimy mass of yeast and bacterial colonies is visible. No direct link has been clearly established with attacks of the grey rot agent *B. cinerea* (Bisiach et al. 1982), although it is not unusual to see grey rot and sour rot cohabiting on some bunches. However, the development of *B. cinerea* is hindered by the release of acetic acid (Hall et al. 2019). Grey rot can open the door to yeasts and acetic bacteria by enzymatically decomposing the epidermis of the berries but is not a necessary condition for the development of sour rot. The first brownish discoloration of the berries after veraison can be mistaken for early infections of *Botrytis cinerea* or *Coniella diplodiella*, the agent of white rot.

4.7.3 Biology and Epidemiology

Sour rot results from the co-infection of berries by several species of yeast (Hall et al. 2019) which convert grape sugars to ethanol, and bacteria which oxidize ethanol to acetic acid (Pinto et al. 2019). This disease only develops when *Drosophila* are present although inoculation with certain combinations of yeast and acetic bacteria has been successfully produced discoloration of the berries and acetic acid in

Fig. 4.62 Drosophila attracted by the release of acetic acid. (**a**) Adult *Drosophila suzukii*. (**b**) *Drosophila suzukii* and pupae. (**c**) Several pupae *Drosophila* sp. on a berry infected with sour rot

the absence of these insects. *Drosophila* are also considered to be the vector of at least some of the sour rot-associated yeasts listed above (Lam and Howell 2015). Yeasts and bacteria carried by these insects (Barata et al. 2012) and/or already present on the surface of the berries (Hall et al. 2019) initiate spontaneous fermentation and the resulting volatiles attract a larger population of *Drosophila* which spread the disease to other injured berries. In addition to biotic factors, skin wounds resulting from hail, sunburn, water stress, or long periods of wet weather have also been shown to be mandatory for the expression of sour rot (McFadden-Smith and Gubler 2015). Skin wounds activate the plant's defence mechanisms, but in the absence of *Drosophila*, the wounds heal and prevent pathogen penetration (Barata et al. 2012). Therefore, it is thought that in the presence of large populations of these insects, attracted by the volatiles produced by yeasts and acetic acid bacteria, the plant's defensive response is too slow to heal the skin wounds and sour rot spreads.

Sour rot never appears before veraison and its severity is linked to hot and humid climatic conditions during the ripening of the grapes. The recent appearance in Europe of *D. suzukii*, also called spotted wing drosophila, is a new source of development for sour rot. The female, very similar to common *Drosophila*, has a well-developed, toothed ovipositor, which allows it to pierce the epidermis of healthy berries. Her eggs hatch into creamy white larvae (Fig. 4.62) that feed on the pulp. This direct perforation of the epidermis of the berries, specific to this species, favours the establishment of yeasts and bacteria associated with sour rot.

Flavonoids Influence Drosophila Suzukii Egg-Laying Activity
(Marcellin-Gros et al. 2024)

Insects' preferences for hosts are regulated by multiple factors that interact in complex ways, often only partially understood. This study employed a comprehensive, untargeted metabolomic approach, combining molecular networking (MN) with supervised Analysis of Variance multiblock Orthogonal Partial Least Squares (AMOPLS), to elucidate the egg-laying preferences of *Drosophila suzukii*. Due to the significant variability in susceptibility among grape cultivars, eight genetically related Vitis vinifera red cultivars (e.g., Ancellotta, Galotta, Gamaret, Gamay, Gamay précoce, Garanoir, Mara, and Reichensteiner) were selected. Behavioural experiments conducted in the laboratory, along with field observations, confirmed substantial differences in susceptibility to *D. suzukii* among these cultivars. The two most attractive cultivars (Gamay précoce and Mara) and the two least attractive cultivars (Galotta and Gamaret) were chosen for further metabolomic analyses of their berry skins. The integration of MN and statistical AMOPLS findings with semi-quantitative detection information allowed to identify flavonoids as potential markers for differences in the attractiveness of the four grape cultivars to *D. suzukii*. Specifically, dihydroflavonols were found to accumulate in unattractive grape cultivars, while attractive ones exhibited higher levels of flavonols. Both flavonoid classes were abundant in the grape skin extracts, underscoring their biological relevance as potential markers. Flavonoids, characterized by specific chromophores with absorption bands in UV wavelengths (around 250 and 350 nm), play a crucial role in berry coloration. Notably, dihydroflavonols absorb shorter wavelengths compared to flavonols, potentially altering reflectance and berry colour, which could impact *D. suzukii* egg-laying behaviour as UV light affects it. Accumulation of dihydroflavonols in unattractive grape varieties may act as a repellent. Moreover, the chemical composition of fruit skins, perceived through specific chemoreceptors, is thought to play a crucial role in *D. suzukii*'s egg-laying behaviour. Various molecules, including glucosinolates, terpenoids, and phenylpropanoids, can act as oviposition stimulants or repellents. While the ability of D. suzukii to sense flavonoids in vivo has not been fully demonstrated, studies

suggest its potential to perceive them. Therefor flavonoid content could significantly influence the visual and organoleptic properties of the berries, serving as initial attractant for *D. suzukii* impacting its egg-laying behaviour.
Polymethoxylated flavonoids (PMFs) found in unattractive grape cultivars
show diverse biological activities, including cytotoxic properties. Their presence in unattractive cultivars suggests a possible role in reducing egg-laying
and hindering larval development of *D. suzukii*, potentially contributing to the
low emergence rate of the vinegar fly in grape varieties enriched in PMFs.

4.7.4   Disease Control

No specific phytosanitary product is registered against yeasts and bacteria. Prophylactic
measures that modify the microclimate of the grapes are the only way to control acid
rot. Varieties with loose clusters or thick-skinned berries are generally less affected by
sour rot. Grey rot-resistant crossbred grape varieties hardly ever develop sour rot. In
susceptible grape varieties splitting the bunches in half after veraison significantly
reduces sour rot and grey mould, as does the application of gibberellic acid during
flowering, which modifies stem structure and fruit set rate (Spring and Viret 2009,
2011). However, this natural phytohormone must be precisely dosed according to the
grape variety, otherwise the entire crop could be lost due to lack of fruit set. All preventive measures must be used to avoid the disease: choosing grape varieties and
clones with loose bunches, controlling grape worms, installing nets to protect against
wasps or hail, regulating yields by avoiding bunch compaction, reducing vigour
through adequate management of nitrogenous manuring and the use cover crops to
produce smaller bunches, and carefully thinning of bunch zones are some of the many
cultivation practices that help reduce the development of the sour rot.

Direct control of *Drosophila* with insecticides is not effective against sour rot, especially after it has already been observed in the vineyard (Kenney and Hall 2021) and is
not recommended because of the residues of active ingredients that can accumulate in
the grapes during ripening. Mass trapping could control the emerging pest *D. suzukii*
and other fruit flies as well. The effectiveness of copper treatments against sour rot is
highly dependent on the situation and weather conditions during grape ripening.

4.8   Rotbrenner (Brenner Disease, Red Fire Disease, Rougeot)

Current name:

- *Pseudopezicula tracheiphila* (Müll.-Thurg.) Korf & W.Y. Zhuang 1986,
  Ascomycota, syn. *Pseudopeziza tracheiphila* Müll.-Thurg 1903
- Angular leaf scorch: *Pseudopezicula tetraspora* Korf, R.C. Pearson &
  W.Y. Zhuang 1986, Ascomycota

Brenner disease, also known as rotbrenner, is found in most European winegrowing regions. Long attributed to physiological problems, it was not until the start of the twentieth century that the symptoms were linked by Müller-Thurgau (1903) to the fungus *Pseudopeziza tracheiphila*—a pathogen which he described and illustrated, including its development and association with the vascular elements of infected leaves. Brenner disease is considered a disease of secondary importance which can cause localised economic losses under favourable weather conditions. Generally, it is confined to well-defined sectors of the vineyard and does not assume an epidemic character on the scale of downy mildew or powdery mildew. The fungus infects every variety of European vine, interspecific hybrids, American vines, and has been reported on *Parthenocissus*. Beyond its occurrence in all the winegrowing regions of Europe, it has since been recorded in Turkey, Algeria, Southern Russia, Brazil and Australia. In the US the first cases of angular leaf scorch, caused by a new species, *Pseudopezicula tetraspora*, were observed in the Great Lakes region of New York State in 1986 (Korf et al. 1986; Pearson et al. 1988). The only difference between the causal agent of this disease and *P. tracheiphila* is that it has smaller apothecia and asci containing four ascospores instead of eight; its symptoms are virtually identical.

4.8.1   Causal Organism

The genus *Pseudopeziza* was first described in Switzerland by Müller-Thurgau in 1903 and renamed *Pseudopezicula* by Korf et al. in 1986. *Pseudopezicula* belongs to the Ascomycetes, the order Helotiales and the family Discinellaceae. The fungus overwinters as mycelium on infected dead leaves. In spring, it produces yellowish-brown apothecia, 0.2–0.4 mm in diameter (Fig. 4.63), arranged in a highly irregular fashion mainly along the principal leaf veins (apothecia of *P. tetraspora*: 0.1–0.3 mm in diameter). At maturity, the apothecia burst open and release a whitish mass containing inoperculate asci (115–145 × 18–28 µm, in *P. tetraspora*: 80–100 × 20–22 µm). An apothecium can contain more than 100 asci (Fig. 4.63), separated by thread-like paraphyses, bent and branched. Each ascus contains eight unicellular ascospores; hyaline, ellipsoidal and reniform (18–22 × 9–11 µm). Large vacuoles (1–2) are generally present in the ascospores. The apothecia develop mainly on the underside of the leaves. Embedded in the foliar tissue initially, they emerge as they mature. Ascospores are the sole source of infection. The asexual form of the fungus has never been observed in natural conditions, although it has been obtained in the laboratory (unicellular conidia, hyaline, elliptical, 2–3 × 1.5–2 µm).

4.8.2   Symptoms

The first symptoms of Brenner disease appear in late spring (May–June) on leaves at the base of the shoots. The yellow patches can be mistaken for the first symptoms of downy mildew. The essential difference is that the yellow patches caused by Brenner disease

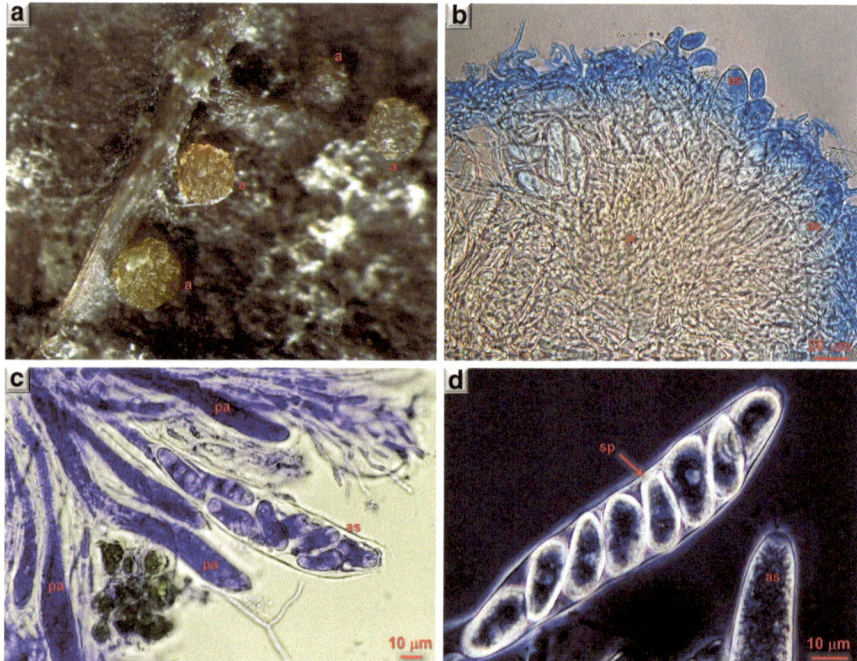

Fig. 4.63 Apothecia of *Pseudopezicula tracheiphila*. (**a**) Apothecia (a) generally form along the veins of desiccated infected leaves. (**b**) Apothecium viewed in semi-thin section dyed cotton blue. Immature asci (as) are visible in the stroma of the apothecium. (**c**) The asci (as) are separated by sterile paraphyses (pa). (**d**) Ascus containing eight reniform ascospores (sp)

are clearly bound by the leaf veins (Fig. 4.64) and there is no sporulation on the underside of the leaves. In a severe infection, foliar symptoms can be observed as far as the 10th or 12th leaf. These patches, subtle at first, enlarge and take on a bright yellow hue in white grape varieties and a reddish hue in red grape varieties. These necrotic areas eventually wither from the centre. Lesions are angular, confined by the main and secondary veins, and generally surrounded by a yellowish border (Fig. 4.65). Affected leaves shrivel completely and fall prematurely in July. Inflorescences can be infected shortly before or during flowering. In contrast to early *Botrytis cinerea* infections which cause the entire inflorescence to die back and detach from the shoot, flowers and petioles affected by Brenner disease shrivel but remain attached to the stem. As in the case of severe coulure (berry shatter), affected bunches have only a few healthy berries (Fig. 4.66). A major attack during flowering can cause total loss of the harvest.

4.8.3   Biology and Epidemiology

The latest findings about the disease cycle of Brenner disease are summarised in Fig. 4.67. The causal agent of Brenner disease is a biotrophic fungus specific to *Vitaceae*. It is a monocyclic disease, having only one infection cycle per year.

Fig. 4.64 First symptoms of *Pseudopezicula tracheiphila* on leaves (**a**) clearly delimited by the leaf veins (inset), which can be confused with oil spots of downy mildew (**b**). In the case of Brenner, sporulation is never seen on the underside of the leaf

Fig. 4.65 Brenner symptoms on white (**a**) and red cultivars (**b**) showing specific leaf colourations

All infections stem from the primary inoculum on dead leaves on the ground (Fig. 4.67 no. 1). Symptoms observed on the leaves in the current year do not transmit the infection to other healthy leaves. Maturation of the apothecia is closely linked to spring rainfall, especially wet weather in April and May. In dry conditions, the fruiting bodies remain embedded in the leaf epidermis, where they are practically invisible. During wet weather or high relative humidity, they hydrate, swell and burst open (Fig. 4.67 no. 2). In infected zones of the vineyard, hedges, walls and forest margins where dead leaves pile up constitute a major source of infection. These zones are often more humid and more conducive to the survival and development of Brenner disease. A temperature of at least 13 °C is required for ascospores to be discharged (Fig. 4.67 no. 3). Discharge reaches a peak in May–June and can continue until August. Ascospores are transported to the vine green organs (Fig. 4.67 no. 4) primarily by the wind

Fig. 4.66 Infected inflorescences and bunches have low levels of fertilisation and produce only a few health berries (**a**). Severe infections can lead to complete withering of bunches (**b**)

Fig. 4.67 Biological development cycle of *Pseudopezicula tracheiphila*. Infections stem from the primary inoculum on dead leaves on the ground represented by the apothecia (1). During wet weather or high humidity, the apothecia hydrate and swell (2), enabling the asci to discharge the ascospores (3) which are dispersed by the wind. The ascospores land on green organs and germinate (4) to form a mycelium which extends into the vascular tissue (5). Mycelial growth on the surface of the leaves facilitates the development of apothecia (6). (Illustration © Virginie Duquette, Gravir un Monde d'Illustration, Switzerland)

and tend to disperse during the drying phase following a rainfall event. These spores can survive for several weeks, even in dry conditions. They can germinate at temperatures of 5–25 °C (optimum 20 °C) with relative humidity above 95%. Unlike downy mildew, Brenner disease does not require a film of water for an infection to occur. The germ tubes penetrate directly through the epidermal tissue without forming appressoria.

The mycelium then grows within the parenchymal cells (Fig. 4.67 no. 5) and colonises the vascular tissue. It continues to develop on the surface of the leaves, forming apothecia when conditions are favourable (Fig. 4.67 no. 6). Brenner disease colonises both the phloem and the xylem (Fig. 4.68). Indeed, the Latin name of the species *P. tracheiphila* indicates its affinity for vascular tissue. The tissues surrounding a necrotic lesion form tyloses (sack-like structures) which prevent the pathogen spreading within plant tissue, thereby causing symptoms to be confined by the main or secondary veins. The period of latency between infection and emergence of symptoms is generally 2–4 weeks.

Microscopic studies of the epidemiology of Brenner disease have shown that the fungus penetrates both the cuticle and the anticline cell wall after producing an infection cushion from which an infectious hypha emerges. This vesicular structure is surrounded a thin mucilaginous envelope which separates the host plasma membrane from the wall of the infectious hypha. The vine defends itself by producing secondary deposits in cell walls adjacent to the infected epidermal cells. At this stage, the infected epidermal cell is dead, the cell walls are damaged, and the cytoplasm is coagulated. In the second stage of disease progression, the biotrophic relationship between pathogen and host evolves into a perthotrophic relationship in which the mycelium grows intracellularly within the spongy parenchyma of the leaf, destroying the host cell. Six days after inoculation, the hyphae of *P. tracheiphila* can be found in the parenchymatous cells of the xylem and phloem. After invasion of the vascular tissues, the adjacent parenchymatous cells form tyloses (Reiss et al. 1997).

Fig. 4.68 The mycelium of *Pseudopezicula tracheiphila* grows inside the vascular system (A) (arrows) of the infected tissues, explaining why symptoms are limited by the leaf veins. Scale bar represents 1 µm

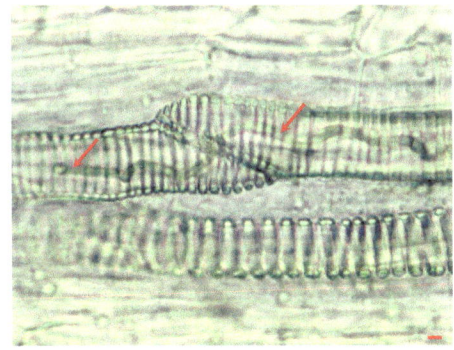

4.8.4   Disease Control

4.8.4.1   Disease Forecasting

Brenner disease appears in a cyclical manner in localised well-defined zones of the vineyard and it is these areas which warrant control measures. No preventative measures are required in any areas of the vineyard where the disease has not been previously observed. Estimating the primary inoculum and the timing of ascospore maturation and release has proved effective in predicting the risk of infection and determining the opportune moment to treat Brenner disease (Pearson et al. 1991).

Estimating the Primary Inoculum

In zones affected by Brenner disease, the primary inoculum can be estimated from the number of dead leaves on the ground producing apothecia and the average number of apothecia per leaf. Each week in spring, infected leaf samples are collected from the ground in different plots that were severely infected the previous year. In each plot, the leaf sample fragments are placed in water at ambient temperature for at least 12 h. To determine the number of apothecia, 50 fragments collected from 50 to 100 leaves are examined under a binocular magnifier (magnification 20–30×) and a tolerance threshold of 5–10% of leaves containing apothecia is applied. Below this level, the risk of infection is low to negligible.

Apothecia Maturation

By harvesting infected leaves in autumn and overwintering them under vineyard conditions, it is possible to identify the moment when the apothecia are ready to discharge the infectious ascospores. In spring, fragments of these leaves are collected once or twice a week to monitor asci differentiation in the apothecia under a microscope. Using this method, it is possible to precisely determine the start of ascospore discharge, which correlates very well with rainfall and the sum of cumulative temperatures above 8 °C since 1 January. When this value reaches 140–160 °C, the next precipitation event can induce infections by the ascospores. At this point, the vine, whose threshold temperature for growth is around 10 °C, reaches the 3–5 leaf stage (BBCH 13–15) in most geographical areas.

Release of Ascospores

Ascospore release is invariably initiated after rainfall. It correlates with drying of the leaf surface, reduction in relative humidity and rise in temperature (Pearson et al. 1991). During the night, ascospore dispersal is virtually zero. Alternating wet and dry periods favour apothecia maturation, although prolonged wet periods inhibit it. About 4 weeks after release of the first ascospores, around 90% of them have been ejected.

4.8.4.2 Disease Control

Brenner Disease needs to be controlled only in contaminated zones of the vineyard or in plots infected the previous year. In particularly dry springs, the risk of infection is negligible. The first treatment is applied in line with the predicted release of ascospores (Reiss and Zinkernagel 1997), which generally corresponds to the 5–6 leaf stage (BBCH15–16). The first phytosanitary treatments are combined with the downy mildew treatment using active ingredients which act jointly against the two diseases, namely those from the strobilurins or quinone outside inhibitors (QoIs), demethylation inhibitors—triazols (DMIs) phthalimide or dithiocarbamate families, or copper.

4.9 Ripe Rot

Current names of fungi:
Four *Colletotrichum* Corda 1831 species complexes:

- *Colletotrichum acutatum* spp. complex:

  - *C. acutatum* J.H. Simmonds 1968, Ascomycota
  - *C. godetiae* Neerg. 1950, Ascomycota
  - *C. limetticola* (R.E. Clausen) Damm, P.F. Cannon & Crous 2012, Ascomycota
  - *C. nymphaeae* (Pass.) Aa 1978, Ascomycota
  - *C. pseudoacutatum* Damm, P.F. Cannon & Crous 2012, Ascomycota

- *Colletotrichum boninense* spp. complex:

  - *C. karsti* You L. Yang, Zuo Y. Liu, K.D. Hyde & L. Cai 2012, Ascomycota
  - *C. phyllanthi* (H. Surendranath Pai) Damm, P.F. Cannon & Crous 2012, Ascomycota

- *Colletotrichum gloeosporioides* spp. complex:

  - *C. aenigma* B.S. Weir & P.R. Johnst.2012, Ascomycota
  - *C. ampelinum* Cavara 1889, Ascomycota
  - *C. fructicola* Prihast., L. Cai & K.D. Hyde 2009, Ascomycota
  - *C. gloeosporioides* (Penz.) Penz. & Sacc. 1884, Ascomycota; syn. *Glomerella cingulate* (Stoneman) Spauld. & H. Schrenk 1903
  - *C. kahawae* J.M. Waller & Bridge 1993, Ascomycota
  - *C. siamense* Prihast., L. Cai & K.D. Hyde 2009, Ascomycota
  - *C. viniferum* L.J. Peng, L. Cai, K.D. Hyde & Z.Y. Liu 2013, Ascomycota

- *Colletotrichum truncatum* spp. complex:

  - *C. capsici* (Syd. & P. Syd.) E.J. Butler & Bisby 1931, Ascomycota

Ripe rot is a berry disease which was originally attributed to a single anamorphic species, *Colletotrichum gloeosporioides,* of which *Glomerella cingulata* is the sexual form. This disease was reported for the first time in the USA (Southworth 1891) but is now recorded worldwide wherever vine is cultivated. However, another species, *C. acutatum*,—along with *C. gloeosporioides*—has been reported to be associated with the disease in several countries (south-eastern USA, Korea, Japan, and Australia; Greer et al. 2011). More recently, molecular systematics have inferred that *C. gloeosporioides* and *C. acutatum* are composed of several cryptic species (Damm et al. 2012; Weir et al. 2012). At present, the genus *Colletotrichum* is divided in 12 phylogenetic lineages (=species complexes) and 23 additional single species (Jayawardena et al. 2016). Fifteen species belonging to four out of these complexes (see species list above) have been reported to be causal agents of ripe rot (Peng et al. 2013; Oo and Oh 2017; Echeverrigaray et al. 2019; Batista et al. 2023).

Ripe rot can cause severe yield losses and is known to reduce wine quality by increasing acidity and bitterness of fruits and wine (Ji et al. 2021a, b, c). At present, this disease is considered problematic only in tropical regions with hot and humid summers.

4.9.1   Causal Organisms

The genus *Colletotrichum* Corda belongs to the Glomerellaceae family (order Glomerellales; class Sordariomycetes; division Ascomycota). This genus, found worldwide but mainly in tropical and subtropical regions, is one of the most important fungal plant pathogens. *Colletotrichum* species (Cano et al. 2004) are characterised by the presence of acervuli (20 × 3–4 µm) that produce masses of yellow-orange conidia, often with dark brown, septate, thick-walled acicular setae (up to 200 µm long). Conidia often harbour appressoria (thick-walled swellings at the end of the hypha or germ tube, enabling the fungus to attach itself to the host surface before penetration) which have different morphologies depending on the species. Previously identified according to their host plant (Sutton 1992), *Colletotrichum* species are now identified by multigenic phylogeny, with most species being cryptic and classified in species complexes.

4.9.2   Symptoms

Ripe rot is characterised by reddish-brown spots, which evolve to form circular lesions at the surface of the berries and can become systemic at harvest time (Ji et al. 2021a, b, c; Greer et al. 2011). Ripe rot (caused by *Colletotrichum* species) and anthracnose (caused by *Elsinoe ampelina*) are sometimes confused in the literature, despite the fact that their symptoms are different (Ye et al. 2023). While berries infected by *E. ampelina* start by developing dark brown spots which evolve to form sunken lesions with a light grey centre and a dark brown margin, those infected by *Colletotrichum* spp. develop reddish-brown circular spots which evolve into concentric lesions producing acervuli with masses of pale orange-pink conidia (Fig. 4.69). Berries may soften, lose turgor, and shrivel

Fig. 4.69 Berries infected by *Colletotrichum* spp. develop reddish-brown circular spots (red arrow) which evolve into concentric lesions producing acervuli (black arrow) with masses of pale orange-pink conidia (white arrow on the right)

Fig. 4.70 Berries infected by *Colletotrichum* spp. may soften, lose turgor and shrivel

(Fig. 4.70); symptoms similar to those produced by *Phomopsis viticola*. Berries altered by *Colletotrichum* spp. may also produce acervuli. *Colletotrichum* spp. pass from biotrophy to necrotrophy, the latter development stage leading to ammonia secretion that alkalises host tissues and leads to the formation of appressoria. This latter stage of development is when typical disease symptoms emerge.

4.9.3 Biology and Epidemiology

According to Bailey and Jeger (1992), both dormant mycelium and spores (conidia or ascospores) can remain in the soil for 1 or 2 years on plant debris or various organic substrates such as mummified berries, pedicels or vine bark. A film of water on the surface of the berries is required for spores to germinate. They penetrate the tissues and gradually invade the berries. Spores are already present after flowering but remain latent until berry ripening. Once infection is established, acervuli production on lesions is relatively rapid. Masses of conidia are dispersed by the wind droplets of water, by insects or by human activity in the vineyard. These conidia are responsible for secondary contamination in late summer. Ascospores are disseminated by the wind during humid periods, which favour infection and rotting of ripe berries when temperatures rise (25–30 °C).

4.9.4 Disease Control

Ripe rot control relies principally on repeated applications of fungicides during the growing season (Ji et al. 2021a, b, c). Lime sulphur is applied early, before bud brake, to reduce overwintering inoculum. Strobilurins or quinone outside inhibitors (QoIs) fungicides are recommended at flowering, while broad-spectrum fungicides (captafol, captan, maneb or benomyl) are applied repeatedly on green to mature bunches. Such intensive treatments are neither ecologically nor economically sustainable and represent a real threat to viticulture in tropical and subtropical regions. However, this disease may be partially controlled by selecting tolerant varieties or rootstocks (Greer et al. 2011; Shiraishi et al. 2007). Rootstocks with poor vigour are more tolerant of ripe rot. Well-aerated vineyards are less susceptible to the disease (reduction of plant density, vigour control and removal of leaves in the fruiting zone). Minimising wounds at the berry surface may also prevent ripe rot developing, as well as preventing infection by other pathogens. Other preventive management measures to control the disease include avoiding sprinkler irrigation (which helps disperse conidia masses), removing early infected organs and plant debris, and working in the vineyard only when plants are not wet.

4.10   Angular Leaf Spot

Current name:

- *Pseudocercospora brachypus* (Ellis & Everh.) X.J. Liu & Y.L. Guo 1992, Ascomycota, syn. *Cercospora brachypus* Ellis & Everh. 1902, *Mycosphaerella angulata* W.A. Jenkins 1942 (teleomorph)

Angular leafspot is caused by the fungal pathogen *Pseudocercospora brachypus*, the anamorph of *Mycosphaerella angulata*. Aptroot (2006) made the connection between the anamorph and the teleomorph of the species. However, the two main taxonomic databases for fungi (Mycobank and Index Fungorum) have not yet been updated to reflect this connection. Consequently, both names are valid and there is no single current name for the species, although the name *Pseudocercospora brachypus* is used in more recent studies (Crous et al. 2013). This disease principally attacks muscadine grapes (*Vitis rotundifolia* Michx.)—the only grape variety that can be successfully cultivated in coastal regions of the south-eastern United States because of its resistance to Pierce's disease caused by the bacteria *Xylella fastidiosa* (Chen and Lamikanra 1997). Angular leaf spot leads to premature defoliation. When the vine defoliates before harvest, fruits fail to reach maturity, leading to partial or total crop loss. Very little has been written about this disease, probably because it is specific to muscadine grapes, resistant cultivars common only in the south-eastern USA, and it can easily be controlled using fungicides, although at high cost when disease severity is high.

4.10.1   Causal Organism

Pseudocercospora brachypus (=*Mycosphaerella angulata*) belongs to the family Mycosphaerellaceae, the order Mycosphaerellales, and the division Ascomycota. The teleomorph *Mycosphaerella angulata* is a species of *Guignardia* (sensu von Arx), producing pseudothecia with clavate asci without hamathecium, each containing eight septate simple hyaline ascospores (11–13 × 5–7 µm), without gelatinous sheath (Aptroot 2006). The anamorph, *Pseudocercospora brachypus*, produces short, narrow conidia (25–60 × 2–3.5 µm) (Crous et al. 2013).

4.10.2   Symptoms

Pseudocercospora brachypus mainly infects leaves of muscadine grapes. The first symptoms to appear are pale yellow spots, more pronounced on the underside of the leaf. During the growing season, these chlorotic spots extend and form

Fig. 4.71 Irregular brown lesions on grapevine leaves caused by *Pseudocercospora brachypus*

irregular brown lesions at their centre (Fig. 4.71). These leaf lesions inhibit photosynthesis and may induce defoliation when they cover the whole leaf, or a substantial part of it.

4.10.3   Biology and Epidemiology

Angular leaf spot infection begins with primary inoculum (ascospores and/or conidia) spread by the wind. Once on the leaves, the spores germinate and penetrate the vine leaves through stomata or wounds. The mycelium invades the leaf tissue and begins to produce clumps of conidiophores on the lower leaf surface. The conidia produced by these conidiophores infect other leaves, constituting the second inoculum. *Pseudocercospora brachypus* (=*Mycosphaerella angulata*) also produces pseudothecia containing asci and ascospores. Both types of spores can restart the life cycle as primary inoculum.

4.10.4   Disease Control

Angular leafspot seems easy to control using fungicides such as dithiocarbamates, phthalimides, strobilurins or DMIs applied mid- to late-season. Preventive measures consist of destroying crop residues, removing nearby wild muscadine plants (sources of primary inoculum), and pruning the canopy. Some muscadine cultivars have been shown to be more resistant than others to angular leaf spot (Chen and Lamikanra 1997).

4.11 Leaf Rust

Current names of fungi:

- *Neophysopella ampelopsidis* (Dietel & P. Syd.) Jing X. Ji & Kakish 2019, Basidiomycota, syn. *Angiopsora ampelopsidis* (Dietel & P. Syd.) Thirum. & F. Kern 1949, *Physopella ampelopsidis* (Dietel & P. Syd.) Cummins & Ramachar 1959, *Phakopsora ampelopsidis* Dietel & P. Syd. 1898.
- *Neophysopella meliosmae-myrianthae* (Henn. & Shirai) Jing X. Ji & Kakish 2019, Basidiomycota, syn. *Aecidium meliosmatis-myrianthi* Henn. & Shirai 1900, *Phakopsora euvitis* Y. Ono 2000, *Neophysopella euvitis* (Y. Ono) R.F. Santos 2020
- *Neophysopella montana* (Y. Ono & Chatasiri) Jing X. Ji & Kakish., 2019, Basidiomycota, syn. *Phakopsora montana* Y. Ono & Chatasiri 2013
- *Neophysopella doipuiensis* Okane & Y. Ono 2020, Basidiomycota
- *Neophysopella tropicalis* Y. Ono, Chatasiri, Pota & Okane 2020, Basidiomycota
- *Neophysopella uva* (Buriticá & J.F. Hennen) Jing X. Ji & Kakish 2019, Basidiomycota
- *Neophysopella muscadinae* (Buriticá) Jing X. Ji & Kakish. 2019, Basidiomycota

At present grapevine leaf rust—like ripe rot—has a tropical to subtropical distribution. This disease is particularly widespread and problematic in Asia and Central and South America (Primiano et al. 2017; Na et al. 2022; Weinert et al. 2003) and in India (Sagar et al. 2023). It appeared in Australia in 2001 but seems to be eradicated since 2007. This disease has also been reported in more temperate regions such as Japan and the south-eastern United States (Okane and Ono 2018; Ono et al. 2020; Santos et al. 2021; Weinert et al. 2003). Initially classified in the genus *Phakopsora* Dietel 1895, rusts infecting grapevines were recently transferred to the genus *Neophysopella* Jing X. Ji & Kakish 2019. Furthermore, leaf rust was originally attributed to a single species, *N. ampelopsidis*. However, four additional species have since been found to be grapevine pathogens. Multigene phylogeny showed that *N. ampelopsidis* included several closely related species, among them *N. meliosmae-myrianthae* (previously *Phakopsora euvitis*). In addition, the sampling of leaf rust-associated fungi on infected grapevines in different continents lead to the description of five new grapevine rust-associated species (Okane and Ono 2018; Ono et al. 2012, 2020; Primiano et al. 2017). A severe outbreak of grapevine leaf rust results in poor shoot growth, reducing the quantity and quality of wines, and leading to an increase in the number of yearly fungicides treatments (Primiano et al. 2017).

4.11.1 Causal Organisms

The genus *Neophysopella* was first classified in the family Crossopsoraceae (Aime and McTaggart 2020) but is now part of the family Neophysopellaceae P. Zhao & L. Cai 2021 (Zhao et al. 2021), in the order Pucciniales and the division

Basidiomycota. Two types of asexual spores are produced by *Neophysopella* spp. associated with leaf rusts on *Vitis* spp. (Ono et al. 2012). The asexual yellow-orange spores forming on the underside of the leaves are called urediniospores (Fig. 4.72). They are produced in structures called uredinia (pustules) and have a subglobose to ovoid shape (20 µm Ø) and an echinulate surface (Rasera et al. 2022). Teliospores, produced later in structures called telia are the overwintering stage of the fungus (they act as chlamydospores). Ontogenetically speaking, teliospores (40–60 × 16–23 µm) are no different from urediniospores. They are also dikaryotic, but more condensed, with thicker walls, and characterised by an absence of the vacuoles and liquid droplets present in urediniospores, hence their dark brown colour. The sexual stage of *Neophysopella* spp. happens in an alternative host plant (Okane and Ono 2018), namely *Meliosma myriantha*.

4.11.2   Symptoms

Leaf rust usually infects grapevine leaves but may eventually infect fruits, stems, and rachises. When disease severity is high, leaves turn to senescence prematurely and drop, leading to poor shoot growth. Typical leaf symptoms are small yellowish-brown, angular necrotic lesions appearing on the upper surface of the leaves. On the underside, lesions are covered with small powdery pustules containing masses of yellow to orange spores (Fig. 4.73).

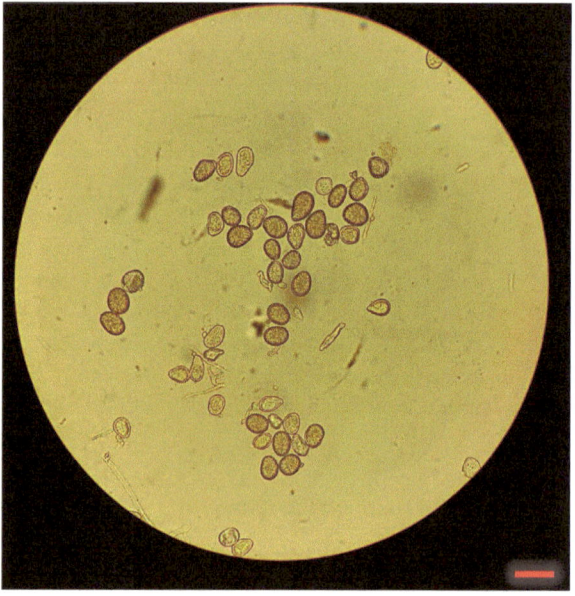

Fig. 4.72 Urediniospores of grapevine rust. Scale bar represents 30 µm. (Photograph © Dr. B.R. Sayiprathap. 2022. Acharya NG Ranga Agricultural University, Lam, Guntur, India)

Fig. 4.73 Symptoms of grapevine leaf rust on vine showing the formation of numerous yellowy-orange pustules (uredinia) containing urediniospores. (Photograph © Dr. B.R. Sayiprathap. 2022. Acharya NG Ranga Agricultural University, Lam, Guntur, India)

4.11.3   Biology and Epidemiology

The life cycle of rusts is very complex and largely unknown for most of the *Neophysopella* spp. associated with grapevine (Okane and Ono 2018). The full cycle involves the production of five types of propagules; two produced on a non-*Vitis* host, in special structures called aecia and pycnidia; two produced on a *Vitis* sp. in structures called uredinia and telia; and basidiospores, which are formed immediately after germination of the teliospores on *Vitis* leaf debris. For *N. euvitis* (Primiano et al. 2017), teliospores (=probasidia)—the spores that overwinter in grapevine leaves litter—produce a promycelium after germination which immediately undergoes karyogamy and meiosis to form basidia containing four haploid basidiospores. These spores, transported by wind or insects, infect *Meliosma myriantha*, its alternative host, in late spring or early summer. A haploid promycelium produced from the germination of a single basidiospore develops in the leaves and produces the pycnidia and receptive hyphae that undergo mating (sexual stage). The resulting diploid mycelium will then produce binucleate aeciospores that are transported by wind and infect *Vitis* spp., usually during late summer through to the end of autumn. In the *Vitis* host species, the fungus produces the typical symptoms of grapevine leaf rust characterised by the production of urediniospores on the underside of the leaves. Seasonal change will generally induce the formation of telia bearing teliospores on the upper surface of grapevine leaves.

Consequently, two host plants are mandatory for *Neophysopella* spp. to complete their full life cycle. Phylogenetic analyses suggest a *Neophysopella* of Asian origin in the genus *Meliosma* as an alternative host (Okane and Ono 2018). However, as the genus *Meliosma* is not present in all continents, *Ampelocissus* or *Parthenocissus* have been proposed as other alternative host genera for *Neophysopella* spp. In favourable climatic conditions such as the tropics, uredinospores can survive all year round and reinfect grapevine without the need for sexual reproduction on a non-*Vitis* host. Further research is needed to understand grapevine rust evolution and determine alternative hosts of the different species of *Neophysopella* spp.

4.11.4   Disease Control

Grapevine leaf rust may be controlled using fungicides commonly deployed to control other fungal diseases. Preventive measures include buying certified propagating material to avoid introducing leaf rust to the vineyard and carefully examining young plants for early leaf rust symptoms before planting. Australia, using both approaches combined with strict border controls and control of people visiting vineyards, successfully eradicated the disease within 6 years of its first appearance in 2001.

4.12   Septoria Leaf Spot (Melanose)

Current name:

• *Septoria ampelina* Berk. & M.A. Curtis 1874, Ascomycota, syn. *Rhabdospora ampelina* (Berk. & M.A. Curtis) Kuntze 1898

McGrew and Pollack (1988) found that *Septoria ampelina* is the causal fungal agent of septoria leaf spot (also called melanose), a disease affecting grapevines and many other crops such as shrubs, cucurbits, and ornamental plants. This fungus was first reported to infect only American *Vitis* species (Boubals and Mur 1983). However, Mitchell et al. (1994), having tested 17 cultivars (American, European, and interspecific hybrids), showed that almost all cultivars developed leaf spot lesions after leaf inoculation with *S. ampelina*. The only two exceptions that appeared resistant to this fungus were two cultivars of *Vitis rotundifolia* ('Cowart' and 'Fry'). Initially seen as a real threat to viticulture in the early 1990 due to the substantial defoliation of plants in Arkansas, this disease has since been reported to be minor and occasional; apparently not sufficiently destructive to be the subject of further studies. Septoria leaf spot is reported to have a higher incidence when the summer is cold. Septoria leaf spot is caused by *Septoria ampelina*, a fungal species belonging to the family Mycosphaerellaceae (order Mycosphaerellales; class Dothideomycetes; division Ascomycota). The original description was made from a specimen affecting leaves of *Vitis vulpinae* in Texas (USA). According to the very succinct description given by Berk. & M.A. Curtis, *S. ampelina* causes intermixed reddish and star-shaped brownish spots on the leaves. This species is mitosporic, producing conidia from amphigenous pycnidia (growing on all sides at once), which are obscurely septate (3–5 septa); conidia are linear and curved, 1–3 × 30–90 μm long (Kleinig 1985).

According to Pearson and Goheen (1998), grapevine leaves affected by *Septoria* present small angular spots (1–2 mm Ø), reddish to brownish or eventually black. These spots become bigger at veraison, with pycnidia sometimes appearing at the centre of the spots. Spots have a well-defined margin surrounded by a diffuse yellow halo. When disease severity is high, plants can lose most of their leaves, leading to 15% or more crop loss at harvest.

Septoria ampelina overwinters in grapevine leaf litter, where it can survive as mycelium for 1 or 2 years. Pycnidia release conidia in spring, which are spread by wind and rain. The optimum temperature for infection is between 16 and 19 °C during wet periods. Infection generally occurs in spring, before the berry skin reaches maturity. Dry, warm weather conditions are not conducive to infection. However, infection can also occur later, in autumn, when temperatures are cooler.

One preventive measure against Septoria leaf spot is to remove leaf debris before winter. Another measure is to prune the vines in such a way as to promote aeration and rapid drying of the leaves, and to examine the leaves carefully for early signs of infection (Pearson and Goheen 1998). The disease can also be controlled with fungicides. According to Utami (1995), the most effective fungicides are benomyl (methyl benzimidazole carbamate MBCs) and tebuconazole (demethylation inhibitors, DMIs), applied either as protectant or systemic treatment.

4.13 Mycotoxins and Secondary Fungal Metabolites in Grape

Current name:

- *Acremonium* spp. Link 1809, Ascomycota
- *Alternaria* spp. Nees 1817, Ascomycota
- *Aspergillus* spp. P. Micheli ex Haller 1768, Ascomycota
- *Aureobasidium pullulans* (de Bary) G. Arnaud 1918, Ascomycota
- *Botrytis cinerea* Pers. 1794, Ascomycota
- *Cladosporium* spp. Link 1816, Ascomycota
- *Colletotrichum* spp. Corda 1831, Ascomycota
- *Fusarium* spp. Link 1809, Ascomycota
- *Greeneria uvicola* (Berk. & M.A. Curtis) Punith. 1974, Ascomycota
- *Penicillium* spp. Link 1809, Ascomycota
- *Trichothecium roseum* (Pers.) Link 1809, Ascomycota
- *Ulocladium* spp. Preuss 1851, Ascomycota

Secondary moulds appearing on berries and bunches shortly before harvest are generally cosmopolitan, ubiquitous fungi which can pose serious health problems, especially in the production of raisins, table grapes and wine grapes. The fungal species involved can vary considerably depending on location and climate conditions. They can significantly impair grape quality by introducing off-flavours or in particular, by producing toxic secondary metabolites (mycotoxins) which can cause health problems when ingested even in low concentrations. These secondary fungal contaminations are generally initiated in the field or during storage. Broadly speaking, these moulds invariably develop in humid conditions and on berries that are damaged (microfissures) or at an advanced stage of ripening (Crandall et al. 2022). With few exceptions, fungi which develop on berries are not generally associated with the vine. Instead, they form part of the mycoflora (mycobiome) of the vineyard environment and/or the atmosphere. Symptoms cannot be linked specifically to one or other of the fungal species which interact during the breakdown of berry tissue. Often it is tricky to draw a clear distinction between grey mould caused by *Botrytis cinerea*, sour rot of fungal (yeasts) or bacterial origin, and secondary rots. When all favourable factors combine (temperatures >20 °C and high relative humidity or rainfall), these different forms of rot, appearing separately or together, contribute to the breakdown and destruction of plant tissue. In certain circumstances, secondary rots can completely alter the harvest, making it unfit for consumption or vinification. The genera *Acremonium, Alternaria, Aspergillus, Cladosporium, Trichothecium, Penicillium, Fusarium, Colletotrichum, Ulocladium, Aureobasidium pullulans* and *Botrytis cinerea* can produce secondary metabolites linked to off-flavours and mycotoxins.

Wines produced from partially rotten harvests can be tainted by off-flavours; of for example, musty aromas caused by the production of 1-octen-3-ol and 1-octen-3-one; of musty-earthy aromas (geosmin); of earthy-camphor aromas (2-methylisoborneol MIB); of aromas of asparagus and pepper (2-isopropyl-3-methoxypyrazine IPMP); or caused by atypical aging or water-nitrogen stress characterised by odours of wet hay (amino-aceto-phenol, sotolon [3-hydroxy-4.5 dimethylfurane-2(5H)-one]), varnish (ethyl acetate), reductive notes (methanethiol, mercaptan), beer (ethanethiol), volatile acidity (ethyl acetate, acetic acid) and astringency (Fig. 4.74). Geosmin (trans-1.10-dimethyl-trans-9-decalol) has been identified as the main molecule responsible for musty-earthy odours in musts and wines (Darriet et al. 2000). It can be produced by bacteria (Streptomyces), algae, cyanobacteria and fungi (*Botrytis cinerea, Aspergillus* sp., *Penicillium* sp., *Trichoderma* sp.). Among the dozens of species of *Penicillium* isolated from infected berries, *P. expansum* is the only one to produce geosmin. When this fungus interacts with *Botrytis cinerea*, the production of geosmin increases its ability to develop and colonise while inhibiting the development of other fungi such as grey mould (La Guerche et al. 2004; Barata et al. 2012). MIB, a terpene alcohol, and IPMP are also produced by different fungi, including species in the genera *Penicillium* and *Aspergillus* as well as *Botrytis cinerea* (Rousseaux et al. 2014). These different off-flavours make the wines unfit for consumption and may vary in intensity depending on the presence of one or other fungal species, the vine cultivar and the vinification method.

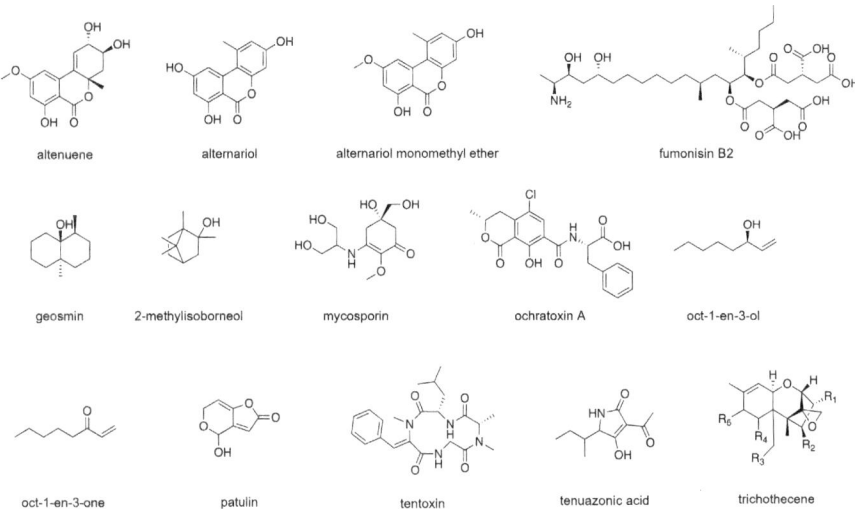

altenuene   alternariol   alternariol monomethyl ether   fumonisin B2

geosmin   2-methylisoborneol   mycosporin   ochratoxin A   oct-1-en-3-ol

oct-1-en-3-one   patulin   tentoxin   tenuazonic acid   trichothecene

Fig. 4.74 Examples of secondary metabolites and mycotoxins produced by different species of filamentous fungi which colonise grape berries during ripening and can cause problems with off-flavours and mycotoxigenic contamination in musts and wine. (Illustrations © Robin Huber, University of Geneva, Switzerland)

Mycotoxins are mostly secondary metabolites of polyketides produced during the development of numerous filamentous fungi, in particular *Aspergillus* spp., *Penicillium* spp., *Fusarium* spp. or *Trichotecium roseum*. These molecules are generally very stable to changes in temperature. Ochratoxin A (OTA) or fumonisin B2 (FB2) produced by *Aspergillus welwitschiae, A. steynii, A. westerdijkiae, A. flavus*, or *A. parasiticus* (Gómez-Albarrán et al. 2021), patulin (PAT) produced by *Aspergillus carbonarius, A. niger* or *A. ochraceus*, and other aflatoxins produced mainly by *A. flavus* (Rousseaux et al. 2014), as well as diverse mycotoxins such as trichothecenes, produced by *T. roseum* or *Fusarium* spp. for example, can constitute a public health problem (Gonçalves et al. 2020). These secondary metabolites are generally toxic to humans. Ochratoxin A, the most toxic of the ochratoxins, is produced by *Aspergillus carbonarius*. Other ochratoxins are produced by species of the genus *Penicillium*, such as *P. verrucosum, P. brevicompactum, Penicillium* in the *nigri* section and *P. expansum* (Amézqueta et al. 2012; Rousseaux et al. 2014). Ochratoxins remain stable in musts and wines. OTA weakens the immune system, attacks the kidneys and urinary tract and is carcinogenic. Patulin, a lactone produced by *Penicillium* sp., causes problems during fermentation by inhibiting the development of the yeast *Saccharomyces cerevisiae*, and is also carcinogenic. However, this toxin is broken down by sulphiting the wine (SO_2) and completely degraded during alcoholic fermentation.

Over 30 species of *Aspergillus* have been isolated from grapes in vineyards around the world, more than 50% of which have been identified as *A. carbonarius* and *A. niger*, two species classified in the *nigri* section of *Aspergillus*. Numerous other mycotoxins, such as the aflatoxins, including the hepatotoxic aflatoxin B1 produced by *Aspergillus* sp., have been characterised as major contaminants of wine in specific climatic conditions (Welke 2019).

Depending on the drying conditions, raisins can also contain mycotoxins associated with *Aspergillus* or *Penicillium*, especially OTA (Şen et al. 2016; Şen and Nas 2013). Grape berries can also be contaminated with species in the genus *Fusarium* spp., known for their ability to produce extremely toxic secondary metabolites, such as trichothecenes or zearalenone. This may occur during the dehydration of grapes for the production of raisins and can seriously impair the edibility of the end product. The main mycotoxins in grapes are beauvericin or fumonisins, produced by *F. proliferatum*, *F. sporotrichoides*, *F. oxysporum* and *F. verticillioides* (Gonçalves et al. 2020). Several mycotoxigenic species belonging to the genus *Alternaria* have also been isolated from grapes. Essentially, these include *Alternaria alternata*, *A. arborescens*, *A. infectoria* and *A. tenuissima*, which can produce altenuene, tenuazonic acid, mycosporine A, tentoxin, alternariol monomethyl ether and alternariol. The latter two mycotoxins are reported to be mutagenic and genotoxic (Mikušová et al. 2014; Tančinová et al. 2016).

Greeneria uvicola, an ascomycete in the order Diaporthales and the family Melanconiellaceae, causes bitter rot of grape, a disease commonly found on muscadine and bunch grapes throughout the world. The fungus overwinters on stem lesions and mummified berries. Although the fungus can infect leaves, tendrils and stems, this disease mainly causes damage to fruit, especially when wet weather persists until harvest. *G. uvicola* is known to attack several species of *Vitis*, including *V. aestivalis*, *V. labrusca*, *V. rotundifolia* and *V. vinifera* (Farr et al. 2011). This species has been described in several countries where vines are exposed to hot and humid climates, such as Japan, Mexico, Greece, Taiwan, China, India, Australia, Costa Rica, Brazil, South Africa and the Eastern United States. *G. uvicola* is the cause of off-flavours in wines produced from infected grapes. The metabolites responsible for these off-flavours are yet to be identified.

4.13.1 Biology and Epidemiology of Involved Fungi

The fungi responsible for secondary bunch rots form part of the vineyard environment or are present in the atmosphere. Their saprophytic opportunistic behaviour places them among the decomposers of organic matter. Their presence is directly linked to the general health status of vine and berries, which is influenced by a hot, humid climate during the later phases of bunch ripening or the overripening phase. Damaged berries are particularly susceptible to fungi. Wounds may arise from insect bites, bird damage, severe powdery mildew infection or berry burst caused by hail or heavy rainfall.

4.13.2 Impact on Berries

Secondary rots are easy to recognise by the characteristic colour of the mycelium sporulating on the berry surface (Fig. 4.75). In a severe grey mould infection, berries are often dotted with aerial mycelium, greyish-white first, then turning

Fig. 4.75 *Botrytis cinerea* (arrow) and *Penicillium* sp./*Aspergillus* sp. complex on a rotten grape berry

blueish-green when the fungus sporulates (*Penicillium* sp., especially *P. expansum*). Colonies of *Trichothecium roseum* are pink to pinkish-orange, while those of *Alternaria* and *Aspergillus* are greyish-brown to black (Fig. 4.76). The initial mycelium of secondary rots can easily be mistaken for the mycelium of *Botrytis cinerea*, which clearly turns grey when it sporulates. These symptoms appear on turgid berries as well as mummified or partially dried berries; even stems in the case of severe infections. Affected bunches are not suitable for harvesting due to their poor appearance and are generally eliminated. Even with mechanical harvesting, triage is difficult and, depending on the proportion of damaged bunches, the secondary metabolite load produced by fungi associated with secondary rots can compromise vinification processes. Raisins can become contaminated during open-air drying (Şen et al. 2016; Şen and Nas 2013).

4.13.3   Biological Control and Detoxification

There are limited opportunities to control secondary rots as fungicidal applications close to harvest expose consumers to undesirable fungicide residues. Apart from maintaining bunch health and protecting them from other fungal diseases and pests, any measures to protect berries from damage play a preventive role in controlling secondary rots.

The grape berry microbiota is a complex community of filamentous fungi, yeasts and bacteria. Although some species within these communities may be mycotoxigenic, the majority constitute a major part of the biocontrol agents described in the literature. Bleve et al. (2006) isolated several natural yeasts present on grapes, such as *Pichia orientalis* (formerly *Issatchenkia orientalis*), *Metschnikowia pulcherrima* or *Candida incommunis*, and showed their potential to significantly inhibit the growth of ochratoxigenic fungi such as *Aspergillus niger* and *A. carbonarius*. Furthermore, *Aureobasidium pullulans* and *Lachancea thermotolerans* can undermine the development *Aspergillus* in the *nigri* section and the biosynthesis of OTA

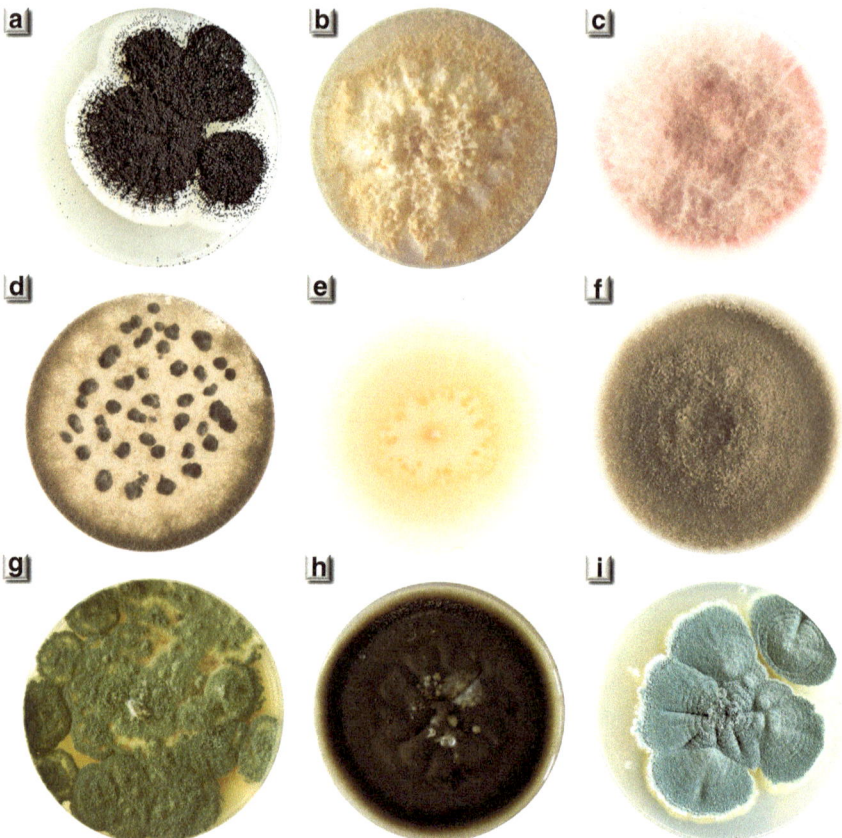

Fig. 4.76 Raisin moulds and rots can be caused by a large spectrum of fungi present in the vineyards or air-borne. Example of different rotting fungal species in pure culture on agar media. (**a**) *Aspergillus niger*. (**b**) *Trichothecium roseum*. (**c**) *Fusarium graminearum*. (**d**) *Botrytis cinerea*. (**e**) *Aureobasidium pullulans*. (**f**) *Alternaria alternata*. (**g**) *Penicillium expansum*. (**h**) *Cladosporium* sp. (**i**) *Penicillium* sp.

(Ponsone et al. 2011). More recently, purification techniques to detoxify mycotoxins have been developed using activated carbon filters and yeasts obtained from the natural berry mycobiome.

Prevention of Mycotoxin Contamination
(Gómez-Albarrán et al. 2021)

Despite the use of prevention and control methods aimed at averting all fungal contamination, it is not always possible to avoid the presence of mycotoxins in food and beverages. Furthermore, the coexistence of several mycotoxins in the same product is problematic because of their cumulative and synergic toxic and carcinogenic effect. It is therefore essential to develop new strategies to eliminate mycotoxins in food. Detoxification involves various methods (physical, chemical or biological) which eliminate mycotoxins without generating toxic products or altering the nutritional value and organoleptic properties of the food. In the case of wine, the detoxification of ochratoxin A (OTA) during the fining process using physical adsorbents such as activated carbon has been tested with positive results, although this treatment generally affects the quality of the wine. Biological detoxification is the most promising approach because it is safe for health and the environment and also retains the organoleptic properties of the food. These biological methods use microorganisms or their enzymes to degrade, transform or adsorb mycotoxins, thereby avoiding toxic effects caused by ingesting contaminated food. In this context, *Hanseniaspora uvarum* U1 has been selected for *in vitro* biocontrol assays. The results showed that this yeast can reduce the growth rate of the main aflatoxigenic and ochratoxigenic species of *Aspergillus* spp. present on grapes that are responsible for producing ochratoxins and aflatoxins. Furthermore, *H. uvarum* U1 seems to be effective in detoxifying aflatoxin B1 (AFB1) and OTA, probably by means of adsorption mechanisms through the cell wall.

References

Abbey JA, Percival D, Abbey L, Asiedu SK, Prithiviraj B, Schilder A (2019) Biofungicides as alternative to synthetic fungicide control of grey mould (*Botrytis cinerea*)—prospects and challenges. Biocontrol Sci Tech 29:207–228. https://doi.org/10.1080/09583157.2018.1548574

Aime MC, McTaggart AR (2020) A higher-rank classification for rust fungi, with notes on genera. Fungal Syst Evol 7:21–47. https://doi.org/10.3114/fuse.2021.07.02

Alonso-Villaverde V, Voinesco F, Viret O, Spring JL, Gindro K (2011) The effectiveness of stilbenes in resistant Vitaceae: ultrastructural and biochemical events during *Plasmopara viticola* infection process. Plant Physiol Biochem 49(3):265–274. https://doi.org/10.1016/j.plaphy.2010.12.010

Amézqueta S, Schorr-Galindo S, Murillo-Arbizu M, González-Peñas E, López de Cerain A, Guiraud JP (2012) OTA-producing fungi in foodstuffs: a review. Food Control 26(2):259–268

Amrani L, Corio-Costet MF (2006) A single nucleotide polymorphism in the β-tubulin gene distinguishing two genotypes of *Erysiphe necator* expressing different symptoms on grapevine. Plant Pathol 55:505–512. https://doi.org/10.1111/j.1365-3059.2006.01390.x

Anant JK, Inchulkar SR, Bhagat S (2018) An overview of copper toxicity relevance to public health. EJPMR 5(11):232–237

Aptroot A (2006). *Mycosphaerella* and its anamorphs: conspectus of *Mycosphaerella* CBS biodiversity series 2(5), ISSN 1571-8859. Centraalbureau voor Schimmelcultures, ISBN 9070351609, 9789070351601, 231 pp

Arya GC, Cohen H (2022) The multifaceted roles of fungal cutinases during infection. J Fungi 8(2):199. https://doi.org/10.3390/jof8020199

Aziz A, Poinssot B, Daire X, Adrian M, Bézier A, Lambert B, Joubert JM, Pugin A (2003) Laminarin elicits defense responses in grapevine and induces protection against *Botrytis cinerea* and *Plasmopara viticola*. Mol Plant-Microbe Interact 16(12):1118–1128. https://doi.org/10.1094/MPMI.2003.16.12.1118. PMID: 14651345

Bailey JA, Jeger MJ (eds) (1992) *Colletotrichum*: Biology, Pathology and Control. CAB International. Wallingford UK

Barata A, Gonzalez S, Malfeito-Ferreira M, Querol A, Loureiro V (2008a) Sour rot-damaged grapes are sources of wine spoilage yeasts. FEMS Yeast Res 8:1008–1017. https://doi.org/10.1111/j.1567-1634.2008.00399.x

Barata A, Seborro F, Belloch C, Malfeito-Ferreira M, Loureiro V (2008b) Ascomycetous yeast species recovered from grapes damaged by honeydew and sour rot. J Appl Microbiol 104(4):1182–1191. https://doi.org/10.1111/j.1365-2672.2007.03631.x

Barata A, Correia Santos S, Malfeito-Ferreira M, Loureiro V (2012) New insights into the ecological interaction between grape berry microorganisms and *Drosophila* flies during the development of sour rot. Microb Ecol 64:416–430

Batista DDC, Vieira WAS, Barbosa MA, Camara MPS (2023) First report of *Colletotrichum siamense* causing grape ripe rot in Brazil. Plant Dis 107:2881. PMID: 37172973

Belhadj A, Telef N, Saigne C, Cluzet S, Barrieu F, Hamdi S, Mérillon JM (2008) Effect of methyl jasmonate in combination with carbohydrates on gene expression of PR proteins, stilbene and anthocyanin accumulation in grapevine cell cultures. Plant Physiol Biochem 46(4):493–499. https://doi.org/10.1016/j.plaphy.2007.12.001

Bernard A (1977) Observations histologiques sur les baies de *Vitis vinifera* au cours de leur croissance. France Viticole 9:137–146

Bessis R (1972) Study on development of peristomatic tissue and stomata of grapes. C R Hebd Séances Acad Sci Sér D 274:2158–2161

Bettinelli P, Nicolini D, Giovannini O, Stefanini M, Hausmann L, Vezzulli S (2023) Breeding for black rot resistance in grapevine: advanced approaches for germplasm screening. Euphytica 219:113. https://doi.org/10.1007/s10681-023-03235-9

Bisiach M, Zerbetto F, Minervini G (1980) Ricerche sulla conservazione invernale di *Botrytis cinerea* su vite. Notizario sulle malattie delle piante 101:13–33

Bisiach M, Zerbetto F, Vercesi A (1982) La protection du vignoble contre *Botrytis cinerea* en Italie du Nord et du Centre. Situation actuelle et perspectives d'avenir. EPPO Bulletin 12(2):87–93. https://doi.org/10.1111/j.1365-2338.1982.tb01685.x

Blakeman JP (1980) Behaviour of conidia on aerial plant surfaces. In: Coley-Smith JR, Verhoeff K, Jarvis WR (eds) The biology of *Botrytis*. Academic Press, New York/London, pp 115–151

Bläser M (1978) Untersuchungen zur Epidemiologie des falschen Mehltaus an Weinrebe, *Plasmopara viticola* (Berk. et Curt. Ex de Bary) Berl. et de Toni. Dissertation Universität Bonn

Bläser M, Weltzien HC (1979) Epidemiologische Studien an *Plamopara viticola* zur Verbesserung der Spritzterminbestimmung. Zeitschrift für Pflanzenkrankheit und Pflanzenschutz 86(8):489–498

Bleve G, Grieco F, Cozzi G, Logrieco A, Visconti A (2006) Isolation of epiphytic yeasts with potential for biocontrol of *Aspergillus carbonarius* and *A. niger* on grape. Int J Food Microbiol 108(2):204–209. https://doi.org/10.1016/j.ijfoodmicro.2005.12.004

Bleyer G, Kassemeyer H-H, Breuer M, Krause R, Viret O, Dubuis P-H, Fabre A-L, Bloesch B, Siegfried W, Naef A, Hubert M (2011) "VitiMeteo"—a future-oriented forecasting system for viticulture. IOBC/WPRS Bull 67:69–77

Bolay A (2005) Les oïdiums de Suisse (Erysiphacées). Cryptogamica Helvetica 20, 176 p

Boubals D, Mur G (1983) Another grapevine disease is raging in Penedes (Spain). Prog Agr Viticult 100:453

Braga ZV, dos Santos RF, Amorim L, Appezzato-da-Gloria B (2020) Histopathological evidence of concomitant sexual and asexual reproduction of *Elsinoë ampelina* in grapevine under subtropical climate. Physiol Mol Plant Pathol 111:101517. https://doi.org/10.1016/j.pmpp.2020.101517

Braun U, Takamatsu S (2000) Phylogeny of *Erysiphe*, *Microsphaera*, *Uncinula* (Erysipheae) and *Cystotheca*, *Podosphaera*, *Sphaerotheca* (Cystotheceae) inferred from rDNA ITS sequences—some taxonomic consequences. Schlechtendalia 4:1–33. https://doi.org/10.25673/90008

Brierly WB (1918) Botrytis cinerea. Kew Bull 1:42

Bristow PR, McNicol RJ, Williamson B (1986) Infection of strawberry flowers by *Botrytis cinerea* and its relevance to grey mould development. Ann Appl Biol 109:545–554. https://doi.org/10.1111/j.1744-7348.1986.tb03211.x

Brooks BA (1908) Observations on the biology of *Botrytis cinerea*. Ann Bot 12(87):479–487

Burril TJ (1886) Grape rots. Amer Pomol Soc Proc 20:47–49

Caffi T, Rossi V, Legler SE, Bugiani R (2011) A mechanistic model simulating ascosporic infections by *Erysiphe necator*, the powdery mildew fungus of grapevine. Plant Pathol 60:522–531. https://doi.org/10.1111/j.1365-3059.2010.02395.x

Caffi T, Legler SE, Rossi V, Bugiani R (2012) Evaluation of a warning system for early-season control of grapevine powdery mildew. Plant Dis 96(1):104–110. https://doi.org/10.1094/PDIS-06-11-0484

Calonnec A, Cartolaro P, Deliere L, Chadoeuf J (2006) Powdery mildew on grapevine: the date of primary contamination affects disease development on leaves and damage on grape. IOBC/WPRS Bull 29(11):67–73

Calonnec A, Cartolaro P, Naulin JM, Bailey D, Langlais M (2008) A host-pathogen simulation model: powdery mildew of grapevine. Plant Pathol 57:493–508. https://doi.org/10.1111/j.1365-3059.2007.01783.x

Cano J, Guarro J, Gené J (2004) Molecular and morphological identification of *Colletotrichum* species of clinical interest. J Clin Microbiol 42(6):2450–2454

Carisse O, Lefebvre A (2011) A model to estimate the amount of primary inoculum of *Elsinoë ampelina*. Plant Dis 95(9):1167–1171. https://doi.org/10.1094/PDIS-11-10-0798

Carisse O, Levasseur A, Provost C (2020) Influence of leaf wetness duration and temperature on infection of grape leaves by *Elsinoë ampelina* under controlled and vineyard conditions. Plant Dis 104:2817–2822. https://doi.org/10.1094/PDIS-02-20-0262-RE

Cesco S, Tolotti A, Nadalini S, Rizzi S, Valentinuzzi F, Mimmo T, Porfido C, Allegretta I, Giovannini O, Perazzolli M, Cipriani G, Terzano R, Pertot I, Pii Y (2020) *Plasmopara viticola* infection affects mineral elements allocation and distribution in *Vitis vinifera* leaves. Sci Rep 10(1):18759. https://doi.org/10.1038/s41598-020-75990-x

Chen XY, Kim JY (2009) Callose synthesis in higher plants. Plant Signal Behav 4(6):489–492. https://doi.org/10.4161/psb.4.6.8359

Chen J, Lamikanra O (1997) Resistance of muscadine grapes to angular leaf spot (*Mycosphaerella angulata* Jenkins) in North Florida. HortScience 32(1):94–95. https://doi.org/10.21273/HORTSCI.32.1.94

Chen TW, He YF, Peng FY, Xiao CM (1980) Studies on the white rot (*Coniothyrium diplodiella* (Speg.) Sacc.) of grapevine II. The environmental conditions and paths of white rot infection. J Plant Prot 7:27–34

Chethana KWT, Zhou Y, Zhang W, Liu M, Xing QK, Li XH, Yan JY (2017) *Coniella vitis* sp. nov. is the common pathogen of white rot in Chinese vineyards. Plant Dis 101:2123–2136. https://doi.org/10.1094/PDIS-12-16-1741-RE

Chitarrini G, Soini E, Riccadonna S, Franceschi P, Zulini L, Masuero D, Vecchione A, Stefanini M, Di Gaspero G, Mattivi F, Vrhovsek U (2017) Identification of biomarkers for defense response to *Plasmopara viticola* in a resistant grape variety. Front Plant Sci 8:1524. https://doi.org/10.3389/fpls.2017.01524

Clifford L, Jiang L, Oghenekome O, Hemanth KNV, Devaiah K, Sheikh MB, Hae Keun Y (2011) Resistance to *Elsinoë ampelina* and expression of related resistant genes in *Vitis rotundifolia* Michx. grapes. Int J Mol Sci 12:3473–3488. https://doi.org/10.3390/ijms12063473

Coley-Smith JR (1980) Sclerotia and other structures in survival. In: Coley-Smith JR, Verhoeff K, Jarvis WR (eds) The biology of *Botrytis*. Academic Press, New York/London, pp 85–114

Coombe BG, Hale CR (1973) The hormone content of ripening grape berries and the effects of growth substance treatments. Plant Physiol 51:629–634. https://doi.org/10.1104/pp.51.4.629

Crandall SG, Spychalla J, Crouch UT, Acevedo FE, Naegele RP, Miles TD (2022) Rotting grapes don't improve with age: cluster rot disease complexes, management, and future prospects. Plant Dis 106:2013–2025. https://doi.org/10.1094/PDIS-04-21-0695-FE

Crous PW, Braun U, Hunter GC, Wingfield MJ, Verkley GJ, Shin HD, Nakashima C, Groenewald JZ (2013) Phylogenetic lineages in *Pseudocercospora*. Stud Mycol 75(1):37–114

Csikós A, Németh MZ, Frenkel O, Kiss L, Váczy KZ (2020) A fresh look at grape powdery mildew (*Erysiphe necator*) A and B genotypes revealed frequent mixed infections and only B genotypes in flag shoot samples. Plan Theory 9(9):1156. https://doi.org/10.3390/plants9091156

Dalbò MA, Weeden NF, Reisch BI (2000) QTL analysis of disease resistance in interspecific hybrid grapes. Acta Hortic. 528:217–222. https://doi.org/10.17660/ActaHortic.2000.528.29

Damm U, Cannon PF, Woudenberg JH, Crous PW (2012) The *Colletotrichum acutatum* species complex. Stud Mycol 73(1):37–113

Darriet P, Pons M, Henry R, Dumont O, Findeling V, Cartolaro P, Calonnec A, Dubourdieu D (2002) Impact odorants contributing to the fungus type aroma from grape berries contaminated by powdery mildew (*Uncinula necator*); incidence of enzymatic activities of the yeast *Saccharomyces cerevisiae*. J Agric Food Chem 50(11):3277–82. https://doi.org/10.1021/jf011527d

Darriet P, Pons M, Lamy S, Dubourdieu D (2000) Identification and quantification of geosmin, a powerful earthy odorant contaminating wines. J Agric Food Chem 48:4835–4838. https://doi.org/10.1021/jf0007683

De Bary HA (1863) Du développement des champignons parasites. Ann Sci Nat Ser 4(t.20):125–126

De Bary HA (1866) Morphologie und physiologie der Pilze, Flechten und Myxomyceten. In: Hoffmeister's Handbuch der Physiol. Botanik, 2Bd., Leipzig

De Bary HA (1874) Über den sogenannten Brenner (Pech) der Reben. Ann Oenol 4:165–167

De Bary HA (1886) Ueber einige Sclerotien und Sclerotinienkrankheiten. Bot Ztg 44:378–398

De Istvanffi G (1903) Etudes microbiologiques et mycologiques sur le rot gris de la vigne (*Botrytis cinerea-Sclerotinia fuckeliana*). Annales de l'Institut Central Ampélologique Royal Hongrois, Budapest

De Luca d'Oro GM, Trippi VS (1987) Effect of stress conditions induced by temperature, water and rain on senescence development. Plant Cell Physiol 28(8):1389–1396. https://doi.org/10.1093/oxfordjournals.pcp.a077430

Délye C, Laigret F, Corio-Costet MF (1997) RAPD analysis provides insight into the biology and epidemiology of *Uncinula necator*. Phytopathology 87(7):670–677. https://doi.org/10.1094/PHYTO.1997.87.7.670

Doss RP, Potter SW, Chastagner GA, Christian JK (1993) Adhesion of nongerminated *Botrytis cinerea* conidia to several substrata. Appl Environ Microbiol 59(6):1786–1791. https://doi.org/10.1128/aem.59.6.1786-1791.1993

Doss RP, Potter SW, Soeldner AH, Christian JK, Fukunaga LE (1995) Adhesion of germlings of *Botrytis cinerea*. Appl Environ Microbiol 61(1):260–265. https://doi.org/10.1128/aem.61.1.260-265.1995

Drayton FL (1932) The sexual function of the microconidia in certain discomycetes. Mycologia 24:345–348

Echeverrigaray SG, Delamare AP, Fontanella G, Favaron F, Stella L, Scariot A (2019) *Colletotrichum* species associated to ripe rot disease of grapes in the "Serra Gaucha" region of Southern Brazil. BIO Web Conf. https://doi.org/10.1051/bioconf/20191201008

Ellis JB (1880) A new *Sphaeria* on grapes. Bull Torrey Bot Club 7(8):90–91

Ellis MA, Madden LV, Wilson LL (1986) Electronic grape black rot predictor for scheduling fungicides with curative activity. Plant Dis 70(10):938–940

Epton HAS, Richmond DV (1980) Formation, structure and germination of conidia. In: Coley-Smith JR, Verhoeff K, Jarvis WR (eds) The biology of Botrytis. Academic, New York, pp 41–83

Esterio M, Osorio-Navarro C, Carreras C, Azócar M, Copier C, Estrada V, Rubilar M, Auger J (2020) *Botrytis prunorum* associated to *Vitis vinifera* blossom blight in Chile. Plant Dis 104(9):2324–2329. https://doi.org/10.1094/PDIS-09-19-2055-SC

Falk SP, Gadoury DM, Cortesi P, Pearson RC, Seem RC (1995) Parasitism of *Uncinula necator* cleistothecia by the mycoparasite *Ampelomyces quisqualis*. Phytopathology 85:794–800. https://doi.org/10.1094/Phyto-85-794

Faretra F, Antonnaci E (1987) Production of apothecia of *Botryotinia fuckeliana* (De Bary) Whetz. Under controlled environmental conditions. Phytopathol Med 26(1):29–35

Farlow WG (1876) Grapevine mildew. Pacific Rural Press 12(18):288

Farlow WG (1882) Notes on some species in the third and eleventh centuries of Ellis's North American fungi. Proc Am Acad Art Sci 18:65–85. https://doi.org/10.2307/25138686

Farr DF, Castlebury LA, Rossman AY, Erincik O (2011) *Greeneria uvicola*, cause of bitter rot of grapes belongs in the Diaporthales. Sydowia-Horn 53(2):185–189

Fontaine MC, Labbé F, Dussert Y, Delière L, Richart-Cervera S, Giraud T, Delmotte F (2021) Europe as a bridgehead in the worldwide invasion history of grapevine downy mildew, *Plasmopara viticola*. Curr Biol 31(10):2155–2166.e4. https://doi.org/10.1016/j.cub.2021.03.009

Freimoser FM, Rueda-Mejia MP, Tilocca B, Migheli Q (2019) Biocontrol yeasts: mechanisms and applications. World J Microbiol Biotechnol 35(10):154. https://doi.org/10.1007/s11274-019-2728-4

Fröbel S, Zyprian E (2019) Colonization of different grapevine tissues by *Plasmopara viticola*—a histological study. Front Plant Sci 10:951. https://doi.org/10.3389/fpls.2019.00951

Gadoury DM, Seem RC, Ficke A, Wilcox WF (2003) Ontogenic resistance to powdery mildew in grape berries. Phytopathology 93(5):547–555

Gadoury DM, Cadle-Davidson L, Wilcox WF, Dry IB, Seem RC, Milgroom MG (2012) Grapevine powdery mildew (*Erysiphe necator*): a fascinating system for the study of the biology, ecology and epidemiology of an obligate biotroph. Mol Plant Pathol 13(1):1–16. https://doi.org/10.1111/j.1364-3703.2011.00728.x

Gäumann E (1951) Pflanzliche Infektionslehre. Verlag Birkhäuser, Basel

Gehmann K (1987) Untersuchungen zur Epidemiologie und Bekämpfung des Falschen Mehltaus der Weinrebe, verursacht durch *Plamopara viticola* (Berk. et Curt. ex de Bary) Berl. et de Toni. University of Hohenheim Dissertation, Germany

Gessler C, Pertot I, Perazzolli M (2011) *Plasmopara viticola*: a review of knowledge on downy mildew of grapevine and effective disease management. Phytopathol Mediterr 50:3–44. https://doi.org/10.14601/Phytopathol_Mediterr-9360

Giménez P, Just-Borras A, Pons P, Gombau J, Heras JM, Sieczkowski N, Canals JM, Zamora F (2023) Biotechnological tools for reducing the use of sulfur dioxide in white grape must and preventing enzymatic browning: glutathione; inactivated dry yeasts rich in glutathione; and bioprotection with *Metschnikowia pulcherrima*. Eur Food Res Technol 249(6):1491–1501. https://doi.org/10.1007/s00217-023-04229-6

Gindro K, Pezet R (1997) Evidence for a constitutive cytoplasmic cutinase in ungerminated conidia of *Botrytis cinerea* Pers.: Fr. FEMS Microbiol Lett 149(1):89–92. https://doi.org/10.1111/j.1574-6968.1997.tb10313.x

Gindro K, Pezet R (2001) Effects of long-term storage at different temperatures on conidia of *Botrytis cinerea* Pers.: Fr. FEMS Microbiol Lett 204:101–104. https://doi.org/10.1111/j.1574-6968.2001.tb10870.x

Gindro K, Pezet R, Viret O (2003) Histological study of the responses of two *Vitis vinifera* cultivars (resistant and susceptible) to *Plasmopara viticola* infections. Plant Physiol Biochem 41(9):846–853. https://doi.org/10.1016/S0981-9428(03)00124-8

Gindro K, Pezet R, Viret O, Richter H (2005) Development of a rapid and highly sensitive direct-PCR assay to detect a single conidium of *Botrytis cinerea* Pers.: Fr in vitro and quiescent forms in planta. Vitis 44(3):139–142. https://doi.org/10.5073/vitis.2005.44.139-142

Gindro K, Alonso-Villaverde V, Voinesco F, Spring JL, Viret O, Dubuis PH (2012) Susceptibility to downy mildew in grape clusters: new microscopical and biochemical insights. Plant Physiol Biochem 52:140–146. https://doi.org/10.1016/j.plaphy.2011.12.009

Gindro K, Lecoultre N, Molino L, de Joffrey JP, Schnee S, Voinesco F, Alonso-Villaverde V, Viret O, Dubuis PH (2014) Development of rapid direct PCR assays to identify downy and powdery mildew and grey mould in *Vitis vinifera* tissues. OENO One 48:261–268. https://doi.org/10.20870/oeno-one.2014.48.4.1697

Gindro K, Schnee S, Lecoultre N, Michellod E, Zufferey V, Spring JL, Viret O, Dubuis PH (2022) Development of downy mildew in grape bunches of susceptible and resistant cultivars: infection pathways and limited systemic spread. Aust J Grape Wine Res 28:572–580. https://doi.org/10.1111/ajgw.12560

Giraud T, Fortini D, Levis C, Lamarque C, Leroux P, Lobuglio K, Brygoo Y (1999) Two sibling species of the *Botrytis cinerea* complex, transposa and vacuma, are found in sympatry on numerous host plants. Phytopathology 89(10):967–973. https://doi.org/10.1094/PHYTO.1999.89.10.967

Giraud T, Fortini D, Levis C, Leroux P, Brygoo Y (1997) RFLP markers show genetic recombination in *Botryotinia fuckeliana (Botrytis cinerea)* and transposable elements reveal two sympatric species. Mol Biol Evol 14(11):1177–1185. https://doi.org/10.1093/oxfordjournals.molbev.a025727

Gómez-Albarrán C, Melguizo C, Patiño B, Vázquez C, Gil-Serna J (2021) Diversity of Mycobiota in Spanish grape berries and selection of *Hanseniaspora uvarum* U1 to prevent mycotoxin contamination. Toxins (Basel) 13(9):649. https://doi.org/10.3390/toxins13090649

Gonçalves RA, Schatzmayr D, Albalat A, Mackenzie S (2020) Mycotoxins in aquaculture: feed and food. Rev Aquac 12:145–175. https://doi.org/10.1111/raq.12310

Gouvernet R (1978) Réflexions sur la pourriture grise de la vigne. Progrès agricole et Viticole 12:361–368

Greer LA, Harper JDI, Savocchia S, Samuelian SK, Steel CC (2011) Ripe rot of south-eastern Australian wine grapes is caused by two species of Colletotrichum: *C. acutatum* and *C. gloeosporioides* with differences in infection and fungicide sensitivity. Aust J Grape Wine Res 17:123–128. https://doi.org/10.1111/j.1755-0238.2011.00143.x

Groves JW, Loveland CA (1953) The connection between *Botryotinia fuckeliana* and *Botrytis cinerea*. Mycologia 45:415–425

Hall M, Loeb G, Cadle-Davidson L, Evans K, Wilcox W (2018) Grape sour rot: a four-way interaction involving the host, yeast, acetic acid bacteria, and insects. Phytopathology 108(12):1429–1442

Hall ME, O'Bryon I, Osier MV (2019) The epiphytic microbiota of sour rot-affected grapes differs minimally from that of healthy grapes, indicating causal organisms are already present on healthy berries. PLoS One 14:e0211378. https://doi.org/10.1371/journal.pone.0211378

Hardie WJ, O'Brein TP, Jaudzems GV (1996) Morphology, anatomy and development of the pericarp after anthesis in grape, *Vitis vinifera* L. Aust J Grape Wine Res 2(2):97–142

Hausmann L, Rex F, Töpfer R (2017) Evaluation and genetic analysis of grapevine black rot resistance. Acta Hortic 1188:285-290. https://doi.org/10.17660/ActaHortic.2017.1188.37

Haviland DR, Bettiga LJ, Varela LG, Roncoroni JA, Smith RJ, Westerdahl BB, Bentley WJ, Daane K M, Ferris H, Gubler WD, Hembree KJ, Ingels CA, Zalom FG, Zasada I (2017) UC IPM pest management guidelines grape. UC ANR Publ. 3448. Online publication. Oakland, CA

He Z, Cui CY, Jiang JX (2017) First report of white rot of grape caused by *Pilidiella castaneicola* in China. Plant Dis 101:1673. https://doi.org/10.1094/PDIS-02-17-0264-PDN

Hill GK (1989) Effect of temperature on sporulation efficiency of oilspots caused by *Plasmopara viticola* (Berk. et Curt. Ex de Bary) Berl. et de Toni in vineyards. Vitic Enol Sci 44:86–90

Hoffman LE, Wilcox WF, Gadoury DM, Seem RC (2002) Influence of grape berry age on susceptibility to *Guignardia bidwellii* and its incubation period length. Phytopathology 92:1068–1076. https://doi.org/10.1094/PHYTO.2002.92.10.1068

Hoffman LE, Wilcox WF, Gadoury DM, Seem RC, Riegel DG (2004) Integrated control of grape black rot: influence of host phenology, inoculum availability, sanitation, and spray timing. Phytopathology 94:641–650. https://doi.org/10.1094/PHYTO.2004.94.6.641

Hopkins EF (1921) The *Botrytis* blight of tulips. Mem Cornell Agric Exp Station 45:315–361

Jailloux F, Willocquet L, Chapuis L, Froidefond G (1999) Effect of weather factors on the release of grape powdery mildew in Bordeaux region. Can J Bot 77(7):1044–1051

Janex-Favre MC, Parguey-Leduc A, Jailloux F (1993) The ontogeny of pycnidia of *Guignardia bidwellii* in culture. Mycol Res 97:1333–1339

Janex-Favre MC, Parguey-Leduc A, Jailloux F (1996) The ontogeny of perithecia in *Guignardia bidwellii*. Mycol Res 100:875–880. https://doi.org/10.1016/S0953-7562(96)80038-4

Jarvis WR (1962) The infection of strawberry and raspberry fruits by Botrytis cinerea Fr. Annals of Applied Biology 50(3):569–575. https://doi.org/10.1111/j.1744-7348.1962.tb06049.x

Jarvis WR (1977) *Botryotinia* and *Botrytis* species: taxonomy, physiology and pathogenicity. Monograph n°15. Research Branch Canada Department of Agriculture, Ontario, Canada

Jayawardena RS, Hyde KD, Damm U, Cai L, Liu M, Li XH, Zhang W, Zhao WS, Yan JY (2016) Notes on currently accepted species of *Colletotrichum*. Mycosphere 7(8):1192–1260

Jermini M, Gessler C (1996) Epidemiology and control of grape black rot in Southern Switzerland. Plant Dis 80:322–325

Ji T, Caffi T, Carisse O, Li M, Rossi V (2021a) Development and evaluation of a model that predicts grapevine anthracnose caused by *Elsinoë ampelina*. Phytopathology 111(7):1173–1183. https://doi.org/10.1094/PHYTO-07-20-0267-R

Ji T, Languasco L, Li M, Rossi V (2021b) Effects of temperature and wetness duration on infection by *Coniella diplodiella*, the fungus causing white rot of grape berries. Plants (Basel) 10(8):1696. https://doi.org/10.3390/plants10081696

Ji T, Salotti I, Dong C, Li M, Rossi V (2021c) Modeling the effects of the environment and the host plant on the ripe rot of grapes, caused by the *Colletotrichum* species. Plants 10(11):2288. https://doi.org/10.3390/plants10112288

Jiao W, Liu L, Zhou R, Xu M, Xiao D, Xue C (2019) Elsinochrome phytotoxin production and pathogenicity of *Elsinoë arachidis* isolates in China. PLoS One 14(6):e0218391. https://doi.org/10.1371/journal.pone.0218391

Karimi B, Masson V, Guilland C, Leroy E, Pellegrinelli S, Giboulot E, Maron PA, Ranjard L (2021) La biodiversité des sols est-elle impactée par l'apport du cuivre ou son accumulation dans les sols des vignes ? Synthèse des connaissances scientifiques. Etude et gestion des sols 28:71–92

Kast WK (1997) A step by step risk analysis (SRA) used for planning sprays against powdery mildew (OiDiag-system). Vitic Enol Sci 52:230–321

Keller M, Viret O, Cole FM (2003) *Botrytis cinerea* infection in grape flowers: defense reaction, latency, and disease expression. Phytopathology 93:316–322

Kennelly MM, Gadoury DM, Wilcox WF, Magarey PA, Seem RC (2005) Seasonal development of ontogenic resistance to downy mildew in grape berries and rachises. Phytopathology 95(12):1445–1452. https://doi.org/10.1094/PHYTO-95-1445. PMID: 18943556

Kenney P, Hall M (2021) Reducing sour rot spray applications initiated after symptom development does not impact disease control. Catalyst 5:22–28

Kleinig H (1985) Parasitische Pilze an Gefäßpflanzen in Europa. Von W. Brandenburger. 1248 Seiten, 403 Abb. auf 150 Bildtafeln. Fischer, Stuttgart, Biologie in unserer Zeit 15:128–128. https://doi.org/10.1002/biuz.19850150412

Koike ST (1997) Gray mold. In: Davis RM, Subbarao KV, Raid RN, Kurtz EA (eds) Compendium of lettuce disease. APS Press, St. Paul, pp 22–23

Koledenkova K, Esmaeel Q, Jacquard C, Nowak J, Clément C, Ait Barka E (2022) *Plasmopara viticola* the causal agent of downy mildew of grapevine: from its taxonomy to disease management. Front Microbiol 13:889472. https://doi.org/10.3389/fmicb.2022.889472

Korf RP, Pearson RC, Zhuang WY, Dubos B (1986) *Pseudopezicula* (Helotiales, Peziculoideae), a new discomycete genus for pathogens causing an angular leaf scorch disease of grapes ("Rotbrenner"). Mycotaxon 26:457–471

Kortekamp A, Wind R, Zyprian E (1997) The role of callose deposits during infection of two downy mildew-tolerant and two-susceptible *Vitis* cultivars. VITIS-J Grape Res 36(2):103–104. https://doi.org/10.5073/vitis.1997.36.103-104

Kuanyshev N, Adamo GM, Porro D, Branduardi P (2017) The spoilage yeast *Zygosaccharomyces bailii*: foe or friend? Yeast 34:359–370. https://doi.org/10.1002/yea.3238

La Guerche S, Garcia C, Darriet P, Dubourdieu D, Labarère J (2004) Characterization of *Penicillium* species isolated from grape berries by their internal transcribed spacer (ITS1) sequences and by gas chromatography-mass spectrometry analysis of geosmin production. Curr Microbiol 48(6):405–411

Lam SSTH, Howell KS (2015) *Drosophila*-associated yeast species in vineyard ecosystems. FEMS Microbiol Lett 362(20):fnv170. https://doi.org/10.1093/femsle/fnv170

Leach R, Moore KG (1966) Sporulation of *Botrytis fabae* on agar cultures. Trans Br Mycol Soc 49(4):593–561. https://doi.org/10.1016/S0007-1536(66)80008-6

Lebleux M, Alexandre H, Romanet R, Ballester J, David-Vaizant V, Adrian M, Tourdot-Maréchal R, Rouiller-Gall C (2023) Must protection, sulfites versus bioprotection: a metabolomic study. Food Res Int 173(Pt 2):113383

Lee SC, Ristaino JB, Heitman J (2012) Parallels in intercellular communication in oomycete and fungal pathogens of plants and humans. PLoS Pathol 8(12):e1003028. https://doi.org/10.1371/journal.ppat.1003028

Li D, Wan Y, Wang Y, He P (2008) Relatedness of resistance to anthracnose and to white rot in Chinese wild grapes. Vitis 47(4):213–2015

Li Z, Dang H, Yuan X, He J, Hu Z, Wang X (2018) Morphological characterization and optimization of conditions for conidial production of *Elsinoë ampelina*, the causal organism of grapevine anthracnose. J Phytopathol 166:420–428. https://doi.org/10.1111/jph.12702

Li Z, Zhang S, Han R, Zhang H, Li K, Wang X (2019) Infection process and host responses to *Elsinoë ampelina*, the causal organism of grapevine anthracnose. Eur J Plant Phytopathol 155:571–582. https://doi.org/10.1007/s10658-019-01793-0

Li BY, Luan BH, Shi J, Wang SL, Tian YY, Nie LX, Wang YZ (2020) Sensitivity of *Coniella diplodiella* to pyraclostrobin in Jiaodong area and comparison with other fungicides. Chin J Pestic Sci 22:959–966. https://doi.org/10.16801/j.issn.1008-7303.2020.0147

Li Z, dos Santos RF, Gao L, Chang P, Wang X (2021) Current status and future prospects of grapevine anthracnose caused by *Elsinoë ampelina*: an important disease in humid grape-growing regions. Mol Plant Pathol 22:899–910. https://doi.org/10.1111/mpp.13076

Li P, Tan X, Liu R, Rahman FU, Jiang J, Sun L, Fan X, Liu J, Liu C, Zhang Y (2023) QTL detection and candidate gene analysis of grape white rot resistance by interspecific grape (*Vitis vinifera* L. × *Vitis davidii* Foex.) crossing. Hortic Res 10(5):uhad063. https://doi.org/10.1093/hr/uhad063

Lopez Pinar A, Rauhut D, Ruehl E, Buettner A (2017) Quantification of the changes in potent wine odorants as induced by bunch rot (*Botrytis cinerea*) and powdery mildew (*Erysiphe necator*). Front Chem 5:57. https://doi.org/10.3389/fchem.2017.00057

Luttrell ES (1948) Physiologic specialization in *Guignardia bidwellii*, cause of black rot of *Vitis* and *Parthenocissus* species. Phytopathology 38:716–723

Maas JL (1969) Effect of time and temperature of storage on viability of Botrytis convolute conidia and sclerotia. Plant Dis Rep 53:141–144

Magarey RD, Coffey BE, Emmett RW (1993) Anthracnose of grapevine, a review. Plant Prot Q 8(3):106–110

Marcellin-Gros R, Hévin S, Chevalley C, Boccard J, Hofstetter V, Gindro K, Wolfender JL, Kehrli P. (2024) An advanced metabolomic approach on grape skins untangles cultivar preferences by Drosophila suzukii for oviposition. Front Plant Sci 15:1435943. https://doi.org/10.3389/fpls.2024.1435943

Marès H (1857) Manuel pour le soufrage des vignes maladies, emploi du soufre, ses effets. Montpellier (France)

Martinez F, Dubos B, Fermaud M (2005) The role of saprotrophy and virulence in the population dynamics of *Botrytis cinerea* in vineyards. Phytopathology 95(6):692–700. https://doi.org/10.1094/PHYTO-95-0692

McClellan WD, Hewitt WB (1973) Early *Botrytis* rot of grapes: time of infection and latency of *Botrytis cinerea* Pers. in *Vitis vinifera* L. Phytopathology 63:1151–1157

McFadden-Smith W, Gubler WD (2015) Sour rot. In: Wilcox WF, Gubler WD, Uyemoto JK (eds) Compendium of grape diseases, pests, and disorders, 2nd edn. American Phytopathological Society Press, St. Paul, pp 87–90

McGrew JR, Pollack FG (1988) Septoria leaf spot. In: Pearson RC, Goheen AC (eds) Compendium of grape diseases. Amer Phytopathol Soc Press, St. Paul, p 31

Melzer V (1924) L'ornementation des spores des russules. Bull Soc Mycol France 40:78–81

Micheli PA (1729) Nova Plantarum Genera: juxta Turnefortii methodum deposita, Firenze

Mikušová P, Sulyok M, Šrobárová A (2014) *Alternaria* mycotoxins associated with grape berries in vitro and in situ. Biologia 69:173–177. https://doi.org/10.2478/s11756-013-0306-z

Millardet A (1887) Instruction pratique pour le traitement du mildiou et du rot de la vigne. In: Masson G, Feret et Fils (eds) Bordeaux, France

Mitchell JK, Patterson WK, Ford RH (1994) Susceptibility of American, European, and interspecific hybrid grape cultivars to the fungus *Septoria ampelina*. HortScience 29(1):31–32

Molitor D, Berkelmann-Löhnertz B (2011) Simulating the susceptibility of clusters to grape black rot infections, depending on their phenological development. Crop Prot 30(12):1649–1654

Molitor D, Beyer M (2014) Epidemiology, identification and disease management of grape black rot and potentially useful metabolites of black rot pathogens for industrial applications—a review. Ann Appl Biol 165:305–317

Molitor D, Baus O, Berkelmann-Löhnertz B (2011) Protective and curative grape black rot control potential of pyraclostrobin and myclobutanil. J Plant Dis Prot 118(5):161–167

Molitor D, Fruehauf K, Baus O, Berkelmann-Löhnertz B (2012) A cumulative degree-day-based model to calculate the duration of the incubation period of *Guignardia bidwellii*. Plant Dis 96:1054–1059. https://doi.org/10.1094/PDIS-11-11-1005-RE

Molitor D, Augenstein B, Mugnai L, Rinaldi PA, Sofia J, Hed B, Dubuis PH, Jermini M, Kührer E, Bleyer G (2016) Composition and evaluation of a novel web-based decision support system for grape black rot control. Eur J Plant Pathol 144:785–798. https://doi.org/10.1007/s10658-015-0835-0

Müller K, Rabanus A, Kotte W (1923) Biologische Versuche mit der Rebenperonospora zur Ermittlung der Inkubationszeit. Weinberg und Keller 2:65–71

Müller-Thurgau H (1903) Der rote Brenner des Weinstockes. Centralblatt für Bakteriologie, Parasitenkunde und Infektionskrankheiten 2(10):1–38

Murria S, Kaur N, Arora N, Mahal AK (2018) Field reaction and metabolic alterations in grape (Vitis vinifera L.) varieties infested with anthracnose. Sci Hortic 235:286–293. https://doi.org/10.1016/j.scienta.2018.03.016

Na DH, Lee JS, Shin HD, Ono Y, Choi YJ (2022) Emergence of a new rust disease of Virginia Creeper (*Parthenocissus quinquefolia*) through a host range expansion of *Neophysopella vitis*. Mycobiology 50(3):166–171. https://doi.org/10.1080/12298093.2022.2076366

Nair NG, Allen RN (1993) Infection of grape flowers and berries by *Botrytis cinerea* as a function of time and temperature. Mycol Res 97(8):1012–1014

Nair NG, Nadtotchei A (1987) Sclerotia of *Botrytis* as source of a primary inoculum for bunch rot of grapes in New South Wales. Aust J Exp Agric 35:1177–1180

Nair NG, Guilbaud-Oulton S, Barchia I, Emmett R (1995) Significance of carry over inoculum, flower infection and latency on the incidence of *Botrytis cinerea* in berries of grapevines at harvest in New South Wales. Aust J Exp Agric 35:1177–1180

Nakagawa S, Komatsu H, Yuda E (1980) A study of micro-morphology of grape berry surface during their development with special reference to stoma. J Japan Soc Hortic Sci 49(1):1–7

Nicholson RL, Hammerschmidt R (1992) Phenolic compounds and their role in disease resistance. Annu Rev Phytopathol 30(1):369–389. https://doi.org/10.1146/annurev.py.30.090192.002101

Okane I, Ono Y (2018) Phylogenetic study of indigenous grapevine leaf rust fungi in North America and biological identity of an invasive grapevine leaf rust fungus in Brazil. Mycoscience 59(2):99–104. https://doi.org/10.1016/j.myc.2017.07.005

Ollat N, Carde J-P, Gaudillère J-P, Barrieu F, Diakou-Verdin P, Moing A (2002) Grape berry development: a review. OENO One 36(3):109–131

Onesti G, Gonzales-Dominguez E, Rossi V (2016) Accurate prediction of black rot epidemics in vineyards using a weather-driven disease model. Pest Manag Sci 71(12):2321–2329. https://doi.org/10.1002/ps.4277

Onesti G, Gonzales-Dominguez E, Rossi V (2017) Production of *Guignardia bidwellii* conidia on grape leaf lesions is influenced by repeated washing events and by alternation of dry and wet periods. Eur J Plant Pathol 147:949–953. https://doi.org/10.1094/PHYTO-07-16-0255-R

Ono Y, Chatasiri S, Pota S, Yamaoka Y (2012) *Phakopsora montana*, another grapevine leaf rust pathogen in Japan. J Gen Plant Pathol 78:338–347. https://doi.org/10.1007/s10327-012-0401-y

Ono Y, Okane I, Chatasiri S, Pota S, Unartngam J, Ayawong C, Nguyen HD, Le CTM (2020) Taxonomy of Southeast Asian-Australasian grapevine leaf rust fungus and its close relatives. Mycol Progr 19:905–919. https://doi.org/10.1007/s11557-020-01607-2

Oo MM, Oh SK (2017) Identification and characterization of new record of grape ripe rot disease caused by *Colletotrichum viniferum* in Korea. Mycobiology 45(4):421–425. https://doi.org/10.5941/MYCO.2017.45.4.421

Padgett M, Morrison JC (1990) Changes in grape berry exudates during fruit development and their effect on mycelial growth of *Botrytis cinerea*. J Am Soc Hortic Sci 115(2):269–273. https://doi.org/10.21273/JASHS.115.2.269

Pearson RC, Goheen AC (eds) (1998) Septoria leaf spot compendium of grape diseases. The American Phytopathological Soc Press (APS Press), St Paul, Minnesota. p 31

Pearson RC, Smith FD, Dubos B (1988) Angular leaf scorch, a new disease of grapevines in North America caused by *Pseudopezicula tetraspora*. Plant Dis 72(9):796–800

Pearson RC, Siegfried W, Bodmer M, Schüepp H (1991) Ascospore discharge and survival in *Pseudopezicula tracheiphila*, causal agent of rotbrenner of grape. J Phytopathol 132(3):177–185. https://doi.org/10.1111/j.1439-0434.1991.tb00110.x

Peng L, Sun T, Yang Y, Cai L, Hyde KD, Bahkali AH, Liu Z (2013) *Colletotrichum* species on grape in Guizhou and Yunnan provinces, China. Mycoscience 54:29–41. https://doi.org/10.1016/j.myc.2012.07.006

Persoon CH (1801) Synopsis Methodica Fungorum, Gottingae, pp 690–691

Pertot I, Caffi T, Rossi V, Mugnai L, Hoffmann C, Grando MS, Gary C, Lafond D, Duso C, Thierry D, Mazzoni V, Anfora G (2017) A critical review of plant protection tools for reducing pesticide use on grapevine and new perspectives for the implementation of IPM in viticulture. Crop Prot 97:70–84. https://doi.org/10.1016/j.cropro.2016.11.025

Petrak F, Sydow H (1927) Repertorium Specierum Novarum Regni Vegetabilis. Beihefte 42: Die Gattungen der Pyrenomyzeten, Sphaeropsideen und Melanconieen

Pezet R, Pont V (1986) Infection florale et latence de *Botrytis cinerea* dans les grappes de *Vitis vinifera* (var. Gamay). Revue Suisse Vitic Arboric Hortic 18(5):317–322

Pezet R, Viret O, Perret C, Tabacchi R (2003) Latency of *Botrytis cinerea* Pers.: Fr. and biochemical studies during growth and ripening of two grape berry cultivars, respectively susceptible and resistant to grey mould. J Phytopathol 151:208–214. https://doi.org/10.1046/j.1439-0434.2003.00707.x

Pezet R, Viret O, Gindro K (2004) Plant-microbe interaction: the *Botrytis* grey mould of grapes, biology, biochemistry, epidemiology, and control management. Adv Plant Physiol 7:71–116

Pinto L, Malfeito-Ferreira M, Quintieri L, Silva AC, Baruzzi F (2019) Growth and metabolite production of a grape sour rot yeast-bacterium consortium on different carbon sources. Int J Food Microbiol 296:65–74

Plesken C, Pattar P, Reiss B, Noor ZN, Zhang L, Klug K, Huettel B, Hahn M (2021) Genetic diversity of Botrytis cinerea revealed by multilocus sequencing, and identification of B. cinerea populations showing genetic isolation and distinct host adaptation Front Plant Sci 5:12:663027. doi:https://doi.org/10.3389/fpls.2021.663027

Ponsone ML, Chiotta ML, Combina M, Dalcero A, Chulze S (2011) Biocontrol as a strategy to reduce the impact of ochratoxin A and *Aspergillus* section *Nigri* in grapes. Int J Food Microbiol 151:70–77. https://doi.org/10.1016/j.ijfoodmicro.2011.08.005

Primiano IV, Loehrer M, Amorim L, Schaffrath U (2017) Asian grapevine leaf rust caused by *Phakopsora euvitis*: an important disease in Brazil. Plant Pathol 66:691–701. https://doi.org/10.1111/ppa.12662

Rasera JB, Amorim L, Appezzato-da-Glória B (2022) Diverse effects of temperature on in vivo and in vitro germination of urediniospores of *Neophysopella tropicalis*. Eur J Plant Pathol 162:501–507. https://doi.org/10.1007/s10658-021-02413-6

Reiss K, Zinkernagel V (1997) Epidemiological research on ascospore discharge of *Pseudopezicula tracheiphila* (Müll.-Thurg.) Korf & Zhuang, the cause of rotbrenner in grapes. Zeitschrift für Pflanzenkrankheiten und Pflanzenschutz 104(2):113–125

Reiss K, Gutmann M, Bartscherer HC, Zinkernagel V (1997) *Pseudopezicula tracheiphila* on *Vitis vinifera*: microscopical studies on the infection mechanism. J Plant Dis Prot 104(5):483–491

Rex F, Fechter I, Hausmann L, Töpfer R (2014) QTL mapping of black rot (*Guignardia bidwellii*) resistance in the grapevine rootstock 'Börner' (*V. riparia* Gm183 × *V. cinerea* Arnold). Theor Appl Genet 127(7):1667–1677. https://doi.org/10.1007/s00122-014-2329-4

Ribéreau-Gayon J, Ribéreau-Gayon P, Séguin G (1980) Botrytis cinerea. In: Coley-Smith JR, Verhoeff K, Jarvis WR (eds) The biology of Botrytis. Academic, New York/London, pp 251–274

Rinaldi P, Skaventzou M, Rossi M, Comparini C, Sofia J, Molitor D, Mugnai L (2013) *Guignardia bidwellii*: epidemiology and symptoms development in Mediterranean environment. J Plant Pathol 95:S1.83–S1.84

Rinaldi PA, Paffetti D, Comparini C, Broggini GAL, Gessler C, Mugnai L (2017) Genetic variability of *Phyllosticta ampelicida*, the causal agent of black rot disease of grapevine. Phytopathology 107:1406–1416

Rossi V, Caffi T, Giosuè S, Bugiani RA (2008) Mechanistic model simulating primary infections of downy mildew in grapevine. Ecol Model 212:480–491

Rossi V, Caffi T, Legler SE (2010) Dynamic of ascospore maturation and discharge in *Erysiphe necator*, the causal agent of grape powdery mildew. Phytopathology 100(12):321–1329

Rossi V, Onesti G, Legler SE, Caffi T (2014) Use of systems analysis to develop plant disease models based on literature data: grape black-rot as a case-study. Eur J Plant Pathol 141:427–444

Rousseaux S, Diguta CF, Radoi-Matei F, Alexandre H, Guilloux-Benatier M (2014) Non-*Botrytis* grape-rotting fungi responsible for earthy and moldy off-flavors and mycotoxins. Food Microbiol 38:104–121

Rouxel M, Mestre P, Comont G, Lehman BL, Schilder A, Delmotte F (2013) Phylogenetic and experimental evidence for host-specialized cryptic species in a biotrophic oomycete. New Phytol 197(1):251–263. https://doi.org/10.1111/nph.12016

Rouxel M, Mestre P, Baudoin A, Carisse O, Delière L, Ellis MA, Gadoury D, Lu J, Nita M, Richard-Cervera S, Schilder A, Wise A, Delmotte F (2014) Geographic distribution of cryptic species of *Plasmopara viticola* causing downy mildew on wild and cultivated grape in eastern North America. Phytopathology 104(7):692–701. https://doi.org/10.1094/PHYTO-08-13-0225-R

Saccardo PA (1882) Sylloge Fungorum Omnium Hucusque Cognitorum. Pyrenomycetum 1:20–21

Saccardo PA (1884) Sylloge Fungorum Omnium Hucusque Cognitorum. Sphaeropsidearum et Melanconiearum 3:310

Sagar N, Sayiprathap BR, Jamadar MM, Patibanda AK, Madhu GS, Harish D, Sathish K (2023) First report of *Phakopsora euvitis* causing leaf rust disease on grapevine (*Vitis labrusca* L.) in India. Plant Dis. https://doi.org/10.1094/PDIS-04-23-0651-PDN

Santos RF, Ciampi-Guillardi M, Amorim L, Massola Júnior NS, Spósito MB (2018) Aetiology of anthracnose on grapevine shoots in Brazil. Plant Pathol 67:692–706. https://doi.org/10.1111/ppa.12756

Santos RF, Primiano IV, Amorim L (2021) Identification and pathogenicity of *Neophysopella* species associated with Asian grapevine leaf rust in Brazil. Plant Pathol 70:74–86. https://doi.org/10.1111/ppa.13274

Schultz HR (1992) An empirical model for the simulation of leaf appearance and leaf area development of primary shoots of several grapevine (*Vitis vinifera* L.) canopy-systems. Sci Hortic 52:179–200. https://doi.org/10.1016/0304-4238(92)90020-D

Şen L, Nas S (2013) Identification of ochratoxigenic fungi and contextual change on dried raisins (Sultanas). J Food Agricul Environ 11(3–4):155–161

Şen L, Ocak İ, Nas S, Şevik R (2016) Effects of different drying treatments on fungal population and ochratoxin A occurrence in sultana type grapes. Food Addit Contam Part A 33(9):1444–1455. https://doi.org/10.1080/19440049.2016.1217066

Shear CL (1929) The life history of *Sphaceloma ampelina* de Bary. Phytopathology 19(7):673–679

Shiraishi M, Fukutomi M, Akai S (1970) Effect of temperature on the conidium germination and appressorium formation of Botrytis cinerea. Ann Phytopathol Soc Jpn 36:234–236

Shiraishi M, Koide M, Itamura H, Yamada M, Mitani N, Ueno T, Nakaune R, Nakano M (2007) Screening for resistance to ripe rot caused by *Colletotrichum acutatum* in grape germplasm. Vitis 46:196–201

Sommer F (1982) Postharvest handling practices and postharvest diseases of fruits. Plant Dis 66(5):359–364

Southworth EA (1891) Ripe rot of grapes and apples. J Mycol 6(4):164–173. https://doi.org/10.2307/3752659

Spegazzini CL (1878) Ampelomiceti Italici, ossia enumerazione, diagnosi e storia dei principali parassiti della vite. Rivista Viticolt. Enologia 2:339

Spotts RA (1977) Effect of leaf wetness duration and temperature on the infectivity of *Guignardia bidwellii* on grape leaves. Phytopathology 67(11):1378–1381

Spotts RA (1980) Infection of grape by *Guignardia bidwellii*—factors affecting lesion development, conidial dispersal, and conidial populations on leaves. Phytopathology 70(3):252–255

Spring JL, Viret O (2009) Influence des techniques d'éclaircissage sur le rendement, la morphologie des grappes, la pourriture et la qualité des vins de Pinot noir. Revue suisse de Viticulture, Atboric Hortic 41(2):95–101

Spring JL, Viret O (2011) Influence des techniques d'éclaircissage sur le rendement, la morphologie des grappes et la sensibilité à la pourriture du cépage Gamay. Revue Suisse Vitic Arboric Hortic 43(5):280–287

Stark-Urnau M, Kast WK (1999) Development of ontogenetic resistance to powdery mildew in fruit of differently susceptible grapevines (cvs. Trollinger and Lemberger). Mitteilungen Klosterneuburg 49:186–189

Sun L, Zhang M, Ren J, Qi J, Zhang G, Leng P (2010) Reciprocity between abscisic acid and ethylene at the onset of berry ripening and after harvest. BMC Plant Biol 10:257. https://doi.org/10.1186/1471-2229-10-257

Sutton BC (1992) The genus *Glomerella* and its anamorph *Colletotrichum*. In: Bailey JA, Jeger MJ (eds) *Colletotrichum* biology, pathology, and control. CAB International, Wallingford, pp 1–26

Szabó M, Csikász-Krizsics A, Dula T, Farkas E, Roznik D, Kozma P, Deák T (2023) Black rot of grapes (*Guignardia bidwellii*)—a comprehensive overview. Horticulturae 9(2):130. https://doi.org/10.3390/horticulturae9020130

Tan KK, Epton HAS (1973) Effect of light on the growth and sporulation of *Botrytis cinerea*. Trans Br Mycol Soc 61:145–157

Tančinová D, Mašková Z, Rybárik ĽubomíR, Felšöciová S, Císarová M (2016) Colonization of grapes berries by *Alternaria* sp. and their ability to produce mycotoxins. Potravinarstvo Slovak. J Food Sci 10(1):7–13. doi:https://doi.org/10.5219/553

Taylor AA, Tsuji JS, Garry MR, McArdle ME, Goodfellow WL, Adams WJ, Menzie CA (2020) Critical review of exposure and effects: implications for setting regulatory health criteria for ingested copper. Environ Manag 65:131–159. https://doi.org/10.1007/s00267-019-01234-y

Utami LB (1995) Fungicides used to control *Septoria ampelina* Berk & Curt. leaf spot of *Vitis labrusca* L. cv. 'concord'. PhD thesis, Ball State University. http://cardinalscholar.bsu.edu/handle/20.500.14291/185385

Van Kan JA, Shaw MW, Grant-Downton RT (2014) *Botrytis* species: relentless necrotrophic thugs or endophytes gone rogue? Mol Plant Pathol 15(9):957–961. https://doi.org/10.1111/mpp.12148

Vance CP, Kirk TK, Sherwood RT (1980) Lignification as a mechanism of disease resistance. Annu Rev Phytopathol 18(1):259–288. https://doi.org/10.1146/annurev.py.18.090180.001355

Vanderplank JE (1963) Plant diseases: epidemics and control. Academic, New York/London. 349p

Vannini A, Chilosi G (2013) *Botrytis* infection: grey mould and noble rot. In: Mencarelli F, Tonutti P (eds) Sweet, reinforced and fortified wines. Wiley, Oxford, pp 159–169. https://doi.org/10.1002/9781118569184.ch11

Viala P (1885) Les maladies de la vigne. In: Coulet C, Delahaye A, Lecrosnier E (eds) Montpellier, Paris, France

Viala P (1893) Les maladies de la vigne, troisième édition. In: Coulet C, Masson G (eds) Montpellier, Paris, France

Viala P, Ravaz L (1886) Mémoire sur une nouvelle maladie de la vigne, le Black rot (pourriture noir). Annales de l'École National d'Agriculture de Montpellier 2:17–58

Viala P, Ravaz L (1892) Sur la dénomination botanique du black rot. Bulletin de la Société mycologique de France 8(2):63

Viret O, Keller M, Jaudzems V, Cole FM (2004) *Botrytis cinerea* infection of grape flowers: light and electron microscopical studies of infection sites. Phytopathology 94(8):850–857

Von Schweinitz LD (1834) Synopsis Fungorum in America Boreali media degentium. Secundum Observationes. Trans Am Philos Soc 4:270–271

Walker JC (1926) *Botrytis* neck rot of onions. J Agric Res 33:893–928

Walker AS, Gautier AL, Confais J, Martinho D, Viaud M, Le Pêcheur P, Dupont J, Fournier E (2011) *Botrytis pseudocinerea*, a new cryptic species causing gray mould in French vineyards in sympatry with *Botrytis cinerea*. Phytopathology 101(12):1433–1445. https://doi.org/10.1094/PHYTO-04-11-0104

Wang XH, Zheng SS, Huang T, Su LM, Zhao YH, Souders CL 2nd, Martyniuk CJ (2018) Fluazinam impairs oxidative phosphorylation and induces hyper/hypo-activity in a dose specific manner in zebrafish larvae. Chemosphere 210:633–644. https://doi.org/10.1016/j.chemosphere.2018.07.056

Webb RW (1921) Studies in the physiology of the fungi, XV germination of the spores of certain fungi in relation to hydrogen ion concentration. Ann Mo Bot Gard 8:282–341

Weinert MP, Shivas RG, Pitkethley RN, Daly AM (2003) First record of grapevine leaf rust in the Northern Territory, Australia. Aust Plant Pathol 32:117–118. https://doi.org/10.1071/AP02060

Weir BS, Johnston PR, Damm U (2012) The *Colletotrichum gloeosporioides* species complex. Stud Mycol 73(1):115–180

Welke JE (2019) Fungal and mycotoxin problems in grape juice and wine industries. Curr Opin Food Sci 29:7–13. https://doi.org/10.1016/j.cofs.2019.06.009

Wheeler S, Loveys B, Ford C, Davies C (2009) The relationship between the expression of abscisic acid biosynthesis genes, accumulation of abscisic acid and the promotion of *Vitis vinifera* L. berry ripening by abscisic acid. Aust J Grape Wine Res 15:195–204. https://doi.org/10.1111/j.1755-0238.2008.00045.x

Wicht B, Petrini O, Jermini M, Gessler C, Broggini GAL (2012) Molecular, proteomic and morphological characterization of the ascomycete *Guignardia bidwellii*, agent of grape black rot: a polyphasic approach to fungal identification. Mycologia 104:1036–1045. https://doi.org/10.3852/11-242

Williamson B, Tudzynski B, Tudzynski P, van Kan JAL (2007) *Botrytis cinerea*: the cause of grey mould disease. Mol Plant Pathol 8:561–580. https://doi.org/10.1111/j.1364-3703.2007.00417.x

Windholtz S, Nioi C, Coulon J, Masneuf-Pomarede I (2023) Bioprotection by non-*Saccharomyces* yeasts in oenology: evaluation of O_2 consumption and impact on acetic acid bacteria. Int J Food Microbiol 405:110338. https://doi.org/10.1016/j.ijfoodmicro.2023.110338

Wong FP, Burr HN, Wilcox WF (2001) Heterothallism in *Plasmopara viticola*. Plant Pathol 50:427–432. https://doi.org/10.1046/j.1365-3059.2001.00573.x

Ye B, Zhang J, Chen X, Xiao W, Wu J, Yu H, Zhang C (2023) Genetic diversity of *Colletotrichum* spp. causing grape anthracnose in Zhejiang, China. Agronomy 13(4):952. https://doi.org/10.3390/agronomy13040952

Yin X, Li T, Jiang X, Tang X, Zhang J, Yuan L, Wei Y (2013) Suppression of grape white rot caused by *Coniella vitis* using the potential biocontrol agent *Bacillus velezensis* GSBZ09. Pathogens 11(2):248. https://doi.org/10.3390/pathogens11020248

Zhang K, Zhang N, Cai L (2013) Typification and phylogenetic study of *Phyllosticta ampelicida* and *P. vaccinii*. Mycologia 105:1030–1042

Zhang Y, Fan X, Li Y, Sun H, Jiang J, Liu C (2020) Restriction site-associated DNA sequencing reveals the molecular genetic diversity of grapevine and genes related to white rot disease. Sci Hortic 261:108907. https://doi.org/10.1016/j.scienta.2019.108907

Zhao P, Zhang ZF, Hu DM, Tsui KM, Qi XH, Phurbu D, Gafforov Y, Cai L (2021) Contribution to rust flora in China I, tremendous diversity from natural reserves and parks. Fungal Divers 110:1–58. https://doi.org/10.1007/s13225-021-00482-w

Zhou N, Chen Q, Carroll G, Zhang N, Shivas RG, Cai L (2015) Polyphasic characterization of four new plant pathogenic *Phyllosticta* species from China, Japan, and the United States. Fungal Biol 119:433–446

Ziliotto F, Corso M, Rizzini FM, Rasori A, Botton A, Bonghi C (2012) Grape berry ripening delay induced by a pre-véraison NAA treatment is paralleled by a shift in the expression pattern of auxin- and ethylene-related genes. BMC Plant Biol 12:185. https://doi.org/10.1186/1471-2229-12-185

Open Access This chapter is licensed under the terms of the Creative Commons Attribution 4.0 International License (http://creativecommons.org/licenses/by/4.0/), which permits use, sharing, adaptation, distribution and reproduction in any medium or format, as long as you give appropriate credit to the original author(s) and the source, provide a link to the Creative Commons license and indicate if changes were made.

The images or other third party material in this chapter are included in the chapter's Creative Commons license, unless indicated otherwise in a credit line to the material. If material is not included in the chapter's Creative Commons license and your intended use is not permitted by statutory regulation or exceeds the permitted use, you will need to obtain permission directly from the copyright holder.

Chapter 5
Wood Diseases

5.1 Esca Syndrome

Current names:

- *Phaeomoniella chlamydospora* (W. Gams, Crous, M.J. Wingf. & L. Mugnai) Crous & W. Gams 2000, Ascomycota
- *Phaeoacremonium minimum* (Tul. & C. Tul.) Gramaje, L. Mostert & Crous 2015, Ascomycota, syn. *Phaeoacremonium aleophilum* W. Gams, Crous, M.J. Wingf. & Mugnai 1996, *Togninia minima* (Tul. & C. Tul.) Berl. 1900
- *Fomitiporia mediterranea* M. Fischer 2002, Basidiomycota
- *Cadophora luteo-olivacea* (J.F.H. Beyma) T.C. Harr. & McNew 2003, Ascomycota

Esca is one of the oldest diseases of the vine, already described in Greek and Roman texts. It is only at the beginning of the twentieth century that this disease attracted the attention of researchers and winegrowers. Vine decline attributable to esca has been gaining ground for almost three decades and has become particularly worrying for winegrowers. Generally observed on old vines, esca now increasingly manifests on young plants harbouring the same dominant fungal species as old vines. Historically, grapevine decline has been referred to by different names: 'folletage', 'Petri disease', 'black measles', 'black goo', 'wood rustle' or 'esca'. The distinction between esca and other wood diseases is not always clear, as there are many species of fungi isolated from diseased wood. Some species described in the past, such as the Basidiomycetes *Stereum hirsutum*, *Stereum necator* and *Phellinus igniarius* (synonym of *Fomes igniarius*) do not seem to play a decisive role in the esca syndrome, although they are present in the affected stumps. Esca is caused by the action of a succession of fungi, some of which are primary colonisers, followed by other opportunistic or secondary fungi (Valtaud et al. 2009). Surico (2009), however, redefined esca syndrome as being caused by three fungal species: *Phaeoacremonium minimum*, *Phaeomoniella chlamydospora* and *Fomitiporia mediterranea*.

© The Author(s) 2025   313
O. Viret, K. Gindro, *Science of Fungi in Grapevine*,
https://doi.org/10.1007/978-3-031-68663-4_5

Cadophora luteo-olivacea was later added to the list of fungal pathogens associated with esca (Gramaje et al. 2011; Maldonado-González et al. 2020). Nevertheless, several recent studies question this definition of esca because the wood of plants suffering from esca appears to host fungal species associated with more than one GTD (Bruez et al. 2016; Del Frari et al. 2019; Hofstetter et al. 2012).

Esca is caused by a complex interaction of factors whose incidences vary from one vineyard to another, depending on climate, soil type, grape variety and management practices (Gramaje et al. 2018; Claverie et al. 2020). Multiple fungal pathogens repeatedly isolated from symptomatic plants are associated with esca but their interactions with each other and with the grapevine are not well understood (Graniti et al. 2000). Fungi have long been implicated in esca due to their consistent presence in affected plants. However, the systematic investigation of both healthy and diseased vines under identical growing conditions reveals no discernible differences in the fungal communities extracted from asymptomatic and symptomatic plants (Hofstetter et al. 2012; Monod 2024; Monod et al. 2023). The mere presence of individual species traditionally linked with esca is not enough to explain the presence of symptoms; nor does the relative abundance of these species vary between asymptomatic and symptomatic plant samples. The precise role of these species and of the trunk fungal communities more generally is therefore an issue warranting further investigation. In a subsequent step, the connection between esca prevalence and a range of abiotic factors in the same network of vineyards displaying varying degrees of disease impact was investigated. In particular, correlations between the presence of esca and specific climatic and pedological factors associated with water availability to the plant were found. A greater incidence of esca (foliar symptoms and apoplexy) is observed in vineyards with a high soil water holding capacity (SWHC) and during rainy years, particularly when it rains in late spring (Monod et al. 2023). Analysis of the vine trunk mycobiome has revealed a remarkably diverse fungal community with weak differentiation at vineyard or regional level. Although several taxa are overrepresented in asymptomatic plants, this is not true for symptomatic plants (Monod 2024). Neither do key taxa typically implicated in esca appear to have any significant association with plant health status. Multiple factors are likely to be involved in complex diseases like tree decline, and establishing a causal relationship between the presence of taxa and health is a challenge. The concept of endophytes living innocuously in plants challenges our traditional understanding of plant infection processes and of how causal taxa should be identified (Mishra et al. 2021). The poorly defined range of symbiotic associations and their implications depend upon environmental conditions, and may transition between commensalism, mutualism and pathogenicity (Mishra et al. 2021). A number of fungi associated with plants can promote health or cause disease, depending on the specific strain or their location (Selosse et al. 2004; Schulz and Boyle 2005; Busby et al. 2016). Microbial community assembly is influenced by cooperative and competitive interactions among the myriad microbial members that perform functions for plant health as a whole (Gao et al. 2017). Microbiome composition is also described as a mechanism for creating novel traits that enhance the plant's ability to thrive in different environmental conditions (Trivedi et al. 2020). Several hypotheses may explain why it is not always possible to establish a link between the presence of fungal species and the expression of symptoms. The ability of an endo-phyte to either grow asymptomatically within its host or to induce disease depends not only on its adaptation to the host but also on the variable virulence of the endophyte, the response of the host's defence mechanisms and the prevailing biotic and abiotic environmental conditions (Schulz and Boyle 2005; Delaye et al. 2013; Busby et al. 2016).

Species conventionally linked with esca are prevalent in the recovered mycobi-ome, with some such as *Phaeomoniella chlamydospora* ranking among the most abundant. The abundance of the species linked to esca implies their adaptation to the trunk environment and suggests that their interaction with the plant is not at least exclusively pathogenic. If the species typically associated with esca were responsi-ble for symptoms this would imply a transition from a mutualistic to a pathogenic relationship at a certain stage. If the relationship were purely pathogenic, the mere presence of these species would induce symptoms, which is not the case in many asymptomatic plants harbouring said species. A similar abundance of esca-linked species in both symptomatic and asymptomatic plants is also observed in other stud-ies (Hofstetter et al. 2012; Bruez et al. 2014; Del Frari et al. 2019; Geiger et al. 2022); these species are also known to be widely distributed in many grape-growing areas worldwide (Bertsch et al. 2013). Several endophytes considered to be latent pathogens subsequently acted as pathogens under changed environmental condi-tions (Viret et al. 1993; Viret and Petrini 1994; Delaye et al. 2013).

An esca outbreak can be caused by certain species functioning as opportunistic pathogens on the host plant. The shift from a commensal to a pathogenic species can be viewed as an unbalanced symbiosis (Kogel et al. 2006). Suboptimal environmental conditions may stress the host plant and weaken its defence status, creating favourable conditions for the development of disease (Schulz and Boyle 2005). Both the predis-positions of the host and the prevailing environmental conditions influence the equi-librium between host and endophyte (Schulz and Boyle 2005). Endophyte modification is context-dependent since environmental factors (temperature, pH, humidity) are pivotal in determining the prevalence and activity of fungal species, influencing both their growth and the production of mycotoxins (Busby et al. 2016; Magan and Medina 2016; Giorni et al. 2019). The ecological function of an endophyte (neutrality, patho-genicity, pathogen antagonism or pathogen facilitation) may be modified by environ-mental factors such as soil properties, nutrient status and climatic conditions, or host-related factors such as physiological status (Romeralo et al. 2022).

5.1.1 Causal Organisms

5.1.1.1 Phaeomoniella chlamydospora

Phaeomoniella chlamydospora, whose sexual form is unknown, is an Ascomycota of the order Phaeomoniellales and the family *Celotheliaceae* (Chen et al. 2015). The growth of its mycelium is slow (diameter of colonies less than 1 cm after 1 week at

20 °C on agar medium). The greenish mycelium quickly turns olive-grey-to-black. Conidia are very small (1.13 ± 0.13 μm × 2.72 ± 0.30 μm), hyaline, oblong-ellipsoid, obovate-to-straight, grouped in a glomerulus and produced at the end of ampullate and elongate phialides, terminated by a collar. Globose-to-subglobose, olivaceous-to-dark-green chlamydospores (7–15 μm long), isolated or in short chains, are formed in variable numbers.

5.1.1.2 *Phaeoacremonium minimum*

Phaeoacremonium minimum is a member of the Ascomycota of the order Togniniales and the family Togniniaceae (Réblová et al. 2004). Conidia are very small (1.18 ± 0.20 μm × 3.56 ± 0.35 μm), hyaline, oblong-ellipsoid-to-allantoid, and emerge at the tip of simple phialides, usually reduced to the single, subcylindrical-to-narrowly-ellipsoidal conidiogenous cell. Its mycelium is slow-growing (<1 cm after 1 week at 20 °C on agar medium), beige-to-honey-coloured with a yellow pigment diffusing into the medium. The sexual form of this species has been observed in Australia and California, but not on the European continent.

5.1.1.3 *Fomitiporia mediterranea*

Fomitiporia mediterranea is a Basidiomycota of the order Hymenochaetales, family Hymenochaetaceae. Also called 'tinder', 'amadou' or 'white rot', this fungus is characterised by whitish hyphae and forms carpophores on affected stumps that adhere to the wood (Fig. 5.1), flat and fully resupinate, pulvinate, with a lighter margin. Beige-to-brown in colour, the carpophores can reach several centimetres in diameter and several millimetres in thickness on the vine, and are generally located at the head of the stock.

5.1.1.4 *Cadophora luteo-olivacea*

Cadophora luteo-olivacea is a member of the Ascomycota of the order Helotiales and the family Ploettnerulaceae (Harrington and Mcnew 2003). This fungus is characterised by short conidiophores, usually unbranched, developing lateral or terminal phialides, mostly monophialidic, hyaline with collarettes. Conidia are hyaline, with up to three guttules, ovoid or oblong-ellipsoidal, 3.5–7.5 μm long × 2–3 μm wide (Gramaje et al. 2011).

Fig. 5.1 Sexual fruiting
bodies of *Fomitiporia
mediterranea* with a
brownish carpophore
embedded in the
grapevine wood

Fungal Community

(Hofstetter et al. 2012; Del Frari et al. 2019; Eichmeier et al. 2018; Gramaje
and Armengol 2011)

Comprehensive mycological studies of fungal communities were conducted
on grapevine plants. The first of these (Hofstetter et al. 2012) was on a plot of
15-to-30-year-old cv. Chasselas grafted on the rootstock 3309. In a symptomatic
vineyard, plants showing foliar symptoms of esca for the first time were compared to seemingly healthy plants and to nursery plants grafted with material from
the same plot. The fungal community of grapevine wood from this single plot
consisted of 158 species identified by molecular methods and statistically analysed. Of the 107 symptomatic and asymptomatic adult plants analysed in total,

only 11 were free of the three species associated with esca. The fungal community of esca-symptomatic vines did not differ significantly from that of healthy vines, with both having an almost-identical isolation frequency for *Phaeomoniella chlamydospora*, *Phaeoacremonium* spp. and *Eutypa lata*. Anamorphs of Botryosphaeriaceae and *Phomopsis* spp. were more frequent in diseased plants than in healthy ones but were also present in nursery plants. The study by Del Frari et al. (2019) also investigated the mycobiome composition of an esca-symptomatic adult vineyard, sampling wood from different parts of the stem and in the canes of symptomatic and asymptomatic plants. Their results are consistent with those of Hofstetter et al. (2012), with an even higher fungal diversity (289 taxa) as well as esca-associated fungi (*P. chlamydospora* and *Fomitiporia* spp.) and other pathogenic species (Botryosphaeriaceae and *Eutypa* spp.) being detected in both plant types with comparable frequency and abundance.

Studies dealing with nursery grapevine plants are controversial. Gramaje and Armengol (2011) reported that *P. chlamydospora* was ubiquitous in nursery plants, while Hofstetter et al. (2012) and Eichmeier et al. (2018) claimed that both it as well as *Fomitiporia mediterranea* and *Phaeoacremonium* species were absent from them. In the latter study, *P. chlamydospora* was isolated only after the first growing season. In Hofstetter et al. (2012), Nectriaceae spp. (*Cylindrocarpon* anamorphs) responsible for black-foot disease (Halleen et al. 2006) were only present in young grafted plants (>60% of plants), while *Cadophora luteo-olivacea*, a species which according to Gramaje and Armengol (2011) was involved in young vine decline, was frequently isolated from nursery plants in studies conducted by Hofstetter et al. and Eichmeier et al. The latter two studies suggest that nursery plants are not necessarily the cause of esca problems in adult grapevines. The fungal community of nursery plants differs from that of adult plants. The grapevine fungal community will change over time depending on the environmental fungal reservoir of latent pathogens and true endophytes present in each terroir. The soil appears to be a major source of inoculum for young vine decline (Agustí-Brisach et al. 2013). Moreover, factors influencing vine physiology appear to significantly influence the expression of esca symptoms. In general, short-term or long-lasting stress seems to favor esca, whether it is of osmotic, mineral and/or climatic origin.

5.1.2 Symptoms

The symptoms of esca appear sporadically from July onwards on isolated vines or in well-defined areas of the vineyard. There is a sudden form of vine decline leading to apoplexy of the plant, as well as a slow form of chronic decline essentially marked by foliar symptoms on certain branches or over the entire plant. With apoplexy, the vines bud and develop normally until summer. Current hypotheses indicate that heavy rainfall followed by hot, dry weather is what causes the leaf blades to begin to dry out. The necroses quickly enlarge and the entire shoot or plant dries out completely within a few days (Fig. 5.2). Unlike other perennial plants that exhibit

Fig. 5.2 Symptoms of esca. Apoplexy of red (**a**) and white (**b**) grape varieties occurring in a few days. Details: Completely desiccated bunches

symptoms of apoplexy, where the root system is typically the causative factor, in esca-diseased grapevines, the root system remains intact regardless of the foliar symptoms. This dramatic and irreversible form of dieback is the easiest to identify.

The slow form of the disease manifests mainly on the foliage (Fig. 5.3). On white grape varieties the leaf blades turn pale and then yellow in an irregular manner between the veins and occasionally at the edges. On red grape varieties these colorations are somewhat reddish. These areas subsequently dry out, with only the main veins remaining green-to-yellowish-green (Fig. 5.4).

The lower leaves of the twigs are affected first, followed by the entire shoot. These leaf symptoms fluctuate, disappearing for a year or more and then reappearing. Berries may be punctuated with small blue-black spots (Fig. 5.5), a symptom termed 'black measles' by analogy with measles in humans. These leaf symptoms are not unique to esca. In the slow form, leaf discoloration can be mistaken for physiological problems such as magnesium deficiency or extreme water stress, while in the apoplectic form it can be mistaken for root rot caused by *Armillaria mellea*.

The interior of vines affected by either the apoplectic or slow form of esca exhibits large patches of dry wood in longitudinal section from the pruning wounds (Fig. 5.6). These necrotic areas are either hard or soft in consistency and are separated from the healthy wood by a dark-brown-to-black border (Fig. 5.7). *F. mediterranea* causes a soft whitish rot also referred to as 'tinder'.

5.1.3   Biology and Epidemiology

Our knowledge of the biology and epidemiology of the different fungi associated with esca remains incomplete, making them difficult to study. We do know, however, that the main lignicolous fungal species described to date in connection with esca are

Fig. 5.3 Slow form of esca: partial discoloration of leaves on white (**a**) and red (**b**) grape varieties

likely to play a decisive role in vine decline. Most scientists agree that the fungi associated with esca are in fact latent pathogens and not true endophytes (Mostert et al. 2000). We also note that apoplectic symptoms are triggered directly after the vine is subjected to physiologically stressful climatic conditions. These fungi form a very large number of anemochorous spores capable of colonising the wood of the vine through wounds of any kind. A morphological characteristic of vine wood is that it does not cover pruning wounds. Unlike other woody plants, vines are characterized by a rhytidome on the exterior, consisting of dead tissue formed from various layers of cork or suber, replacing the juvenile bark or cortex. In other perennial plants, the bark is separated from the wood by a continuous cambium and phloem encircling the entire circumference of the stem. The secondary growth of vine therefore depends on a discontinuous cambium located between the phloem and the xylem. Pruning wounds thus lead irreparably to the formation of dead wood, which provides favourable conditions for the development of lignicolous and saprophytic fungi.

Esca agents produce numerous microscopic conidia on affected plants. According to current knowledge, *P. minimum* sporulates only in summer and cannot generally infect pruning wounds. This species is isolated in small quantities from vines affected by esca. By contrast, *P. chlamydospora* can produce spores all year round and directly infect pruning wounds. The occasional presence of these two pathogens in young vines along with other fungal species suggests that grapevines, like all woody plants, naturally contain microscopic organisms within their tissues (Fig. 5.8). A coloniser of pruning wounds, the tinder *F. mediterranea* plays a decisive role in the degradation of wood. Generally absent from young grapevine plants,

Fig. 5.4 Complete desiccation of leaves: only the main veins still retain a yellowish-green hue

it is always isolated from soft parts of decaying wood, and occasionally produces brownish carpophores embedded in the wood of infected stumps (Fig. 5.1). These three fungi and perhaps other fungal species and/or bacteria (Bruez et al. 2015) also generate molecules that have a more-or-less-active phytotoxic effect. Recent studies have shown the extreme diversity of the fungi capable of inhabiting grapevines (Jayawardena et al. 2018) even after hot-water treatment (Eichmeier et al. 2018) to prevent phytoplasmas like *flavescence dorée* and *bois noir*. The role of these microorganisms has yet to be clarified, especially as regards the composition of fungal and bacterial communities not affecting the plant, as well as their relationship with biotic and abiotic factors that may favour certain species and lead to vine decline.

Plant Physiology and Climatic Conditions
(Bortolami et al. 2019, 2021)
These authors have shown that the development of leaf symptoms of esca is associated with a disruption of vessel integrity that could be related to the translocation from the trunk to the leaves of elicitors or toxins produced by pathogenic fungi (Bortolami et al. 2019). Because physiology controls grapevine water use, they tested the hypothesis of a strong interaction between esca

and drought, both known to affect water transport in the plant. Contrary to this theory, drought was found to completely suppress foliar symptoms of esca (Bortolami et al. 2021). Both esca and drought alter water transport and carbon balance, but in completely different ways. These results highlight the complexity and unpredictability of interactions between the different stresses hypothesised to cause grapevine mortality.

Fig. 5.5 'Black measles' symptoms on the berries due to esca

5.1.4 Disease Control

There are no direct means of controlling esca syndrome. Only prophylactic measures can limit the incidence of the disease, the risks of contamination and the sources of inoculum. Among vine cultural practices, winter pruning carried out to regulate yield and control trunk structure is a determining factor for the sustainability and homogeneity of vineyards. Vine growth is dictated by the phenomenon of acrotonia, which results in the development of increasingly long branches if the latter are not cut back. Decades ago, in the Charentes region, the Guyot-Poussard pruning technique described by Lafon (1921) was proposed as a means of preventing apoplexy by careful pruning. Recent experiments and observations in the main wine-growing countries of Europe show that the vine's reaction to pruning, and in particular the depth of desiccation of the wood caused by the pruning wound, varies according to grape variety. The timing and method of pruning can also play an important role from a health and physiological perspective. The earlier the pruning, the longer the period of wound receptivity. Vines pruned early have been shown to be more susceptible to dieback than those pruned later. Studies by Sun et al. (2006,

Fig. 5.6 Patches of necrotised dry wood forming from pruning wounds. (**a**) Vine with location of pruning wounds (arrows). (**b**) Longitudinal section of the same vine showing the patches of dry wood formed from pruning wounds (arrows)

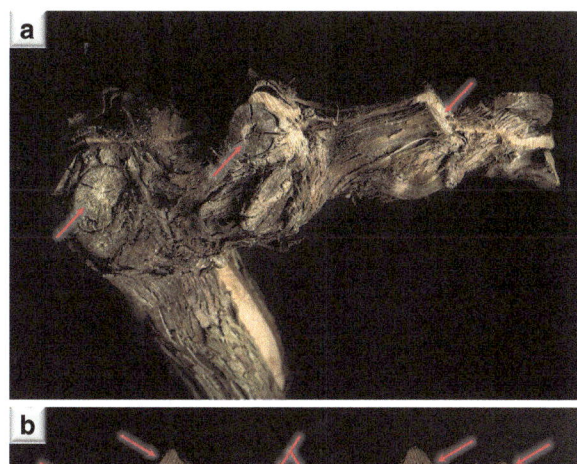

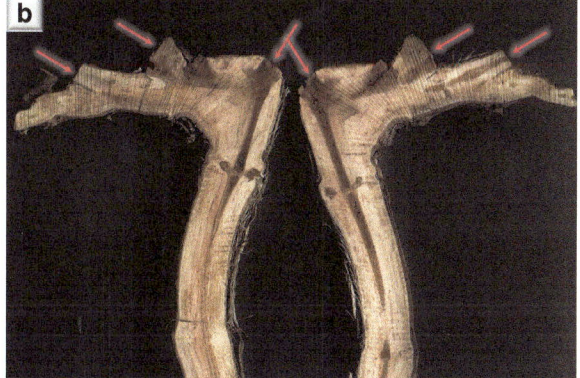

Fig. 5.7 Symptoms of esca in the wood: necrotic zones of hard or soft consistency, each delimited by a dark-brown border (arrows). *hw* healthy wood, *sw* soft wood

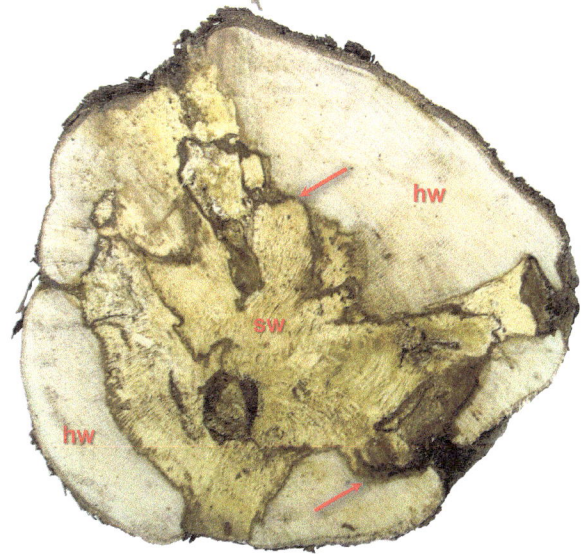

Fig. 5.8 Presence of
fungal hyphae (arrows)
inside the conducting
vessels of a vine affected
by esca. Detail: Hyphae
growing in a xylem vessel.
Scale bars 100 μm

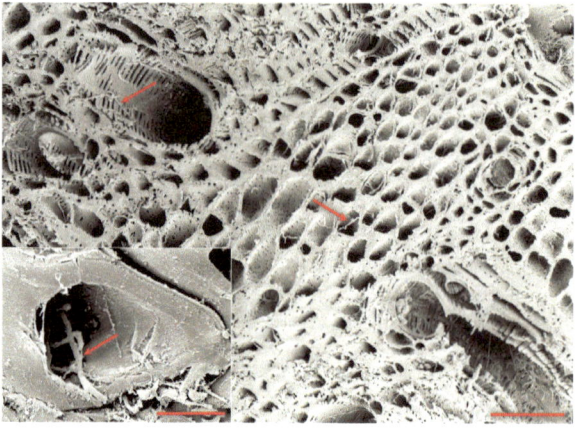

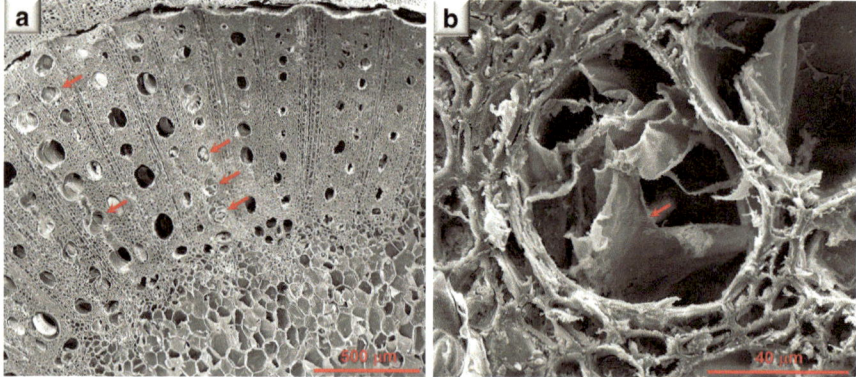

Fig. 5.9 The formation of tylose in the conducting vessels of grapevine leads to cavitation, reduces hydraulic conductance but is not degraded by fungi. (**a**) Embolised xylem vessels with tylose formation (arrows). (**b**) Detail of tylosis (arrow) in a xylem vessel

2008), which show that grapevines respond very strongly to injury by producing large amounts of tylose (Fig. 5.9) during the summer and pectin gels during the winter in order to close damaged vessels, may explain these observations. Fungi do not degrade tylose, but pectin gels are an excellent substrate for them. In general, pruning must be done with great care. Guyot-Poussard pruning, which consists in orienting the pruning wounds on the same side to preserve an uninterrupted flow of sap, recommends forming stocks with two sap flows to guarantee the continuous circulation of the raw and elaborated sap. Generally speaking, the larger the pruning wounds, the greater the proportion of dead wood, which provides a favourable environment for the development of fungi. Large pruning wounds should therefore be avoided, as should the flush pruning that is possible with pneumatic or electric pruning shears. Training pruning according to the Guyot-Poussard principle serves as a basis for constructing vines that can later be trained in single or double Guyot, or in permanent spur cordons. Although it does not constitute a direct means of fighting esca or wood diseases, careful pruning ensures continuous sap flow and reduces the spaces that favour the growth of fungi inside the vines.

Disinfection of pruning wounds with healing putty, or with injection pruning shears by covering the blade and the pruning wound with fungicides, remains contentious. The effectiveness of these practices is not proven and they involve the use of systemic fungicides with a broad spectrum of efficacy. At present, there is no registered chemical active ingredient for controlling esca, except antagonistic fungi. The use of preparations based on *Trichoderma* sp., *Fusarium* spp., or other antagonists have shown limited efficacy. Re-cutting affected plants from the trunk regrowth can be successful provided that the lower part of the trunk is healthy. Another approach involves the curettage and regrafting of vines, a method noted for its time-consuming and challenging nature (Bruez et al. 2021; Dewasme et al. 2023).

Removing affected stumps and destroying them by grinding and composting limits the sources of inoculum. Removed diseased stumps should be stored away from rain and at a distance from the vineyard to minimise the risk of spreading the spores found on the affected vines. Nurseries do not have sufficiently versatile fungicides that can penetrate into the wood. Hot-water treatment of young grapevines, used against *flavescence dorée* and *bois noir* phytoplasma, alters the composition of the fungal community in the wood, but does not destroy esca agents completely (Casieri et al. 2009; Eichmeier et al. 2018). All the research conducted to date confirms the importance of hygiene measures at the various stages of vine plant production by nurseries. These hygiene measures reduce the incidence of pathogenic fungi, but do not fully suppress plant microorganisms. Furthermore, a recent study (Ramsing et al. 2021) showed that rootstocks with large xylem diameters are the most susceptible to *Phaeomoniella chlamydospora* and *Phaeoacremonium minimum*. Choosing rootstocks with narrow xylem diameters may thus be a promising solution for a more sustainable viticulture.

5.2 *Botryosphaeria* **Dieback (Black Dead-Arm)**

Current names:

- *Botryosphaeria obtusa* (Schwein.) Shoemaker 1962, Ascomycota
- *Botryosphaeria dothidea* (Moug. ex Fr.) Ces. & De Not. 1863, Ascomycota
- *Diplodia corticola* A.J.L. Phillips, A. Alves & J. Luque 2004, Ascomycota
- *Diplodia mutila* (Fr.) Mont. 1834, Ascomycota, syn. *Botryosphaeria stevensii* Shoemaker 1964
- *Dothiorella iberica* A.J.L. Phillips, J. Luque & A. Alves 2005, Ascomycota
- *Dothiorella sarmentorum* (Fr.) A.J.L. Phillips, Alves & Luque 2005, Ascomycota, syn. *Dothiorella americana* J.R. Úbez-Torres, F. Peduto & W.D. Gubler 2011
- *Lasiodiplodia crassispora* T.I. Burgess & Barber 2006, Ascomycota
- *Lasiodiplodia gilanensis* Abdollahz., Javadi & A.J.L. Phillips 2010,, Ascomycota, syn. *Lasiodiplodia missouriana*, J.R. Úrbez-Torres, F. Peduto & W.D. Gubler 2011

- *Lasiodiplodia mahajangana* Begoude, Jol. Roux & Slippers 2009, Ascomycota, syn. *Lasiodiplodia exigua* Linaldeddu, Deidda & A.J.L. Phillips 2014
- *Lasiodiplodia mediterranea* Linaldeddu, Deidda & Berraf-Tebbal 2015, Ascomycota
- *Lasiodiplodia parva* A.J.L. Phillips, A. Alves & Crous 2008, Ascomycota
- *Lasiodiplodia pseudotheobromae* A.J.L. Phillips, A. Alves & Crous 2008, Ascomycota
- *Lasiodiplodia theobromae* (Pat.) Griffon & Maubl. 1909, Ascomycota
- *Lasiodiplodia viticola* J.R. Úrbez-Torres, F. Peduto & W.D. Gubler 2011, Ascomycota
- *Neofusicoccum australe* (Slippers, Crous & M.J. Wingf.) Crous, Slippers & A.J.L. Phillips 2006, Ascomycota
- *Neofusicoccum luteum* (Pennycook & Samuels) Crous, Slippers & A.J.L. Phillips 2006, Ascomycota
- *Neofusicoccum macroclavatum* (T.I. Burgess, Barber & Hardy) T.I. Burgess, Barber & Hardy 2006, Ascomycota
- *Neofusicoccum mediterraneum* Crous, M.J. Wingf. & A.J.L. Phillips 2007, Ascomycota
- *Neofusicoccum parvum* (Pennycook & Samuels) P.W. Crous, Slippers & A.J.L. Phillips 2006, Ascomycota
- *Neofusicoccum ribis* (Slippers, P.W. Crous & M.J. Wingfield) P.W. Crous, Slippers & A.J.L. Phillips 2006, Ascomycota
- *Neofusicoccum viticlavatum* (Van Niekerk & Crous) Crous, Slippers & A.J.L. Phillips 2006, Ascomycota
- *Neofusicoccum vitifusiforme* (Van Niekerk & Crous) Crous, Slippers & A.J.L. Phillips 2006, Ascomycota
- *Phaeobotryosphaeria porosa* (Van Niekerk & Crous) Crous & A.J.L. Phillips 2008, Ascomycota, syn. *Diplodia porosa* Van Niekerk & Crous 2004
- *Phyllosticta ampelicida* (Engelm.) Aa 1973, Ascomycota, syn. *Guignardia bidwellii* (Ellis) Viala & Ravaz 1892
- *Spencermartinsia viticola* (A.J.L. Phillips & J. Luque) A.J.L. Phillips, A. Alves & Crous 2008, Ascomycota, syn. *Botryosphaeria viticola* A.J.L. Phillips & J. Luque 2005

The term 'black dead arm' (BDA) was first used by Lehoczky (1974a) to describe symptoms in grapevine wood in association with a single species of fungus, *Diplodia mutila* (syn. *Botryosphaeria stevensii*). In 1978, symptoms similar to those of BDA were described on grapevines in Italy but were associated with *Botryosphaeria obtusa* by Cristinzio (1978). Still in Italy but several years later, Rovesti and Montermini (1987) isolated *D. mutila* from necrotic grapevine wood exhibiting symptoms similar to those previously described by Cristinzio, but associated these symptoms with esca rather than with BDA and were able to confirm by pathogenicity tests that the wood necroses were indeed caused by *D. mutila*. In the early 2000s, Larignon and Dubos (2001) reported the presence of BDA on grapevines in the Bordeaux region of France, associating the disease with *B. dothidea* and *B. obtusa*.

In addition, Larignon et al. (2001) associated BDA with a mild and a severe form, a distinction that has been debated because of their respective similarities to the "tiger stripe" and apoplectic symptoms characteristic of esca (Mugnai et al. 1999). A decade later, a Spanish vine survey demonstrated the impossibility of differentiating BDA foliar symptoms *per se*, as they commonly occurred along with both *Eutypa* dieback and esca symptoms (Luque et al. 2009). Based on the most recent studies, most researchers agree that BDA leaf symptoms are probably just one of many manifestations of esca (Surico et al. 2006). These symptoms may also result from unknown causes yet to be determined. Thus, to avoid confusion, until the status of BDA is clarified, researchers should be cautious when reporting this disease in vineyards (Úrbez-Torres 2011).

In addition, the disease known as '*Botryosphaeria* canker of grapevines' was first named by Leavitt (1990) to describe the specific symptoms caused by *Lasiodiplodia theobromae* in Californian vineyards. This disease is characterised by the presence of perennial cankers in spurs, cordons and/or trunks. Research carried out in the United States showed that Botryosphaeria canker—the name adopted for this disease—was not caused by *L. theobromae* alone, and that 21 species of Botryosphaeriaceae, including the species thought to be involved in BDA *sensus stricto* (*B. dothidea*, *D. mutila* and *B. obtusa*), were involved. The name '*Botryosphaeria* dieback', instead of 'black dead-arm' or '*Botryosphaeria* canker', was therefore proposed by Úrbez-Torres (2011) to describe the different symptoms of grapevine wood disease caused by Botryosphaeriaceae species. Úrbez-Torres also suggests that if used, the term 'BDA' should be restricted to the symptoms caused by *D. mutila* in the original description of the disease.

5.2.1 Causal Organisms

To date, at least 25 species of Botryosphaeriaceae belonging to 9 genera have been identified as grapevine wood pathogens (see above list of species; Linaldeddu et al. 2015; Úrbez-Torres 2011). All these species are Ascomycotas belonging to the order Botryosphaeriales and the family Botryosphaeriaceae. At present, three species are considered responsible for BDA *sensus stricto*: *Botryosphaeria obtusa*, *Botryosphaeria dothidea* and *Diplodia mutila* (syn. *Botryosphaeria stevensii*). Of these three, *B. obtusa* was reported as one of the main fungi associated with black dead-arm of grapevine (Auger et al. 2004). First described by Schweinitz in 1832 as *Sphaeria obtusa* and later reassigned to the genus *Physalospora* by Cooke in 1892, this species was finally renamed *Botryosphaeria obtusa* by Shoemaker in 1964. The fungus lives in dead wood or bark, where it produces pycnidia (<500 μm in diameter) embedded in a dark-brown-to-black-stroma with ostioles. The wall of the pycnidia is thick, composed of sclerotised cells. The holoblast-like conidiogenous cells each release an ovoid conidium, first hyaline and then rapidly turning dark-brown, truncated at the base (18–26 × 9–12 μm). The wall of the conidia is thick (ca. 0.5 μm) and rough. Asci (bitunicate, 90–120 × 17–23 μm) and ascospores (hyaline, fusoid, usually unicellular, sometimes septate, wider in the middle than at the ends, 25–33×7–12 µm) are formed in perithecia also embedded in a dark-brown-to-black stroma. Botryosphaeriaceae species are extremely similar in terms of their morphological characteristics, and only molecular identification methods can reliably determine the different species of this family.

5.2.2 Symptoms

Grapevine symptoms associated with Botryosphaeriaceae are numerous and highly variable. Wedge-shaped cankers, brown stripes below the bark, progressive budbreak failure, leaf spots, leaf wilting, bud necrosis and plant dieback may also be associated with *Eutypa* dieback, since they are "tiger-stripe" foliar symptoms.

The term 'BDA' currently relates to three species (*B. dothidea*, *D. mutila* and *B. obtusa*); however, BDA-related species not only cause different symptoms of grapevine decline in different countries and continents (Phillips 2000; Taylor et al. 2005; Úrbez-Torres et al. 2006) but also in different winegrowing regions of the same country (van Niekerk et al. 2004). According to Úrbez-Torres (2011), it is clear that the name BDA is misapplied to one or more diseases, and that the use of this name should be strictly limited to cases coinciding precisely with the original disease concept: diffuse chlorosis of leaves followed by wilting, black streaks of wood in the xylem, caused by *D. mutila*. Furthermore, the term 'BDA' should not be applied to disease symptoms such as sectorial necrosis, light-brown discoloration, progressive budbreak failure, plant dieback of wood or 'tiger stripe' patterns on leaves—symptoms that might result from other grapevine trunk diseases.

5.2.3 Biology and Epidemiology

Until recently, scientists considered Botryosphaeriaceae to be saprophytes, secondary colonisers or weak pathogens in grapevine wood (Phillips 2000). This might be why the role played by Botryosphaeriaceae species in grapevine wood diseases has hitherto been underestimated. Another reason could be the difficulty in differentiating the vascular and foliar symptoms caused by Botryosphaeriaceae from those caused by other grapevine trunk disease pathogens, mainly esca (Luque et al. 2009) and *Eutypa* dieback (Úrbez-Torres 2011). In recent decades, studies have mainly focused on the latter two diseases, which did not help us to understand the role played by Botryosphaeriaceae in grapevine trunk diseases (Úrbez-Torres 2011). Consequently, there has been little research into the epidemiology of Botryosphaeriaceae species associated with grapevine trunk diseases. Early studies suggested that, as esca-associated species, Botryosphaeriaceae species enter grapevines primarily through pruning wounds (Leavitt 1990) and through graft unions between rootstocks and scions (Lehoczky 1974b). Spores are probably released in wet and humid conditions. They are often isolated from symptomatic wood together with esca and *Eutypa* dieback species (Hofstetter et al. 2012; Del Frari et al. 2019).

Toxins Produced by Botryosphaeriaceae
(Martos et al. 2008; Félix et al. 2019; Nazar Pour et al. 2020; Reveglia et al. 2021)

In recent years, an increasing number of Botryosphaeriaceae have been associated with the decline of grapevines worldwide. Five species (*Botryosphaeria dothidea*, *Botryosphaeria obtusa*, *Dothiorella viticola*, *Neofusicoccum luteum* and *N. parvum*) isolated from affected grapevines in Spain were tested for their ability to synthesise toxins in liquid culture (Martos et al. 2008). The phytotoxicity of *Botryosphaeria obtusa* and *Neofusicoccum parvum* peaked after 14 days of growth, while that of the other species only emerged after 21 days of culture. All fungi produced low-molecular-weight lipophilic compounds with phytotoxic properties. In addition, *N. luteum* and *N. parvum* produced low-molecular-weight phytotoxins specific to them. This led to an extensive study on the phytotoxicity of *N. luteum* and *N. parvum*. Culture filtrates and corresponding extracts of these two species exhibited phytotoxic properties in a wide range of biotests. Gas chromatography analyses of the O-methyl-glucosides of the phytotoxic exopolysaccharides of *N. parvum* revealed these substances to be mainly composed of glucose, mannose and galactose. More recently, other types of phytotoxic compounds such as (*R*)-mellein, protocatechuic alcohol and spencertoxin have been isolated from Botryosphaeriaceae (Reveglia et al. 2021) (Fig. 5.10). These results suggest that these toxic metabolites may account for the virulence to grapevine of these two fungal species. Recent studies (Félix et al. 2019; Nazar Pour et al. 2020) have also shown that temperature affects the phytotoxicity and cytotoxicity of Botryosphaeriaceae.

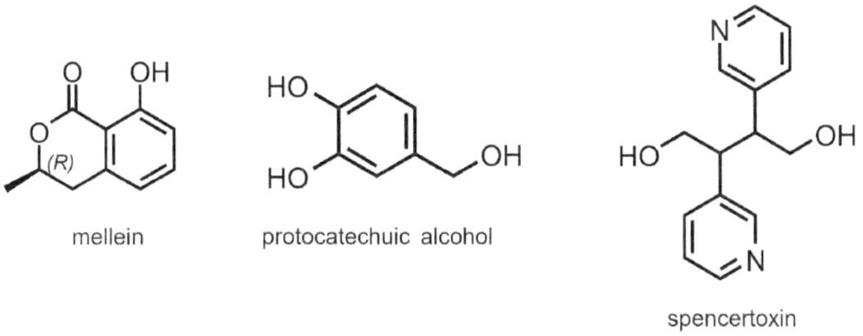

Fig. 5.10 Chemical structure of (*R*)-mellein, protocatechuic alcohol and spencertoxin, three phytotoxic compounds isolated from Botryosphaeriaceae

5.2.4   Disease Control

Long considered to be weak pathogens of grapevine trunks, this family has not been the subject of many epidemiological studies, which has limited the development of control methods. As with esca, prophylactic measures such as careful pruning, the protection of pruning wounds, and chemical and hot-water treatment of nursery material may reduce the propagation of Botryosphaeriaceae (see Úrbez-Torres 2011).

5.3   *Eutypa* Dieback (Eutypiosis)

Current names:

- *Eutypa lata* (Pers.: Fr.) Tul. & C. Tul. 1863, Ascomycota, syn. *E. armeniacae* Hansf. & Carter, *Libertiella blepharis* A. L. Smith
- *Eutypa laevata* (Nitschke) Sacc. 1882, Ascomycota

Because it was the most frequently isolated Diatrypaceae species from the wood of *Eutypa* dieback-symptomatic plants (Carter 1957; Kuntzmann et al. 2010; Munkvold et al. 1994; Rolshausen et al. 2008; Rudelle et al. 2005; Travadon and Baumgartner 2015), a single ascomycete species, *Eutypa lata*, was long thought to be responsible for *Eutypa* dieback. A wood-rotting fungus, *Eutypa lata* is not specific to grapevines and has been identified on over 80 species of woody plants belonging to around 30 different botanical families (Carter et al. 1983). *Eutypa* dieback affects most of the fruit species grown in temperate and Mediterranean climates (apricot, almond, lemon, plum, peach, apple, pear, quince, blackcurrant, gooseberry, walnut, hazelnut, fig, persimmon, elderberry) and in Australia (Bolay and Carter 1985). In Europe, the disease is frequently found on grapevine and apricot. Because *E. lata* can only infect the woody tissues of its host through wounds, it tends to develop on plants undergoing regular and severe pruning, as is the case with grapevines. Recent studies have shown that *E. lata* is not the only Diatrypaceae associated with wood cankers of grapevines (Díaz et al. 2011; Luque et al. 2012; Trouillas et al. 2010, 2011; Trouillas and Gubler 2010), but it remains to be determined whether these species are pathogenic or not. However, *E. laevata* and an unidentified species of *Eutypa* have been shown to be the most frequently isolated species from grapevine plants suffering from *Eutypa* dieback in eastern North America (Rolshausen et al. 2014).

5.3.1   Causal Organism

Eutypa lata is an Ascomycota, of the order Xylariales and the family Diatrypaceae. It was first identified as *E. armeniacae*, the agent of apricot apoplexy, in 1957 (Carter 1957). *Eutypa* dieback was then formally identified in vineyards in

Switzerland in the 1970s. *E. lata* forms black subcortical stromata reaching several centimetres in length on the surface of dead wood (Fig. 5.11). These stromata contain a multitude of perithecia, globose fruiting bodies 300–500 μm in diameter, extended by an obtuse conical ostiole opening into a pore. Usually arranged in a single layer, the perithecia contain a multitude of cylindro-claviform, unitunicate, very long pedicellate asci measuring 30–35 × 4–5 μm in their fertile part. Each ascus contains eight allantoid ascospores, brownish-yellow and 7–11 × 1.5–2 μm in length. *Libertella blepharis*, the anamorph of *E. lata*, can be seen on the surface of necrotic tissue, forming black pycnidia around 2 mm in diameter. Given sufficient moisture, the pycnidia exude ivory-to-yellowish-orange cirrhus containing innumerable filiform, more-or-less-arcuate hyaline conidia (18–25 × 1 μm).

5.3.2 Symptoms

The typical external symptom of vines suffering from *Eutypa* dieback is a significant reduction of vegetative growth during the early stage of the growing season (stunted shoots with short internodes) (Fig. 5.12) and chlorotic leaves, brownish-yellow in white grape varieties and somewhat reddish in red grape varieties, with marginal necroses (Bertsch et al. 2013; Munkvold et al. 1994).

The leaves are distinctly smaller than those of healthy plants, chlorotic, curled, deformed, or shredded (Fig. 5.13). In the most severe cases, marginal necrosis appears, whereupon the leaves dry up completely and fall off. The initial leaf symptoms can easily be mistaken for those caused by the fanleaf virus. Bunches appear relatively normal until flowering, after which they exhibit poor fruit setting, often resulting in Millerand age (Fig. 5.14), or complete abortion and desiccation.

Cutting longitudinally through affected vines or cordons reveals one or more well-demarcated necroses, each starting from a pruning wound and penetrating deep into the wood. The necrotic tissues are hard and greyish-brown-to-purplish-brown in colour, depending on the cultivar (Fig. 5.15).

Fig. 5.11 Subcortical stromata of *Eutypa lata* on the surface of the wood containing a multitude of perithecia (arrow). Detail: perithecia in cross section in the subcortical stroma

Fig. 5.12 Symptoms of
Eutypa dieback. (**a**) and
(**b**) Stunted vegetative
development and yellowish
foliage. (**c**) Growth
anomaly of the shoots,
with very short internodes
(above)

Fig. 5.13 Foliar symptom
of *Eutypa* dieback:
chlorotic, curled
appearance of the affected
leaves compared to a
healthy leaf (right)

Fig. 5.14 Millerandage
(uneven fruit-set) of grape
bunches affected by *Eutypa*
dieback

Fig. 5.15 Delimited necroses in the wood from a pruning wound or lesion affected by *Eutypa* dieback showing numerous similarities to esca or Botryosphaeria wood symptoms

According to several authors, *E. lata* might also play a role in esca disease (Armengol et al. 2001; Bertsch et al. 2013). Esca-associated species sometime coincide with *E. lata* in wood lesions of plants expressing *Eutypa* dieback symptoms (Wagschal et al. 2008). The results of a field survey conducted to characterise the decline of mature grapevine plants in Spain (Luque et al. 2009) have shown that wood cankers of more than one disease often coincide in the same plant. In the case of esca, plants suffering from *Eutypa* dieback may eventually die. Moreover, the symptoms expressed by *Botryosphaeria* dieback (Úrbez-Torres 2011) make it difficult to distinguish between major grapevine wood diseases: dwarf shoots and wedge-shaped wood necroses are attributable to *Eutypa* and *Botryosphaeria* dieback alike (Taylor et al. 2005).

5.3.3 Biology and Epidemiology

The disease is spread solely by ascospores from perithecia buried in the stroma developing on the dead wood of many hosts. Perithecia appear only in areas of high annual rainfall. When mature, the stroma retains fertile perithecia for at least 5 years. Ascospores are ejected from the perithecia during and immediately after rainfall. Wind ensures their dissemination over long distances.

Ascospores of *E. lata* can germinate at temperatures between 1 and 45 °C, the optimum range being 22–25 °C. Between 13 and 27 °C, ascospores need 16 h to germinate. They remain viable for up to 2 months after release. The function of conidia is unknown. They do not germinate in water or on the surface of an agar medium and have no infectivity. *E. lata and E. laevata* are wound parasites. The ascospores use the vascular tissues of the wood as a portal of entry. With the colonisation of antagonistic saphrophytic microorganisms, wound susceptibility gradually decreases. The infectious process is slow, with symptoms appearing several years after contamination of the pruning wounds.

Toxins
(Tey-Rulh et al. 1991; Afifi et al. 2004; Rudelle et al. 2005; Wagschal et al. 2008)

The hyphae of *Eutypa lata* are found only in the wood of the vine and not in the current year's twigs. The presence of symptoms in the herbaceous parts in the absence of mycelium suggested the existence of toxin(s). Analysis of the metabolites produced by *E. lata* in liquid culture revealed the presence of eutypine (4-hydroxy-3-(3-methyl-3-buten-1-ynyl) benzaldehyde). On grapevine seedlings, this toxin induces necrosis proportional to the applied dose. Eutypine can be detected in the inflorescences and raw sap of infected vines and is absent from healthy vines. Grapevines can convert eutypine into an alcohol, eutypinol, which is non-toxic. More recently, Wagschal et al. (2008) isolated additional metabolites produced by *E. lata*: eutypinic acid, cyclisation products of eutypinol and eutypinic acid, siccayne, eulatinol, epoxidised chromanones and two highly oxygenated cyclohexene oxides as eutypoxide B, allenic epoxy-cyclohexane and eulatachromene (Fig. 5.16). The phytotoxicity of *E. lata* most likely results from these structurally related compounds, with each compound probably having a different level of toxicity and different molecular targets within the plant cell. *E. lata* alters the structure of the xylem at the site of infection and in the current year's shoots. The presence of the pathogen induces anatomical and physiological changes in the cells associated with the vessels bordering the xylem. Treatment of healthy seedlings with eutypine partially reproduced these changes and the death of xylem-associated cells.

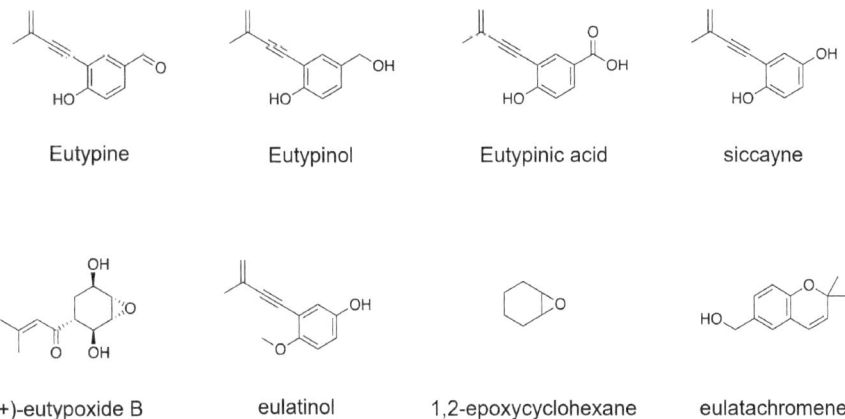

Fig. 5.16 Chemical structure of different secondary metabolites isolated from *Eutypa lata*

5.3.4   *Disease Control*

As with esca and *Botryosphaeria* dieback, there is no direct way to control *Eutypa* dieback. Only prophylactic measures can limit the incidence of the disease. To counter *Eutypa* dieback preventively, the pruning period is crucial. The earlier pruning takes place, the longer the wounds are receptive and coincide with the release of ascospores. Elimination of the affected vine stocks limits the sources of inoculum if stumps are stored away from rain and away from the vineyard to minimise the risk of spreading *E. lata* ascospores.

5.4 Excoriosis (*Diaporthe* or *Phomopsis* Dieback, Cane and Leaf Spot)

Current names:

- *Diaporthe ampelina (Berk. & M.A. Curtis)* R.R. Gomes C. Glienke & Crous 2013, Ascomycota, syn. *Diaporthe neoviticola* Udayanga, Crous & K.D. Hyde 2012, *Phomopsis viticola* (Sacc.) Sacc. 1915, *Phomopsis ampelina* (Berk. & M.A. Curt.) Grove 1919, *Phomopsis viticola* var. *ampelopsidis* (Berk. & M.A. Curt.) Grove 1919.
- *Diaporthe ambigua* Nitschke 1870, Ascomycota, syn. *Phoma ambigua* (Nitschke) Sacc., Grevillea 1872
- *Diaporthe amygdali* (Delacr.) Udayanga, Crous & K.D. Hyde 2012, Ascomycota, syn. *Phomopsis amygdali* (Delacr.) J.J. Tuset & M.T. Portilla
- *Diaporthe baccae* L. Lombard, G. Polizzi & Crous 2014, Ascomycota
- *Diaporthe celeris* Guarnaccia, Woodhall & Crous 2018, Ascomycota
- *Diaporthe eres* Nitschke 1870, Ascomycota
- *Diaporthe fukushii* (Tanaka & S. Endô) Dissan., A.J.L. Phillips & K.D. Hyde 2017, Ascomycota, syn. *Phomopsis fukushii* Tanaka & S. Endô 1927
- *Diaporthe foeniculina* (Sacc.) Udayanga & Castl. 2014, Ascomycota, syn. *Phomopsis foeniculina* (Sacc.) 1880, *Diaporthe neotheicola* A.J.L. Phillips & J.M. Santos 2009
- *Diaporthe hispaniae* Guarnaccia, Armengol & Crous 2018, Ascomycota
- *Diaporthe hongkongensis* R.R. Gomes, Glienke & Crous 2013, Ascomycota
- *Diaporthe hungariae* Guarnaccia, Armengol & K.Z. Váczy 2018, Ascomycota
- *Diaporthe phaseolorum* (Cooke & Ellis) Sacc. 1882, Ascomycota
- *Diaporthe perjuncta* Niessl 1876, Ascomycota
- *Diaporthe rudis* (Fries) Nitschke 1870, Ascomycota
- *Diaporthe sojae* Lehman 1923, Ascomycota, syn. *Diaporthe phaseolorum* var. *sojae* (Lehman) Wehm 1933, *Diaporthe longicolla* (Hobbs) J.M. Santos, Vrandecic & A.J.L. Phillips 2011
- *Diaporthe kyushuensis* Kajitani & Kanem 2000, Ascomycota, syn. *Diaporthe vitimegaspora* (K.C. Kuo & L.S. Leu) Rossman & Udayanga 2015, *Phomopsis vitimegaspora* K.C. Kuo & L.S. Leu 1998.

As with other groups of fungi associated with grapevine wood diseases, molecular techniques have greatly modified the taxonomy and species concepts of the genus *Diaporthe* (syn. *Phomopsis*) (Gao et al. 2017; Udayanga et al. 2015). *Phomopsis* dieback is a major disease of grapevines, causing severe losses due to basal-shoot breakage, stunting, dieback, loss of vigour, reduced cluster formation and fruit rot (Wilcox 2015). Canes exhibit irregularly shaped brown-to-black necrotic lesions and clusters exhibit necrosis of the rachis and brown, shrivelled berries shortly before harvest (Pearson and Goheen 1994). Together with *D. ampelina* (syn. *Phomopsis viticola*), historically the most common species known to cause this disease, *D. amygdali* has been confirmed as a severe pathogen of grapevine (Mostert et al. 2001; Van Niekerk et al. 2005). *Phomopsis* dieback is most severe in humid temperate regions, occurring throughout the growing season (Erincik et al. 2001). More recently, Úrbez-Torres et al. (2013) provided strong evidence for the role of *D. ampelina* as a canker-causing organism but also reported the presence of *D. ambigua*, *D. eres* and *D. foeniculina* in Californian vineyards. In addition, *D. ampelina* is the causal agent of grapevine swelling arm, also induced by *D. kyushuensis*. Cane bleaching is another symptom of grapevine caused by *D. perjuncta* and *D. ampelina*. *Diaporthe eres* was found to be a weak-to-moderate pathogen causing wood-canker of grapevine (Baumgartner et al. 2013). In China, the four species reported to cause *Phomopsis* dieback are *D. eres*, *D. hongkongensis*, *D. phaseolorum* and *D. sojae*. A recent study on *Diaporthe* spp. pathogenic to *V. vinifera* in Europe (Guarnaccia et al. 2018) added *D. hungariae*, *D. celeris*, *D. hispaniae*, *D. rudis*. and *D. baccae* to the list of species responsible for *Phomopsis* and related dieback symptoms.

5.4.1 Causal Organisms

The genus *Diaporthe* is a member of the Ascomycetes, order Diaporthales, family Diaporthaceae. *D. ampelina* (*P. viticola*), the most virulent species and the one most frequently associated with Phomopsis dieback (Úrbez-Torres et al. 2013), forms black (0.2–0.5 mm diameter), unilocular, initially submerged discoid pycnidia on the surface of infected twigs; they become globose and emerge from the epidermis at maturity, releasing conidia through an apical ostiole. Cylindrical, enteroblastic conidiophores of the phialid type emit conidia in yellowish cirrhus or in gelatinous masses. There are two types of hyaline and unicellular conidia (α and β) depending on the conditions of sporulation: the conidia α ($7–10 \times 2–4$ µm-long, ellipsoidal-to-fusiform, bigutulate, germinating) and the conidia β ($18–30 \times 0.5–1$ µm-long, filiform, curved and non-germinating), whose role is still unknown.

5.4.2 Symptoms

The most characteristic symptoms of *Phomopsis* dieback in grapevine are similar to those of *Botryosphaeria* dieback and include perennial cankers in the grapevine trunk and lack of budbreak in infected spurs (Úrbez-Torres et al. 2013). *Phomopsis* dieback is primarily caused by *D. ampelina*, the most virulent *Diaporthe* species on grapevine. This species has long been known as the causal agent of the disease referred to as 'Phomopsis cane and leaf spot' in the United States and 'excoriosis' in Europe (Gramaje et al. 2018).

In early summer, the centres of young shoots, bunch stalks and petioles exhibit dark chlorotic spots which increase in size over the season to form elongated blackish necroses (Fig. 5.17). On severely affected vines internodes are short,

Fig. 5.17 Elongated blackish necroses (cane spot) caused by *Diaporthe ampelina* at the base of the vine branches (**a**), progressing towards a deep excoriation (**b**)

Fig. 5.18 Symptoms of leaf spot caused by *Diaporthe ampelina* (excoriosis) with the presence of yellowish patches and central leaf blade necroses

excoriations are not limited to the base of the shoots and the lower buds do not break, which prevents the formation of spurs during the following winter pruning. Affected inflorescences are no longer properly fed and abort rapidly. Symptoms progress after successive rains in spring. The plants are weakened, crop quantity and quality are reduced, pruning the following winter is difficult and plant perennity is compromised over the long term.

Leaf infections are frequent, causing round yellow spots at the base of the leaf blade and on the veins, with black spots in the centre (Fig. 5.18). Badly affected leaves dry out and the leaf blade falls off, while the petiole usually remains attached to the shoots. Berries may also be affected, turning bluish-purple after veraison. The epidermis of the berry becomes covered with darker dots arranged in concentric circles that constitute the fruiting bodies of the fungus (Fig. 5.19).

Fig. 5.19 Symptoms of excoriosis (*Diaporthe* dieback) on a grape bunch. (**a**) violet-blue berries appearing after veraison, covered with fungal fruiting bodies (pycnidia) which can be confused with black rot or white rot diseases. (**b**) Detail: pycnidia of *Diaporthe ampelina* on a berry

Fig. 5.20 The base of the woody shoots affected by excoriosis shows a greyish bark (below) compared to the healthy wood (above) and contains small black pustules, the pycnidia (arrows)

Phomopsis dieback is clearly visible during winter pruning: the affected shoots are discoloured, the epidermis takes on a grey-white appearance and contains a multitude of small black pustules emerging from the bark. Cane bleaching is commonly associated with *D. ampelina* and *D. perjuncta*. The base of the shoots exhibits isolated or confluent greyish spots bordered with black, 0.2–5 cm long, which can be confused with thrips (*Drepanothrips reuteri*) damages (Fig. 5.20). The fungus invades almost all of the shoot tissues, especially the medullary rays and parenchyma, causing extensive necrosis (Fig. 5.21).

Usually, the bark bursts open, revealing one or more crevices giving the wood its excoriated appearance. The presence of the species responsible for swelling-arm symptoms, *D. ampelina* and *D. kyushuensis*, can be easily detected by placing pieces of shoot in a humid chamber (glass container, plastic bag): after a few days incubating at room temperature, the fruiting bodies emit yellowish-white cirrhus (Fig. 5.22).

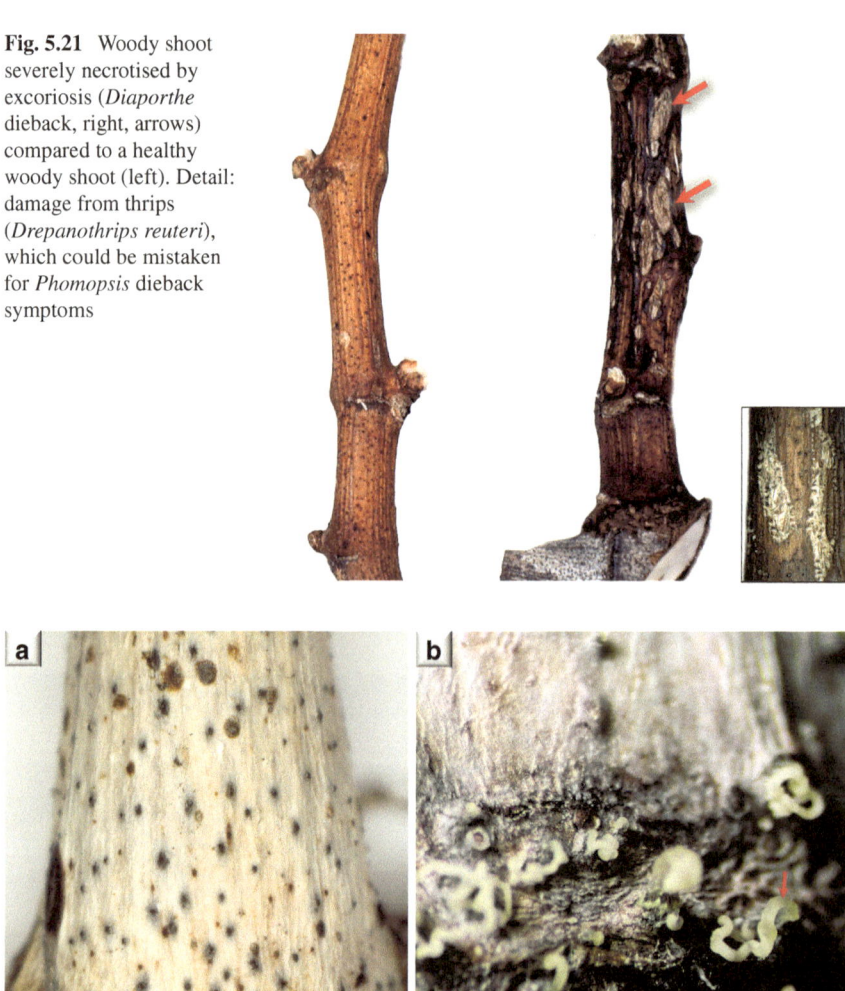

Fig. 5.21 Woody shoot severely necrotised by excoriosis (*Diaporthe* dieback, right, arrows) compared to a healthy woody shoot (left). Detail: damage from thrips (*Drepanothrips reuteri*), which could be mistaken for *Phomopsis* dieback symptoms

Fig. 5.22 Fruiting bodies of *Diaporthe ampelina*. (**a**) Pycnidia on the surface of the wood. (**b**) Yellowish-white cirrhus (arrow) emerging from the pycnidia under high moisture

5.4.3 Biology and Epidemiology

Diaporthe ampelina overwinters as pycnidia in the bark or as mycelium in the buds at the base of shoots. Prolonged rainy periods and cool temperatures are the main requirements for the epidemiological development of the pathogen. As soon as budbreak occurs and provided that there is sufficient moisture, pycnidia emerge from the epidermis and release conidia which are carried by rainwater to the young shoots. A minimum of 12 h of wetting is required for the spores to infect healthy tissue. Infections are dependent on temperature and the duration of leaf wetness (Erincik et al. 2003). Tissue susceptibility is greatest when shoots are emerging and elongating (BBCH 11 and 14). Spores germinate between 5 and 36 °C (optimum temperature 16–22 °C) and can infect tissues within a few hours. Symptoms appear 20–30 days after infection. The fungus becomes less active in summer, although infections are possible over the entire vegetation period. Excoriosis is a progressive disease with outbreaks, the extent of which depends on the inoculum linked to the presence of the disease in the previous year and on varietal sensitivity. Müller-Thurgau, Kerner, Cabernet-Sauvignon, Grenache, Palomino, Muscadelle, Sultana, Zinfandel are susceptible, while Riesling, Silvaner, Merlot, Cabernet franc, Pinot, Barbera, Chardonnay, Thomson seedless are less so (Úrbez-Torres et al. 2013). The natural spread of spores in space is very limited, with conidia being released in a relatively adhesive gelatinous mass. Management systems that include a large proportion of wood over 2 years old, such as permanent cordon or spur, are particularly vulnerable to *Diaporthe* dieback, especially in cultivars and clones with fertile buds at the base of the shoots, which necessitate spur pruning. Grafting can be a major source of contamination in the nursery, with grafts taken from infected shoots able to spread the disease on a large scale and over great distances.

5.4.4   Disease Control

Prophylactic measures are important in the control of *Diaporthe* dieback, and consist in detecting the presence of pycnidia at the base of the shoots and eliminating infected material. Although shredding the shoots in the field encourages their rapid decomposition and has not been reported as a potential source of inoculum, depositing whole shoots on the ground as a preventive measure against erosion in steep vineyards can be a significant source of contamination. Manual or mechanical extraction of vine shoots from the vineyard for crushing and composting ensures the elimination of any pycnidia present. In vineyard nurseries, grafted shoots should be selected from pycnidia-free material. Routine disinfection with oxyquinoline has been shown to be effective against *Diaporthe* dieback. If early symptoms are present, as much of the affected material as possible should be removed during pruning. Heavily infected vines can be replaced or reformed from regrowth. Where direct control is required, one to two applications of fungicide are recommended, the first at BBCH 09–10, the second at 11–14. Active ingredients used against downy mildew are generally effective against excoriosis, as is wettable sulphur. If possible, fungicides should be applied before rainfall, which triggers sporulation of the fungus.

5.5 Black Foot Disease

Current names:

- *Campylocarpon fasciculare* Schroers Halleen & Crous 2004, Ascomycota
- *Campylocarpon pseudofasciculare* Halleen & Crous 2004, Ascomycota
- *Cylindrocladiella parva* (P.J. Anderson) Boesew. 1982, Ascomycota
- *Cylindrocladiella lageniformis* Crous, M.J. Wingf. & Alfenas 1993, Ascomycota
- *Cylindrocladiella peruviana* (Bat., J.L. Bezerra & M.P. Herrera) Boesew. 1982, Ascomycota
- *Dactylonectria alcacerensis* (A. Cabral, Oliveira & Crous) L. Lombard & Crous 2014, Ascomycota
- *Dactylonectria novozelandica* (A. Cabral & Crous) L. Lombard & Crous 2014, Ascomycota
- *Dactylonectria pauciseptata* (Schroers & Crous) L. Lombard & Crous 2014, Ascomycota
- *Dactylonectria pinicola* L. Lombard & Crous 2014, Ascomycota
- *Dactylonectria riojana* Berlanas Andrés-Sodupe, Ojeda & Gramaje 2019, Ascomycota
- *Dactylonectria torresensis* (A. Cabral, Rego & Crous) L. Lombard & Crous 2014, Ascomycota
- *Dactylonectria vitis* (A. Cabral, Rego & Crous) L. Lombard & Crous 2014, Ascomycota
- *Ilyonectria destructans* (Zinssm.) Rossman, L. Lombard & Crous 2015, Ascomycota
- *Ilyonectria estremocensis* A. Cabral, Nascim. & Crous 2012, Ascomycota
- *Ilyonectria europaea* A. Cabral, Rego & Crous 2011, Ascomycota
- *Ilyonectria liriodendri* A. Cabral, Nascim. & Crous 2012, Ascomycota
- *Ilyonectria lusitanica* A. Cabral, Rego & Crous 2011, Ascomycota
- *Ilyonectria novozelandica* A. Cabral & Crous 2012, Ascomycota
- *Ilyonectria pseudodestructans* A. Cabral, Rego & Crous 2011, Ascomycota
- *Ilyonectria robusta* (A.A. Hildebr.) A. Cabral & Crous 2011, Ascomycota
- *Ilyonectria venezuelensis* A. Cabral & Crous 2011, Ascomycota
- *Ilyonectria vivaria* Berlanas, López-Manzanares, Bujanda & Gramaje 2019, Ascomycota
- *Ilyonectria vredehoekensis* L. Lombard & Crous 2012, Ascomycota
- *Neonectria liriodendri* Halleen, Rego & Crous 2006, Ascomycota
- *Neonectria macrodidyma* Halleen, Schroers & Crous 2004, Ascomycota
- *Neonectria quercicola* B. Mora-Sala, A. Cabral, J. Armengol & P. Abad-Campos 2018, Ascomycota
- *Neonectria obtusispora* (Cooke & Harkness) Rossman, L. Lombard & Crous 2014, Ascomycota
- *Thelonectria blackeriella* M.L. Raimondo & A. Carlucci 2017, Ascomycota
- *Thelonectria olida* (Wollenw.) P. Chaverri & C. Salgado 2011, Ascomycota

First reported in France in 1961 and dubbed "gangrene" of grapevine (Maluta and Larignon 1991), black-foot is a soilborne disease whose fungal pathogens have frequently been isolated from nursery soils as well as nursery plants (Agustí-Brisach et al. 2014; Berlanas et al. 2017; Langenhoven et al. 2018). Like Petri disease, black-foot disease primarily attacks young vines in the nursery or after planting. This pathology is present in most vineyards worldwide and can cause significant, usually irreversible damage. The name of the disease refers to the brownish-black appearance of the rootstock heel and categorizes it as a vine trunk disease rather than a root disease, despite its impact on the entire root system.

5.5.1 Causal Organisms

Black-foot disease was originally thought to be caused primarily by two species, *Ilyonectria destructans* (asexual form *Cylindrocarpon destructans*) and *Neonectria obtusispora* (asexual morph *Cylindrocarpon obtusisporum*) (Agustí-Brisach and Armengol 2013). Since then, numerous other *Cylindrocarpon* asexual morphs (see species list above) all belonging to the family Nectriaceae (Hypocreales, Ascomycota) have been isolated from wood affected by black-foot disease (Carlucci et al. 2017; Berlanas et al. 2020). As with grapevine-wood diseases, the presence of these species in wood can be confirmed by isolating the fungi from infected wood fragments placed on agar medium. On starch-based agar, the aerial, flocculent mycelium of *I. destructans* grows relatively slowly, with the culture changing from a whitish-grey to a light-brown and then to a reddish-brown. The reverse side of the culture is reddish-brown with a yellowish margin. Elongate, straight or slightly curved, cylindrical, rounded-ended, multi-celled macroconidia (1–3 and occasionally 5 septa, measuring 45–52×6.5–7.5 μm) form on the lateral phialides in an isolated manner. Oval-to-elliptical microconidia (6–10×3.5–4 μm) as well as chlamydospores are also observed. Globose, red-to-reddish-brown perithecia (170–350×150–320 μm) surround cylindrical asci (53–83×4.5–10 μm), each of which contains eight elliptical-to-fusiform hyaline ascospores (10–13×3–3.5 μm) with a septum.

5.5.2 Symptoms

Mainly affecting young vines, black-foot disease can impact vines up until the tenth year after planting. Symptoms first appear on the aerial parts of the vine, with foliar symptoms very similar to those caused by Petri disease or abiotic disorders such as frost, water stress or nutrient deficiencies (Gramaje et al. 2018). Bud break is delayed or absent, vegetation is stunted, and leaves turn yellow and dry up during the summer, indicating root-system dysfunction.

When the base of affected plants is cleared the roots are seen to be poorly developed, tracing and growing horizontally. The first layer of roots around the root-stock heel is partially or completely necrotic, grey in colour and dark-brown-to-black once the bark is removed. A second layer of roots usually develops above the main roots at the base of the plant. These roots may remain healthy and enable the vine to survive for a more-or-less long period of time. The wood of plants affected by black foot exhibits a partial or complete necrosis of the rootstock circumference from the heel, which may extend to the graft point (Fig. 5.23). Diseased vines may be isolated or in clusters of several plants in the same row. Although randomly distributed, black-foot disease appears to be favoured by periodically wet soil conditions.

5.5.3 Biology and Epidemiology

The fungi associated with black-foot disease are saprophytes that can infect young plants or develop under temporarily unfavourable soil and climatic conditions. Fungal species associated with black-foot are cosmopolitan root fungi, which colonise a very large number of generally weakened plants. n viticulture, the vast majority of fungi are opportunistic and can frequently be isolated from wood as part of the grapevine's overall mycobiome.

Fig. 5.23 Symptom of black foot: blackish necrosis of wood of a young grapevine plant, starting from the base of the rootstock (arrows)

5.5.4   Disease Control

There are no direct control measures for black-foot disease of the vine. Prevention involves creating a favourable soil structure by avoiding excessive compaction and water accumulation. The choice of land on which to establish a vineyard or nursery is thus decisive, with poorly drained, compacted and wet soils being unsuitable. All measures that guarantee good soil structure help to prevent the development of black foot. In nurseries, the hot-water treatment used against *flavescence dorée* and *bois noir* phytoplasma seems to be effective against black foot (Halleen et al. 2007). In some countries, the grafting site is often covered with soil to stop the adjacent tissue drying up. This practice should be avoided as it is likely to increase the incidence of black-foot fungi in the grafting area, given that these fungi are soilborne. Fruit overload in the first few years of production is a source of stress for young vines and should be avoided so as to reduce the likelihood of endophytic wood fungi becoming pathogenic (Gramaje et al. 2018). The factors responsible for the emergence and expression of black-foot disease remain poorly understood.

Phytotoxins Produced by Fungi-Associated Grapevine Trunk Diseases
(Andolfi et al. 2011)

Over 60 species of fungi belonging to the Botryosphaeriaceae family along with species of the genera *Cadophora*, *Cryptovalsa*, *Cylindrocarpon*, *Diatrype*, *Diatrypella*, *Eutypa*, *Eutypella*, *Fomitiporella*, *Fomitiporia*, *Inocutis*, *Phaeoacremonium* and *Phaemoniella* have been isolated worldwide from grapevines showing symptoms of decline. To understand the mechanisms of these die-offs and their associated symptoms, the toxins produced by the fungi involved in these diseases have been isolated and characterised chemically and biologically. To date, only a few molecules have been studied. A synthesis of the work on toxins produced by the main fungi associated with grapevine wood diseases, *Eutypa lata*, *Phaeomoniella chlamydospora*, *Phaeoacremonium minimum* and some taxa belonging to the Botryosphaeriaceae family, leads to the following conclusions. Fungi produce toxins, specific or otherwise, belonging to different classes of natural compounds. These substances have been chemically characterised, at least in part, but their mode of action remains to be elucidated, except for eutypine, eutypinol, eulatinol and other metabolites produced by *Eutypa lata*. In phytotoxicity tests on leaves, callus or protoplasts, all the bioactive substances produced by *P. chlamydospora*, *P. aleophilum* and the few species of Botryosphaeriaceae were found to be toxic to grapevine or other plants, and could reproduce the original symptoms of wood diseases to some extent. At present, these findings are not considered sufficient proof that the substances play a decisive role in the death of grapevines. They may, however, cause primary damage such as browning of tissues, necrosis of parenchymatic cells, exudation of blackish gum in the vessels and occlusion of xylem elements in the wood in which they are produced, after which the molecules could then act remotely from where they were synthesised and accumulate in the leaves.

References

Afifi M, Monje MC, Legrand V, Roustan JP, Nepveu F (2004) Metabolisation of eutypine by plant tissues: an HPLC determination. Anal Chim Acta 513:21–27. https://doi.org/10.1016/j.aca.2003.10.018

Agustí-Brisach C, Armengol J (2013) Black-foot disease of grapevine: an update on taxonomy, epidemiology and management strategies. Phytopathol Mediterr 52(2):245–261. https://doi.org/10.14601/Phytopathol_Mediterr-12662

Agustí-Brisach C, Gramaje D, Garcíia-Jiménez J, Armengol J (2013) Detection of black-foot and Petri disease pathogens in natural soils of grapevine nurseries and vineyards using bait plants. Plant Soil 364:5–13. https://doi.org/10.1007/s11104-012-1333-1

Agustí-Brisach C, Mostert L, Armengol J (2014) Detection and quantification of *Ilyonectria* spp. associated with black-foot disease of grapevine in nursery soils using multiplex nested PCR and quantitative PCR. Plant Pathol 63:316–322. https://doi.org/10.1111/ppa.12093

Andolfi A, Mugnai L, Luque J, Surico G, Cimmino A, Evidente A (2011) Phytotoxins produced by fungi associated with grapevine trunk diseases. Toxins 3:1569–1605. https://doi.org/10.3390/toxins3121569

Armengol J, Vicent A, Torné L, García-Figueres F, García-Jeménez J (2001) Fungi associated with esca and grapevine declines in Spain: a three-year survey. Phytopathol Mediterr 40:S325–S329

Auger J, Esterio M, Ricke G, Pérez I (2004) Black dead arm and basal canker of *Vitis vinifera* cv. Red Globe caused by *Botryosphaeria obtusa* in Chile. Plant Dis 88(11):1286–1286. https://doi.org/10.1094/PDIS.2004.88.11.1286A

Baumgartner K, Fujiyoshi PT, Travadon R, Castlebury LA, Wilcox WF, Rolshausen PE (2013) Characterization of species of *Diaporthe* from wood cankers of grape in eastern North American vineyards. Plant Dis 97(7):912–920. https://doi.org/10.1094/PDIS-04-12-0357-RE

Berlanas C, López-Manzanares B, Gramaje D (2017) Estimation of viable propagules of black-foot disease pathogens in grapevine cultivated soils and their relation to production systems and soil properties. Plant Soil 417:467–479. https://doi.org/10.1007/s11104-017-3272-3

Berlanas C, Ojeda S, López-Manzanares B, Andrés-Sodupe M, Bujanda R, del Pilar Martínez-Díaz M, Díaz-Losada E, Gramaje D (2020) Occurrence and diversity of black-foot disease fungi in symptomless grapevine nursery stock in Spain. Plant Dis 104(1):94–104. https://doi.org/10.1094/PDIS-03-19-0484-RE

Bertsch C, Ramírez-Suero M, Magnin-Robert M, Larignon P, Chong J, Abou-Mansour E, Spagnolo A, Clément C, Fontaine F (2013) Grapevine trunk diseases: complex and still poorly understood. Plant Pathol 62:243–265. https://doi.org/10.1111/j.1365-3059.2012.02674.x

Bolay A, Carter MV (1985) Newly recorded host of *Eutypa lata* (= *E. armeniacae*) in Australia. Plant Protect Q 1(1):10–12

Bortolami G, Gambetta GA, Delzon S, Lamarque LJ, Pouzoulet J, Badel E, Burlett R, Charrier G, Cochard H, Dayer S, Jansen S, King A, Lecomte P, Lens F, Torres-Ruiz JM, Delmas CEL (2019) Exploring the hydraulic failure hypothesis of esca leaf symptom formation. Plant Physiol 181(3):1163–1174. https://doi.org/10.1104/pp.19.00591

Bortolami G, Gambetta GA, Cassan C, Dayer S, Farolfi E, Ferrer N, Gibon Y, Jolivet J, Lecomte P, Delmas CEL (2021) Grapevines under drought do not express esca leaf symptoms. Proc Natl Acad Sci U S A 118(43):e2112825118. https://doi.org/10.1073/pnas.2112825118

Bruez E, Vallance J, Gerbore J, Lecomte P, Da Costa JP, Guerin-Dubrana L, Rey P (2014) Analyses of the temporal dynamics of fungal communities colonizing the healthy wood tissues of esca leaf-symptomatic and asymptomatic vines. PLoS One 9(5):e95928. https://doi.org/10.1371/journal.pone.0095928

Bruez E, Haidar R, Alou MT, Vallance J, Bertsch C, Mazet F, Fermaud M, Deschamps A, Guerin-Dubrana L, Compant S, Rey P (2015) Bacteria in a wood fungal disease: characterization of bacterial communities in wood tissues of esca-foliar symptomatic and asymptomatic grapevines. Front Microbiol 27(6):1137. https://doi.org/10.3389/fmicb.2015.01137

Bruez E, Baumgartner K, Bastien S, Travadon R, Guérin-Dubrana L, Rey P (2016) Various fungal communities colonise the functional wood tissues of old grapevines externally free from grapevine trunk disease symptoms. Aust J Grape Wine Res 22:288–295. https://doi.org/10.1111/ajgw.12209

Bruez E, Cholet C, Thibon C, Redon P, Lacampagne S, Martignon T, Giudici M, Darriet P, Gény L (2021) Influence of curettage on esca-diseased *Vitis vinifera* L. cv. Sauvignon blanc plants on the quality of musts and wines. OENO One 55(1):171–182. https://doi.org/10.20870/oeno-one.2021.55.1.4479

Busby PE, Ridout M, Newcombe G (2016) Fungal endophytes: modifiers of plant disease. Plant Mol Biol 90(6):645–655. https://doi.org/10.1007/s11103-015-0412-0

Carlucci A, Lops F, Mostert I, Halleen F, Raimondo M (2017) Occurrence fungi causing black foot on young grapevines and nursery rootstock plants in Italy. Phytopathol Mediterr 56(1):10–39. https://doi.org/10.14601/Phytopathol_Mediterr-18769

Carter MV (1957) *Eutypa armeniacae* Hansf. & Carter, sp. Nov. an airborne vascular pathogen of *Prunus armeniaca* L. in southern Australia. Aust J Bot 5(1):21–35. https://doi.org/10.1071/BT9570021

Carter MV, Bolay A, Rappaz F (1983) An annotated list and biography of *E. armeniacae*. Review of. Plant Pathol 62(7):251–258

Casieri I., Hofstetter V, Viret O, Gindro K (2009) Effet du traitement à l'eau chaude sur les champignons associés aux jeunes plants de vigne. Rev Suisse Vitic Arboric Hortic 41(4):219–224

Chen KH, Miadlikowska J, Molnár K, Arnold AE, U'Ren JM, Gaya E, Gueidan C, Lutzoni F (2015) Phylogenetic analyses of eurotiomycetous endophytes reveal their close affinities to Chaetothyriales, Eurotiales, and a new order—Phaeomoniellales. Mol Phylogenet Evol 85:117–130. https://doi.org/10.1016/j.ympev.2015.01.008

Claverie M, Notaro M, Fontaine F, Wery J (2020) Current knowledge on grapevine trunk diseases with complex etiology: a systemic approach. Phytopathol Mediterr 59(1):29–53

Cristinzio G (1978) Gravi attacchi di *Botryosphaeria obtusa* su vite in provincia di Isernia. Inf Fitopatol 6:21–23

Del Frari G, Gobbi A, Aggerbeck MR, Oliveira H, Hansen LH, Ferreira RB (2019) Characterization of the wood mycobiome of *Vitis vinifera* in a vineyard affected by esca. Spatial distribution of fungal communities and their putative relation with leaf symptoms. Front Plant Sci 12(10):910. https://doi.org/10.3389/fpls.2019.00910

Delaye L, García-Guzmán G, Heil M (2013) Endophytes versus biotrophic and necrotrophic pathogens-are fungal lifestyles evolutionarily stable traits? Fungal Divers 60(1):125–135. https://doi.org/10.1007/s13225-013-0240-y

Dewasme C, Mary S, Roby JP (2023) Regrafting and curettage of esca-affected vines: useful control techniques but difficult to implement. https://doi.org/10.20870/IVES-TR.2023.7604

Díaz GA, Prehn D, Latorre BA (2011) First report of *Cryptovalsa ampelina* and *Eutypella leprosa* associated with grapevine trunk diseases in Chile. Plant Dis 95(4):490. https://doi.org/10.1094/PDIS-12-10-0919

Eichmeier A, Pečenka J, Hakalova E, Baránek M, García SA, Leon M, Armengol J, Gramaje D (2018) High-throughput amplicon sequencing-based analysis of active fungal communities inhabiting grapevine after hot-water treatments reveals unexpectedly high fungal diversity. Fungal Ecol 36:26–38. https://doi.org/10.1016/j.funeco.2018.07.011

Erincik O, Madden LV, Ferree DC, Ellis MA (2001) Effect of growth stage on susceptibility of grape berry and rachis tissues to infection by *Phomopsis viticola*. Plant Dis 85(5):517–520. https://doi.org/10.1094/PDIS.2001.85.5.517

Erincik O, Madden LV, Ferree DC, Ellis MA (2003) Temperature and wetness-duration requirements for grape leaf and cane infection by *Phomopsis viticola*. Plant Dis 87(7):832–840. https://doi.org/10.1094/PDIS.2003.87.7.832

Félix C, Meneses R, Gonçalves MFM, Tilleman L, Duarte AS, Jorrín-Novo JV, Van de Peer Y, Deforce D, Van Nieuwerburgh F, Esteves AC, Alves A (2019) A multi-omics analysis of the grapevine pathogen *Lasiodiplodia theobromae* reveals that temperature affects the expression of virulence- and pathogenicity-related genes. Sci Rep 9:1–12. https://doi.org/10.1038/s41598-019-49551-w

Gao Y, Liu F, Duan W, Crous PW, Cai L (2017) *Diaporthe* is paraphyletic. IMA Fungus 8(1):153–187. https://doi.org/10.5598/imafungus.2017.08.01.11

Geiger A, Karacsony Z, Golen R, Vaczy KZ, Geml J (2022) The compositional turnover of grapevine-associated fungal communities is greater among intraindividual microhabitats and terroir than among healthy and esca-diseased plants. Phytopathology 112:1029–1035

Giorni P, Bertuzzi T, Battilani P (2019) Impact of fungi co-occurrence on mycotoxin contamination in maize during the growing season. Front Microbiol 10:1265. https://doi.org/10.3389/fmicb.2019.01265

Gramaje D, Armengol J (2011) Fungal trunk pathogens in the grapevine propagation process: potential inoculum sources, detection, identification, and management strategies. Plant Dis 95(9):1040–1055. https://doi.org/10.1094/PDIS-01-11-0025

Gramaje D, Mostert L, Armengol J (2011) Characterization of *Cadophora luteo-olivacea* and *C. melinii* isolates obtained from grapevines and environmental samples from grapevine nurseries in Spain. Phytopathol Mediterr 50:S112–S126. https://doi.org/10.14601/PHYTOPATHOL_MEDITERR-8723

Gramaje D, Úrbez-Torres JR, Sosnowski MR (2018) Managing grapevine trunk diseases with respect to etiology and epidemiology: current strategies and future prospects. Plant Dis 102(1):12–39. https://doi.org/10.1094/PDIS-04-17-0512-FE

Graniti A, Surico G, Mugnai L (2000) Esca of grapevine: a disease complex or a complex of diseases. Phytopathol Mediterr 39(1):16–20

Guarnaccia V, Groenewald JZ, Woodhall J, Armengol J, Cinelli T, Eichmeier A, Ezra D, Fontaine F, Gramaje D, Gutierrez-Aguirregabiria A, Kaliterna J, Kiss L, Larignon P, Luque J, Mugnai L, Naor V, Raposo R, Sandor E, Vaczy KZ, Crous PW (2018) *Diaporthe* diversity and pathogenicity revealed from a broad survey of grapevine diseases in Europe. Persoonia 40:135–153. https://doi.org/10.3767/persoonia.2018.40.06

Halleen F, Fourie PH, Crous P (2006) A review of black foot disease of grapevine. Phytopathol Mediterr 45:S55–S67

Halleen F, Fourie PH, Crous P (2007) Control of black foot disease in grapevine nurseries. Plant Pathol 56:637–645. https://doi.org/10.1111/j.1365-3059.2007.01613.x

Harrington TC, McNew DL (2003) Phylogenetic analysis places the *Phialophora*-like anamorph genus Cadophora in the Helotiales. Mycotaxon 87:141–151

Hofstetter V, Buyck B, Croll D, Viret O, Couloux A, Gindro K (2012) What if esca disease of grapevine were not a fungal disease? Fungal Divers 54:51–67. https://doi.org/10.1007/s13225-012-0171-z

Jayawardena RS, Purahong W, Zhang W, Wubet T, Li X, Liu M, Zhao W, Hyde KD, Liu J, Yan J (2018) Biodiversity of fungi on *Vitis vinifera* L. revealed by traditional and high-resolution culture-independent approaches. Fungal Divers 90:1–84. https://doi.org/10.1007/s13225-018-0398-4

Kogel KH, Franken P, Hückelhoven R (2006) Endophyte or parasite—what decides? Curr Opin Plant Biol 9(4):358–363. https://doi.org/10.1016/j.pbi.2006.05.001

Kuntzmann P, Villaume S, Larignon P, Bertsch C (2010) Esca, BDA and eutypiosis: foliar symptoms, trunk lesions and fungi observed in diseased vinestocks in two vineyards in Alsace. Vitis 49(2):71–76

Lafon R (1921) Modifications à apporter à la taille de la vigne dans les Charentes, taille Guyot-Poussard mixte et double. L'apoplexie, traitement préventif (méthode Poussard), traitement curatif. Work published under the direction of MM. James HENNESSY, Sénateur de la Charente and Jean HENNESSY, Député de la Charente, Roumégous et Déhan Publishers, Montpellier, 96 pages

Langenhoven SD, Halleen F, Spies CFJ, Stempien E, Mostert L (2018) Detection and quantification of black foot and crown and root rot pathogens in grapevine nursery soils in the Western Cape of South Africa. Phytopathol Mediterr 57(3):519–537. https://doi.org/10.14601/Phytopathol_Mediterr-23921

Larignon P, Dubos B (2001) Le Black Dead Arm. Maladie nouvelle à ne pas confondre avec l'esca. Phytoma 538:26–29

Larignon P, Fulchic R, Cere L, Dubos B (2001) Observation on black dead arm in French vineyards. Phytopathol Mediterr 40:S336–S342

Leavitt GM (1990) The occurrence, distribution, effects and control of *Botryodiplodia theobromae* on *Vitis vinifera* in California, Arizona and northern Mexico. Ph.D. dissertation, University of California, Riverside, CA, USA

Lehoczky J (1974a) Black dead-arm disease of grapevine caused by *Botryosphaeria stevensii* infection. Acta Phytopathol 9:319–327

Lehoczky J (1974b) Necrosis of nurseried grapevine grafts of *Botryosphaeria stevensii* infection. Acta Phytopathol 9:329–331

Linaldeddu BT, Deidda A, Scanu B, Franceschini A, Serra S, Berraf-Tebbal A, Boutiti MZ, Jamâa MLB, Phillips AJL (2015) Diversity of Botryosphaeriaceae species associated with grapevine and other woody hosts in Italy, Algeria and Tunisia, with descriptions of *Lasiodiplodia exigua* and *Lasiodiplodia mediterranea* sp. nov. Fungal Divers 71:201–214. https://doi.org/10.1007/s13225-014-0301-x

Luque J, Martos S, Aroca A, Raposo R, Garcia-Figueres F (2009) Symptoms and fungi associated with declining grapevine plants in northern Spain. J Plant Pathol 91(2):381–390. https://doi.org/10.14601/Phytopathol_Mediterr-1621

Luque J, Garcia-Figueres F, Legorburu FJ, Muruamendiaraz A, Armengol J, Trouillas FP (2012) Species of Diatrypaceae associated with grapevine trunk diseases in eastern Spain. Phytopathol Mediterr 51(3):528–540. https://doi.org/10.14601/PHYTOPATHOL_MEDITERR-9953

Magan N, Medina A (2016) Integrating gene expression, ecology and mycotoxin production by *Fusarium* and *Aspergillus* species in relation to interacting environmental factors. World Mycotoxin J 9(5):673–684. https://doi.org/10.3920/WMJ2016.2076

Maldonado-González MM, Del Pilar Martínez-Diz M, Andrés-Sodupe M, Bujanda R, Díaz-Losada E, Gramaje D (2020) Quantification of *Cadophora luteo-olivacea* from grapevine nursery stock and vineyard soil using droplet digital PCR. Plant Dis 104(8):2269–2274. https://doi.org/10.1094/PDIS-09-19-2035-RE

Maluta DR, Larignon P (1991) Pied-noir: mieux vaut prévenir. Viticulture 159:71–72

Martos S, Andolfi A, Luque J, Mugnai L, Surico G, Evidente A (2008) Production of phytotoxic metabolites by five species of Botryosphaeriaceae causing decline of grapevines, with special interest in the species *Neofusicoccum luteum* and *N. parvum*. Eur J Plant Pathol 121(4):451–461. https://doi.org/10.1007/s10658-007-9263-0

Mishra S, Bhattacharjee A, Sharma S (2021) An ecological insight into the multifaceted world of plant-endophyte association. Crit Rev Plant Sci 40(2):127–146. https://doi.org/10.1080/07352689.2021.1901044

Monod V (2024) Deciphering the multifaceted determinants of esca incidence across a vineyards network. PhD Thesis of the University of Neuchâtel, Faculty of Science, Switzerland

Monod V, Zufferey V, Wilhelm M, Viret O, Gindro K, Croll D, Hofstetter V (2023) A systemic approach allows to identify the pedoclimatic conditions most critical in the susceptibility of a grapevine cultivar to esca/*Botryosphaeria* dieback. bioRxiv 2023.05.23.541976. https://doi.org/10.1101/2023.05.23.541976

Mostert L, Crous PW, Petrini O (2000) Endophytic fungi associated with shoots and leaves of *Vitis vinifera*, with specific reference to the *Phomopsis viticola* complex. Sydowia 52(1):46–58

Mostert L, Crous PW, Kang JC, Phillips AJL (2001) Species of *Phomopsis* and a *Libertella* sp. occurring on grapevines with specific reference to South Africa: morphological, cultural, molecular and pathological characterization. Mycologia 93:146–167. https://doi.org/10.1080/00275514.2001.12061286

Mugnai L, Graniti A, Surico G (1999) Esca (black measles) and brown wood-streaking: two old and elusive diseases of grapevines. Plant Dis 83(5):404–418. https://doi.org/10.1094/PDIS.1999.83.5.404

Munkvold GP, Duthie JA, Marois JJ (1994) Reductions in yield and vegetative growth of grapevines due to *Eutypa* dieback. Phytopathologia 84:186–192. https://doi.org/10.1094/Phyto-84-186

Nazar Pour F, Ferreira V, Félix C, Serôdio J, Alves A, Duarte AS, Esteves AC (2020) Effect of temperature on the phytotoxicity and cytotoxicity of Botryosphaeriaceae fungi. Fungal Biol 124(6):571–578. https://doi.org/10.1016/j.funbio.2020.02.012

Pearson RC, Goheen C (1994) Phomopsis cane and leaf spot. In: Hewitt WB, Pearson RC (eds) Compendium of grape diseases. APS Press, St Paul, MI, pp 17–18

Phillips AJL (2000) Excoriose, cane blight and related diseases of grapevines: a taxonomic review of the pathogens. Phytopathol Mediterr 39:341–356. https://doi.org/10.14601/Phytopathol_Mediterr-1583

Ramsing CK, Gramaje D, Mocholí S, Agustí J, Cabello Sáenz de Santa María F, Armengol J, Berbegal M (2021) Relationship between the xylem anatomy of grapevine rootstocks and their susceptibility to *Phaeoacremonium minimum* and *Phaeomoniella chlamydospora*. Front Plant Sci 12:726461. https://doi.org/10.3389/fpls.2021.726461

Réblová M, Mostert L, Gams W, Crous PW (2004) New genera in the Calosphaeriales: *Togniniella* and its anamorph *Phaeocrella*, and *Calosphaeriophora* as anamorph of *Calosphaeria*. Stud Mycol 50:533–550

Reveglia P, Billones-Baaijens R, Millera Niem J, Masi M, Cimmino A, Evidente A, Savocchia S (2021) Production of phytotoxic metabolites by Botryosphaeriaceae in naturally infected and artificially inoculated grapevines. Plan Theory 10:802. https://doi.org/10.3390/plants10040802

Rolshausen P, Baumgartner K, Bergemann S, Fujiyoshi PT, Gubler W, Wilcox W (2008) Geographical diversity of the grapevine pathogen *Eutypa lata* in North American vineyards. Phytopathology 98:S135

Rolshausen PE, Baumgartner K, Travadon R, Fujiyoshi P, Pouzoulet J, Wilcox WF (2014) Identification of *Eutypa* spp. causing *Eutypa* dieback of grapevine in eastern North America. Plant Dis 98(4):483–491. https://doi.org/10.1094/PDIS-08-13-0883-RE

Romeralo C, Martín-García J, Martínez-Álvarez P, Muñoz-Adalia EJ, Gonçalves DR, Torres E, Witzell J, Diez JJ (2022) Pine species determine fungal microbiome composition in a common garden experiment. Fungal Ecol 56:101137. https://doi.org/10.1016/j.funeco.2021.101137

Rovesti L, Montermini A (1987) Un deperimento della vite causato da *Sphaeropsis malorum* diffuso in provincia di Reggio Emilia. Inf Fitopatol 1:59–61

Rudelle J, Octave S, Kaid-Harche M, Roblin G, Fleurat-Lessard P (2005) Structural modifications induced by *Eutypa lata* in the xylem of trunk and canes of Vitis vinifera. Funct Plant Biol 6:537–547. https://doi.org/10.1071/FP05012

Schulz B, Boyle C (2005) The endophytic continuum. Mycol Res 109(6):661–686. https://doi.org/10.1017/S095375620500273X

Selosse MA, Baudoin E, Vandenkoornhuyse P (2004) Symbiotic microorganisms, a key for ecological success and protection of plants. C R Biol 327(7):639–648. https://doi.org/10.1016/j.crvi.2003.12.008

Sun Q, Rost TL, Matthews MA (2006) Pruning-induced tylose development in stems of current-year shoots of *Vitis vinifera* (Vitaceae). Am J Bot 93(11):1567–1576. https://doi.org/10.3732/ajb.93.11.1567

Sun Q, Rost TL, Matthews MA (2008) Wound-induced vascular occlusions in *Vitis vinifera* (Vitaceae): Tyloses in summer and gels in winter. Am J Bot 95(12):1498–1505. https://doi.org/10.3732/ajb.0800061

Surico G (2009) Towards a redefinition of the diseases within the esca complex of grapevine. Phytopathol Mediterr 48(1):5–10. https://doi.org/10.14601/Phytopathol_Mediterr-2870

Surico G, Mugnai L, Marchi G (2006) Older and more recent observations on esca: a critical overview. Phytopathol Mediterr 45:S68–S86. https://doi.org/10.14601/Phytopathol_Mediterr-1847

Taylor A, Hardy GESJ, Wood P, Burgess T (2005) Identification and pathogenicity of *Botryosphaeria* species associated with grapevine decline in Western Australia. Australas Plant Pathol 34:187–195. https://doi.org/10.1071/AP05018

Tey-Rulh P, Philippe I, Renaud JM, Tsoupras G, De Angelis P, Fallot J, Tabacchi R (1991) Eutypine, a phytotoxin produced by *Eutypa lata* the causal agent of dying-arm disease of grapevine. Phytochemistry 30(2):471–473. https://doi.org/10.1016/0031-9422(91)83707-R

Travadon R, Baumgartner K (2015) Molecular polymorphism and phenotypic diversity in the generalist, wood-decay fungus *Eutypa lata*. Phytopathology 105:255–164. https://doi.org/10.1094/PHYTO-04-14-0117-R

Trivedi P, Leach JE, Tringe SG, Sa T, Singh BK (2020) Plant–microbiome interactions: from community assembly to plant health. Nat Rev Microbiol 18:607–621. Available from: http://www.nature.com/articles/s41579-020-0412-1

Trouillas FP, Gubler WD (2010) Host range, biological variation, and phylogenetic diversity of *Eutypa lata* in California. Phytopathology 100(10):1048–1056. https://doi.org/10.1094/PHYTO-02-10-0040

Trouillas FP, Urbez-Torres JR, Gubler WD (2010) Diversity of diatrypaceous fungi associated with grapevine canker diseases in California. Mycologia 102(2):319–336. https://doi.org/10.3852/08-185. PMID: 20361500

Trouillas FP, Pitt WM, Sosnowski MR, Huang RJ, Peduto F, Loschiavo A, Savocchia S, Scott ES, Gubler WD (2011) Taxonomy and DNA phylogeny of Diatrypaceae associated with *Vitis vinifera* and other woody plants in Australia. Fungal Divers 49:203–223. https://doi.org/10.1007/s13225-011-0094-0

Udayanga D, Castlebury LA, Rossman AY, Chukeatirote E, Hyde KD (2015) The *Diaporthe sojae* species complex: phylogenetic re-assessment of pathogens associated with soybean, cucurbits and other field crops. Fungal Biol 119(5):383–407. https://doi.org/10.1016/j.funbio.2014.10.009

Úrbez-Torres JR (2011) The status of Botryosphaeriaceae species infecting grapevines. Phytopathol Mediterr 50:S5–S45. https://doi.org/10.14601/Phytopathol_Mediterr-9316

Úrbez-Torres JR, Leavitt GM, Voegel TM, Gubler WD (2006) Identification and distribution of *Botryosphaeria* spp. associated with grapevine cankers in California. Plant Dis 90(12):1490–1503. https://doi.org/10.1094/PD-90-1490

Úrbez-Torres JR, Peduto F, Smith RJ, Gubler WD (2013) Phomopsis dieback: a grapevine trunk disease caused by *Phomopsis viticola* in California. Plant Dis 97(12):1571–1579. https://doi.org/10.1094/PDIS-11-12-1072-RE

Valtaud C, Larignon P, Roblin G, Fleurat-Lessard P (2009) Developmental and ultrastructural features of *Phaeomoniella chlamydospora* and *Phaeoacremonium aleophilum* in relation to xylem degradation in esca disease of the grapevine. J Plant Path 91(1):37–51. https://doi.org/10.4454/jpp.v91i1.622

Van Niekerk JM, Crous PW, Groenewald JZ, Fourie PH, Halleen F (2004) DNA phylogeny, morphology and pathogenicity of *Botryosphaeria* species on grapevines. Mycologia 96(4):781–798. https://doi.org/10.1080/15572536.2005.11832926

Van Niekerk JM, Groenewald JZ, Farr DF, Fourie PH, Halleen F, Crous P (2005) Reassessment of Phomopsis species on grapevines. Aust Plant Pathol 34:27–39. https://doi.org/10.1071/AP04072

Viret O, Petrini O (1994) Colonization of beech leaves (*Fagus sylvatica*) by the endophyte *Discula umbrinella* (teleomorph: *Apiognomonia errabunda*). Mycol Res 98(4):423–432. https://doi.org/10.1016/S0953-7562(09)81200-8

Viret O, Scheidegger C, Petrini O (1993) Infection of beech leaves (*Fagus sylvatica*) by the endophyte *Discula umbrinella* (teleomorph: *Apiognomonia errabunda*): low-temperature scanning electron microscopy studies. Can J Bot 71:1520–1527

Wagschal I, Abou-Mansour E, Petit AN, Clément C, Fontaine F (2008) Chapter 16: Wood diseases of grapevine: a review on eutypa dieback and esca. In: Barka EA, Clément C (eds) Plant-microbe interactions. Research Signpost, pp 367–391. ISBN: 9788130802121

Wilcox WA (2015) *Phomopsis* cane and leaf spot. In: Wilcox WF, Gubler WD, Uyemoto JK (eds) Compendium of grape diseases, disorders, and pests, 2nd edn. APS, St. Paul, MN. https://doi.org/10.1094/9780890544815

Chapter 6
Root Diseases

6.1 *Armillaria* Root Rot

Current name:

Armillaria mellea (Vahl) Kumm. 1871, Basidiomycota, syn. *Armillariella mellea* (Vahl) Karst. 1881.

Armillaria is a globally distributed genus of basidiomycetes that includes plant-pathogenic species of economic importance for forestry and agronomic systems. In recent decades this genus has garnered particular attention for its role in the decomposition of woody plants as well as for its involvement in mycorrhizal symbioses with a number of plants (Koch and Herr 2021). Species of this genus spread through root contact or through rootlike structures called rhizomorphs (flattened and sclerotised hyphal bundles), which may form networks hundreds of metres long and can penetrate the root bark of hosts (Calamita et al. 2021). Among *Armillaria* species, *A. mellea* is considered a primary parasite of stressed woody plants and is the species responsible for grapevine root rot. As with many other fungal species, climate change is thought to exacerbate the damage caused by this species, especially with agricultural intensification (La Porta et al. 2008). Widespread in temperate and tropical climates, the fungus *A. mellea* grows on several hundred different host plants. This root pathogen can be very virulent to grapevines. Its persistence is due to its saprophytic nature, which allows it to live for several years on wood debris in the soil. Wooden stakes can also act as a substrate and allow it to infect adjacent vine roots. In general, *A. mellea* is not a problem in vineyards where preventive measures have been taken before planting. If soil preparation has been neglected, however, it can cause significant economic losses. The proximity of a forest can also favours its appearance. *Armillaria* carpophores appear in groups at the foot of affected vines in autumn, when conditions are favourable. They are found in particular around the stumps of trees felled in the forest or at the base of infected trunks (Fig. 6.1).

O. Viret, K. Gindro, *Science of Fungi in Grapevine*,
https://doi.org/10.1007/978-3-031-68663-4_6

Fig. 6.1 *Armillaria* carpophores in a clump at the foot of a tree in a forest (**a**) and detail view of carpophores in a vineyard (**b**)

6.1.1   Causal Organism

Armillaria mellea is a fungus of the lineage Basidiomycota, order Agaricales and family Physalacriaceae. In keeping with its common name of 'honey fungus', in autumn it forms honey-brown-to-dark-brown, reddish-to- olivaceous (darker towards the centre) carpophores of medium size, with a slightly scaly cap 5 to 10–15 cm in diameter, spherical and closed by a white veil (Fig. 6.2), which then spreads. The lamellae are uneven, white then yellowish, and produce white spores. The stem, 10–20 cm long, is slender and rigid, the same colour as the cap, its top streaked with a whitish collar above and speckled with white below. *A. mellea* is an extremely polymorphic species with highly variable features. This fungus also forms rigid rhizomorphs. In nature, these are often dark-brown-to-black, leathery, flat and naked, growing under the bark of the host plant (Fig. 6.3). In culture, rhizomorphs remain very pale and extremely hairy (Fig. 6.4), and their structure is distinguished by a cortex composed of thin-walled, cylindrical hyphae growing parallel to the surface of the medium. Inside the rhizomorph, the pith consists of large physalohyphae. An adult rhizomorph is made up of hundreds of tubular hyphae. On woody plants, rhizomorphs grow up the trunk from the soil between the bark and the wood. These mycelial structures also enable the fungus to spread in the soil litter and to survive long-term in extreme conditions.

6.1.2   Symptoms

Vines affected by *Armillaria* root rot show decreased vigour and stunted, pale or yellowing leaves that progressively lose their turgor (Fig. 6.5), as well as a higher number of side shoots and stunted, unlined shoots (Baumgartner and Rizzo 2002).

This disease is never uniform across a plot. Branch growth is markedly reduced and the leaves generally fall prematurely. After a period of water stress, the disease

Fig. 6.2 *Armillaria mellea* young carpophores with their white veil partially obscuring the gills of the mushrooms

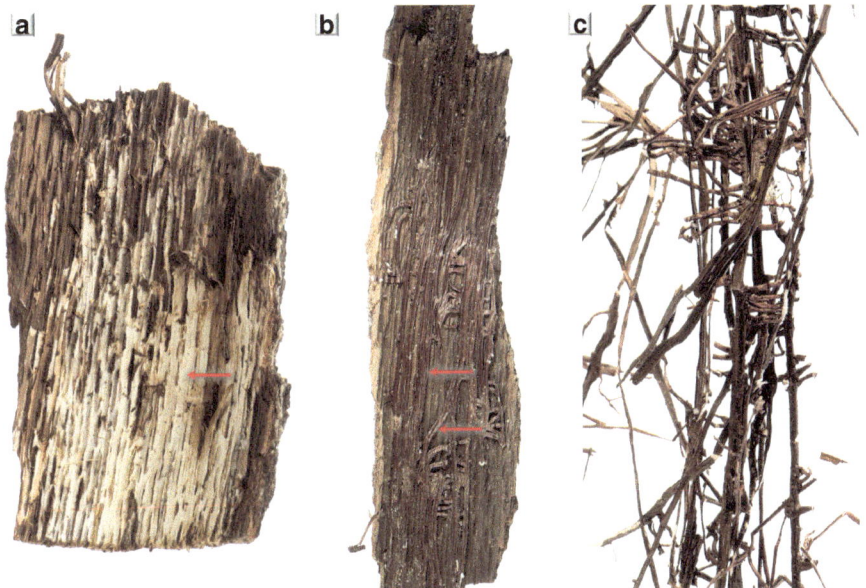

Fig. 6.3 Creamy-white mycelium (arrow) (**a**) and dark-brown sclerified rhizomorphs (**b**) of *Armillaria* (arrows) under the bark. (**c**) Rhizomorphs detached from the bark

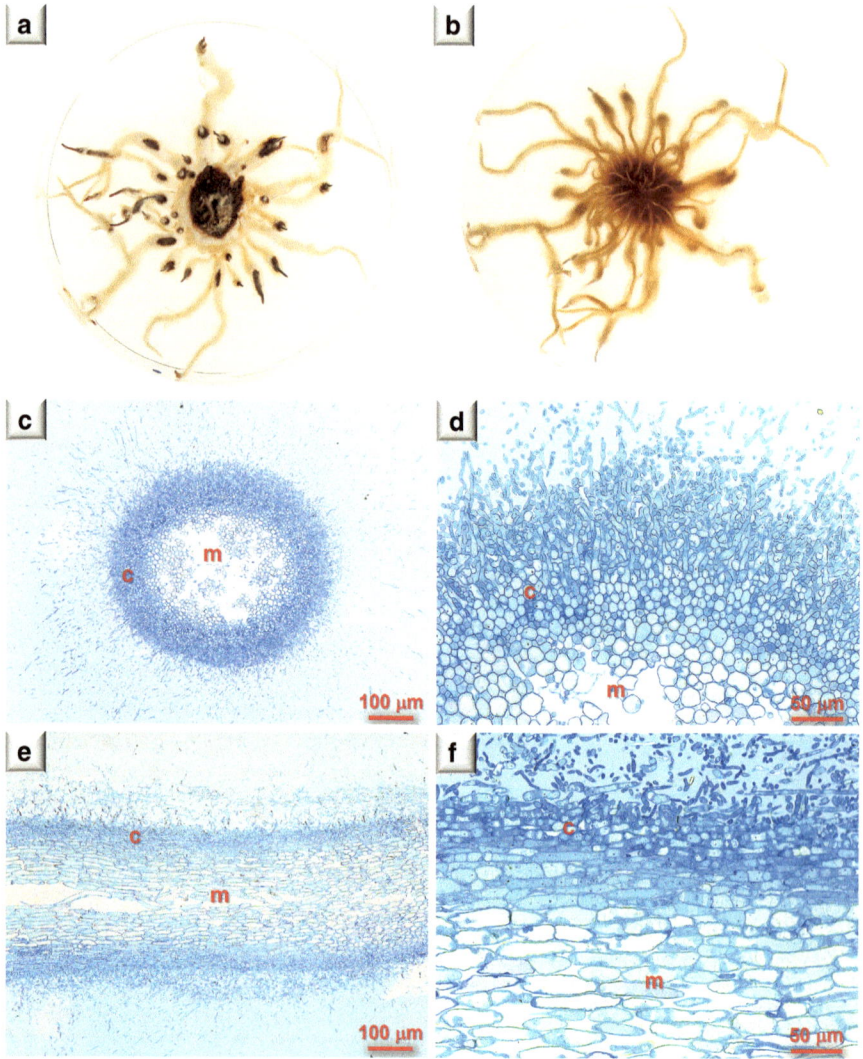

Fig. 6.4 Anatomy of *Armillaria mellea* rhizomorphs. (**a**) Rhizomorphs cultured on an agar medium, top view. (**b**) Rhizomorphs cultured on an agar medium, bottom view. (**c, d**) Cross section of rhizomorphs. (**e, f**) Longitudinal section. *c* cortex, *m* medulla

sometimes manifests in the sudden apoplexy of the plants. Infected plants may fail for several years, producing no or very few grapes and exhibiting a stunted vegetation indicative of dysfunction of the vascular system. *Armillaria mellea* infects the roots and degrades the woody tissues, leading to vine dieback. Clearing the base of affected plants reveals an abundant whitish mycelium completely surrounding the base of the trunk or main roots (Fig. 6.6). The mycelium may be on the surface or under the filaments of the vine bark. A characteristic odour of button mushrooms

Fig. 6.5 Grapevine
dieback (left) due to
Armillaria; the foliage
fades and shoots remain
stunted

Fig. 6.6 Signs bearing
witness to the presence of
Armillaria on a vine
showing a whitish-grey
mycelium on a main root

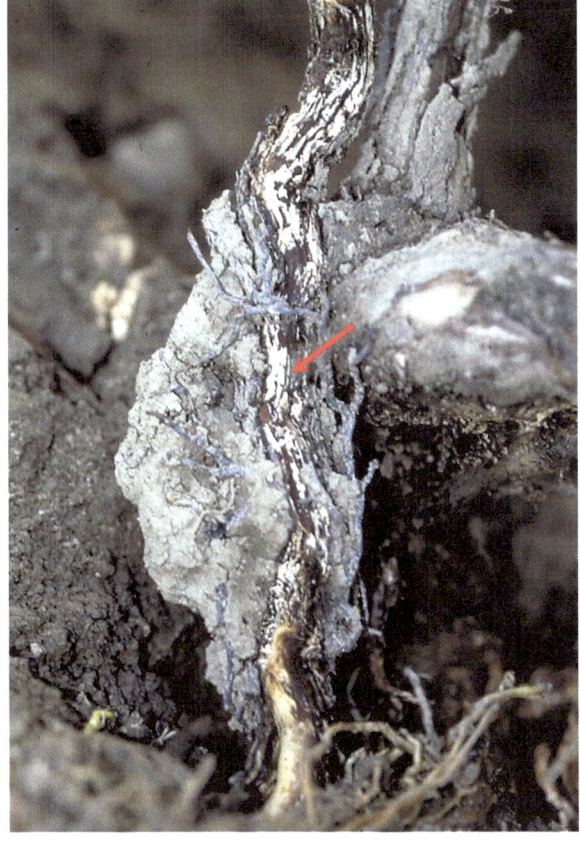

Fig. 6.7 Carpophore of *Armillaria mellea* at the foot of an affected vine

Fig. 6.7 Carpophore of *Armillaria mellea* at the foot of an affected vine

emanates from the affected tissues. Infected roots are dark, easily detached from the soil, with a fibrous consistency. Dark-brown rhizomorphs can be found in the roots and soil. In autumn, the appearance of carpophores clustered at the base of the vines is strong evidence of *Armillaria* root rot (Fig. 6.7).

6.1.3   Biology and Epidemiology

Fanning out under the bark and rhizomorphs, the creamy-white mycelium can infect stumps up to several tens of centimetres above the ground. Rhizomorphs may also be epiphytic and grow in the soil, in which case they assume a fairly cylindrical shape in cross section. The entire cycle of infection takes place in the soil. The fungus hibernates as a saprophyte on woody waste. When in their vicinity, it attaches to living roots via a mucilage and then directly penetrates the bark and xylem before invading the plant tissue. The genus *Armillaria sensu lato* and, more specifically, the species *A. mellea*, *Desarmillaria tabescens* and *A. lutea*, are fungal species that have been studied due to their bioluminescence.

Bioluminescence of the Genus Armillaria
(Baumgartner et al. 2011; Calamita et al. 2021; Purtov et al. 2017)

Bioluminescence is present in several fungal genera all of which belong to the order Agaricales (Basidiomycota). The reason for fungal bioluminescence remains a mystery. Several hypotheses have been advanced that ascribe an ecological function to this characteristic, specifically, that of attracting invertebrate organisms or fungivorous predators involved in spore dispersion (Calamita et al. 2021). Another hypothesis is that the release of light rather than heat might be a side-effect of the detoxification of peroxides formed during the degradation of wood lignin (Baumgartner et al. 2011). *A. mellea* is only partially luminous, as its carpophores are non-luminescent. Comparison of the elements involved in the bioluminescence process in both the luminous and non-luminous organs of this species (Purtov et al. 2017) led to the discovery that not all enzymes and substrates required for bioluminescence are present in fruiting bodies. Because the synthesis of the precursor of luciferin (hispidin) and hispidin 3-hydroxylase (the penultimate enzyme in the luminescent pathway that catalyses the conversion of hispidin to 3-hydroxyhispidin, the substrate of luciferin) is blocked in the carpophores, *A. mellea* is only partially bioluminescent, despite the presence in the fruiting bodies of luciferase, the enzyme using luciferin as a substrate and making it luminescent by oxidation in the presence of oxygen.

6.1.4 Disease Control

Direct treatment by carbon disulphide fumigation has long been prohibited. Control is only prophylactic and must be carried out before planting. Tillage is used to remove root residues and other woody debris from the previous crop. Preventive measures consist in creating a favourable soil structure by avoiding excessive compaction and temporary accumulations of water. Plots must be carefully selected and tilled to avoid *Armillaria* root rot because once an outbreak of this disease emerges in a producing vineyard it is very difficult to control. Young replacement plants often perish after a few years because the soil cannot be tilled deeply to eliminate all residues of the infected rootlets.

6.2 *Dematophora* Root Rot (White or Woolly Root Rot)

Current name:

Dematophora necatrix R. Hartig 1883, Ascomycota, syn. *Rosellinia necatrix* Berl. ex Prill. 1904, *Hypoxylon necatrix* (Berl. ex Prill.) P.M.D. Martin 1976.

Dematophora root rot is a disease caused by a soilborne fungus not specific to grape-vines, *Dematophora necatrix*. To date, the Systematic Botany and Mycology Laboratory of the United States Department of Agriculture has established a list of 437 host plants of this fungus (Pasini et al. 2016). The list includes mainly woody plants, but also herbaceous species. Many of the host plants—apple, pear, orange, avocado, almond, cherry, coffee, olive and grapevine, among others—are of significant economic importance. *D. necatrix* causes sizeable economic losses in orchards and vineyards. In general, this root rot has long been considered a secondary pathogen of minimal economic importance, due to its saprophytic behaviour; however, studies have shown that on some host plants, including grapevine, the fungus can be virulent and may play a crucial role as a primary pathogen (Pérez-Jiménez 2006; Schena et al. 2008).

6.2.1   Causal Organism

A member of the Ascomycetes, *Dematophora necatrix* belongs to the order Xylariales and the family Xylariaceae. This fungal species forms perithecia 0.9–1.8 mm in diameter and 1.4–1.6 mm in height embedded in a black stroma on the surface of infected roots. The young perithecia are gelatinous in appearance, initially honey-coloured and gradually turning brownish-black. These structures contain filiform (155–230 × 7–10 µm), unitunicate asci with ellipsoid and cymbi-form (boat-shaped) ascospores (31–49 × 5.4–10 µm). The ascospores are initially hyaline with numerous vacuoles and granules and become dark brown at maturity. In culture, the fungus produces a fluffy white mycelium turning dark-grey-to-black and with a crusty appearance. The life cycle of *D. necatrix* also encompasses two different types of asexual spores: chlamydospores and conidiospores. Produced only in exceptional environmental conditions, chlamydospores are rarely observed in natural or artificial conditions (Pérez-Jiménez et al. 2003). Nearly spherical and measuring 15 mm in diameter, these spores are produced by condensation of the pyriform swellings of the protoplasm and the subsequent formation of a cell wall. The conidia are borne at the tip of the synnemata of the conidiogenous cells, which are produced either from sclerotia or from brown mycelial masses. The synnemata range between 0.5–1.5 mm in length. Conidia are solitary, unicellular, hyaline, elliptical (3–5 × 2.5–3 µm), and borne both apically and laterally to conidiogenous cells (Petrini 1993).

6.2.2   Symptoms

A vineyard's vegetative symptoms of decline due to *Dematophora* root rot are identical to those caused by *Armillaria* root rot: the aerial organs are poorly developed, the vine is stunted and some shoots die during growth. In other cases the plant is stricken by apoplexy and dies. These symptoms are easily confused with those of other root rots, as well as with wood diseases or abiotic or biotic problems of the vine (chlorosis, mineral deficiencies). When clearing the underground parts of the vines, a cottony greyish-white spider-web-like mycelial network covering the roots and penetrating the surrounding soil particles can be seen. In the bark, it forms cottony white rhizomorphs arranged in a spider's web, which later turn grey-green-to-black. The fruiting bodies of *D. necatrix* are embedded in a blackish mass on the root surface and are difficult to identify with the naked eye. Appearing only temporarily during the fungal development cycle, they may be absent during root removal and disposal of dead stock.

6.2.3   Biology and Epidemiology

Dematophora necatrix is a polyphagous facultative parasite capable of surviving for several years in the soil in the absence of a host. The fungus is dispersed in the soil either by mycelium and mycelial filaments, or along infected roots. Upon contact with a healthy root, the mycelial network proliferates, first covering the root surface in a diffuse manner, then forming aggregates on the root surface. Root penetration by *D. necatrix* occurs at these sites, through natural openings (lenticels), wounds, or directly through the formation of a penetrating sclerotia. Primary infection occurs simultaneously at several random positions along the longitudinal axis of the roots. Hyphae invade and penetrate the primary and secondary xylem, progressively destroying the vascular system (Pliego et al. 2012). The mycelial growth of *D. necatrix* is dependent on temperature, oxygen, moisture, organic matter content, pH and soil microflora. Heavy soils with a high water-retention capacity, a pH between 5 and 7, a temperature of 20–25 °C and/or a high organic-matter content are favourable environments for the development of this pathogen. The epidemiological role of the three spore types produced by *D. necatrix* remains unclear, as the mycelium and aggregated organs appear capable of performing the entire cycle of infection (Pliego et al. 2012).

6.2.4   Disease Control

The prophylactic control measures recommended for *Armillaria* root rot also apply to *Dematophora* root rot. No effective direct chemical or biological control method has been identified to date. Given that the soil is a highly complex biological system, the multiple interactions between microorganisms together with the various pedoclimatic factors provide scant opportunities to find appropriate solutions for all situations in the quest to control soil fungi.

6.3   *Roesleria* Root Rot

Current name:

Roesleria subterranea (Weinm.) Redhead 1985, Ascomycota, syn. *Roesleria hypogaea* Thüm. & Pass 1887.

Roesleria subterranea is a poorly studied soilborne fungus whose systematic placement has long remained uncertain (Kirchmair et al. 2008). It owes its French name, *pourridié morille* (lit. 'morel root rot'), to its fruiting bodies, which resemble tiny morel mushrooms on dead roots (Fig. 6.8). Described as occurring on several genera of fruit trees (e.g. *Cydonia*, *Malus*, *Pyrus*, *Prunus*) and on other woody plant genera (e.g. *Rosa*, *Salix*, *Populus*, *Tilia*), it is also found on the roots of apoplectic or dying vines. Although long considered a facultative parasite of the vine, its decline being attributed to a weakness or physiological imbalance of the plants, this fungus has also been shown to infect healthy roots (Höfer 1992). Since publication of the study of Neuhauser et al. (2011), it has been considered a primary and emerging pathogen of grapevine. *Roesleria* root rot dieback has been reported in northern vineyards in Austria, Germany, Hungary, Scotland, Luxembourg, France, Canada, as well as in New Zealand (Neuhauser et al. 2011).

6.3.1   Causal Organism

Originally called *Roesleria hypogea* by Thümen in 1877, *Roesleria subterranea* is an Ascomycete of the order Helotiales and the family Helotiaceae (Kirchmair et al. 2008). The fruiting bodies of *R. subterranea* are solitary, nailhead-shaped apothe-

Fig. 6.8 Nail-shaped fruiting bodies of *Roesleria subterranea* on a root. Inset: enlarged view of fruiting bodies

cium stromatocarps 2–20 mm in length, comprising a cluster of compact parallel hyphae formed on infected roots and rootlets. The apothecia contain cylindrical asci up to 50 × 10 μm in size, each containing 8 elliptical, hyaline-to-pale-greyish-green ascospores (5–6 × 4–5 μm). In pure culture, the mycelium can be recognised by its characteristic green pigmentation.

6.3.2   Symptoms

The vegetative symptoms of *Roesleria* root rot are indistinguishable from those of other root rots: stunting and partial discoloration of foliage, and atrophied branches that eventually dry up during the summer. The infection process may take several years but systematically results in the death of the infected vines. Apoplexy of individual branches or whole plants is frequently observed. When the roots are extracted, the rot is easily identified by the absence of rootlets and the presence of fungal fruiting bodies on the main roots, small nailhead-shaped carpophores atop a several-millimetre-tall stipe containing the apothecia (Fig. 6.8). The symptoms of the aerial parts of the plant resemble those of other root diseases, or even abiotic or biotic problems of the vine (chlorosis, mineral deficiencies, wood diseases).

6.3.3   Biology and Epidemiology

R. subterranea can live as a saprophyte on dead wood, but also infects healthy plants. Fruiting bodies of the fungus are usually produced from autumn to spring, or even year-round if conditions are favourable (cool, wet summers, moist soils). The mycelium can survive several years in the soil on woody debris or in organic matter. As with other root fungi, root infection begins with the invasion of the cortex and vascular system. *Roesleria* root rot can survive highly variable temperature (−3 to +35 °C, optimum range 15–20 °C), pH (2.5–8.5) and soil-moisture (water-retention capacity 10%–80%) conditions, which explains its adaptation to viticultural soils in northern regions (Höfer 1992).

6.3.4   Disease Control

The prophylactic control measures recommended for *Armillaria* and *Dematophora* root rot also apply to *Roesleria* root rot. No effective direct chemical or biological control methods have been identified to date.

6.4   *Phymatotrichopsis* Root Rot (Cotton Root Rot)

Current name:

Phymatotrichopsis omnivora (Shear) Hennebert 1973, Ascomycota, syn. *Phymatotrichum omnivorum* (Shear) Duggar 1916.

Phymatotrichopsis root rot of grape is commonly referred to as cotton root rot, Texas root rot or *Phymatotrichum* root rot. Like other root rots, this disease has a very wide range of hosts, including over 2000 species of dicotyledons. In Europe the causal fungus has been wrongly listed as harmful organism under *Trechispora brinkmannii* (or *Sistotrema brinkmannii*) a basidiomycete fungus. Evidence has since been produced that *P. omnivora* is an ascomycete species (Marek et al. 2009). This root rot is destructive to many economically important crops, including grapes on clay-limestone soils with high pH's (between 7.0 and 8.5), as is the case in central and southern Arizona where it is responsible for significant economic losses. Present in the southwestern United States, northern Mexico, Libya and Venezuela, it has also been reported in Europe, although not yet on grapevines; however, to our knowledge, no recent study has focused on the fungal community of grapevine roots in Europe. Given its wide host range, *P. omnivora* could become established in parts of the European Union where soil and climatic conditions are favourable for the fungus, and be spread mainly by human-assisted means (EFSA PLH Panel, Bragard et al. 2019).

6.4.1   Causal Organism

Phymatotrichopsis omnivora is an Ascomycete of the order Pezizales and family Rhizinaceae (Marek et al. 2009). The teleomorph of this fungal species remains unknown. The hyphae of *P. omnivora* form a loose web of large branched cells on the surface of the roots of the host with certain more rigid, acicular (i.e. needle-like) hyphae, some of which have cruciform ramifications. These rigid hyphae form mycelial cords resembling rhizomorphs, with a melanised rind consisting of parenchymatous polygonal cells. Almost hyaline when young, this web turns cinnamon-brown once mature. The fertile hyphae are dotted irregularly about this mycelial web and produce spheroid-to-ellipsoid-shaped spores (20–28 × 15–20 µm). *P. omnivora* may also form characteristic spherical conidiophores with botryoid blastoconidia borne separately on denticles. The conidial stage forms a continuous powdery zone, sometimes in the form of a rind, on the ground.

6.4.2   Symptoms

As with other root rots, a wide range of symptoms can be expressed by vines affected by *Phymatotrichopsis* root rot. These include yellowing, scorching or browning of the leaves or even total defoliation of the plants, and, over the long term, the progressive decline or sudden death of the plants. The symptoms, which can be confused with those of other root rots or wood diseases, vary according to the rootstock and scion combinations, the stage of growth of the vine when infection occurs, the altitude of the vineyard, the density of fungal inoculum in the soil and the pedoclimatic conditions, as well as the volume of infected roots and the speed at which they are destroyed by *P. omnivora* (Hu 2020).

6.4.3   Biology and Epidemiology

Phymatotrichopsis omnivorum can be spread over long distances by conidia, through nursery stock or contaminated soil. It is transmitted from vine to vine by mycelial strands that develop along roots and in the soil. The fungus overwinters in infected root residues of grapevines or other host plants, or as sclerotia that can persist at various soil depths for decades. Sclerotia can be activated by root exudate and germinate as fungal hyphae that directly attack the root bark.

6.4.4   Disease Control

Phymatotrichopsis root rot is difficult to manage and previous research on cotton, nut trees and landscape plants has shown various methods to have varying degrees of success in its control. An integrated approach is needed to reduce the occurrence and severity of the disease in grapevines (Hu 2020). The best strategy is to avoid establishing new vineyards on land planted with cotton or alfalfa with a history of the disease, to select tolerant rootstocks, to apply cultivation practices that increase soil organic matter and lower soil pH below 6.5 and to increase microbial diversity and soil quality. Fungicide treatments with flutriafol (not permitted in countries like Germany, Austria or Switzerland) or other fungicides (Appel and McBride 2018) are of limited efficacy owing to their difficulty in reaching fungal particles in the soil.

6.5   *Verticillium* Wilt

Current names:

- *Verticillium dahliae* Kleb. 1913, Ascomycota
- *Verticillium alboatrum* Reinke and Berthold 1879, Ascomycota

As for the other root diseases, Verticillium wilt occurs in nurseries and young plantings, caused by poor soil quality, wood residues in the soil, or water-saturated conditions leading to asphyxia of the roots. Verticillium in grapevine has been reported in

Germany (Thate 1960; Böning et al. 1960), Austria (Nieder 1980), Greece (Zachos and Panagopoulos 1963), Italy (D'Ercole 1970), Turkey (Kapkin and Ari 1982), Chile (Álvarez and Sepúlveda 1977), New Zealand (Canter-Visscher 1970; Mundy 2015), California (Schnathorst and Goheen 1977), and more recently in China (Zhang et al. 2009) and Crete (Ligoxigakis 2000). Verticillium wilt on grapevine can be considered a weak root pathogen that occurs in specific soil conditions. Sometimes reported in the complex of grapevine wood diseases, it is mentioned worldwide as a ubiquitous soilborne pathogen on over 200 wild host plants of economic importance.

6.5.1   Causal Organism

Verticillium wilt is caused by *Verticillium dahliae* (Ascomycota, Plectosphaerellaceae), an ubiquitous soilborne fungus. Longstanding confusions in differentiating *V. dahliae* from *V. alboatrum* have been clarified by sequencing technologies revealing a wide diversity within the species *V. dahlia*e in terms of its ability to mute depending on the substrate and biotic conditions. Colonies grow fairly quickly, initially exhibiting a white, partly aerial mycelium with a regular margin. Later, the culture turns black from the centre outwards from the development of microsclerotia. Conidia ($2.5–6 \times 1.5–3.0$ μm) are generally single-celled, ellipsoidal, hyaline, formed at the end of pointed conidiogenous cells. Microsclerotia formed on senescing diseased tissues are irregular in shape and size ($50–200 \times 15–50$ (-100) μm), dark-brown-to-black and almost globose.

6.5.2   Symptoms

During the growing season, when sap demand of the canopy is at its highest, leaves begin to yellow, wilt and collapse as early-summer temperatures increase. Leaf symptoms are limited by the veins or appear on the whole limb if the leaf petiole is infected. Depending on the severity of the disease, symptoms occur on isolated shoots of some plants, or entire plants dry up. This apoplectic form can be confused with the symptoms of grapevine wood diseases such as esca or black dead arm. Cross- and longitudinal sections of the infected shoots exhibit a brownish-red discoloration of the vessels and streaking of the wood caused by reduced sap conductivity in the vessels. Unlike wood diseases, the rootstock and basal part of the trunk exhibit brown-to-black necrosis decreasing upwards towards the grafting point. Cross-sections of diseased main roots and wood parts exhibit occluded brownish vessels which after a while exude mucilaginous material. Often, vines are only affected in part and strong new growth can appear. Wilted leaves generally remain attached, and fruit clusters at the base of affected canes shrivel and dry up. Symptom expression is strongly linked to water supply, demand and growth intensity. Vines infected only in part by the disease may recover completely by the following year.

6.5.3   Biology and Epidemiology

The fungus can infect the plant's roots and colonise its vascular system, spreading systematically throughout the entire plant. It can survive for several years as micro-sclerotia on plant debris or free in the soil. The occurrence of the disease has been linked to soils previously planted with susceptible crops such as potatoes, melons, tomatoes, alfalfa, sunflower, oilseed rape and cotton, or covered with weeds (*Solanum nigrum, Conyza canadensis, Senecio vulgaris, Chenopodium album,* etc.). Microsclerotia can survive for a decade or more in the soil and form conidia that infect host plants through the root system. Mycelium growth inside the vascular system forms occlusions and intermittently interrupts the sap flow (tyloses), leading to symptoms of wilt. The disease occurs predominantly in cool, wet spring conditions in clay-rich, water-retentive soils with a high organic-matter content, and on young vines with a less-well-developed root system. There have been reports of the recovery of weakly diseased plants.

6.5.4   Disease Control

There are no direct control measures for verticillium wilt of vine. Prevention consists in avoiding susceptible previous crops such as potatoes, tomatoes and alfalfa, and in creating a favourable soil structure. Excessive compaction and water accumulation encourage *Verticillium*. It is essential to avoid incorporating compost or other organic manure in the planting holes.

6.6   *Phytophthora* Crown and Root Rot

Current names:

- *Phytophthora cactorum* (Lebert and Cohn) J. Schröt. 1914, Oomycota.
- *Phytophthora cinnamomi* Rands 1922, Oomycota.
- *Phytophthora cryptogea* Pethybr. and Laff. 1919, Oomycota.
- *Phytophthora drechsleri* Tucker 1931, Oomycota.
- *Phytophthora megasperma* Drechsler 1931, Oomycota.
- *Phytophthora nicotianae* Breda de Haan 1896, Oomycota, syn. *Phytophthora parasitica* Dastur 1913.

Phytophthora crown and root rot is present throughout the grape-producing areas of the world without species specificity for grapevine. The disease can be considered a minor one due to its weak incidence and sporadic occurrence. *P. cinnamomi* was initially identified as the causal organism of root rot on grapevine in Australia (McGechan 1966), India (Agnihothrudu 1968) and South Africa (Van der Merwe et al. 1972). In

Chile (Latorre et al. 1997), other species of *Phytophthora* (*P. cinnamomi, P. cryptogea* and *P. drechsleri*) were later isolated from the diseased roots of table-grape vines, having caused similar symptoms with no distinction in virulence after artificial inoculations. The different species of *Phytophthora* described as pathogens on grape today include *P. cinnamomi, P. cactorum, P. nicotianae, P. cryptogea, P. megasperma* and *P. drechsleri* (Wilcox et al. 2015). As is the case with other soilborne diseases on crops such as pome fruits (Sutton et al. 2014), *Phytophthora* infection of grapevine roots may stem from complex interactions between different species of *Phytophthora* under favourable environmental conditions.

6.6.1   Causal Organism

The genus *Phytophthora* belongs to the class Oomycetes, the order Peronosporales and the family Peronosporaceae, and produces oospores as survival structures which germinate to form sporangia that contain motile zoospores. After encystment on the surface of the host, zoospores penetrate the host's tissues with a hypha, colonising the host's internal tissues and later forming sexual oospores. The penetration of the nucleus of the antheridium (male component) in the oogonium (egg-containing female component) forms the oospores.

Various Phytophthora species have been described as pathogens on grapes, including *P. cinnamomi, P. cactorum, P. nicotianae, P. cryptogea, P. megasperma and P. drechsleri. P. cinnamomi* is one of the most devastating plant pathogens in the world, with up to 5000 host-plant species, including economically important agricultural, forestry and horticultural plants (Hardham and Blackman 2018). It has been reported as most virulent on vines. In culture, *P. cinnamomi* produces coralloid hyphae with regular swellings and globose, thin-walled, mainly terminal chlamydospores, often in grape-like clusters of 3–10. Sporangia production ($75 \times 40\,\mu m$, ovoid, obpyriform or ellipsoid and non-papillate) is stimulated in soil extracts. The sporangia are formed by internal or external proliferation or by sympodial development of the sporangiophore immediately below empty sporangia. They can germinate directly or release 10–30 motile zoospores. Antheridia ($19 \times 17\,\mu m$) are amphigynous. Oogonia (21–58 μm) are round, hyaline-to-yellowish-brown and smooth-walled. Oospores (19–54 μm in diameter) are round and hyaline-to-yellowish-brown.

6.6.2   Symptoms

The disease appears on individual vines or on circular areas of several plants in poorly drained, temporarily saturated, heavy clay-rich soils. These stress conditions inhibit the plant's metabolism, causing a yellowing of the leaves followed by premature defoliation. High disease pressure causes vine death. Where there is root rot the primary roots are brown under the bark, with extensive necrosis inside the wood. The absence of rootlets prevents the absorption of water and nutrients. In the case of crown rot, necroses develop on the basal part of the trunk, appearing brownish-black under the bark and periderm. Where disease pressure is high, both root- and crown rot cause apoplectic death of the vines. Symptoms appear mainly after excessive irrigation or heavy rainfall on water-retentive soils. Recovery is possible if environmental conditions become less hospitable to the pathogen. Reliable diagnosis of the presence of *Phytophthora* spp. in early yellowing and wilting vines requires the root system to be dug up and fungi to be isolated from the border zone between necrotic and healthy wood tissues on cornmeal agar, which is a specific culture medium to oomycetes.

6.6.3 Biology and Epidemiology

Phytophthora spp. can survive as mycelium in host tissues or as oospores in the soil. Oospores ensure long-term survival for months to several years. Sporangia and encysted zoospores play a role in the short-term contamination of roots. *Phytophthora* species are considered weakly saprophytic compared with other root fungal pathogens. Infections occur through the release of zoospores by sporangia. Soil moisture for sporangia production is highly species-specific and is always encouraged by water-saturated conditions regardless of the species. Soil temperature is not a limiting factor for sporangia production and zoospore release: depending on the species of *Phytophthora* this can even occur below 5 °C, although it generally coincides with the beginning of vegetative growth of the vines in spring, when temperatures rise above 10 °C. Most of the contaminating particles are concentrated in the top few centimetres of the soil on plant residues. The zoospores reach the vine roots through the movement of water due to flooding or saturated conditions; contamination can also occur when water that contains zoospores splashes on the vine crown. In contrast to crown rot in apple trees, the direct penetration of the bark by zoospores has not been described in grapevine, possibly because of the different morphological configuration of the external bark tissues. In grapevine, the outside of the trunk consists of the so-called rhytidome—a layer of dead phloem and old periderm that forms strips of dead bark eliminated annually from woody organs—which probably protects the inner living bark from the zoospores. Crown rot symptoms might therefore stem from the systemic growth of mycelium from the roots upwards through the vascular system.

6.6.4 Disease Control

The disease only develops in vines that are periodically exposed to flooding or prolonged periods of excessive soil moisture. Young vines and nurseries are particularly affected by root rot in saturated water conditions. To date, *Phytophthora* root rot has never been described in northern European vineyards planted for the most part on south- to south-west-facing slopes to avoid water saturation. In more-vulnerable areas, the avoidance of poorly drained soils or the use of drains is recommended. In irrigated vineyards, placing drips at a minimum distance of 30 cm from each trunk minimises the risk of periodic saturation of the base of the plants. Although sensitivity of the rootstocks to *Phytophthora* spp. has not been extensively studied, findings from South Africa have indicated that the rootstocks Paulsen 1045, Paulsen 1103 and St George are highly resistant to *P. cinnamomi* (Wilcox et al. 2015). In California, the extensive use of phylloxera-resistant rootstocks such as SO4, 5BB, 5C, 420A, 110R, 3309 and 101-14 which have never been assessed for sensitivity to soilborne pathogens such as *Phytophthora* resulted in a significant increase in root rot in new plantings (Gubler et al. 2004).

References

Agnihothrudu V (1968) A root rot of grapes in Andhra Pradesh. Curr Sci 37:292–294

Álvarez AM, Sepúlveda RG (1977) *Verticillium dahliae* asociado com síntomas de "Amarillamento" en vid. Agricultura Technical (Chile) 37:138–140

Appel DN, McBride SA (2018) Management strategies for cotton root rot disease in Texas Winegrapes. Texas A&M Agrilife Extension publication EPLP-040

Baumgartner K, Rizzo DM (2002) Spread of *Armillaria* root disease in a California vineyard. Am J Enol Vitic 53:197–203. https://doi.org/10.5344/ajev.2002.53.3.197

Baumgartner K, Coetzee MPA, Hoffmeister D (2011) Secrets of the subterranean pathosystem of *Armillaria*. Mol Plant Pathol 12(6):515–534. https://doi.org/10.1111/j.1364-3703.2010.00693.x

Böning K, Mallach N, Spraw F, Wagner F (1960) 33. Deutsche Pflanzenschutz Tagung in Freiburg/ Br. Pflanzenschutz 12:11–12

Calamita F, Imran HA, Vescovo L, Mekhalfi ML, La Porta N (2021) Early identification of root rot disease by using hyperspectral reflectance: the case of pathosystem grapevine/*Armillaria*. Remote Sens (Basel) 13:2436. https://doi.org/10.3390/rs13132436

Canter-Visscher TW (1970) *Verticillium* wilt of grapevine, a new record in New Zealand. N Z J Agric Res 13(2):359–361. https://doi.org/10.1080/00288233.1970.10425408

D'Ercole N (1970) Tracheomycosis of grapevine caused by *Verticillium dahliae*. Inform Fitopatol:3–5

EFSA PLH Panel (EFSA Panel on Plant Health), Bragard C, Dehnen-Schmutz K, Di Serio F, Gonthier P, Jacques MA, Jaques Miret JA, Justesen AF, MacLeod A, Sven Magnusson C, Milonas P, Navas-Cortes JA, Parnell S, Potting R, Reignault PL, Thulke HH, Van der Werf W, Yuen J, Zappala L, Jeger M, Vloutoglou I, Bottex B, Vicent Civera A (2019) Scientific opinion on the pest categorisation of *Phymatotrichopsis omnivora*. EFSA J 17(3):5619. https://doi.org/10.2903/j.efsa.2019.5619

Gubler WD, Baumgartner K, Browne GT, Eskalen A, Latham SR, Petit E, Bayramian LA (2004) Root diseases of grapevines in California and their control. Aust Plant Pathol 33:157–165. https://doi.org/10.1071/AP04019

Hardham AR, Blackman LM (2018) Pathogen profile update *Phytophthora cinnamomi*. Mol Plant Pathol 19(2):260–285. https://doi.org/10.1111/mpp.12568

Höfer M (1992) Untersuchung über *Roesleria hypogea* Thüm. & Pass. Als Erreger des Wurzelschimmels der Weinrebe. Geisenheimer Berichte 13:1–137

Hu J (2020) Phymatotrichopsis root rot of grape. The Division of Agriculture, Life and Veterinary Sciences and Cooperative Extension, University of Arizona. AZ1859

Kapkin A, Ari M (1982) A new host of *Verticillium dahliae* Kleb. in Turkey. J Turk Phytopathol 11:77

Kirchmair M, Neuhauser S, Buzina W, Huber L (2008) The taxonomic position of *Roesleria subterranea*. Mycol Res 112(Pt 10):1210–1219. https://doi.org/10.1016/j.mycres

Koch RA, Herr JR (2021) Global distribution and richness of *Armillaria* and related species inferred from public databases and amplicon sequencing datasets. Front Microbiol 12:733159. https://doi.org/10.3389/fmicb.2021.733159

La Porta N, Capretti P, Thomsen IM, Kasanen R, Hietala AM, von Weissenberg K (2008) Forest pathogens with higher damage potential due to climate change in Europe. Can J Plant Pathol 30(2):177–195. https://doi.org/10.1080/07060661.2008.10540534

Latorre BA, Wilcox WF, Banados MP (1997) Crown and root rots of table grapes caused by Phytophthora spp. in Chile. Vitis 36(4):195–197. https://doi.org/10.5073/vitis.1997.36.195-197

Ligoxigakis EK (2000) Host of Verticillium dahliae in Kriti (Greece). EPPO Bull 30(2):235–238. https://doi.org/10.1111/j.1365-23382000tb00886.x

Marek SM, Hansen K, Romanish M, Thorn RG (2009) Molecular systematics of the cotton root rot pathogen, Phymatotrichopsis omnivora. Persoonia 22:63–74. https://doi.org/10.3767/003158509X430930

McGechan JK (1966) Phytophthora cinnamomi responsible for a root rot of grapevines. Aust J Sci 28:354

Mundy DC (2015) Ecology and control of root diseases in New Zealand: a review. N Z Plant Pathol 68:396–404. https://doi.org/10.30843/nzpp.2015.68.5818

Neuhauser S, Huber L, Kirchmair M (2011) Is Roesleria subterranea a primary pathogen or a minor parasite of grapevines? Risk assessment and a diagnostic decision scheme. Eur J Plant Pathol 130(4):503–510. https://doi.org/10.1007/s10658-011-9769-3

Nieder G (1980) Holzzerstörende Pilze der Rebe. Ursachen für Kümmerwuchs und Schlagtreffen (Apoplexie). Pflanzenarzt 33:7–9

Pasini L, Prodorutti D, Pastorelli S, Pertot I (2016) Genetic diversity and biocontrol of Rosellinia necatrix infecting apple in northern Italy. Plant Dis 100(2):444–452. https://doi.org/10.1094/PDIS-04-15-0480-RE

Pérez-Jiménez RM (2006) A review of the biology and pathogenicity of Rosellinia necatrix—the cause of white root rot disease of fruit trees and other plants. J Phytopathol 154:257–266. https://doi.org/10.1111/j.1439-0434.2006.01101.x

Pérez-Jiménez RM, Zea-Bonilla T, López-Herrera CJ (2003) Studies of Rosellinia necatrix perithecia found in nature on avocado roots. J Phytopathol 151:660–664. https://doi.org/10.1046/j.0931-1785.2003.00782.x

Petrini LE (1993) Rosellinia species of the temperate zone. Sydowia 44:169–281

Pliego C, López-Herrera C, Ramos C, Cazorla FM (2012) Developing tools to unravel the biological secrets of Rosellinia necatrix, an emergent threat to woody crops. Mol Plant Pathol 13(3):226–239. https://doi.org/10.1111/j.1364-3703.2011.00753.x

Purtov KV, Petushkov VN, Rodionova NS, Gitelson JI (2017) Why does the bioluminescent fungus Armillaria mellea have luminous mycelium but nonluminous fruiting body? Dokl Biochem Biophys 474(1):217–219. https://doi.org/10.1134/S1607672917030176

Schena L, Nigro F, Ippolito A (2008) Integrated management of Rosellinia necatrix root rot on fruit tree crops. In: Ciancio A, Mukerji K (eds) Integrated management of diseases caused by fungi, phytoplasma and bacteria. Integrated management of plant pests and diseases, vol 3. Springer, Dordrecht. https://doi.org/10.1007/978-1-4020-8571-0_7

Schnathorst WC, Goheen AC (1977) A wilt disease of grapevines (Vitis vinifera) in California caused by Verticillium dahliae. Plant Dis Rep 61:909–913

Sutton TB, Aldwinckle HS, Agnello AM, Walgenbach JF (2014) Compendium of apple and pear diseases and pests. APS Press, St. Paul, MN. https://doi.org/10.1094/9780890544334

Thate R (1960) Die Apoplexie der Rebe: eine Verticilliose. 33. Deutsche Pflanzenschutz-Tagung der Biologischen Bundesanstalt für Land-und Forstwirtschaft in Freiburg/Br

Van der Merwe JJH, Joubert DJ, Matthee FN (1972) Phytophthora cinnamomi root rot of grapevine in the Western Cape. Phytophylactica 4:133–136

Wilcox WF, Gubler WD, Uyemoto JK (2015) Compendium of grape diseases, disorders and pests, 2nd edn. APS Press, St. Paul, MN. https://doi.org/10.1094/9780890544815

Zachos DG, Panagopoulos CG (1963) Une hydromycose de la vigne due au Verticillium alboatrum Reinke et Berth. Ann Inst Phyto Benaki 5:303–305

Zhang L, Zhang GL, Qian X, Li GY (2009) First report of Verticillium wilt of grapevine (Vitis vinifera) caused by Verticillium dahliae in China. Plant Dis 93(8):841. https://doi.org/10.1094/PDIS-93-8-0841C

Chapter 7
Grafting and Fungi in the Nursery

7.1 Grafting

Ever since the phylloxera outbreak in the late nineteenth century that devastated the vineyards of Europe, grafting has been considered one of the most effective means of biological pest control in the history of viticulture. After long time of field experiments on the perspectives to replace *Vitis vinifera* by American varieties, especially in France (Millardet 1877), the most conclusive results have been obtained by grafting traditional European varieties, on the field or on table, to establish sustainable healthy vineyards (Bender and Vermorel 1890). The root form of the louse responsible for phylloxera, *Daktulosphaira vitifoliae*, attacks the roots of the European vine *Vitis vinifera*, causing the rapid decline of the latter. The accidental importation of this insect from the northern USA fundamentally altered vine propagation practices, causing simple propagation by cuttings to be replaced worldwide by grafting, which operation consists in joining a graft of the European vine (*Vitis vinifera*) onto an American rootstock (*Vitis* spp.) (Fig. 7.1). The choice of rootstock is based on soil type, in particular limestone content, drought resistance, and conferred graft vigour (Table 7.1). The method of cultivating American vines to yield rootstock wood, called mother vine fields, varies from country to country, although two main methods are widespread, viz., prostrate and trellised growth. Grapevine plants are produced by vine nurserymen who graft and market rooted grafts after stratification and 1 year's growth in the nursery. Another alternative is to produce the vine plants in pots and forgo the year in the nursery.

Grafting-on offers the possibility of adapting the plant material by changing vine variety in 1 year. It is done on mature rooted grafts that are in production. Compared to replanting, this method enables the use of the established root system and the trellis infrastructure that is in place and ensures a harvest the year after the grafting-on. Whether using the principle of chip-budding (Chaudière 2013), T-budding, cleft-grafting or herbaceous grafting (Chaudière 2017), grafting-on involves the overlapping of the cambiums to form a uniform callus, as with bench grafting.

© The Author(s) 2025
O. Viret, K. Gindro, *Science of Fungi in Grapevine*,
https://doi.org/10.1007/978-3-031-68663-4_7

Fig. 7.1 Phylloxera (*Daktulosphaira vitifoliae*). (**a**) Leaf form of phylloxera with galls on the underside of a leaf of an American vine. (**b**) Phylloxera gall in cross section. The orange-yellow female (arrow) and the numerous laid eggs can be seen

Table 7.1 Characteristics of the most common rootstocks, crossings, total limestone resistance (%), vigour and drought susceptibility (according to Carbonneau (1985))

Rootstock	Crossings	Total limestone (%)	Vigour	Drought susceptibility
101-14 MGt	*V. riparia* × *V. rupestris*	0–20	Low-to-medium	Low resistance/susceptible
110 Richter	*V. berlandieri* × *V. rupestris*	0–30	High	Highly resistant
1103 Paulsen	*V. berlandieri* × *V. rupestris*	0–30	Very high	Resistant
125AA (Kober)	*V. riparia* × *V. berlandieri*	0–35	Medium	Resistant
140 Ruggieri	*V. berlandieri* × *V. rupestris*	0–90	Very high	Highly resistant
161-49 C	*V. riparia* × *V. berlandieri*	0–50	Low-to-medium	Low resistance
3309 C	*V. riparia* × *V. rupestris*	0–22	Medium	Low resistance
333 EM	*V. vinifera* × *V. berlandieri*	0–60	High	Highly resistant
41B	*V. vinifera* × *V. berlandieri*	>50	High	Low resistance
420 A	*V. riparia* × *V. berlandieri*	0–40	Low-to-medium	Low resistance
44-53 M	*V. riparia* × (*V. cordifolia* × *V. rupestris*)	0–15	Medium	Highly resistant
5BB	*V. riparia* × *V. berlandieri*	0–40	Medium-to-high	Low resistance
5C	*V. riparia* × *V. berlandieri*	0–40	Medium-to-high	Low resistance
99 Richter	*V. berlandieri* × *V. rupestris*	0–30	High	Resistant

(continued)

Table 7.1 (continued)

Rootstock	Crossings	Total limestone (%)	Vigour	Drought susceptibility
Fercal	V. vinifera × V. berlandieri x V. longii	>60	Medium-to-high	Low resistance
Gravesac	161-49 C × 3309 C	0–15	Medium-to-high	Moderate
Riparia Gloire	V. riparia	0–15	Low	Susceptible
Rupestris du Lot	V. rupestris	0–14	High	Low resistance / Susceptible
SO₄	V. riparia × V. berlandieri	0–35	Medium-to-high	Resistant

7.2 Production Stages of the Young Vine Plants

Although the principle behind grafting remains the same, each nurseryman produces the rooted grafts ('rootings') using the techniques providing the highest success rate based on their available infrastructure and plant material. Though not described exhaustively, the procedure set out below charts the main steps for producing a rooted graft, from the harvesting of the canes to the planting of the rootings. The scions are selected from specific plots during the summer. They are harvested during the winter according to variety and are chosen based on their size and the quality of shoot lignification. The rootstocks, for their part, are harvested after leaf fall. It is important that the scions and rootstocks only be harvested once they are well lignified.

The rootstocks are prepared in two stages: first, the axillary shoots, lateral shoots and tendrils are removed. The second stage, disbudding, consists in removing all buds to prevent their development. The canes are then calibrated to the desired length (e.g. 42, 50 or 65 cm). The scions are cut to an approximate length of 5 cm with a single latent bud (Fig. 7.2).

Rootstocks and scions are rehydrated in a water bath for at least 12 h, then immersed in a disinfectant bath containing a fungicide or alternative product (oxyquinoline, various products of natural origin such as essential oil of oregano) for around 5 h. The active compound penetrates the wood, thus also filling the intercellular cavities of the xylem. The canes are then cold-stored (at 3 °C) in micro-perforated bags at a humidity ranging between 70% and 80% so that they remain hydrated. Storage is generally for 2–3 months. Grafting can be done by hand (English grafting), or more commonly with the help of grafting machines that make an incision allowing the graft to be joined to the rootstock. The grafted vines thus formed are dipped in paraffin wax maintained at 75 °C (Fig. 7.3) to ensure that the graft union is protected against desiccation and mould during forcing. A hormone paraffin wax generally contains auxin, which is used as a growth regulator and encourages the formation of scar tissue or callus. The treated canes are placed in boxes or a large-volume container, cold-stored at approx. 5 °C and

Fig. 7.2 Preparation of scions cut to an approximate length of 5 cm with a single latent bud. (© Dutruy nursery, Founex, Switzerland)

Fig. 7.3 Paraffin-dipping
of grafted plants. (**a**)
Liquid paraffin maintained
at 75 °C, in which the
grafted plants are dipped.
(**b**) Paraffin-dipped grafted
plants

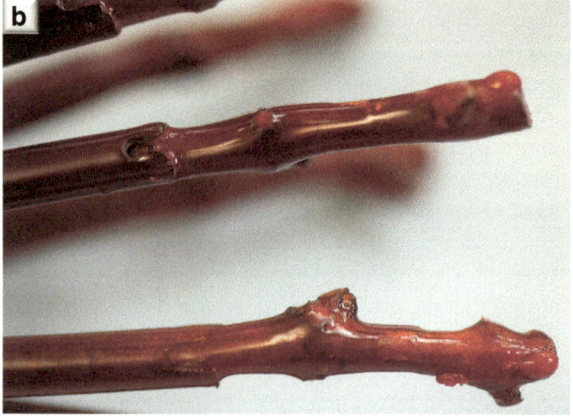

covered with black plastic until forcing, enabling callus formation. Forcing is for around 10 days in a heated room whose temperature is raised by stages up to 30 °C at 90% relative humidity.

After callus formation around the grafting point and development of the primary shoot from the graft bud, the rooted grafts can be removed from the forcing boxes. They are prepared and sorted for planting in a nursery. The rootings are re-dipped in a non-hormone paraffin wax, then planted in the ground on plastic-covered mounds (Fig. 7.4).

Fig. 7.4 After being forced in a box, the rooted grafts are planted at a depth of around 8 cm (**a**) on mounds of earth covered with black plastic (**b**). (**c**) In the nursery, the rooted grafts develop one or two shoots at the top of the graft. (Photo (**a**) © Dutruy nursery, Founex, Switzerland)

After undergoing their complete growing cycle, the plants are uprooted in late autumn, visually sorted (one or two lignified shoots, a minimum of three large roots spread over the entire periphery of the rootstock) and tested via the 'nudge' technique to ascertain the quality of the union at the grafting point. They are then rehydrated for 6–8 h in water at 18 °C for acclimatisation purposes, after which the roots are clipped and the shoots pruned back to two buds before the plants are dipped in paraffin wax. Depending on the country and regulations, the plants are treated in hot water at 50 °C for 45 min to guarantee the absence of flavescence dorée and bois noir phytoplasmas. The roots and base of the vine are then dipped in an aqueous solution of crosslinked potassium acrylate and acrylamide copolymers to limit desiccation and ensure better development after planting in the vineyard. The plants are acclimatised at ambient air temperature (approx. 22 °C) for 10–11 h, then placed in a cold, humid chamber in micro-perforated plastic bags before being sold and finally planted in the vineyard in spring.

7.3   Union of Rootstocks and Scions

Among the numerous production stages of a rooted graft, the most important is the union of the rootstock and graft to obtain a snug-fitting cambial zones. Coating the grafting zone with a hormone wax is effective in improving callogenesis (callus formation), with the callus playing a crucial role in establishing a regular flow of sap after emergence from dormancy. The callus (Fig. 7.5) consists of non-differentiated scar tissue stemming from the proliferation of the cambium around the grafting point, which subsequently transforms into conductive tissue. The quality of the union, size of the callus, and root formation and distribution (Fig. 7.6) are all essential factors for good vegetative recovery after planting in the vineyard. Grafting is a delicate operation requiring a great deal of know-how. A poorly executed step can adversely affect the quality of the rooted graft, leading to limited graft-junction tissue and a predisposition to cork formation such as found in dead wood, or in necroses that partially block sap flow and may offer a niche for the development of saprophytic or lignivorous fungi. Early decline and the rapid progression of wood diseases call into question certain grafting methods, although the causal link between decline and fungi has not been established to date (Hofstetter et al. 2012; Geiger et al. 2022).

Reprogramming of Primary Metabolism and Induction of Stylbene Synthesis at the Graft Interface
(Prodhomme et al. 2019)

Grafting is a considerable stress on plants, stimulating wounding responses and healing processes (viz., the production of non-differentiated callus tissue and the differentiation of the latter into xylem and phloem). The profile of the primary and secondary metabolites (19 amino acids, 5 primary metabolism and 30 secondary metabolism compounds) and the quantification of the activity of two enzymes—phenylalanine ammonia lyase (PAL) and neutral invertase (NI)—in the scion and rootstock tissues and at the graft interface of homo- (genotype grafted with itself) and hetero-grafts (two different genotypes) of grapevine were analysed 1 month post-graft. Graft interface tissues were compared with the surrounding rootstock and scion tissues.

Primary-metabolite profiling at the grapevine graft interface revealed that the graft-interface metabolome is reprogrammed to sustain callus-cell proliferation through the accumulation of glutamine and γ-aminobutyric acid, and the reduction of phenylalanine, arginine, tyrosine and lysine concentrations. Overall, flavonol concentration decreased and stilbene concentration increased at the graft interface compared to the surrounding tissues for all scion/rootstock combinations. Genotype-specific differences in the profile of the flavanols and stilbenes in the rootstock tissues were found, and although the concentration of these metabolites at the graft interface often reflects these genotype-specific differences, certain specific metabolite accumulation patterns were observed. A small number of stilbenes exhibited scion- and rootstock-specific accumulation patterns, which might suggest that hetero-grafting of two different genotypes triggers specific changes in secondary metabolism and plant defence responses at the graft interface. This study revealed the modification of primary metabolism to sustain callus cell formation and the stimulation of stilbene synthesis at the graft interface, and showed how these processes are modified by hetero-grafting. Knowledge of the metabolites and/ or enzymes necessary for successful graft union formation allows the identification of markers that could be used for selection and improvement purposes.

Fig. 7.5 Formation of the callus or scar tissue around the graft union, produced by proliferation of the cambium. *c* callus

Fig. 7.6 Rooted grafts after forcing in boxes, showing good vegetative recovery on the top of the graft (**a**) and well developed calluses (**b** and **c**). Upon leaving the nursery, the root mat is well developed and the lignified shoots are cut back. The plants have been dipped in paraffin wax and are ready to be marketed (**d**). *c* callus, *r* roots. (Photos (**a**, **b** and **d**) © Guillaume nursery, Charcenne, France. Photo (**c**) © Dutruy nursery, Founex, Switzerland)

7.4 Grafting Systems

Before the phylloxera crisis, grapevines were cultivated ungrafted and propagated by cutting or layering. From 1860 onwards, we witness a large number of grafting experiments, with all but the cleft-grafting method for *in situ* ligneous grafting and English bench cleft grafting by the nurseries ultimately being rejected. This relatively complicated method persists into the 1970s, in parallel with various attempts at English herbaceous grafting which meet with qualified success. Thus, over more than a century of history, several grafting systems were developed until the advent of the partially or fully automated omega (Ω) inlay graft (Fig. 7.7). At present this is the most commonly used graft, with the patenting of the omega grafting machine in 1975 by Wagner nurseries in Germany (Fig. 7.8).

Developed in the seventeenth century, single or double English cleft grafting is done by hand on a bench and involves the union of wood rigorously of the same diameter via a single or double bevelled incision in the rootstock combined with an inversely symmetrical cutting of the scion for interlocking (Fig. 7.9). V-grafting (also called cleft grafting or inlay grafting) is a grafting system in which the scion is cut to a point and the rootstock to an inversely symmetrical shape (Fig. 7.9). Old-fashioned grafting techniques such as mortise grafting, dating from the 1920s, have recently been mechanised (Celerina Plus machine developed by the VCR Rauscedo nursery, Italy) or, more recently, the similar but four-point-star-shaped F2 graft (Hebinger nurseries in Germany). The latter two techniques aim to double the contact surface between scion and rootstock to encourage callus welding.

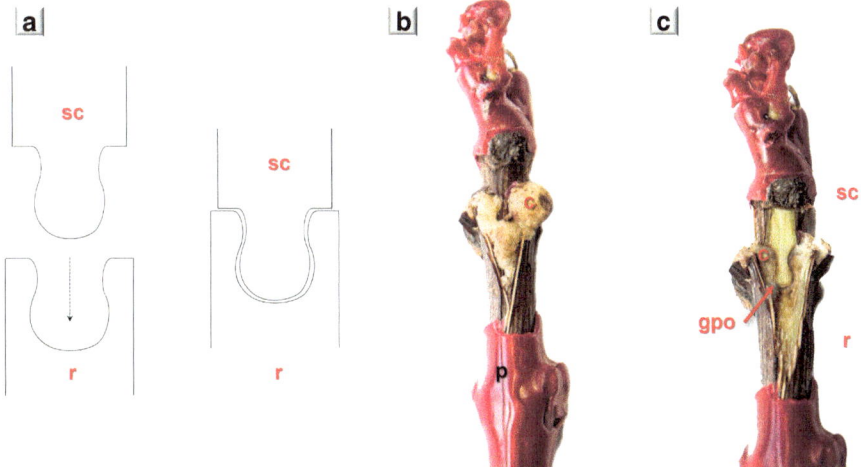

Fig. 7.7 Omega grafting. (**a**) Principle of omega grafting (joining of the scion and rootstock that have been mechanically cut by an omega-shaped punch). (**b**) Rooted graft (rooting) with a well-formed callus (after removal of the wax at the grafting point). (**c**) Longitudinal section at the omega-shaped grafting point. *c* callus, *gpo* omega-shaped grafting point, *p* paraffin wax, *r* rootstock, *sc* scion

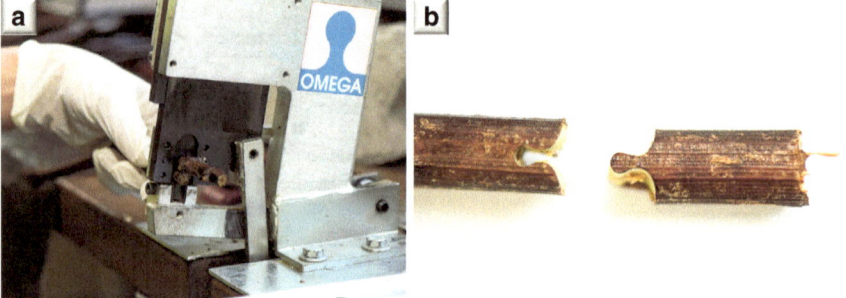

Fig. 7.8 Mechanised omega graft. (**a**) H. Wagner omega grafting machine. (**b**) Cut-out scion and rootstock which will be slotted together

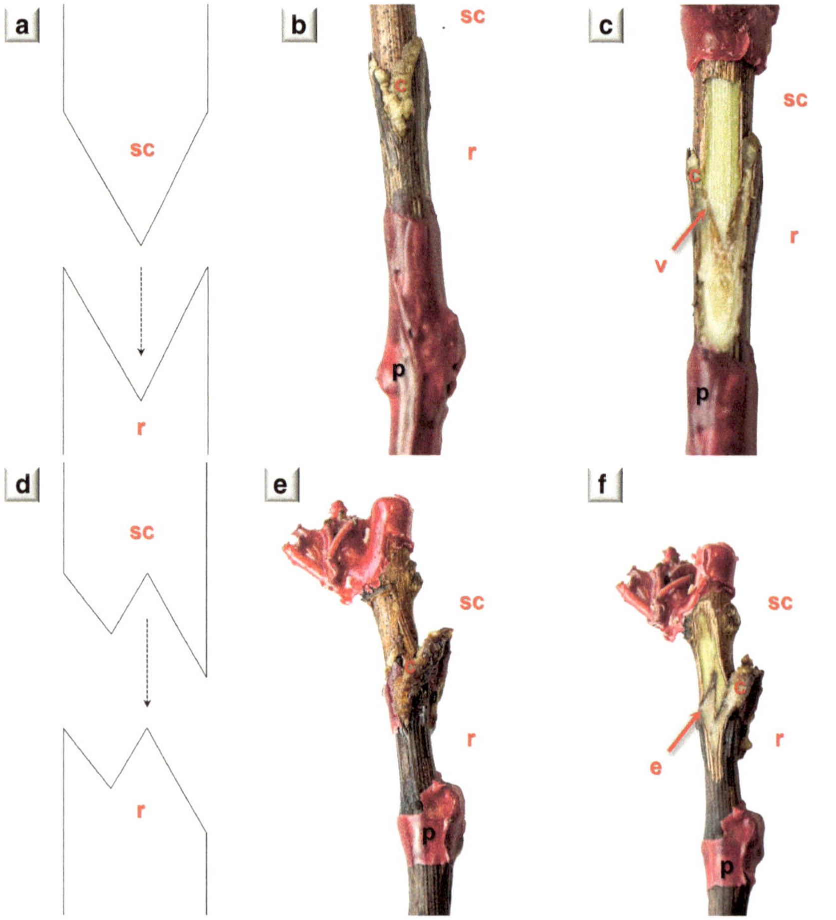

Fig. 7.9 English-style (double) V-grafts. (**a**) V-graft (union of scion and rootstock cut to a point and double-bevelled). (**b**) V-shaped grafted vine with well-developed callus. (**c**) Longitudinal section of a V-graft. (**d**) Principle of double English grafting using complex complementary notches. (**e**) Callus formation in an English graft. (**f**) Longitudinal section at the English double grafting point. *c* callus, *e* English graft zone, *p* paraffin wax, *r* rootstock, *sc* scion, *v* V-graft zone

Graft-Union Bulge and Callus

Grafting exposes the tissue-generating layer (cambium) of the scion and rootstock. It is therefore essential for the generating layers to be perfectly aligned, to allow the start of scar-tissue development either side of the graft union from the respective cambiums. Two types of structure may be produced, depending on the plants: a graft-union bulge and/or callus.

- If the wound remains exposed (with no protection against light and/or dehydration), a graft-union bulge produced by the secondary meristems (cambium and phellogen) develops, e.g. in fruit trees. This bulge is therefore composed of wood, phloem, phellogen and cork, and as its name indicates, does no more than cover the wound.
- If the wound is protected (paraffin wax, grafting wax) a mass of non-differentiated cells produced by cell multiplication—the callus—forms. In the plant kingdom we speak of totipotent cells, which are capable of initiating the formation of all possible plant tissues. Through the action of specific plant hormones, including auxins, the progressive differentiation of callus cells results in the development of neocambium and then of the various functional tissues, including the elements circulating the raw and elaborated sap. The latter form the vascular links between the phloem and xylem of the scion and rootstock, capable of ensuring a lasting union and flows of raw and elaborated sap that are essential for long-term grapevine development.

Regardless of the grafting method used (omega, cleft, English graft), a poor-quality union is highly likely to cause the decline of the plant, even after several years. An analysis of these three grafting systems shows that the omega graft provides for a union of high quality, although a certain amount of cork and necrotic wood are formed in the periphery of the junction zone. With English grafts, there is a high amount of contact surface between the scion and rootstock, and hence a larger amount of necrotic wood and cork than in omega grafts; the union is more fragile, holding only at the callus formation points. The 'V' system exhibits a generally less developed callus that in omega grafting as well as a more-fragile quality of union, requiring careful paraffin waxing; however, there is a slightly lower amount of cork and necrotic wood in the junction zone (Fig. 7.10) than in the case of English grafting.

X-Ray Tomography: A Promising Tool for Evaluating the Selection of High-Quality Grafted Vines

(Renault-Spilmont et al. 2021)

The production of grafted vines is a complex process ranging from the grafting to the final sorting in the nursery. Before they can be put on the market, grafted vines must meet various criteria, including a manual graft-union test called the 'nudge test' which evaluates the mechanical strength of the union. This test depends on individual conducting it and is thus subjective. The aim of the study is to evaluate the possibility of utilising internal quantitative and objective criteria to select high-quality grafted vines in place of the current criteria. The analytic parameters were first optimised to allow for the study of a large number of grapevine plants whilst maintaining sufficient resolution for image analysis. This study was carried out on two batches with different degrees of grafting. Grafted vines were selected at random in these two combinations to be scanned before being assessed with the regulatory criteria, including the nudge test. Numerous variables concerning the graft union and the anatomy of the scions and rootstocks were measured for all the vines. The batches of plants which passed or failed the nudged test were then compared. Two internal criteria associated with wood production and volume of air and necrosis in the omega interface appeared to be discriminatory, and hence relevant. The quantity of xylem produced post-grafting by the graft is greater in the plants passing the nudge test. Similarly, the grafts passing this test have significantly less air and fewer necroses in the union zone and a threshold can be defined to separate these vines, which could be particularly useful from a practical perspective. X-ray tomography could facilitate or replace manual sorting of the grafted vines in the medium term. In a wider context, imaging methods, as non-destructive tools, can furnish key information for understanding all the complex events of the grafting process. These innovative techniques are also opening up new practical prospects for helping nurserymen to identify the key points of the callusing process or assess different grafting methods.

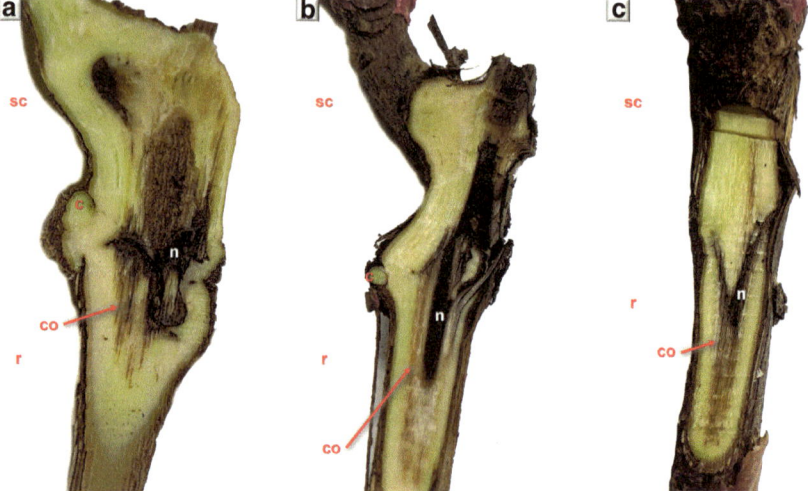

Fig. 7.10 Formation of cork and necrotic wood at the graft point as a function of the assembly system: omega (**a**), English graft (**b**), 'V' (**c**). In the three systems, necrotic wood forms at the graft point and sap circulation is focused at the periphery. *c* callus, *co* cork, *n* wood necrosis, *r* rootstock, *sc* scion

Once the plants have sprouted, the differentiation of the callus cells into conducting tissues is a key stage for proper resumption of raw sap (xylem) and processed sap (phloem) flows (Fig. 7.11). The sap flow in the rooted grafts can be observed in the coloration, which shows that the sap flows bypass the central zone of the graft point and follow the differentiated margin in the callus zone, whatever the grafting system (Fig. 7.12).

All irregularities at the graft point (difference in diameter, both parts out of line with each other, one-sided assembly) lead to the imperfect union of the scion and rootstock which will irretrievably jeopardise the durability of the grapevine plant, regardless of the grafting system.

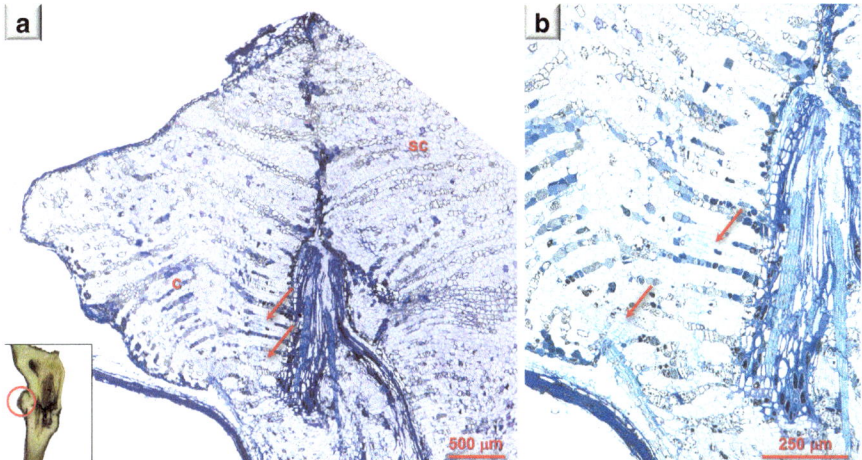

Fig. 7.11 Differentiation of callus cells into steles of conducting bundles (arrows). (**a**) Semi-thin section of the callus zone stained with azure blue (1%), sodium tetraborate (1%) and methylene blue (1%) observed by optical microscopy (inset: sampling zone). (**b**) Detail of the conducting vessels providing for the sap flow between the scion and rootstock. *c* callus, *sc* scion

Fig. 7.12 Longitudinal sections at the graft union of year-old rooted grafts, sprouted then placed in a 0.1% solution of Methylene Blue for 24 h after washing of the roots (**a**) based on the graft joining system. Regardless of the grafting system, sap flow (in blue) is concentrated around the graft union. (**b**) Omega graft; (**c**) V-graft; (**d**) English graft; (**e**) omega graft (2 years old plant)

7.5 Fungi in the Young Vine Plants

7.5.1 Development of Disinfection Processes

The battle against fungal pathogens in nurseries goes back to the 1950s. At this time, *Botrytis cinerea* (Botrytis bunch rot or grey mould) posed a huge number of problems in the propagation of grapevine plants by grafting. The presence of *B. cinerea* on the propagation material severely limited the success rate in nurseries (Fig. 7.13).

Owing to the climatic conditions, the basic plant material harboured a significant quantity of fungal sclerotia since it had been massively colonised during the viticultural season. In 1964, the diseases caused by *Botrytis* blight in German nurseries exceeded 30% (Becker 1966) and were estimated at over one million dollars financial losses a year in the 1970s (Becker and Hiller 1977). The fungus affects the green organs and buds of the young vine plants during forcing, as well as the callus. To increase the success rate, G. Auer (Trasadingen, Schaffhouse, Switzerland) introduced the use of oxyquinoline (syn.: 8-quinolinol; quinophenol) in vineyard nurseries in 1950, a compound derived from quinoline, hydroxylated at carbon 8 (Fig. 7.14).

Effervescent oxyquinoline tablets commonly used in human medicine as an antiseptic for throat disinfection were diluted in water, then sprayed onto young grafted plants from bud-break during the forcing stage. The qualitative improvement in the plants obtained, the reduced production costs and increased success rates resulted in the generalised use of this disinfectant, which evolved into the commercial product Chinosol® based on 8-quinolinol sulphate. The research of Becker and

Fig. 7.13 Damage on the rooted grafts caused by *Botrytis cinerea* (arrows), leading to significant loss of material for the nurserymen. Inset: Pure-culture isolate of *Botrytis cinerea*. (Photos © Nora Viret, Agroscope, Switzerland)

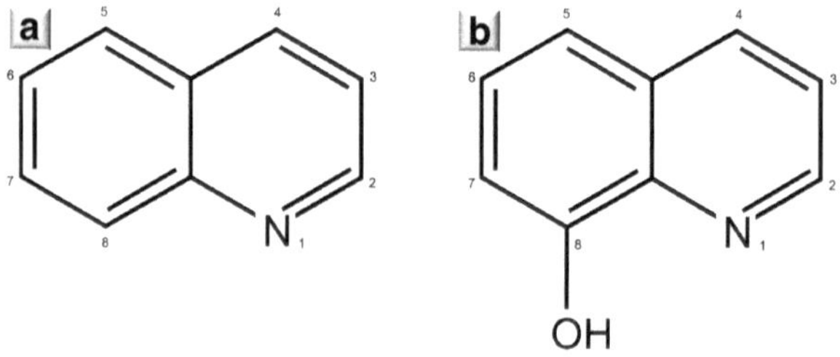

Fig. 7.14 Structure of
quinoline (**a**), from which
oxyquinoline, hydroxylated
at carbon 8, is derived (**b**)

Hiller (1977) allowed the determination and optimisation of concentrations as well as cane-dipping durations and temperatures, revolutionising the approach to graft disinfection before storage as well as the use of paraffins containing fungicides and growth hormones.

Alternatives to Oxyquinoline-Based Products for the Treatment of Nursery Canes: Effectiveness on the Development of Fungal Communities

(Viret 2020)

   The disinfection of grapevine canes for the purpose of grafting is a major challenge to ensure propagating material that is free from fungal pathogens and for the success of the entire rooted-graft production process. Oxyquinoline was commonly used in the majority of vineyard nurseries until its withdrawal from the market in 2016, instigating the search for alternatives. Three alternative products (essential oils of oregano and sweet orange and a patented, hydrogen peroxide-based product) were evaluated in comparison with the non-disinfected control and the reference product. The fungal communities present inside the grapevine wood were isolated (976 isolates) and identified (104 different species) by means of molecular methods (ITS region sequencing) before grafting, during grafting, after stratification and when the canes are dug up in the nursery. The initial fungal community in the untreated controls consists of 17 fungal species (129 isolates). In the controls, the number of species increases during the process until after stratification and returns to the initial level when the canes are dug up in the nursery. The reference product (oxyquinoline) permits a significant reduction (75.7% effectiveness) in the number of isolates until grafting, whilst the other disinfection products have proven ineffective and actually increase the number of isolates, except for the essential oils, until grafting. The absence or very low presence of significant pathogens of the vine such as the species responsible for wood diseases (e.g. *Phaemoniella chlamydospora*) or *Botrytis cinerea* in the controls calls into question the need for cane disinfection when hygiene conditions are optimal during all the stages of the propagation process (Waite and Morton 2007; Kızıldeniz et al. 2022).

7.5.2 Mycobiota of Young Vines

Various strategies and treatments have been developed to limit the progression and presence of certain fungi—some of which are considered responsible for wood diseases (WDs)—during the stages of creating a rooted graft. The fungal diseases of vine wood, such as the decline of young vines (Petri disease) or esca lead to major losses in recently planted vineyards as well as in established vineyards in the majority of wine-producing countries (Bertelli et al. 1998; Scheck et al. 1998; Mugnai et al. 1999; Cottral and Pascoe 2000; Whiting et al. 2001; Giménez-Jaime et al. 2006; Ferreira et al. 2017). Until now, numerous studies on the fungi associated with *Vitis vinifera* essentially focused on the detection and identification of fungal species present in the necrotic wood such as *Phaeomoniella chlamydospora* and species of the *Phaeoacremonium* and *Fomitiporia* genera, considered responsible for esca (Crous et al. 1996; Larignon and Dubos 1997; Armengol et al. 2001; Rumbos and Rumbou 2001). More recently, Giménez-Jaime et al. (2006) showed that the decline of young vines was associated with the same fungi as those of esca, with different fungal species predominating depending on the geographic zone. Broadly speaking, the presence of these fungi in young plants is acknowledged as a cause of reduced vigour and yield losses in recently established vineyards (Morton 1999). Nevertheless, our knowledge about the interactions between these fungi in the wood is still very limited, as is our knowledge of their role as primary pathogens which can act simultaneously or successively (Graniti et al. 2002). Furthermore, like almost all other woody plants, *V. vinifera* harbours a large and long-underestimated number of fungal endophytes (Schweigkofler and Prillinger 1999; Hofstetter et al. 2012; Monod 2024). These endophytes can play an important role in the balance of the fungal community living in grapevine, either as latent pathogens or as protective endophytes, certain of which can also be utilised as biocontrol agents (Król and Machowicz-Stefaniak 2008). Identifying and locating fungi associated with the grapevine in its early growth phase is an essential stage for understanding the potential emergence of grapevine wood diseases. The findings of Casieri et al. (2009) showed that a very large number of fungal species, some of which are considered to be pathogens of grapevine or other plant species, are present in the young plants in the nursery, their abundance appearing to be linked with grape variety. This can be explained in part by the size and greater number of xylem vessels of certain varieties, leading to the faster movement of fungi in the plant as well as a more homogeneous fungal community. The chemical composition (tannin content) of the wood and the defence responses (phytoalexins) of the different grape varieties can also interfere with the growth of certain fungal species.

Microbiota of Plants
(Buchs et al. 2018)

Plants do not develop in a sterile manner in nature. Rather, they harbour a community of microorganisms (fungi, bacteria, viruses), even in the absence of specific symptoms. These microorganisms interact with the plants at root and leaf level. Depending on their different types of interaction with the host, we distinguish between beneficial microorganisms (symbionts), neutral organisms (commensalists) and harmful organisms (pathogens). Taken as a whole, these microorganisms may be said to constitute a microbiota. Just as humans or animals harbour a microbiota on their skin's surface or in their digestive tracts, microbiota have been shown to exist in plants. For several years now, and thanks to the boom in sequencing techniques and to bio-informatic methods, it has been possible to describe in detail the microbiota found on the different parts of a plant. For example, bacteria (*Methylobacterium* sp.) of the genus Rhizobia have been observed in the petioles of Gamay vines not showing any symptoms of disease. The majority of Rhizobia are bacteria living in symbiosis with the roots of plants, where they fix atmospheric nitrogen. These same Methylobacteria have been described by other researchers in the fluid issuing from the xylem of the vine. Hence, these bacteria could be endophytes colonising the interior of the xylem vessels of the vine in the same way as the pathogenic bacteria *Xylella fastidiosa* that causes Pierce's disease. In the case of Methylobacterium, the bacterium possesses an enzyme, β-galactosidase, enabling the hydrolysis of galactane, a polysaccharide found in the cell walls of plants and providing them with a positive interaction.

The microbiota of plants have attracted the attention of researchers, given that they could influence plant growth and the plant's resistance to abiotic and biotic stress. Some fungi, for example, could facilitate vine root access to nutrients and contribute positively to the establishment of the young plants. Understanding the microbiota of the vine and other perennial plants is a highly complex matter, particularly as concerns growth and pathogen resistance. The symbiotic relationship may be at three different levels: that of the plant, the endophytic fungus and a virus, as in the case of *Panicum trichoides* (tropical panicgrass), where the association between the endophytic fungus and the plant allows the latter to grow in high-temperature conditions (Màrquez et al. 2007). The authors have also observed that this heat resistance was the result of the endophytic fungus being infected by a mycovirus. In fact, when the fungus is not infected by the said mycovirus, its association with the plant does not endow the latter with thermal resistance.

7.5.3 Hygiene Techniques for Young Vine Plants

Vine production via the classic grafting methods and on a commercial scale first appeared over 130 years ago. This system remained artisanal until the mid-1950s, with the advent of the first certification programmes aiming to obtain virus-free mother plants. The need to increase the production scale on an industrial model and the production of plant material based on minimal morphological standards first appeared in the late 1960s. During the 1970s, research led to the acquisition of knowledge on semi-automatic grafting, process hygiene, the use of plant-growth regulators and the understanding of the physiological elements regarding the compatibility of rootstock and graft, the formation of calluses, and rooting (Grohs et al. 2017). Up until 2000, certification systems and propagation processes changed very little. Propagation techniques were gradually reviewed with the aim of reducing the incidence of wood diseases (complex of esca, eutypa dieback, dead arm, Petri's disease, blackfoot disease) and viral, bacterial and phytoplasma diseases. More recently, the concepts of producing sustainably and reducing the use of traditional plant-protection inputs have taken hold at the vineyard nursery scale, where innovative solutions are pursued.

7.5.3.1 Hot-Water Treatment (HWT)

Treatment of rootstock wood, grafts or rooted grafts in hot water at 50 °C for 45 min was developed on grapevines affected by Pierce's disease (Goheen et al. 1973; Davis et al. 1978) and flavescence dorée (Caudwell et al. 1997). This treatment completely eliminates the phytoplasmas of the infected wood provided that a temperature of 50 °C can be maintained throughout the entire treatment and throughout the dip tank (Fig. 7.15). Variations in temperature of more than one degree are sufficient to reduce the effectiveness of the treatment. Dipping the wood in hot water—obligatory in the regions affected by the disease—is a method that is strongly recommended to vine nurserymen. It has also been demonstrated that HWT affects the eggs of the flavescence dorée vector (*Scaphoideus titanus*) and can reduce hatching rate by over 90% (Caudwell et al. 1990; Linder et al. 2010). This technique is therefore strongly recommended to ensure that the propagating material is free of phytoplasmas.

The temperature and duration of treatment represent critical values beyond which damage can be detected on the woody tissue of the vine. The buds of the canes undergoing hot-water treatment at 50 °C exhibit localised cellular degeneration of varying degrees depending on grape variety, whilst bud damage is complete at 60 °C (Fig. 7.16). HWT at 50 °C does not damage conductive tissues or significantly disrupt hydraulic conductivity. By contrast, at 60 °C hydraulic conductivity is reduced, leading to a significant embolism in the xylem ducts as well as tissue damage (Fig. 7.17).

The impact of hot-water treatment (HWT) on the fungal community in limiting the incidence of wood diseases (WDs) remains contentious. Some studies show a decrease in the number of fungal pathogens, and hence better control of wood

Fig. 7.15 Hot-water treatment of grapevine plants in a specific device for maintaining the temperature precisely at 50 °C to destroy the phytoplasma and eggs of the flavescence dorée vector. (© Multiplants nursery, Vétroz, Switzerland)

diseases (Fourie and Halleen 2004), whilst, according to Whiting et al. (2001), HWT has no effect on fungi. The coloration of the vascular system and frequency of isolation of *Phaeomoniella chlamydospora* (number of times a particular species is isolated with respect to the total number of isolates) are influenced by the HWT, which does not eliminate the pathogen completely (Rooney and Gubler 2001).

Bearing in mind the great fungal diversity in nursery plants, we have yet to learn how this fungal community may be affected by HWT, and whether the latter has a specific incidence on the species recognised as pathogens. Casieri et al. (2009) showed that the number of species isolated is always lower in the HWT plants, whatever the grape variety examined. The number of fungal species isolated from 1-year-old plants remains high, whether they are recognised as pathogens or not. These species include secondary colonisers with a high potential for lignocellulolytic activity, as well as a high number of saprophytic species. The grouping of the species isolated from each grape variety by taxonomic class allows to gain an overview of the changes taking place post-HWT within the fungal communities of the examined grape varieties (Fig. 7.18). Broadly speaking, a decrease in phytopathogens is noted at the same time as an increase in the frequency of isolation of the fungal species capable of degrading cellulose, hemicellulose and lignin. The species of fungi most sensitive to HWT would appear to be eliminated, making way

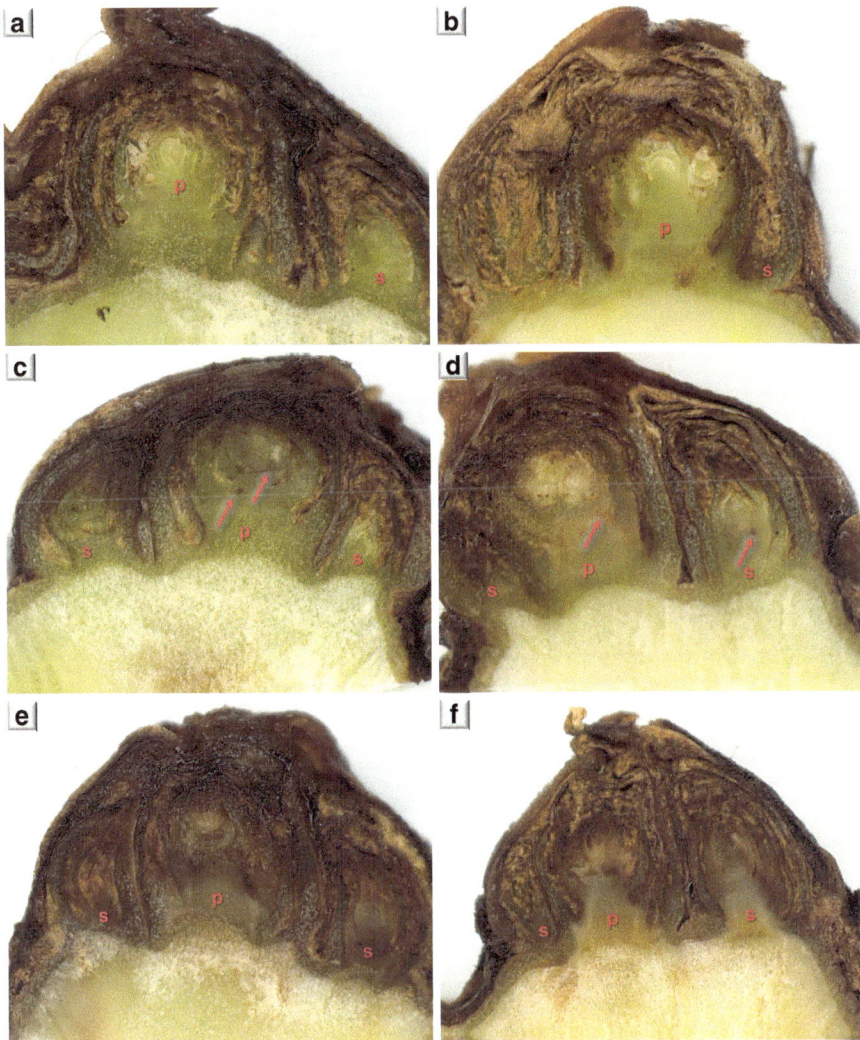

Fig. 7.16 Appearance of *Vitis vinifera* cv Cabernet Sauvignon and Merlot buds from hot-water-treated (HWT) woody cuttings compared to the untreated controls. (**a–c**) Cabernet Sauvignon. (**a**) Untreated control. (**b**) HWT at 50 °C. (**c**) HWT at 60 °C with complete damage of the primary and secondary buds. (**d–f**) Merlot. (**d**) Untreated control. (**e**) HWT at 50 °C. (**f**) HWT at 60 °C with complete damage of the primary and secondary buds. The arrows show the damage in the tissues (browning). *p* primary buds, *s* secondary buds

for other saprophytic species that are generally secondary colonisers. For example, the fungal community present in *Vitis vinifera* cv. Gamay is distinguished by the preponderance of Zygomycetes and a major reduction in Dothideomycetes and Sordariomycetes, classes of fungi encompassing the majority of pathogenic and endophytic species (Casieri et al. 2009). HWT might have certain *a priori* advantages

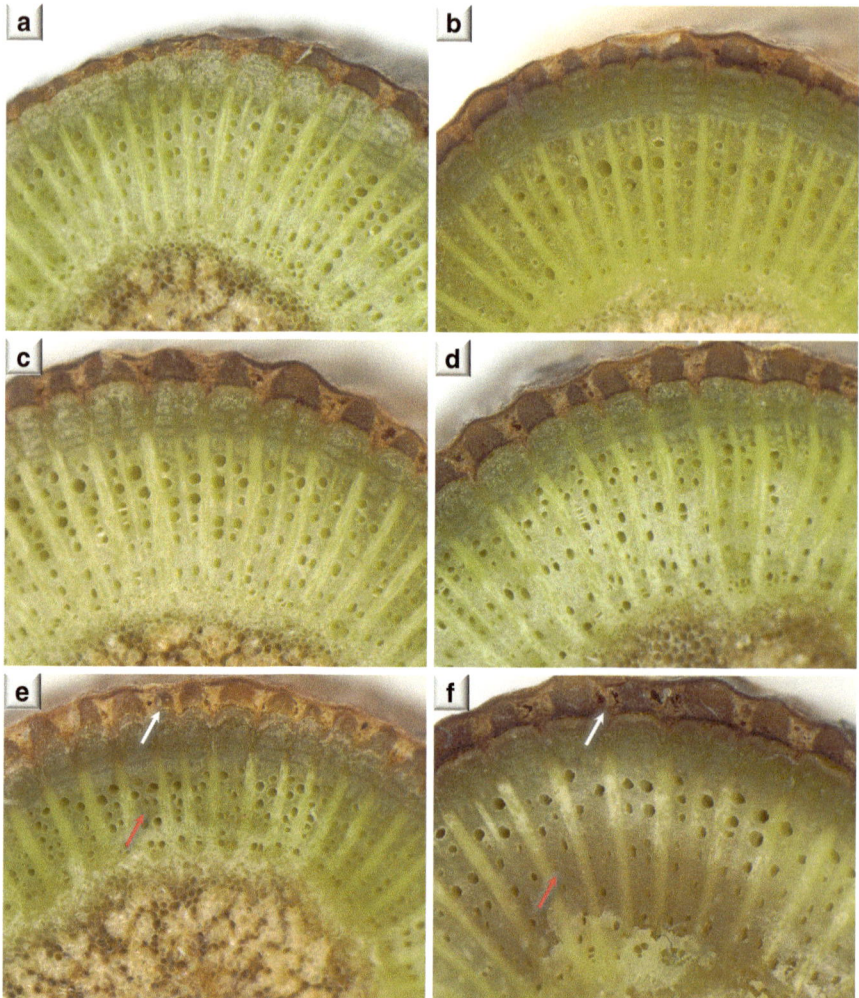

Fig. 7.17 Cross-sections of woody cuttings of *Vitis vinifera* cv Cabernet Sauvignon and Merlot undergoing hot-water treatment (HWT). (**a–c**) Cabernet Sauvignon. (**a**) Untreated control. (**b**) HWT at 50 °C. (**c**) HWT at 60 °C. (**d–f**) Merlot. (**d**) Untreated control. (**e**) HWT at 50 °C. (**f**) HWT at 60 °C. The red arrows show the damage to the conductive tissues (browning), the white arrows show the rupture spaces between the bark and vascular system

for the hygiene treatment of young plants since it reduces the frequency of isolation of several plant-pathogen fungal species and of certain pathogenic species that are strictly limited to grapevine. Nevertheless, the niches freed up by HWT are colonised by saprophytic species with lignocellulolytic activity, and these might subsequently play a major role in the expression of wood diseases.

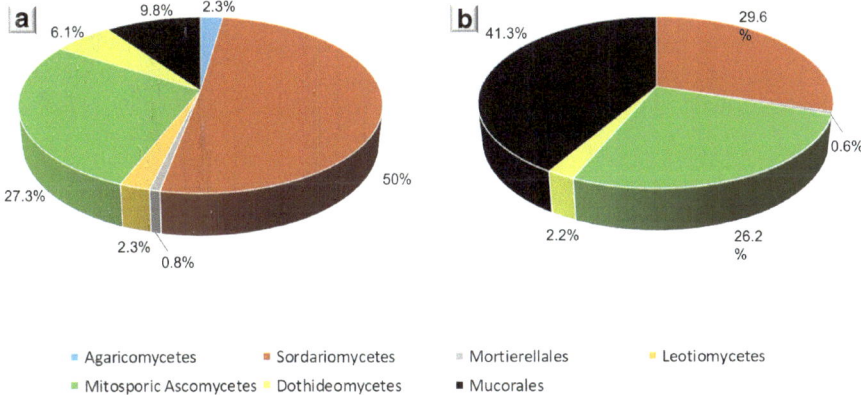

Fig. 7.18 Effect of hot-water treatment on the fungal community isolated from grapevine plants (*Vitis vinifera* cv Gamay) grouped by taxonomic class, from 1 year in the nursery onwards (Casieri et al. 2009). (**a**) Untreated plants; (**b**) HWT plants

7.5.3.2 Plant-Protection Products Used to Replace Oxyquinoline

Several natural products as well as aqueous solutions of organic fungicides have been evaluated before cold storage, before grafting, and prior to planting in the field. Good results were obtained against *P. chlamydospora* and *P. aleophilum* using carbendazime and dodecyl dimethyl ammonium chloride in the hydration baths of the propagation material (Gramaje et al. 2009). Hydrogen peroxide did not demonstrate significant efficacy. Bronicide, a mixture of halogenated alcohols and water, proved to be a good disinfectant, although it significantly reduces the nursery success rate. Benomyl (carbamate family), dodecyl dimethyl ammonium chloride and captan (phthalimide family) were also used successfully and without phytotoxicity (Fourie and Halleen 2006). Rego et al. (2009) demonstrated the efficacy of cyprodinil and fludioxonil against *Cylindrocarpon* spp. and the Botryosphaeriaceae family when dipping vine cuttings before grafting, thereby improving the quality of the planting material. Battiston et al. (2021) conducted tests to evaluate the efficacy of copper-based treatments formulated with hydroxyapatite (HA-Cu) and applied only on the rootstock whilst the scion was inoculated with a strain of *Phaeoacremonium minimum*. The results of this study were promising, showing a reduction in infection by *P. minimum* in the propagation material treated with HA-Cu formulations as well as movement of the HA-Cu formulations within plant tissues and their persistence over time.

Used as a disinfectant, ozone was studied as a means of controlling *Phaeoacremonium aleophilum* on propagation material to prevent esca (Pierron et al. 2015). *In vitro* tests on spores and after inoculation *in planta* suggest that ozonated water completely suppresses their germination. Moreover, 9 weeks after inoculation, fungal development was significantly reduced *in planta* by 50%, showing the use of ozonated water to have promise for limiting infection of the vine by *P. aleophilum* in nurseries. A further example is the possible use of chitosan (Nascimento et al. 2007), which has demonstrated a promising inhibiting effect on the mycelial growth of the main fungi involved in grapevine wood diseases, including *Phaeomoniella chlamydospora*, *Fomitiporia* sp. and *Botryosphaeria* sp. by foliar spray.

7.5.3.3 Biocontrol

Several species of *Trichoderma* were evaluated for their ability to curb the development of certain fungi associated with grapevine wood diseases, due to their hyperparasitic nature. These studies first focused on the treatment of pruning wounds, then on their effectiveness during the propagation material preparation stages in the nursery. Several species of *Trichoderma* seem to provide worthwhile efficacy in inhibiting the growth of fungi considered responsible for grapevine wood diseases or black foot in the field or nursery, such as *T. asperellum* and *T. gamsii* (Di Marco et al. 2022); *T. asperelloides, T. atroviride, T. harzianum, T. koningii, T. tomentosum, T. canadense* and *T. viticola* (Halleen et al. 2001; Langa-Lomba et al. 2022; Pollard-Flamand et al. 2022); *T. atroviride* (Pertot et al. 2016; van Jaarsveld et al. 2021). At present, formulated products are used preventively against infections and the development of grapevine wood diseases in vineyard nurseries. Grafted plants treated with different products containing, or not containing, an association of *Trichoderma* species, such as Trichodex Rootshield®, have enabled a marked decrease in the presence of *Phaeomoniella chlamydospora* in the wood necroses of young grapevine plants (Di Marco and Osti 2007). According to Mondello et al. (2018), this efficacy has more to do with the reinforcement of the plant's defences than with a direct curative effect on the fungus. Nevertheless, the *Trichoderma* sp. preventive application trials in the nursery have not succeeded in putting a stop to contamination due to wood diseases (Mondello et al. 2018). Only when the fungus is applied to the roots (dip treatment) does it serve to reinforce the morpho-physiological traits of the vine, thereby increasing the latter's tolerance to diseases (Di Marco and Osti 2007).

References

Armengol J, Vicent A, Torne L, García-Figueres F, García-Jiménez J (2001) Fungi associated with esca and grapevine declines in Spain: three-year survey. Phytopathol Mediterr 40:325–329. https://doi.org/10.1023/A:1008638409410

Battiston E, Compant S, Antonielli L, Mondello V, Clément C, Simoni A, Di Marco S, Mugnai L, Fontaine F (2021) In planta activity of novel copper(II)-based formulations to inhibit the esca-associated fungus P*haeoacremonium minimum* in grapevine propagation material. Front Plant Sci 12:649694. https://doi.org/10.3389/fpls.2021.649694

Becker H (1966) Untersuchungen über die Wirkung von Chinosol auf *Botrytis cinerea* in der Rebenveredelung. Wein Wiss 21:232–245

Becker H, Hiller M (1977) Hygiene in modern bench-grafting. Am J Enol Vitic. https://doi.org/10.5344/ajev.1974.28.2.113

Bender E, Vermorel V (1890) Le vigneron moderne, établissement et culture des vignes nouvelles. Bibliothèque du Progrès Agricole et Viticole, C. Coulet, Montpellier and G. Masson, Paris, p 446

Bertelli E, Mugnai L, Surico G (1998) Presence of *Phaeoacremonium chlamydosporum* in apparently healthy rooted grapevine cuttings. Phytopathol Mediterr 37:79–82

Buchs N, Braga-Lagache S, Uldry AC, Brodard J, Debonneville C, Reynard JS, Heller M (2018) Absolute quantification of grapevine red blotch virus in grapevine leaf and petiole tissues by proteomics. Front Plant Sci 9:1735. https://doi.org/10.3389/fpls.2018.01735

Carbonneau A (1985) The early selection of grapevine rootstocks for resistance to drought conditions. Am J Enol Vitic 36:195–198. https://doi.org/10.5344/ajev.1985.36.3.195

Casieri L, Hofstetter V, Viret O, Gindro K (2009) Fungal communities living in the wood of different cultivars of young *Vitis vinifera* plants. Phytopathol Mediterr 48(1):73–83

Caudwell A, Larrue J, Valat C, Grenan S (1990) Les traitements à l'eau chaude des bois de vigne atteints de la flavescence dorée. Prog Agric Vitic 107:281–286

Caudwell A, Larrue J, Boudon-Padieu E, Mclean GD (1997) Flavescence dorée elimination from dormant wood of grapevines by hot-water treatment. Aust J Grape Wine Res 3:21–25. https://doi.org/10.1111/j.1755-0238.1997.tb00112.x

Chaudière F (2013) La greffe en chip-budding sur vigne âgée: méthode et résultats de surgreffages réalisés de 2008 à 2012 dans le Médoc. Rev Suisse Vitic Arboric Hortic 45(6):358–365

Chaudière F (2017) La greffe anglaise herbacée de la vigne. Méthode et résultats des essais 2013b à 2016 dans l'Hérault. Rev Suisse Vitic Arboric Hortic 49(3):160–167

Cottral EH, Pascoe IG (2000) Developments in grapevine trunk diseases research in Australia. Phytopathol Mediterr 39:68–75. https://doi.org/10.14601/PHYTOPATHOL_MEDITERR-1534

Crous PW, Gams W, Wingfield MJ, Wyk PS (1996) *Phaeoacremonium* gen. nov. associated with wilt and decline diseases of woody hosts and human infections. Fungal Biol 88(5):786–796. https://doi.org/10.1080/00275514.1996.12026716

Davis MJ, Purcell AH, Thomson SV (1978) Pierce's disease of grapevines: isolation of the causal bacterium. Science 199:75–77. https://doi.org/10.1126/science.199.4324.75

Di Marco S, Osti F (2007) Effect of biostimulant sprays on *Phaeomoniella chlamydospora* and esca proper infected vines under greenhouse and field conditions. Phytopathol Mediterr 48(1):47–58. https://doi.org/10.14601/Phytopathol_Mediterr-1851

Di Marco S, Metruccio EG, Moretti S, Nocentini M, Carella G, Pacetti A, Battiston E, Osti F, Mugnai L (2022) Activity of *Trichoderma asperellum* strain ICC 012 and *Trichoderma gamsii* strain ICC 080 toward diseases of esca complex and associated pathogens. Front Microbiol 12:813410. https://doi.org/10.3389/fmicb.2021.813410

Ferreira JS, Wyk PS, Calitz FJ (2017) Slow dieback of grapevine in South Africa: stress-related predisposition of young vines for infection by *Phaeoacremonium chlamydosporum*. S Afr J Enol Vitic 20:43–46. https://doi.org/10.21548/20-2-2228

Fourie PH, Halleen F (2004) Proactive control of petri disease of grapevine through treatment of propagation material. Plant Dis 88:1241–1245. https://doi.org/10.1094/PDIS.2004.88.11.1241

Fourie PH, Halleen F (2006) Chemical and biological protection of grapevine propagation material from trunk disease pathogens. Eur J Plant Pathol 116:255–265. https://doi.org/10.1007/s10658-006-9057-9

Geiger A, Karacsony Z, Golen R, Vaczy KZ, Geml J (2022) The compositional turnover of grapevine-associated fungal communities is greater among intraindividual microhabitats and terroir than among healthy and esca-diseased plants. Phytopathology 112:1029–1035

Giménez-Jaime A, Aroca A, Raposo R, García-Jiménez J, Armengol J (2006) Occurrence of fungal pathogens associated with grapevine nurseries and the decline of young vines in Spain. J Phytopath 154:598–602. https://doi.org/10.1111/j.1439-0434.2006.01153.x

Goheen AC, Nyland G, Lowe SK (1973) Association of a rickettsia-like organism with Pierce's disease of grapevines and alfalfa dwarf and heat therapy of the disease in grapevines. Phytopathology 63:341–345

Gramaje D, Aroca A, Raposo R, García-Jiménez J, Armengol J (2009) Evaluation of fungicides to control petri disease pathogens in the grapevine propagation process. Crop Prot 28(12):1091–1097. https://doi.org/10.1016/j.cropro.2009.05.010

Graniti A, Surico G, Mugnai L (2002) Esca of grapevine: a disease complex or a complex of diseases? Phytopathol Mediterr 39:16–20. https://doi.org/10.14601/Phytopathol_Mediterr-1539

Grohs DS, Kurtz Almança AA, Fajardo TVM, Halleen F, Miele A (2017) Advances in propagation of grapevine in the world. Rev Bras Frutic 39(4):e-760. https://doi.org/10.1590/0100-29452017760

Halleen F, van der Vyver J, Fourie P, Schreuder W (2001) Effects of Trichoderma treatments on the occurrence of decline pathogens in the roots and rootstocks of nursery grapevines. Phytopathol Mediterr 40(4):473–478. https://doi.org/10.14601/Phytopathol_Mediterr-1619

Hofstetter V, Buyck B, Croll D, Viret O, Couloux A, Gindro K (2012) What if esca disease of grapevine were not a fungal disease? Fungal Divers 54:51–67. https://doi.org/10.1007/s13225-012-0171-z

Kızıldeniz T, Movila M, Michelotti Bettoni M, Abdullateef S, Candar S (2022) Grapevine propagation method with two temperature controlling process. Vitic Stud 2(1):45–53. https://doi.org/10.52001/vis.2022.10.45.53

Król E, Machowicz-Stefaniak Z (2008) Biotic effect of fungal communities inhabiting grapevine phyllosphere on *Phoma negriana*. Biologia 63:466–470. https://doi.org/10.2478/s11756-008-0074-3

Langa-Lomba N, Martín-Ramos P, Casanova-Gascón J, Julián-Lagunas C, González-García V (2022) Potential of native Trichoderma strains as antagonists for the control of fungal wood pathologies in young grapevine plants. Agronomy 12(2):336. https://doi.org/10.3390/agronomy12020336

Larignon P, Dubos B (1997) Fungi associated with esca disease in grapevine. Euro J Plant Path 103:147–157. https://doi.org/10.1023/A:1008638409410

Linder C, Schaub L, Klötzli-Estermann F (2010) Efficacité du traitement à l'eau chaude contre les oeufs de *Scaphoideus titanus*, vecteur de la flavescence dorée de la vigne. Rev Suisse Vitic Arboric Hortic 42(2):132–135

Márquez LM, Redman RS, Rodriguez RJ, Roossinck MJ (2007) A virus in a fungus in a plant: three-way symbiosis required for thermal tolerance. Science 315(5811):513–515. https://doi.org/10.1126/science.1136237

Millardet A (1877) La question des vignes américaines au point de vue théorique et pratique. Féret et Fils, Bordeaux, p 83

Mondello V, Songy A, Battiston E, Pinto C, Coppin C, Trotel-Aziz P, Clément C, Mugnai L, Fontaine F (2018) Grapevine trunk diseases: a review of fifteen years of trials for their control with chemicals and biocontrol agents. Plant Dis 102(7):1189–1217. https://doi.org/10.1094/PDIS-08-17-1181-FE

Monod V (2024) Deciphering the multifaceted determinants of esca incidence across a vineyards network. PhD Thesis, University of Neuchâtel, Switzerland

Morton L (1999) On the trail of black goo. In: Morton L (ed) Black goo—symptoms and occurrence of grapevine declines. IAS/ICGTD proceedings 1998. International Ampelography Society, Fort Valley, VA, pp 56–77

Mugnai L, Graniti A, Surico G (1999) Esca (black measles) and brown wood-streaking: two old and elusive diseases of grapevines. Plant Dis 83(5):404–418. https://doi.org/10.1094/PDIS.1999.83.5.404

Nascimento T, Rego C, Oliveira H (2007) Potential use of chitosan in the control of grapevine trunk diseases. Phytopathol Mediterr 46(2):218–224. https://doi.org/10.14601/Phytopathol_Mediterr-2156

Pertot I, Prodorutti D, Colombini A, Pasini L (2016) *Trichoderma atroviride* SC1 prevents *Phaeomoniella chlamydospora* and *Phaeoacremonium aleophilum* infection of grapevine plants during the grafting process in nurseries. BioControl 61:257–267. https://doi.org/10.1007/s10526-016-9723-6

Pierron RJG, Pages M, Couderc C, Compant S, Jacques A, Violleau F (2015) In vitro and in planta fungicide properties of ozonated water against the esca-associated fungus *Phaeoacremonium aleophilum*. Sci Hortic 189:184–191. https://doi.org/10.1016/j.scienta.2015.03.038

Pollard-Flamand J, Boulé J, Hart M, Úrbez-Torres JR (2022) Biocontrol activity of *Trichoderma* species isolated from grapevines in British Columbia against Botryosphaeria dieback fungal pathogens. J Fungi 8(4):409. https://doi.org/10.3390/jof8040409

Prodhomme D, Valls Fonayet J, Hévin C, Franc C, Hilbert G, de Revel G, Richard T, Ollat N, Cookson SJ (2019) Metabolite profiling during graft union formation reveals the reprogramming of primary metabolism and the induction of stilbene synthesis at the graft interface in grapevine. BMC Plant Biol 19(1):599. https://doi.org/10.1186/s12870-019-2055-9

Rego C, Nascimento T, Cabral A, Silva M, Oliveira H (2009) Control of grapevine wood fungi in commercial nurseries. Phytopathol Mediterr 48(1):128–135. https://doi.org/10.14601/Phytopathol_Mediterr-2881

Renault-Spilmont AS, Carrere C, Hmedi Y, Mathieu G (2021) X-ray tomography: a promising tool to assess the selection of good quality grafted vines. Internet J Viticult Enol. www.infovine.com

Rooney S, Gubler WD (2001) Effect of hot water treatments on eradication of *Phaeomoniella chlamydospora* and *Phaeoacremonium inflatipes* from dormant grapevine wood. Phytopathol Mediterr 40(4):467–472

Rumbos I, Rumbou A (2001) Fungi associated with esca and young grapevine decline in Greece. Phytopathol Mediterr 40:S330–S335. https://doi.org/10.36253/phyto-4911

Scheck H, Vasquez S, Fogle D, Gubler WD (1998) Grape growers report losses to black foot and grapevine decline. Calif Agric 52:19–23. https://doi.org/10.3733/ca.v052n04p19

Schweigkofler W, Prillinger H (1999) Molekulare Identifizierung und phylogenetische Analyse von endophytischen und latent pathogenen Pilzen der Weinrebe. Mitteilungen Klosterneuburg 49:65–78

van Jaarsveld WJ, Halleen F, Bester MC, Pierron RJ, Stempien E, Mostert L (2021) Investigation of Trichoderma species colonization of nursery grapevines for improved management of black foot disease. Pest Manag Sci 77:397–405. https://doi.org/10.1002/ps.6030

Viret N (2020) Alternatives aux produits à base d'oxyquinoléine pour le traitement des bois de pépinière: efficacité sur le développement des communautés fongiques, 2020. Bachelor's thesis, HES-SO University of Applied Sciences and Arts Western Switzerland, School of Viticulture and Oenology, Changins

Waite H, Morton LT (2007) Hot water treatment, trunk diseases and other critical factors in the production of high-quality grapevine planting material. Phytopathol Mediterr 46:5–1. https://doi.org/10.14601/Phytopathol_Mediterr-1857

Whiting EC, Khan A, Gubler WD (2001) Effect of temperature and water potential on survival and mycelial growth of *Phaeomoniella chlamydospora* and *Phaeoacremonium spp*. Plant Dis 85(2):195–201. https://doi.org/10.1094/PDIS.2001.85.2.195

Chapter 8
Disease Control

8.1 Fungicides in Grapevine

8.1.1 Historical Background

At a global level, the overwhelming majority of wine- and table grape varieties are susceptible to various potentially destructive fungal diseases and must be protected by preventive applications of fungicides. In the second half of the twentieth century several destructive pathogens, most notably powdery mildew, downy mildew and black rot, were introduced into Europe, triggering severe crises with a collapse in production and vineyard area. Intensive research led to the discovery of effective control methods, viz., dusting with sulphur to control powdery mildew (Marès 1856) and the application of copper in the form of Bordeaux mixture to control downy mildew (Millardet 1887). Until the mid-twentieth century, these two multi-site active substances of mineral origin were the only plant-protection products available to winegrowers. Since the 1950s, advances in chemistry have led to the development of organic fungicides from combinatorial chemistry, opening up new prospects in the battle against fungal diseases. Dithiocarbamates (ziram, thiram, mancozeb, metiram, propineb) begin appearing in the mid-1950s. Folpet appears in the early 1960s. In the 1970s, systemic fungicides from the group of benzimid-azoles (benomyl, thiophanate-methyl, carbendazim) used to control grey mould (*Botrytis cinerea*), which penetrate the plant's tissues and are transported in the plant by sap flows, were added to these contact products with preventive action. These were followed in the 1980s by the phenylamides (metalaxyl, benalaxyl, oxa-dixyl, ofurace) to control downy mildew. These active substances arouse considerable interest, their curative effect on existing infections and their systemic transport within the different organs of the plant foreshadowing a less intense, more targeted control approach than the preventive control previously practised. Unfortunately, close on the heels of their introduction, these hopes were shattered by the appearance of resistant populations of *Botrytis* and downy mildew. In the battle against

© The Author(s) 2025
O. Viret, K. Gindro, *Science of Fungi in Grapevine*,
https://doi.org/10.1007/978-3-031-68663-4_8

powdery mildew—long limited to wettable sulphur—sterol synthesis inhibitors (DMIs, SBIs) were developed at the same time as the dicarboximides (iprodione, vinclozolin, procymidone) used to control grey mould. These two groups of fungicides also select for resistant powdery mildew and Botrytis populations. In the 1990s, the strobilurins (azoxystrobin, trifloxystrobin, kresoxim-methyl, pyraclostrobin) synthesised according to the model of secondary metabolites of *Strobilurus tenacellus*, a forest fungus, provide a new mode of action with a broad spectrum of efficacy against *inter alia* downy and powdery mildew. Unfortunately, they very quickly select for resistant strains of downy mildew.

From the 1970s onwards, the undesirable side-effects of synthetic organic fungicides are noted, and voices are raised to limit their use. Within the context of the registration process, assessment of the effects of plant-protection products on the environment and human health gains in importance, and increasingly precise and stringent standards are introduced. In particular, in 2009 the European Union sets stricter requirements with elimination criteria (Directive 2009/128/EC). This new directive leads to the ongoing withdrawal of nearly two-thirds of the active substances previously authorised in Europe, with the retention of approx. 400 active substances.

8.1.2 Groups of Active Substances

The majority of fungicides interfere with the vital functions of pathogenic fungi, such as respiration, sterol biosynthesis and cell division. As well as posing risks for humans and non-target organisms, they may lead to selection for resistant fungal strains, depending on their mode of action. Modern fungicides act very specifically on the metabolism of pathogenic fungi, usually at very low effective concentrations. These unisite modes of action are sought after to limit potentially harmful side-effects in humans, other organisms and the environment in general. By contrast, multisite products such as copper, sulphur, dithiocarbamates or phthalimides are toxic for aquatic organisms or predators Moreover, due to its widespread and regular use, copper that does not biodegrade will steadily increase levels of heavy metals in the soil.

8.1.2.1 Mode of Action, Effects and Resistance Risk

The choice of active substances is key for the effective control of fungal diseases and forms an integral part of the control strategies pursued. Depending on the specific conditions at plot level, the susceptibility of the grape varieties and the seasonal dynamics of the grapevine pathogens, this choice is not always an easy one. Spraying schedules cannot be completely determined in advance: the broad brushstrokes can be laid down, but the growth of the vegetation and progress of the disease epidemiology require a certain flexibility to home in on the most suitable active substances based on their mode of action and their interaction with the plant tissues. In general, contact fungicides, penetrating, loco-systemic or translaminar fungicides, and systemic fungicides are distinguished.

- Contact fungicides act only on the surface of the plant organs which they cover, and hence only where the droplets generated by the spraying device have been deposited. Copper, wettable sulphur and folpet are examples of contact products.
- Penetrating or loco-systemic or translaminar fungicides are absorbed by the plant tissues, but only migrate locally within plant cell structures, at most from one side of the leaf to the other, or towards the adjacent organs. This is the case for numerous synthetic active substances such as demethylation inhibitors (DMIs or SBI for sterol biosynthesis inhibitors), strobilurins (quinone outside inhibitors, QoIs), or carboxylic acid amides (CAAs).
- Systemic fungicides boast the advantage of penetrating the plant tissues and then theoretically being transported throughout the entire plant. This is for example the case for aluminium fosetyl and the phenylamides. During the vine's vigorous growth stages, the active substance deposited by the spraying device is rapidly diluted by the increase in plant biomass, limiting the sought-after fungicidal effect. Furthermore, transport towards the bunches is often limited.

Fungicides act preventively, curatively or eradicatively. The great majority of active substances have a preventive effect, i.e. they protect the plant before infection by a pathogen occurs. This is the case for all contact products.

Curative active substances act by penetrating plant tissues and destroying fungal particles at the incubation stage, i.e. during the period from the colonisation of the plant tissues by the pathogen until the development of new infectious spores. At the time of a plant-protection treatment the successive infections of polycyclic pathogens such as downy or powdery mildew are never all at the same stage of development, especially when conditions are favourable and the infectious cycles are close together. Consequently, the efficacy of the curative mode of action is often overestimated by practitioners. The poorer the quality of fungicide application and the denser the vegetation, the more critical the limit of efficacy. Usually, these situations coincide with a period of rapid vine development when new leaves are exposed to pathogens on a daily basis. These biological processes can limit the curative effect, which generally also partially inhibits pathogen sporulation (anti-sporulant effect).

The eradicative effect of an active substance theoretically leads to the elimination of all fungal particles present on the plant tissues of the vine. At present, only sulphur in powdered form is able to ensure this effect against powdery mildew, provided that it enters the gaseous phase through sublimation of the solid form. This property of sulphur powder is achieved via passage from solid to gaseous form without passing through the liquid state, a phenomenon which is conditional upon strong luminosity and temperatures above 25 °C. Above 30–35 °C sulphur vapours can be phytotoxic for vines, with specific tolerances for different grape varieties. Sulphur vapours are directly absorbed by fungal particles and interfere with fungal metabolism by blocking cell respiration, which in turn inhibits the synthesis of proteins and nucleic acids.

Quite a few commercial fungicides contain several active substances with different modes of action, the aim being to control pathogens with a several-pronged approach and ensure the efficacy of the treatment. This combinatorial approach is dictated by the complexity of the biology of the pathogens and their infection dynamic associated with weather and climatic conditions, vine vegetative development, and vine susceptibility to pathogens and cultivation measures (vigour, canopy management). The use of combined products also combats the resistance of fungi to active substances, as well as strictly limiting the number of annual applications of unisite fungicides. When designing treatment schedules, it is imperative to respect the maximum number of applications of substances from the same chemical group, so as to guarantee the alternation of modes of action and prevent resistance. This principle is followed over the long term in the avoidance of repetition of the same strategy over several years, in particular when controlling pathogens with a high risk of resistance like *Botrytis cinerea*.

The risk of resistance is not only influenced by the mode of action of the active substances, but also by the biological properties of the pathogens in question. For grapevine pathogens, the greatest risk factor concerns *Botrytis cinerea* combined with the dicarboximides or benzimidazoles, and downy mildew combined with Qols or phenylamides. For these same pathogens, the use of the DMIs, phenylpyr-roles and anilinopyrimidines constitutes an average risk and the use of copper, sulphur dithiocarbamates, phthalimides or chlorothalonil constitutes a low risk (Brent and Hollomon 2007). Lastly, the resistance risk is additionally influenced by agronomic factors associated with the specificities of the crop and cultivation practices. This risk is relatively high for vines because they are a perennial crop treated at regular intervals of during the season to control downy and powdery mildew.

Below, the different groups of main active substances relevant to grapevines are presented according to their Fungicide Resistance Action Committee (FRAC) classification codes.

8.1.2.2 Methyl Benzimidazole Carbamates (MBCs, FRAC Code 01)

The group of Methyl benzimidazole carbamates (MBCs) contains active substances belonging to two chemical groups: the benzimidazoles (such as benomyl and car-bendazim) and the thiophanates (such as thiophanate-methyl). Their mode of action targets β-tubulin assembly, and hence inhibits mitosis. In grapevine they are used to control grey mould (*Botrytis cinerea*). The benzimidazoles are of very limited usefulness in viticulture due to the generalised resistance of numerous fungal populations to them. If benzimidazoles are used, the principles of resistance management must be implemented without fail in order to prevent a widespread selection of resistant fungal populations.

8.1.2.3 Dicarboximides (FRAC Code 02)

The dicarboximides appeared in the early 1970s (Fujinami et al. 1971) in the form of the three main active substances iprodione, vinclozolin and procymidone. Although their mode of action is not always fully clarified, the dicarboximides appear to inhibit a receptor in the osmoregulation signalling pathway (Yamaguchi and Fujimura 2005). The pathogen targeted in viticulture is grey mould (*Botrytis cinerea*). Due to the widespread cross-resistance among the active substances of this group their use is generally very limited.

8.1.2.4   Demethylation Inhibitors: Triazols (DMIs, SBIs for Sterol Biosynthesis Inhibitors, FRAC Code 03)

Demethylation inhibitors (DMIs) inhibit the synthesis of class I sterols which act on the enzyme lanosterol 14α-demethylase (Erg11p/CYP51). The sterols are essential components of eukaryotic cell membranes. The DMIs constitute one of the most powerful groups of fungicides for controlling numerous economically important fungal pathogens. The majority of DMIs form part of the triazole chemical group with 27 active substances, among these difenoconazole, tebuconazole, myclobutanil, flusilazole and penconazole. The first DMIs appeared in the early 1980s and until recently new active substances such as mefentrifluconazole have been regularly introduced on the market. There are significant differences in the activity spectra of the different active substances. In viticulture, DMIs are used to control powdery mildew (*Erysiphe necator*) and black rot (*Phyllosticta ampelicida*). Resistance is present in various fungal pathogens and although not disruptive, it represents a progressive shift in susceptibility which gradually weakens the efficacy of DMIs. Several resistance mechanisms are known, including different target-site mutations in the CYP51 gene (such as V136A), in the promoter region of CYP51 (Parker et al. 2014), or the overexpression of ABC transporters. In Europe, a majority of the triazoles are candidates for substitution within the context of the registration process of active ingredients owing to persistence and toxicity in the environment. Their progressive withdrawal from the market is foreseeable.

8.1.2.5   Phenylamides (PAs, FRAC Code 04)

Appearing in the late 1970s, the group of phenylamides (PAs) includes benalaxyl, benalaxyl-M, metalaxyl and metalaxyl-M (=mefenoxam). Benalaxyl and metalaxyl are racemic mixtures and only the 'R' enantiomers, benalaxyl-M and metalaxyl-M, are active against Oomycetes such as downy mildew (*Plasmopara viticola*). PAs interfere with RNA polymerase 1 and thus inhibit ribosomal RNA synthesis. Phenylamides belong to the rare systemic active substances. Being liposoluble, they penetrate plant tissues rapidly and are distributed acropetally in the plant via the xylem (Kennelly et al. 2007). PAs prevent spore germination and limit germinative-tube formation, mycelium development and sporulation. A still-unknown mutation triggers a specific resistance which is widespread and limits the efficacy of this group. Phenylamides are used to control downy mildew (*P. viticola*), often in a mixture with a multisite active substance.

8.1.2.6   Amines: Piperidines (FRAC Code 05)

Piperidines are class II sterol-biosynthesis inhibitors belonging to the group of amines. Among others, they include the active substances fenpropidin and spiroxamine, used to control powdery mildew (*E. necator*) of grapevine. The piperidines inhibit two different enzymes (erg24 and erg2) in the sterol biosynthesis pathway, resulting in a low resistance risk. Fenpropidin may cause phytotoxicity symptoms under specific conditions and in sensitive cultivars.

8.1.2.7   Succinate Dehydrogenase Inhibitors (SDHI, FRAC Code 07)

Succinate dehydrogenase inhibitors block mitochondrial respiration (Keon et al. 1991). SDHIs encompass over 20 active substances belonging to a variety of chemical groups. Some such as boscalid have been used since the early 2000s, while others such as fluopyram and fluxapyroxad are more recent. SDHIs are used in viticulture to control grey mould (*B. cinerea*) and powdery mildew (*E. necator*). The resistance risk associated with different mutations at the site of action ranges from medium to high. In Europe, fluxapyroxad is a candidate for substitution within the context of the registration process of the active substances due to its excessive persistence in the soil.

8.1.2.8   Anilinopyrimidines (APs, FRAC Code 09)

The anilinopyrimidines were launched on the market in the early 1990s. The group contains three active substances—cyprodinil, pyrimethanil and mepanipyrim—which are used in viticulture to control grey mould (*B. cinerea*). Although their mode of action is not fully elucidated, APs are thought to inhibit methionine amino acid biosynthesis (Fritz et al. 2003). The risk of resistance is considered moderate and in Europe the presence of resistant strains of *Botrytis cinerea* is generally low. In Europe, cyprodinil is a candidate for substitution due to its high toxicity for aquatic organisms as well as its persistence in water.

8.1.2.9   Strobilurins or Quinone Outside Inhibitors (QoIs, FRAC
Code 11)

The strobilurins or QoIs are on the market since the 1990s. The antifungal substances strobilurin A and B were isolated from the Basidiomycete fungus *Strobilurus tenacellus* as secondary metabolites in the late 1970s (Anke et al. 1977); however, these molecules were not sufficiently photostable for use as fungicides. By modifying their chemical structure to render them photostable and increase their efficacy, chemists obtained azoxystrobin, which was marketed first. The QoIs have a very broad spectrum of action and a high efficacy. They are used in viticulture to control *inter alia* downy (*P. viticola*) and powdery mildew (*E. necator*), black rot (*P.*

ampelicida), rotbrenner (*Pseudopezicula tracheiphila*) and excoriosis (*Diaporthe ampelina*) of grapevine. The QoIs contain over 20 active substances from different chemical groups. The most widely used active substances in viticulture are azoxystrobin, trifloxystrobin, kresoxim-methyl, famoxadone and fenamidone. QoIs inhibit mitochondrial respiration by preventing electron transport at the enzyme cytochrome b, thereby blocking the production of energy (Becker et al. 1981). Unfortunately, resistant strains quickly emerged for various pathogens. Resistance is linked to a point mutation of the target gene cytochrome b (G143A) which leads to a high resistance factor. Resistance selection was at times very rapid, as in the case of *Plasmopara viticola*. For downy and powdery mildew, resistance is very widespread for azoxystrobin, trifloxystrobin, kresoxim-methyl, famoxadone and fenamidone.

8.1.2.10 Phenylpyrroles (PPs, FRAC Code 12)

In viticulture, the group of phenylpyrroles is represented by the active substance fludioxonil, used to control grey mould. Fludioxonil inhibits the transmission of a signal in osmotic regulation by interacting with a mitogen-activated protein (MAP) histidine kinase. There is medium-to-low risk of sporadic resistance, the mechanism of which remains unknown to date. Fludioxonil is a candidate for substitution due to its toxicity for aquatic organisms.

8.1.2.11 Azanaphtalenes (AZNs, FRAC Code 13)

The azanaphtalenes consist of two active substances used to control powdery mildew (*E. necator*): quinoxyfen and proquinazid. Although their mode of action is unknown, the hypothesis of signal transmission inhibition has been posited. AZN-resistant strains have been detected in *Erysiphe necator* with moderate resistance factors. Quinoxyfen is a candidate for substitution due to its persistence, bioaccumulation and toxicity for the environment.

8.1.2.12 Keto-Reductase Inhibitors (KRIs, FRAC Code 17)

Keto-reductase inhibitors comprise two substances, fenhexamid and fenpyrazamine, both of which are used to control grey mould (*B. cinerea*). KRIs are class III sterol-biosynthesis inhibitors that interfere with the C4-demethylation of the enzyme 3-keto reductase. The resistance risk is low-to-medium and strains of resistant Botrytis are present at low-to-medium frequencies.

8.1.2.13 Quinone Inside Inhibitors (QiIs, FRAC Code 21)

The group of quinone inside inhibitors contains cyazofamid and amisulbrom, both of which are active against downy mildew (*P. viticola*). QiIs inhibit mitochondrial respiration by interacting with cytochrome b. The mode of action is different from that of the strobilurins (QoIs), and there is no cross-resistance. The risk of resistance is considered medium-to-high. Resistant strains are described and present in the majority of vineyards worldwide.

8.1.2.14   Dinitroanilines (FRAC Code 29)

The group of dinitroanilines contains one active substance, fluazinam. Thanks to its broad spectrum of action, it is used in viticulture to control downy (*P. viticola*) and powdery mildew (*E. necator*), rotbrenner (*P. tracheiphila*), excoriosis (*D. ampelina*) and *Botrytis*. Fluazinam is an uncoupler of oxidative phosphorylation, which inhibits cell respiration. The risk of resistance risk is low; nevertheless, fluazinam can trigger skin allergies and is moderately toxic for beneficial organisms like *Typhlodromus* predatory mites.

8.1.2.15   Carboxylic Acid Amides (CAAs, FRAC Code 40)

The group of carboxylic acid amides contains, among others, the active substances benthiavalicarb, dimethomorph, mandipropamid, iprovalicarb and valifenalate, all used to control downy mildew (*P. viticola*). The mode of action is the inhibition of a cellulose synthase involved in the biosynthesis of the cell walls of Oomycetes (Blum et al. 2010). Resistance to CAAs is present in the vast majority of vineyards at variable frequencies.

8.1.2.16   Benzamides (FRAC Code 43)

A member of the benzamide group, fluopicolide is used to control downy mildew (*P. viticola*). Although its mode of action is not known precisely, it is thought to interfere with spectrin-type proteins in the cytoskeleton of Oomycetes (Toquin et al. 2006). Resistant isolates have been detected and the risk of resistance has been rated as moderate.

8.1.2.17   Oxysterol-Binding Protein Homologue Inhibitors (OSBPIs, FRAC Code 49)

The family of oxysterol-binding protein homologue inhibitors includes the active substances oxathiapiprolin and fluoxapiprolin (Pasteris et al. 2016). The oxysterol-binding proteins are involved in lipid transfer between membranes. Their inhibition can impact different cell processes such as signal transmission, maintenance of cell-membrane integrity and the formation of essential complex lipids for the cell. OSBPIs are used to control downy mildew (*P. viticola*). The risk of resistance is considered to be medium-to-high and site-of-action mutations have been described.

8.1.2.18 Benzophenones (FRAC Code 50)

The group of benzophenones contains the active substances metrafenone and pyrio-fenone, which interact with the cytoskeleton of fungi, inhibiting the functions of actin and myosin fibres (Nave et al. 2008). Benzophenones are used to control powdery mildew (*E. necator*). The risk of resistance is rated as moderate, and resistant isolates have been detected in Europe.

8.1.2.19 Phenyl-Acetamide (FRAC Code U06)

Cyflufenamid is used to control powdery mildew (*E. necator*) (Sano et al. 2007). It forms part of the phenyl-acetamide group whose mode of action remains unknown. No resistance has been described to date.

8.1.2.20 Copper (FRAC Code M01)

Copper has been known since the late nineteenth century in Bordeaux mixture (*bouillie bordelaise*) for its efficacy in controlling downy mildew (*P. viticola*) and Oomycetes in general. Copper is one of the most-used multisite active substances in the world, despite its persistence in the soil and its toxicity for aquatic organisms. Copper ions (Cu^{2+}) are absorbed at the cell surface and can replace hydrogen (H^+), potassium (K^+), calcium (Ca^+) and magnesium (Mg^+) ions, leading to penetration of the copper in the cells. Copper's biocide effect is associated with the affinity of copper ions for the imidazole, carboxyl, phosphate, amine and hydroxyl groups, which interfere with protein synthesis and inhibit the pathogen's enzymes via protein denaturation, RNA degradation or DNA modification (Gaetke et al. 2014; Anant et al. 2018). The multiplicity of modes of action makes the probability of selecting for copper-resistant downy mildew strains practically nil. Copper has an inhibitory effect on powdery mildew (*E. necator*), grey mould (*B. cinerea*), acid rot (yeasts and bacteria), anthracnose (*Elsinoë ampelina*), black rot (*P. ampelicida*) and bacteria (bacterial blight, *Xylophilus ampelinus*; Pierce's disease, *Xylella fastidiosa* subsp. *fastidiosa*). It can have a phytotoxic effect on plants, causing leaf necrosis or reduced root growth. Despite this, grapevine is one of the least copper-sensitive plants. Copper is used primarily in the form of oxychloride $Cu_2Cl(OH)_3$, sulphate $CuSO_4$, hydroxide $Cu(OH)_2$ or oxide Cu_2O, the metallic-copper contents of which vary considerably.

8.1.2.21 Sulphur (FRAC Code M02)

Sulphur is the first fungicide in the history of viticulture. This mineral element—needed by plants both for the metabolism of nitrogen and the synthesis of sulphur amino acids—is essentially utilised as a fungicide for controlling powdery mildew

(*E. necator*). It is also effective against excoriosis (*D. ampelina*) and eriophyid mites (grape erineum mite, *Colomerus vitis* and leaf rust mite, *Calepitrimerus vitis*). By blocking cell respiration and inhibiting protein and nucleic acid synthesis, its multisite mode of action depletes overall cell energy levels (Pezet and Pont 1977; Beffa et al. 1987). Thanks to these properties, selection for sulphur-resistant fungal isolates can virtually be ruled out.

Sulphur can be used in wettable or powder form. The vapour phase of the powder form endows it with a preventive action on conidia, a curative action on mycelium and an eradicative action through the desiccation of fungal structures.

Sulphur can be phytotoxic for grapevine through the production of sulphuric acid (H_2SO_4), which burns the plant tissues. This phenomenon is observed at high temperatures when sulphur sublimation occurs, and at high relative humidity, depending on plant water status, grape variety and the sulphur doses applied.

8.1.2.22 Dithiocarbamates (FRAC Code M03)

The dithiocarbamates, which include active substances such as ziram, thiram, mancozeb, metiram and propineb, make their appearance from the mid-1940s in the case of thiram and in the early 1960s for mancozeb. They are broad-spectrum multisite fungicides active against downy mildew (*P. viticola*), rotbrenner (*P. tracheiphila*), black rot (*P. ampelicida*) and excoriosis (*D. ampelina*). The dithiocarbamate group also includes insecticides (carbaryl) and herbicides (tri-allate). Mancozeb is toxic for predator mites. In 2020 the European Food Safety Authority (EFSA) classified mancozeb as toxic for reproduction and an endocrine disruptor, leading to its withdrawal from the European market.

8.1.2.23 Phthalimides (FRAC Code M04)

The phthalimide group includes two active substances, folpet and captan, which are widely used in viticulture and arboriculture respectively. Folpet, which appeared in the early 1960s, is a multisite contact fungicide used to control downy mildew (*P. viticola*), rotbrenner (*P. tracheiphila*) and excoriosis (*P. ampelicida*). A partner of choice in the *Plasmopara viticola* resistance management strategy, folpet is chronically toxic to aquatic organisms and is suspected of being carcinogenic. Folpet degrades rapidly in water, where it has a half-life of less than 3 h.

8.1.2.24 Chloronitriles (FRAC Code M05)

The chloronitrile group contains a single active substance, chlorothalonil, which has been used in viticulture since the 1970s. Chlorothalonil is a broad-spectrum multisite fungicide used in viticulture to control downy mildew (*P. viticola*), rotbrenner (*P. tracheiphila*) and excoriosis (*P. ampelicida*), and can cause skin allergies.

Following the re-evaluation of the health risk of chlorothalonil in Europe in 2019 and its classification as a probable carcinogen, attention has focused on the metabolites of chlorothalonil and their recurrent presence in the subterranean waters of agricultural regions as well as in drinking water. As a result, it was withdrawn from the European market.

8.1.2.25   Quinones (FRAC Code M09)

Dithianon is a broad-spectrum multisite fungicide from the group of quinones/ anthraquinones that has been used in viticulture since the late 1960s to control downy mildew (*P. viticola*), rotbrenner (*P. tracheiphila*) and excoriosis (*P. ampelicida*). Its ecotoxicological profile is promising, apart from an acute toxicity for fish. Dithianon can also cause some skin allergies.

8.1.2.26   Phosphonates (FRAC Code P07)

The phosphonates group includes aluminium-fosetyl and phosphorous acid and salts (potassium phosphonate). Phosphonates are systemic active substances transported acropetally through the xylem. They have a direct fungicidal effect, as well as triggering the plant's defence mechanisms (elicitor effect). Phosphonates are principally used to control downy mildew (*P. viticola*). Resistance risk is deemed to be low, with no confirmed resistance to date. Phosphonates degrade into phosphorous acid, which is highly persistent and readily detectable in wine. For several years now, potassium phosphonate has been used in organic viticulture in Germany to enhance the efficacy of copper in humid conditions.

8.1.3   Natural Fungicides

Natural biofungicides and microorganisms offer sustainable, environmentally friendly alternatives for protecting grapevines from fungal diseases. In Europe in particular, societal pressure to cut back on the use of synthetic plant-protection products in order to reduce the risks for human health and the environment is pushing winegrowers and farmers in general to examine alternatives. The efficacy of natural extracts from plants and fungi, products of mineral origin and micro- and macro-organisms is being intensively researched (Zaker 2016). The use of these alternative biocontrol products offers several benefits. They are usually less toxic for the environment and human health than numerous chemical fungicides, and they often have multisite or alternative modes of action that reduce the risk of resistance in the pathogens. A growing number of 'biocontrol' alternatives are available on the market. It is important to note, however, that the efficacy of these alternative products can vary significantly depending on environmental conditions, pathogen pressure and practices specific to each vineyard. Consequently, an integrated approach combining different disease management methods is still often the best strategy for the appropriate protection of vines considering the disease pressure and the cultivar's sensitivity. The main obstacle to the use of such fungicides is their limited, even insufficient efficacy, which implies significant risk-taking for growers and necessitates more frequent applications, thereby increasing production costs.

8.1.3.1 Plant Extracts (FRAC Code BM01)

Numerous plant extracts are used to protect grapevine against major fungal diseases. Different essential oils such as cinnamon (*Cinnamomum verum*), thyme (*Thymus vulgaris*) or oregano (*Origanum vulgare*), used on their own or in a mixture, exhibit promising efficacy.

In biodynamics, different plant extracts such as horsetail (*Equisetum* sp), common osier (*Salix vinimalis*), willow (*Salix* sp), valerian (*Valeriana officinalis*) and stinging nettle (*Urtica dioica*) are regularly used in the form of preparations generally in mixture with copper and sulphur. Chitosan, which stimulates the natural defence mechanisms of the plant, is a polymer derived from the chitin present in the shells of insects and crustaceans as well as the cell walls of certain fungi. Under high disease pressure and with sensitive grapevine cultivars, these plant extracts show limited efficacy against the principal fungal diseases.

8.1.3.2 Mineral Substances

Apart from copper and sulphur which have been described above, various rock powders such as sulphurated clays, kaolin and talc are used to control grey mould (*B. cinerea*) or downy mildew (*P. viticola*). These rock powders act either through a direct fungicidal effect or by forming a physical barrier. Talc and kaolin are sometimes contaminated by heavy metals, and their use can be problematic for the environment.

8.1.3.3 Microorganisms (FRAC Code BM02)

Various bacterial and fungal species are used to protect the vine through antagonism, i.e., by occupying an ecological niche in competition with the pathogens. *Bacillus subtilis* is a bacterium widely used to control a range of fungal diseases, including grey mould (*B. cinerea*), by producing antimicrobial compounds inhibiting the growth of pathogenic fungi such as downy (*P. viticola*) and powdery mildew (*E. necator*). Different species of fungi of the genus *Trichoderma spp* colonise the roots, soil and/or the phyllosphere and surface of the bunches, competing for space and nutrients with the pathogenic fungi and thereby reducing their impact on the vine. The use of hyperparasites such as the fungus *Ampelomyces quisqualis* has also been attempted as a means of controlling powdery mildew with mitigated success.

8.1.3.4 UV Light

The efficacy of ultraviolet B rays (wavelength 280–290 nm) on ectoparasites like powdery mildew has been demonstrated in various cultivated plants such as strawberry (Onofre et al. 2021) and vine (Gadoury et al. 2023). Natural UVB rays from solar radiation partially damage the DNA of the pathogen, which can regenerate with the help of a photolyase enzyme activated by blue light and UVA rays. Through nocturnal exposure of the vine to UVB or UVC rays (maximum wavelength 254 nm) produced by a specific infrastructure, this DNA restoration mechanism is circumvented, leading to a lethal effect for the pathogen (Gadoury et al. 2023). UVC rays are also known to stimulate the natural defence mechanisms of plants, in particular the phytoalexins of the vine (Marti et al. 2014). Exposing the canopy to UV light at the optimal timing in accordance with the epidemiological development of pathogens poses a significant challenge, as does determining the duration necessary to activate natural defense mechanisms induced by UV rays to ensure sufficient efficacy.

8.2 Control Strategies

Regardless of the crop management techniques chosen, the active control of fungal diseases via the regular application of fungicides remains indispensable for vineyards planted with *Vitis vinifera* grape varieties. Long limited to the use of copper to control downy mildew and sulphur to control powdery mildew, control strategies saw the development of numerous synthetic chemical active substances from the 1950s onwards, leading to the safeguarding of grape production without particular regard for the protection of the environment and user. In the mid-1950s this gave rise to the initial discussions on limiting inputs in farming and viticulture, leading to the formulation of the principles of integrated and organic production (Boller et al. 2004; Wijnands et al. 2012) as well as those of anthroposophy-based biodynamic production. These different approaches target all the economic, environmental and social pillars of sustainability (OIV 2016) by proposing specific practices for the control of fungal diseases and the cultivation of grapevines in general. Compliance with the specifications generally confers the entitlement to use certification labels specific to winegrowing countries and regions.

8.2.1 Integrated Pest Management (IPM)

The concept of integrated protection or control was developed from the 1950s onwards, and primarily targeted pests in field crops (Delucchi 1987). Swiss agronomic research played a key role in the development of this concept (Baggiolini 1990), which is nowadays widely recognised. The principle consists in observing the crop and its environment and in anticipating and identifying the harmful organisms, only applying the plant-protection treatments when there is an actual risk of the diseases or pests developing and prioritising ecological and biological approaches (Fig. 8.1).

In Switzerland, the most spectacular successes have been achieved with the biological control of two spotted spider mites (*Tetranychus urticae*) and red spider mites (*Panonychus ulmi*) with naturally present or introduced predatory mites of the group Phytoseiidae (*Typhlodromus pyri, Amblyseius andersoni, Kampimodromus aberrans, Euseius finlandicus*). A second example is given by the control of the vine mothes (*Lobesia botrana* and *Eupoecilia ambiguella*) through sexual confusion with the aid of synthetic pheromones (Linder et al. 2016). These biological and biotechnological control strategies have rendered the use of acaricides and insecticides permanently superfluous. Integrated control encompasses all prophylactic cultivation measures leading to the creation of a favourable environment for beneficial organisms, with plant protection products only used as a last resort (Fig. 8.1). Fungal disease control strategies are trickier to implement if the decision is made to forgo synthetic active substances. The currently available alternatives involve more frequent intervention owing to the limited effect of the natural active substances, with the increased production costs calling the economic sustainability of the approach into question. The most sensible outcome is the planting of grape varieties with greater resistance to fungal pathogens. This approach helps mitigate the risk of economic losses and provides greater flexibility in disease management.

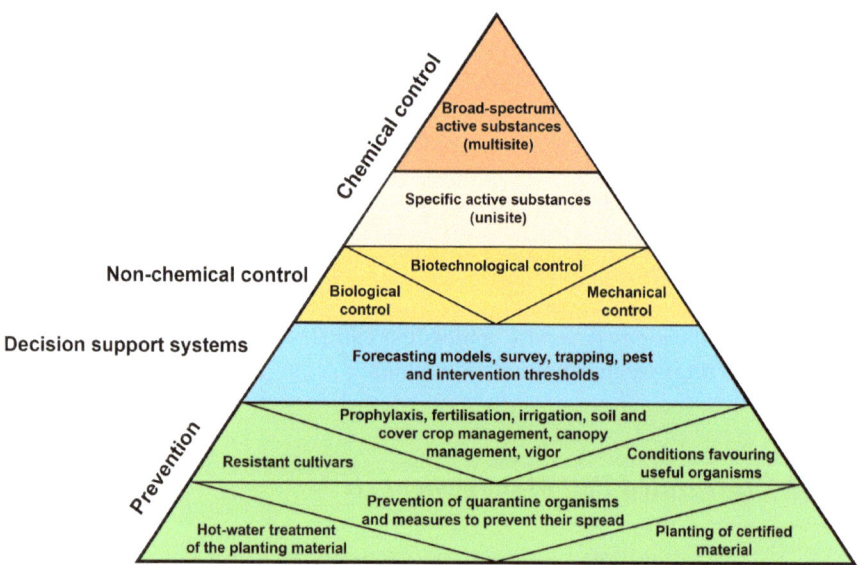

Fig. 8.1 Principles of integrated pest management (IPM) in viticulture prioritising all possible preventive and prophylactic approaches as well as non-chemical control strategies over the use of chemical active substances

The guidelines for ecological and integrated production aim to ensure the production of healthy grapes and quality wine products with a minimum of residues; to protect the health of producers during the handling of inputs; to seek and maintain a high level of biodiversity within and in the environs of the vineyard ecosystem; to prioritise gentle, biological and biotechnological methods, using natural resources and natural regulation mechanisms first and foremost; and to preserve soil fertility over the long term and reduce to a minimum the impacts of vineyard practices on the environment (water, soil, air). Today, this progressive technical management approach that has led to countless ecological and qualitative advances is standard practice in all the world's vineyards, to the point where it is termed conventional viticulture.

As part of its commitment to sustainability, the International Organisation of Vine and Wine (OIV) defines the following in its resolutions:

- integrated production (VITI 01-1999 and CST 01-2004),
- the preservation of biodiversity (VITI 01-2002),
- organic production (ECO 460-2012),
- sustainable grape production (CST 01-2004, CST 01-2008 and VITI 422-2011).

8.2.2 Organic Pest Management

The need to respect the environment is constantly drummed into the public and represents a major social issue for today's farmers and winegrowers. This trend is reflected in the development of organic and biodynamic production based on the use of natural preparations, copper and sulphur. Even with these approaches, however, it remains necessary to treat susceptible varieties of *Vitis vinifera* to control fungal diseases; sometimes the number of applications exceeds those of integrated production, due to the authorised preparations being only partially effective.

In organic production the protection of the vineyard is chiefly based on prophylactic measures, including the choice of site and grape variety, the upkeep of the vine, and the beneficial organisms present. Priority is given to the planting of grape varieties and clones with low susceptibility to fungal diseases and to sensible soil and manure maintenance through the recommendation of permanent grass cover and organic fertilisers. Beneficials are encouraged through the planting of cover crops between grapevine rows, a diversified maintenance strategy for the space between vines, the establishment of hedgerows, and the building and maintenance of dry-stone walls to uphold terraces in steep vineyards. The permitted active substances for controlling fungal diseases are copper products, sulphur, and all non-synthetic chemical preparations such as potassium bicarbonate, fennel oil, sulphurated clays, extracts of horsetail (*Equisetum* sp.) or other plants, laminarin and various antagonists (*Bacillus subtilis*, *Aureobasidium pullulans*, *Gliocladium catenulatum*, etc.). The list of plant-protection product inputs varies according to the specific legislation of the individual wine-producing countries regarding plant-protection product authorisation.

8.2.3   Biodynamic Pest Management

Organic farming has its origins in biodynamic production. An organic-dynamic approach, biodynamics implements a philosophical dimension developed by the anthroposophist and philosopher Rudolf Steiner (1861–1925) that traces its origins back to 1924 and which now has global reach (Paull 2011a, b). The concept of bio-dynamics is based on the autonomy of the individual farm, the reduced use of inputs (plant matter, manure, plant-protection products) and recourse to natural prepara-tions. This respect for and return to nature is a response to the industrialisation of agriculture, which has led to the use of mineral fertilisers and synthetic plant-protection products (Kirchmann et al. 2008). Organic soil amendments are based on compost derived from nine preparations (nos. 500–508) whose recipes were devel-oped by Steiner. Considerable focus is placed on the cosmic aspect, with the consid-eration—variable, depending on applications and trends—of the synodic (waxing and waning moon) and sidereal (signs of the zodiac) rhythms. Depending on the position of the moon with respect to the 12 constellations of the zodiac, the latter define flower, root, leaf, fruit and grain days. This concept gave rise to Maria Thun's sowing calendar, widely disseminated among followers of this approach.

In the fight against fungal diseases, dynamised preparations based on plant extracts are said to exert a homeopathic effect suitable for activating the cosmic forces of the planets to support a strong growth process and limit the development of pathogens. The most common preparations are based on horsetail (*Equisetum* sp.), stinging nettle (*Urtica dioica*), willow (*Salix* sp.) or calcified algae powder (diatomaceous earth) containing silica and generally mixed with copper and sulphur of mineral origin.

8.3   Disease Management

Cultivating *Vitis vinifera*, by far the most common vine species in the world, requires an active battle against pathogens. The use of plant-protection products at regular intervals is essential for ensuring high-quality yields and grapes, regardless of the technical strategies implemented (organic, biodynamic, integrated production, sus-tainable or conventional). Successful control is only indirectly linked to the number of plant-protection treatments applied during the growing season and depends on a number of interacting parameters (Fig. 8.2). Geographic situation, climatic condi-tions, varietal and clonal sensitivity, the physiological state of the vine, type of plant-protection products used, their characteristics and mode of action, the quality of application connected with the type of sprayer used, leaf-wall microclimate and the overall management of the vineyard influence the efficacy of the control measures.

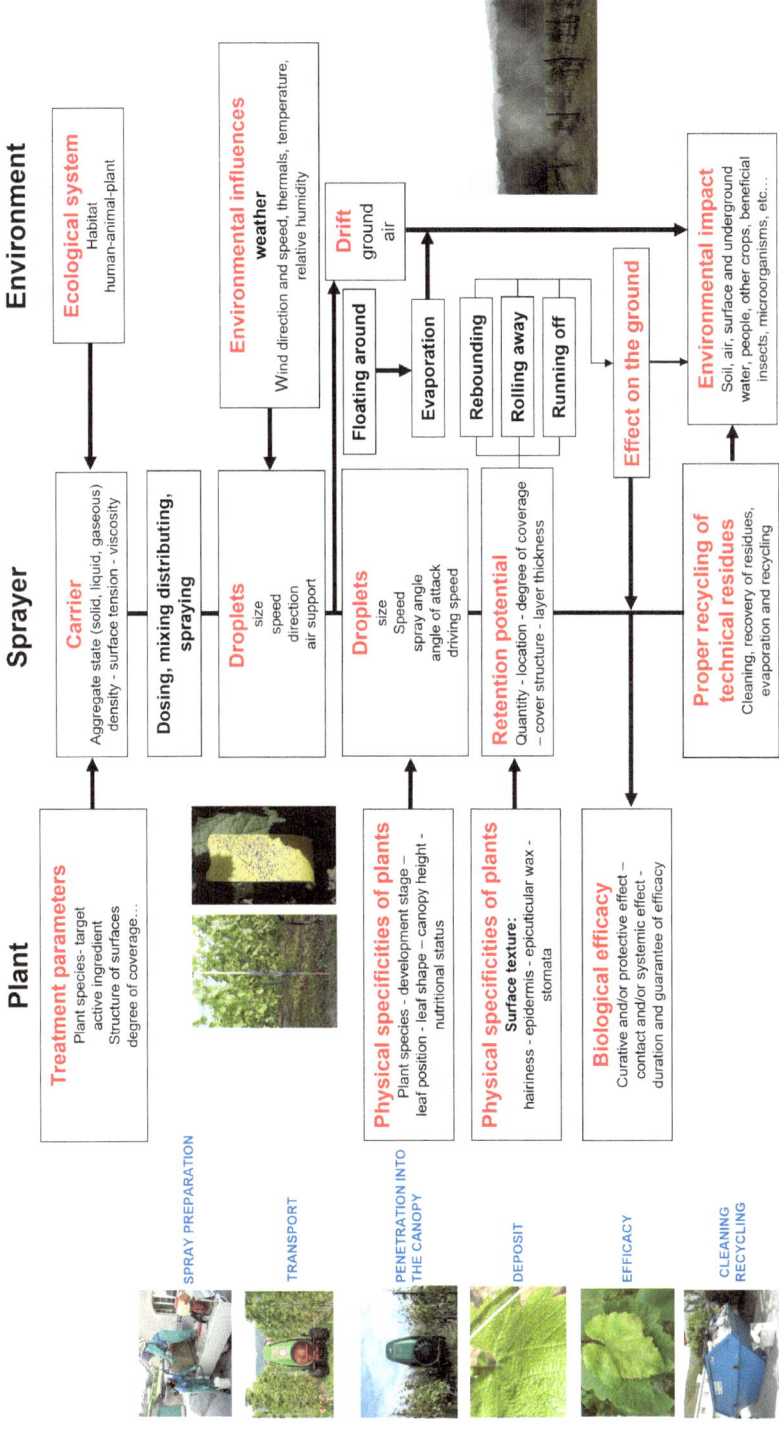

Fig. 8.2 Various interacting parameters (plant, sprayer, environment, weather) influencing disease-control success, from fungicide solution preparation to droplet generation, deposit on the target, followed by cleaning of the spraying device

The application of the plant protection products is a delicate technical operation, and detailed knowledge is required to perform the work accurately, effectively and in an environmentally friendly manner.

For products dissolved in water, the principle consists in producing more-or-less-fine droplets from a flow of liquid, which are projected onto the plant organs to be protected. In this case, the water merely serves to transfer the active substance to the organs of the vine. The aim is to cover the plant surfaces as evenly and uniformly as possible (Fig. 8.3), while avoiding any unintentional dispersal into the environment (Fig. 8.4). Preparations in powdered form, e.g. sulphur used to control powdery mildew, cannot be applied precisely, which leads to an unacceptably diffuse, uncontrolled dispersal in the environment (Fig. 8.5).

Pathogen control decisions must consider plant protection risks in a viticultural (siting of the vineyard, variety, clone), meteorological and environmental context, taking into account issues regarding public health and the protection of water, air and beneficial organisms, as well as harvest quality and economic considerations. These elements are monitored at different levels by specific regulations which may vary from country to country.

Fig. 8.3 Optimal spraying results in an even distribution of droplets on the leaf surface (**a**) that can be viewed on water-sensitive paper at the level of the entire canopy (**b**)

Fig. 8.4 Sprayer calibration aims to prevent undesirable environmental contamination caused by excessive drift

Fig. 8.5 The application of power-formulated plant protection products (e.g. sulphur powder to control powdery mildew) leads to uncontrolled and unacceptable environmental contamination

Consumers expect wine sector production to be as eco-friendly as possible, which implies taking more risks by reducing plant protection inputs and searching for relatively effective alternatives to fungicides derived from combinatorial chemistry (i.e. synthetic fungicides) in particular. In this context, varietal susceptibility is at the heart of the development of global viticulture, which dates back to the late nineteenth century with the creation of interspecific hybrids resistant to the main fungal diseases (downy mildew, powdery mildew, black rot). These grape varieties have been widely cultivated throughout the world, particularly in France, where they accounted for one-third of vineyard surface area in 1960, and in the northern United States, where they are grown for wine and non-winemaking uses (mainly for grape juice). In France, after the banning of six hybrids of *V. labrusca* with a 'foxy' flavour (Jacquez, Noah, Herbemont, Clinton, Isabelle, Othello), hybrids were banned from the registered designation of origin (Appellation d'origine contrôlée or AOC) scheme in the 1950s. In the twenty-first century, the interest in fungal-disease-resistant grape varieties once again made them the focus of debate. To meet society's expectations regarding a significant reduction in the use of plant-protection products, the breeding of new-generation hybrids that are indistinguishable from *V. vinifera* in terms of flavour is essential. On the basis of their main fungal-disease-resistance genes and pathogen pressure, a limited number of phytosanitary treatments may be required. The most robust present-day grape varieties accumulate polygenic resistance to downy mildew, powdery mildew and black rot, and require almost no additional phytosanitary treatments. To prevent any adaptation of the fungal pathogens to the natural defence mechanisms of the vine, it is recommended that a reduced active control strategy be implemented at the times of greatest vulnerability, i.e. around the flowering of the vine, owing to disease pressure.

To ensure that fungal pathogen control is effective, eco-friendly and user-friendly, it must incorporate the following basics:

- **Prophylactic measures** whose aim is to create an inhospitable environment for pathogens and to facilitate penetration of the applied plant protection products in order to ensure optimal cover of the organs requiring protection;

- Pathogen **risk forecasting** based on weather conditions and the receptivity of the plant organs;
- A **suitable choice of active ingredients** based on their mode of action and risk factors for the environment (surface waters, underground waters, persistence, etc.), users and consumers (residues in grapes, must and wine) while considering the risk of pathogen resistance in a long-term strategy (alternating active substances with different modes of action).
- The **dosage of plant-protection product appropriate for the leaf area** to be treated, based on planting density per unit area (hectare), canopy height and width;
- An **application technique** appropriate for the crop (plant density, pruning and trellising system) based on mechanisation options and the vineyard's configuration (slope, access, area, plot fragmentation);
- **Sprayer calibration**: whatever the type of device, aiming for perfection in terms of plant-organ cover and penetration into the leaf wall whilst minimising adverse environmental impacts;
- **Observing all safety measures** in the handling, storage and disposal of plant-protection product residues and packaging.

8.3.1   Prophylactic Measures

Prophylaxis encompasses the range of preventive measures rendering the microclimate as inhospitable as possible to the grapevine pathogens that develop at a more or less rapid pace depending on relative humidity, temperature, or the presence of a film of water on the vegetation.

Theoretically, all measures aiming to reduce the primary inoculum of a pathogenic fungus by eliminating infected organs (dead leaves, bunches, shoots) or burying them in the soil can reduce the risk of infection; however, the impact of these measures depends on the pathogens in question. Generally speaking, fungi always have sufficient inoculum to initiate new infections in which the developmental dynamic and the intensity of the epidemic are dependent on climatic conditions. Downy mildew (*Plasmopara viticola*) and powdery mildew (*Erysiphe necator*) develop regardless of their presence the previous year. In 2021, for example, the downy mildew epidemic in northern Europe led to considerable economic losses due to the rainy weather in spring and summer, whilst in 2022 the very hot, dry conditions meant that the disease remained quite inconspicuous, if not absent. The epidemiological virulence of the pathogens is defined by the relative favourability of the current year's climatic conditions. The infectious cycles of these pathogens are set off by specific survival particles (oospores, cleistothecia, sclerotia, chlamydospores) accumulating over the years. Conversely, for black rot (*Phyllosticta ampelicida*), eliminating the infected and mummified bunches is one of the essential prophylactic measures for reducing the initial inoculum in order to prevent the occurrence of the disease the following year.

When establishing a vineyard, the choice of grape variety is an essential factor in the future management of fungal diseases. Planting resistant interspecific grape varieties simplifies control according to the level of specific resistance to each of the fungal diseases in question. Plant density per unit area and orientation of the vine rows also determines the microclimate of the leaf wall, defining the shade cast between the vine rows and the foliage aeration potential according to the prevailing air currents. Numerous scientific studies on the density of plants in viticulture have been conducted and are ongoing (Keller and Mills 2021). On a global scale, this parameter is dictated by tradition, regional regulation, mechanisation and economic factors (Fig. 8.6). In the vast majority of cases, densities of the order of 3000–6000 plants per hectare spur-pruned or according to the Guyot system are generally best at meeting a vineyard's prophylaxis and cost-efficiency targets.

The orientation and spacing of the rows, the height and architecture of the leaf wall (Fig. 8.7) as a function of latitude, the vineyard's exposure and prevailing valley, plain, sea and thermal currents directly affect the microclimate of the vegetation

Fig. 8.6 Vine planting density is often determined by tradition and regulations. Standard spur- or cordon-pruned vines (6000 plants/ha) (**a**) differ from the traditional dense 'gobelet' growing system (10,000–12,000 plants/ha) still practiced in certain traditional regions (**b**), which significantly impacts the microclimate of the canopy and the conditions for fungal pathogens

Fig. 8.7 The growing system, planting density (standard spur- or cordon-pruned vines) and row orientation as well as canopy and soil management determine the microclimate, which in turn influences the epidemiology of fungal pathogens

cover (Zufferey et al. 2022) as well as the harmfulness potential of the pathogens. Generally speaking, the denser, more shaded and less aerated the foliage, the greater the relative humidity, transpiration and presence of a film of water on the leaves, and the more favourable the conditions are for infection by pathogens. Soil maintenance, in particular the use of companion plants as ground cover, also influences the microclimate of the vineyard due to its impact on vine vigour. Lastly, leaf wall management, encompassing the choice of training system, winter pruning, disbudding, cluster-zone leaf removal and shoot trimming, is a key factor, as it not only significantly influences the microclimate, but also ensures a relatively good degree of penetration of the plant protection product into the canopy.

8.3.2   Decision Support Systems (DSSs), Models and Disease Forecasting

In vineyards planted with *Vitis vinifera* vines that are susceptible to fungal diseases, plant protection products must be applied preventively to the foliage and clusters, whether integrated, organic or biodynamic production methods are used.

For ecological and economic reasons, it is best to intervene as seldom as possible, but at the most opportune moment to avoid yield losses. In the past, a systematic control approach based solely on contact products was practised from budburst up until grape ripening, at regular 8–10-day intervals. The forecasting models developed from new knowledge on the biology of pathogenic fungi directly influenced by climatic conditions (Fig. 8.8) allow to better time the plant protection treatments and to adjust the frequency of interventions to the risks determined. All pathogens

Fig. 8.8 Weather station measuring the key parameters of fungal outbreak built into disease forecasting systems

develop in a cyclical manner. The different stages of the cycles are determined by the climatic conditions influencing the epidemic and disease pressure as a function of the receptivity of the plant organs of the vine. The models are incorporated in weather stations or in decision support systems (DSSs) which always involve a degree of error depending on the threshold parameters built into in the algorithms (Pertot et al. 2016). The forecast consists in anticipating events based on more or less reliable future projections of the weather conditions whose probability of occurrence remains partially random.

8.3.2.1   Meteorological Data and Decision Support Systems (DSSs)

Historically, meteorological data have been used in the battle against fungal diseases. By way of example, Müller's calendar (Müller et al. 1923), which defines the incubation time of downy mildew (*Plasmopara viticola*) as a function of temperature, is still valid (Fig. 8.9). Initial measures using rain gauges and thermometers, and later, manually read thermo-humectograph recorders (Fig. 8.10), electronics and communication technologies have enabled the automation and transmission at regular intervals of the required climate data from the end of the twentieth century

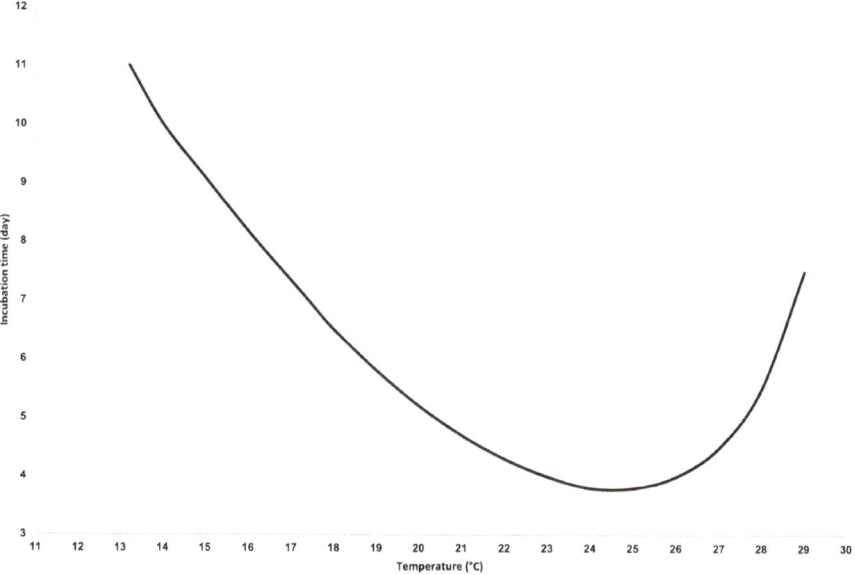

Fig. 8.9 The incubation curve of downy mildew (*Plasmopara viticola*) after Müller et al. (1923) is still valid and implemented in decision support systems

Fig. 8.10 Thermo-humectographs replaced manual recording via thermometers and rain gauges, and were later implemented in automatic weather stations measuring microclimatic conditions. (Photographs © G. Bleyer, Weinbauinstitut Freiburg i. Br., Germany)

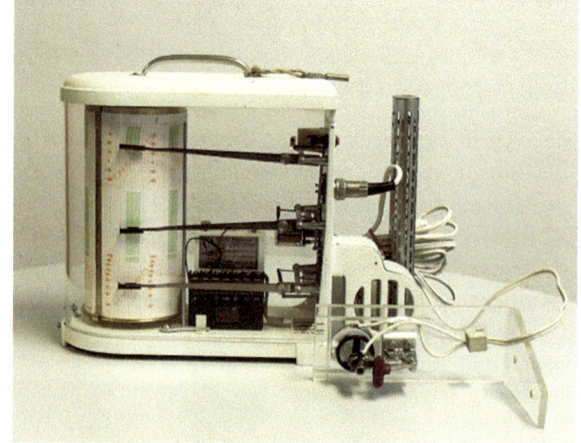

onwards. If in the 1950s information was transmitted via broadcast media, then by fax machines, nowadays the Internet and mobile telephony are the preferred means of communication for forecasting plant health risks. Nowadays there are numerous types of meteorological stations on the market, the great majority of which supply reliable data.

The key parameters for forecasting grapevine fungal disease risk are temperature, rainfall, relative humidity and for downy mildew, foliage wetness duration.

These data are continuously fed into algorithms that estimate the risk of infection, leading to a control strategy. Forecasting model algorithms may be protected and are generally not parameterisable. Statistical empirical models were developed based on historical observations specific to a region (e.g. Strizyk 1983), then mechanistic ones dissecting the interconnected stages of the pathogen development cycle (e.g. Rossi et al. 2008; Caffi et al. 2012; Bleyer et al. 2008), ultimately arriving at artificial intelligence (Baker et al. 2018; Lu et al. 2020) which requires long series of observations and of data to ensure a certain robustness of the predictions.

In Europe, the VitiMeteo decision support system (Viret et al. 2011) for forecasting downy mildew (Bleyer et al. 2008), powdery mildew (Kast 1997), black rot (Molitor et al. 2016), pests (European grapevine moths, *Lobesia botrana* and *Eupoecilia ambiguella*; leafhopper vector of bois noir, *Hyalesthes obsoletus*; grape erineum mite, *Colomerus vitis* and grape leaf rust mite, *Calepitrimerus vitis*) and vegetative growth of the vine (Schultz 1992; Molitor et al. 2014) serves as a benchmark and offers a web service on a global scale. Other DSSs have been developed in the world's main winegrowing regions and pursue the same objective of optimising disease and pest control (Pertot et al. 2016; Velasquez-Camacho et al. 2023). The inclusion of plant-organ sensitivity, particularly with the advent of relatively fungal-disease-resistant grape varieties, is an important control optimisation element that also requires forecasting models.

As a rule, decision support systems require support and continuous training and must be monitored on an ongoing basis to adapt to the development of scientific knowledge at the IT, technical and biological levels.

8.3.2.2 Spore Detection via Artificial Intelligence and Holography

Current risk-forecasting models essentially use meteorological data to predict the development of diseases. To improve the reliability of these forecasts, information on epidemic development in the field could be incorporated in the models (Velásquez-Camacho et al. 2023). However, these observations are time consuming, and it is unrealistic to expect reliable large-scale quantitative data on the presence of disease to be easily obtainable. A useful approach consists in quantifying the inoculum of spores in the atmosphere to improve disease control (Van der Heyden et al. 2021). For a long time, the fact that monitoring the spore in the air by conventional methods requires qualified staff and an abundance of time constituted a major impediment to the development of this approach. New approaches combining spore traps with molecular identification methods such as quantitative polyclonal chain reaction (PCR) (Carisse et al. 2021) or faster methods such as loop-mediated isothermal amplification (LAMP) (Douillet et al. 2022) are being developed for different pathogens. Another very promising approach is the identification and quantification of spores in the atmosphere using conventional optical methods or digital holography incorporating artificial-intelligence-based recognition algorithms (Basso et al. 2023). The use of holographic images leads to more information being obtained than with a conventional image, which is highly useful for identifying spores of fungi that might be very similar under the microscope. With these optical methods it is also possible to forgo the use of a laboratory, which represents a decisive advantage. Moreover, the information is available practically in real time. Considering the concentration of infectious spores in the air presents new opportunities for predicting disease occurrence under field conditions. This allows for the development of more precise control strategies against the main fungal diseases.

8.3.3 Choice of Active Ingredients

The choice of active ingredients is one of the key elements for a successful control of fungal diseases depending on the strategy pursued. The more effective an active ingredient, the less frequently it must be applied. The degree of efficacy against a fungal disease is defined with respect to an identical pedoclimatic situation but without any phytosanitary treatment (EPPO 2017). For fungicides, in the best-case scenarios, efficacy is total between 80% and 100%; it may be partial between 50% and 80% or insufficient below 50% in terms of disease frequency (percentage of infected leaves or bunches) and intensity (average percentage of infected leaf or bunch surface area) compared to the unsprayed control.

The performance of plant protection products is closely linked to their mode of action depending on pathogen pressure. Multisite contact products (combining several modes of action) with a preventive effect are only effective where the active ingredient is deposited, whilst penetrating products (i.e. loco-systemic or translaminar products) and systemic products are absorbed by the plant and migrate in part within it. The latter, however, are more prone to the risk of pathogen resistance. Owing to their highly specific single-site mode of action (directed at a single metabolite or at a specific enzyme of the pathogen), their repeated application may encourage mutations in the pathogens which will then be capable of blocking the modes of action of the active ingredients, causing them to rapidly lose their efficacy.

Most commercial products contain several active ingredients with different modes of action, with the aim of controlling several pathogens simultaneously or tackling them from several angles. This combinational approach is dictated by the biological complexity of the pathogens, which interact with the vegetative development of the vine, weather conditions and agronomical measures. The use of combined products also prevents the risk of resistance of the fungi to the active ingredients. When designing the control strategy, it is essential to respect the maximum number of applications of substances from the same chemical group to guarantee the alternating of modes of action and prevent resistance by reducing the frequency of use of the same single-site products.

Preparing a vineyard pathogen control strategy also involves considering the ecotoxicological profile of the plant-protection products. The protection of surface and underground waters, the secondary effects on beneficial fauna and all the available information on human health encourage the choice of molecules with the lowest possible levels of adverse impacts. For this purpose, risk indices (Labite et al. 2011) for the plant protection products must enable an objective choice to be made through the combination of the different criteria mentioned above.

The active ingredients are of natural origin or derived from combinatorial chemistry. A degree of controversy has crept in with the demonisation of plant protection products or chemistry in general, despite their vital importance for an economically sustainable agricultural sector. By definition, molecules of natural origin are in principle derived from processes in which humans are not involved, whilst so-called synthetic active ingredients derive from natural molecules that have been chemically modified. These molecules do not exist in nature but may be structurally similar to the natural molecules from which they are derived. Paradoxically, nature synthesises a large number of toxic molecules, often secondary metabolites, both in plants (e.g. digitalin, nicotine, pyrethrum) and in fungi (e.g. ergotamine, psilocybin, amanitins, aflatoxins, alkaloids) or in other microorganisms such as microalgae and bacteria (bacteriotoxins, neurotoxins). After the molecular and structural identification of molecules of natural origin and the identification of their bioactivity as part of phytosanitary protection, synthesis and combinatorial chemistry enable their streamlined, stable and reproducible production to transform them into commercial plant protection products (for example, the strobilurins). In the same context, natural molecules are marketed under diverse formulations involving synthetic adjuvants and additives (solvents, dispersants, emulsifiers, surfactants, stabilisers, anti-foaming agents, preservatives, colourants, etc.), which among others ensure their UV-light stability and resistance to oxidation and leaching owing to their ability to adhere to the plant organs. Copper, for example, in the form of hydroxide $Cu(OH)_2$, dihydroxide CuH_4O_2, sulphate $CuSO_4$, oxide Cu_2O, oxychloride $Cu_2Cl(OH)_3$ or oxysulphide, is a metal (Cu^{++} ions) that is present in nature, and hence allowed in organic viticulture, despite the fact that these different formulations involve synthetic products and their toxicity to living organisms has been established (Kumar Anant et al. 2018; Taylor et al. 2020).

Depending on their use, plant protection products are formulated in different ways (FAO 2016), for example as wettable powder (WP), water-dispersible granules (WGs), suspension or flowable concentrate (SC), emulsifiable concentrate (EC), suspo-emulsion (SE), water-soluble concentrate (SL), dispersible concentrate (DC) or water-soluble tablet (ST).

8.3.4 Dosage of Plant-Protection Products

With the exception of the powders, plant protection products are always diluted in water to form a more or less concentrated spray mixture that is sprayed onto the plant organs in the form of droplets that vary in size depending on the technical characteristics of the sprayers. The water is the support or the transporter of the active ingredients and thus determines the degree of cover of the plant surface to be treated. In the ideal scenario, all the surfaces of the vine organs—leaves and bunches—are evenly covered by the spray mixture. The denser the foliage, the more difficult this objective is to achieve (Fig. 8.11). The curve for active ingredient deposit per unit leaf area decreases in all three-dimensional crops (e.g., grapevine, fruit trees, berries-bushes) with the increase in leaf biomass. This reality is explained by the fact that the penetration of the spray mixture into the canopy is limited by the leaf density. Increasing the volume of spray mixture results in the rapid reaching of the runoff-point on the outer leaves of the canopy (Fig. 8.12). By strengthening the

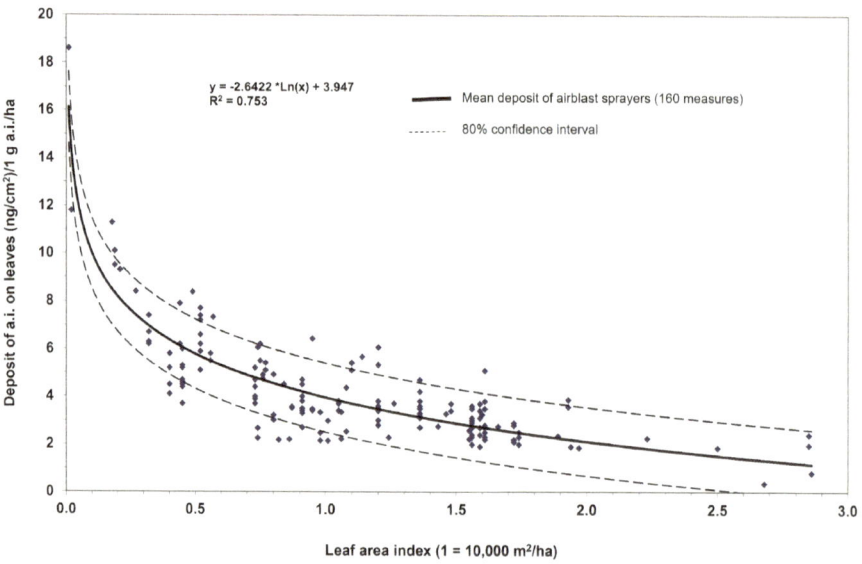

Fig. 8.11 Standardised leaf deposit of active ingredient per cm^2 leaf area as a function of the leaf area index, measured on grapevine. Mean leaf deposit curve and 80% confidence intervals of different commercial airblast sprayers used by growers (N = 160 measures with axial, reverse-axial, tunnel and pneumatic sprayers related to the leaf area index (each point represents the mean deposit value analysed on 10–40 leaves, Data from Viret et al. 2003)

Fig. 8.12 An excessive amount of spray on the leaf surface results in drop-off and loss of active ingredient onto the ground (arrows)

airflow, the outer leaves are pressed against each other and, in the worst-case scenario, may tear apart.

The quantity of spray mixture depends on the type of sprayers (non-air-assisted boom, air -blast, pneumatic or directed air duct, crossflow or tangential, multi-head fan, tunnel sprayers), which generate droplets of varying size. The finer the droplets, the more the volume of spray mixture can be reduced, significantly streamlining work. In viticulture, the recommended quantities of spray mixture for good distribution on trellised vines range, depending on the type of equipment, from 150 to 400 L/ha for boom and air-blast sprayers and from 100 to 200 L/ha for pneumatic sprayers producing finer droplets. Nevertheless, the dose of plant protection product per hectare cannot decrease proportionally to this quantity of water but remains the quantity determined for a given phenological stage as a function of plant biomass to ensure a good efficacy to control diseases.

Only the active ingredient exercises a biocidal effect on pathogens, and its concentration in the water defines the number of active molecules per unit of plant area required to ensure sufficient efficacy. This dimension is particularly sensitive, offering very little flexibility when tackling fungal diseases. The quantity of a plant-protection product to be applied to the organs of the vine to ensure sufficient efficacy is part of the information supplied by the manufacturer within the context of registration. This dose is derived from data obtained under controlled conditions using dose-efficacy curves (Fig. 8.13) for a given pathogen, supplemented by practical field trials.

The efficacy of a phytosanitary treatment is influenced by numerous parameters associated with:

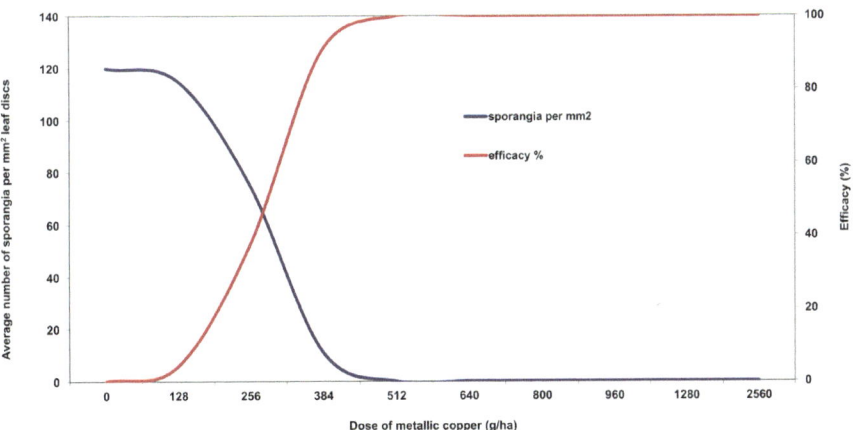

Fig. 8.13 Dose-efficacy curve of increasing the amount of metallic copper reported per ha (copper dihydroxide, 40% metallic copper) to control downy mildew (number of sporangia per mm on sprayed and inoculated grapevine leaf discs)

1. the plant (cultivar's genotype, vegetative development, foliage density, pilosity, epicuticular waxes, bunch morphology, yield, etc.);
2. the pathogen (presence of fungal particles, incubation in the plant, absence or presence of symptoms, fungal spore load, etc.);
3. type of sprayer (formation and size of droplets, air assistance and volume, working pressure, etc.);
4. the weather conditions (temperature, relative humidity, turbulence, wind, etc.).

To minimise the negative impact of these parameters on the success of the disease control, the registered dose always includes a security margin. Registered doses can vary from one Member Country of the EPPO (European and Mediterranean Plant Protection Organization) to another, for the same product and the same application (Rüegg et al. 2001). Both overdosage (Fig. 8.14) and underdosage, combined among others with inappropriate adjustment of the sprayer to the crop or a deliberate reduction of the dose lead to failure from both a phytosanitary and environmental perspective (Fig. 8.15). Copper hydroxide's efficacy against downy mildew (*Plasmopara viticola*), for example, is ensured with a minimum dose of the order of 300 g metallic copper per hectare. Under high disease pressure, below this value, efficacy decreases in proportion to the reduction in the quantity of active ingredient (Fig. 8.13). This principle applies to all fungicides, regardless of their mode of action, when pathogen pressure is high and vine organ sensitivity is at its highest (around the time of flowering when vegetative development reaches its peak).

The reference value for the dose of a fungicide is given per unit area of soil (ha), in kilograms or in litres. This indicator has its origins in field crops (e.g., cereals, leguminous plants) with bidimensional development treated by drawn or mounted horizontal bars. With three-dimensional perennial crops like grapevine or fruit trees, the area or volume of foliage to be treated per hectare varies according to planting density (Table 8.1), pruning system (e.g. cane- or spur pruning, minimal pruning,

Fig. 8.14 Plant-protection product overdose stems from imprecise calculation of dosage rates and improper calibration of the sprayer

Fig. 8.15 Improper calibration of sprayers leads to unacceptable drift and environmental contamination (**a**), especially in the case of fine droplets generated by excessive air flow (**b**)

Table 8.1 Example of the variation in leaf volume (m³ ha⁻¹) as a function of the planting densities for (spur- or cordon-pruned) trellised vines with a fully developed canopy after flowering (BBCH 75–89) for trimming heights of 1.5–1.8 m and a fix canopy depth of 0.5 m

Planting distances [m][a]	Number of plants per ha	Canopy height (H) [m]	Canopy depth (L) [m]	Volume of canopy [m³ ha⁻¹][b]
3 × 1	3333	1.8	0.5	3000
2 × 0.8	6250	1.8	0.5	4500
1.5 × 0.7	9520	1.5	0.5	5000
1.2 × 0.8	10,416	1.5	0.5	6220

[a] *Between rows (D) and between plants in row*
[b] *(H × L × 10,000)/distance between rows (D)*

Guyot, Scott-Henry, double curtain, pergola or lyre pruning) and canopy management (trimming height, partial/total/absent leaf-stripping). If these parameters are not considered, the dose per hectare may be too low for effective control, or conversely too high, contributing to avoidable environmental contamination. Hence, the amount of active ingredient per hectare is determined with respect to the volume or leaf area at the time of application.

As part of their annual cycle, vines develop a sizeable biomass. The leaf wall reaches several thousand cubic or square metres if taking only the outer area of the canopy into account per hectare, formed in several months from dormant buds (Fig. 8.16). Apart from plant density, the dose must also be adapted to vegetation development, so that the quantity of active ingredients is in keeping with the biomass to be protected.

In all the world's countries, the registered dose corresponds to a concentration in kilograms or litres per hectare when the leaf wall is fully developed, but without relation to the effective leaf surface. This reality leaves the manner in which doses are applied and adjusted up to the user's discretion (Landers 2010). In all cases where there is no precise correlation between the dose and the sigmoid curve of vegetative vine growth (Fig. 8.17), a linear adaptation of the reference dose (i.e. the registered dose in full vegetation) as a function of the phenological stage (BBCH

Fig. 8.16 Grapevine develops an impressive leaf biomass from the dormant buds to a canopy of several thousand cubic metres in just a few months, which cannot require the same amount of plant protection product regardless of the growing stage

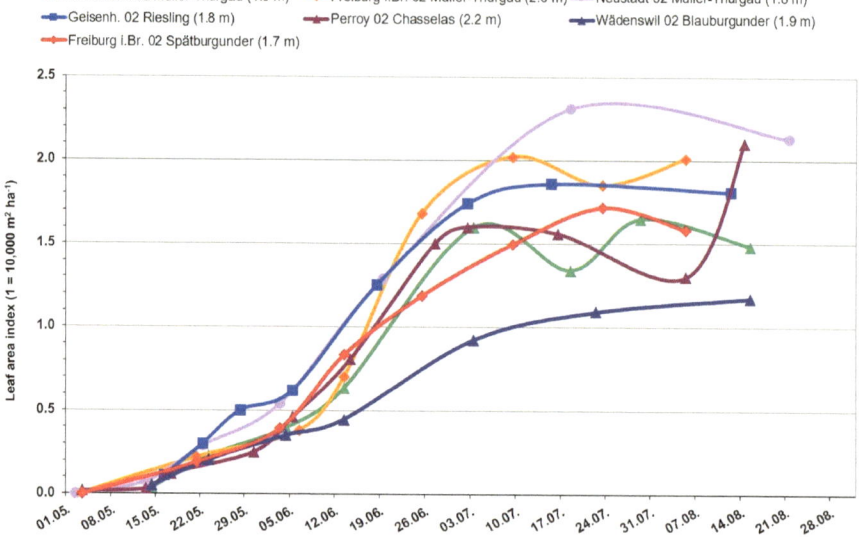

Fig. 8.17 Examples of the sigmoid growth curve of the grapevine canopy of trellised vines (row spacings of 1.7–2 m) for different *Vitis vinifera* cultivars (Müller-Thurgau, Riesling, Chasselas, Pinot noir) and locations in Germany (Freiburg i.Br., Neustadt, Geisenheim) and Switzerland (Perroy, Wädenswil) (data from Siegfried et al. 2007)

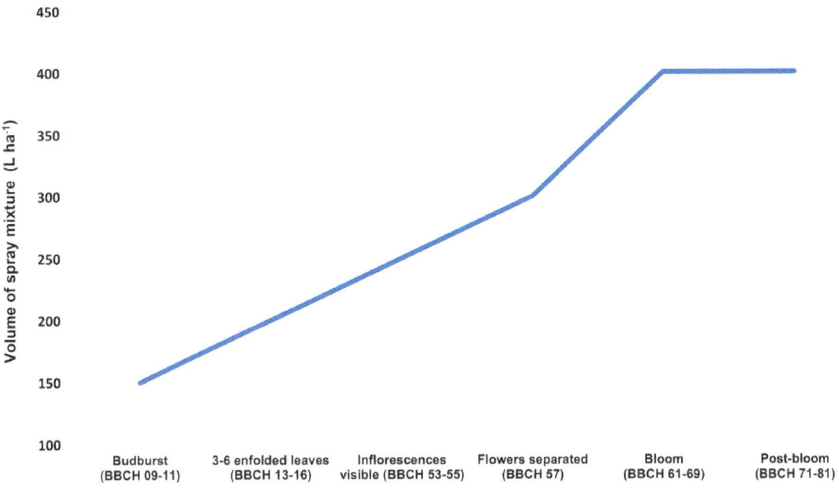

Fig. 8.18 Linear adjustment of the spray volume and dose of active ingredient for conventional boom or air-blast sprayers as a function of growing stage (BBCH)

2001) enables a partial optimisation of this basic notion. According to this approach, the quantity of spray mixture and the dose increases gradually until just after the flowering of the vine, which generally relates to the trimming height. During the post-flowering period, the dose remains stable until the final treatment (Fig. 8.18).

As a rule, the plant-protection product dose may not be reduced linearly without justification and under the pretext of the need to economise but should be determined according to the plant biomass to be treated (Siegfried et al. 2007).

8.3.4.1 Adjustment of Dose to the Canopy Size

Several dose-adjustment models have emerged, enabling an indispensable precision in terms of the optimisation of pathogen control in line with the environmental protection targets (Planas et al. 2022). Four principles for modifying the dose as a function of the canopy have emerged in viticulture, i.e. AGROMETEO or AGMET (Siegfried et al. 2007; Viret et al. 2005, 2011), OPTIDOSE (Davy et al. 2010), DOSAVINA (Gil et al. 2011, 2019) and DOSA3D (Planas et al. 2016; Román et al. 2022). Taking into account leaf wall area (LWA) or canopy volume (vine row volume VRV or tree row volume in fruit orchards TRV) represents a considerable improvement in precision in keeping with the optimisation of pathogen control with respect to a registered single dose per unit area of soil (Planas et al. 2022).

In Switzerland, for example, the AGMET dosage model follows the concept of TRV, with a registered dose corresponding to a leaf volume of 4500 m^3 ha^{-1} which is valid for trellised vine systems (Siegfried et al. 2007; Syngenta 2004). This model measures the height (H in m) and width (W in m) of the leaf wall at the time of treatment. These two values are multiplied together and divided by the row spacing (D in m), then the whole is multiplied by 10,000 m² to obtain the leaf volume per hectare (Fig. 8.19), which is highly correlated with the leaf area (Siegfried et al. 2007). This approach corresponds to the French 'Optidose' system, which in addition adjusts the dose as a function of pathogen's pressure.

In Germany and the European Union, the reference dose is calculated based on the leaf-wall area (LWA) method, but without taking leaf-wall width into account. This is done by multiplying the height (H in m) by 2 (2 sections of foliage to be treated), then by 10,000 m² to define the leaf area per hectare, then dividing everything by the row spacing (D in m). Registration is based on a reference quantity of product per 10,000 m² of leaf area per hectare (Fig. 8.20). The dose to be applied is thus based on this reference and leaf-wall height at the time of treatment, with a possible maximum for 20,000 m² of leaf area per hectare. In fully developed

Fig. 8.19 Measuring vine row volume (VRV) to determine the leaf-volume-adapted dosage of plant protection product incorporated in the AGROMETEO, OPTIDOSE and other dosage systems

$$VRV\ (m^3\,ha^{-1}) = \frac{H\ (m)\ x\ W\ (m)\ x\ 10{,}000\ (m^2)}{D\ (m)}$$

Fig. 8.20 Measuring leaf-wall area (LWA) to determine the leaf-area-adapted dosage of plant protection product as a basis for the registration of grapevine plant-protection products in the European Community

$$LWA\ (m^2\,ha^{-1}) = \frac{H\ (m)\ x\ 2\ x\ 10{,}000\ (m^2)}{D\ (m)}$$

vegetation, a vine planted with 2 m between rows and a 1.5 m leaf wall attains a leaf area of 15,000 m^2 ha^{-1}.

Compared to VRV, which was developed in Switzerland (Siegfried et al. 2007), the LWA approach is less well correlated with the effective leaf surface to be treated, and hence less precise (Rüegg et al. 2001; Planas et al. 2022). Adjusting doses according to canopy volume has been practised with success and enables a 10–30% reduction of plant protection products depending on vine growth conditions and fungal pathogen's pressure. This precise approach, however, can only succeed when sprayers are perfectly calibrated and the spray flow suited to the crop.

8.3.5 Techniques for a Precise Application of Plant Protection Products

Plant-protection product application is influenced by a multitude of interacting factors that determine the success of fungal pathogen control. Calibration of the sprayer and its adjustment to the leaf wall are key for ensuring successful control after the timing of application is determined by risk-forecasting models, the active ingredients are chosen objectively, and the exact dose is calculated as a function of the canopy and plant density per hectare. In other words, treatments must be carried out:

- at the right time,
- with the right product,
- at the right dosage,
- with calibrated equipment that is appropriate for the crop.

A sprayer consists of a tank with varying capacity (ranging from a few liters for handheld sprayers to several hundred or even several thousand liters) containing the spray mixture, a supply of clean water, an agitation system, a pump (piston, membrane, diaphragm, centrifugal or roller pump) and control valves linked by hoses, filtration devices, a pressure regulator and gauge, a fan generating an airflow, and nozzles or adjustable openings allowing the generation of droplets that transport the spray mixture to the organs of the vine. There are different sprayer formats—mounted, drawn or self-propelled. In the world's major winegrowing structures, plant protection treatments are carried out by dedicated vehicles catering specifically to the configuration of the vineyards (Fig. 8.21).

Water serves as the support of the active ingredient, which is transported in solution towards the target. Control efficacy depends on the quantity of active ingredient deposited per unit leaf area and transported by the water droplets. In other words, the quantity of fungicide must be calculated independently of the quantity of water, which depends on sprayer type. The droplets must be small in size to ensure that the spray mixture is distributed evenly over the surface of the organs of the vine, covering the target sufficiently (Fig. 8.22). However, the smaller the droplets, the greater the risk of drift and diffuse dispersion in the environment. The droplets can be

Fig. 8.21 Large over-row air duct sprayer operating over several rows' width

Fig. 8.22 Degree of coverage on water-sensitive paper sheath. Fine droplets (left) ensure better leaf-surface coverage than larger ones (right)

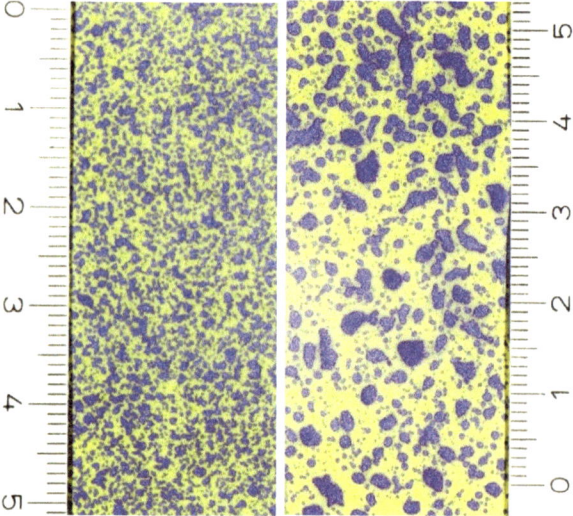

generated hydraulically via a nozzle, or pneumatically by breaking up the spray-mixture stream with air (air shear in pneumatic or directed air duct sprayers).

8.3.5.1 Hydraulic Formation of Droplets Through Nozzles

Droplets are formed by applying pressure of around 5–10 bars (depending on nozzle type and chosen volume of spray mixture) through a nozzle, either without air assistance or via the addition of air before or after the liquid passes through the nozzle.

Many different types of spraying devices have been constructed depending on the country and configuration of the vineyards, each with specificities in producing

Fig. 8.23 Knapsack airless motorised sprayers (**a**) or vertical boom sprayers (**b**) are simple grapevine-spraying devices

droplets to reach the vine canopy more or less precisely (Landers 2010). Two main groups can be distinguished:

1. Vertical-boom non-air-assisted hydraulic sprayers
   With these, only the working pressure (5–10 bar) through a nozzle breaks up the flow of the spray mixture. This is the case with vertical boom sprayers and knapsack sprayers used in steeply sloping, poorly mechanisable vineyards (Fig. 8.23).
2. Air-blast canopy sprayers or air-assisted sprayers
   These devices are assisted by air produced by a fan. With axial flow fans (Fig. 8.24), the air is generated in the axis of the fan which is generally placed at the back, after the flow of the spray mixture through the nozzles, or with inverse axial flow fans, before the flow of the spray mixture through the nozzles. The fan can also be a radial fan, in which the air is driven centrifugally off the axis of the fan. These devices produce a high volume of low-speed air (20–40 m s^{-1}) with a mainly upwards flow, depending on the configuration and orientation of the deflectors.

Tangential or crossflow fans, which consist of vertical ramps the height of the leaf wall, generate a horizontal or oblique airflow (Fig. 8.25). These devices enable improved penetration of the foliage by the spray mixture whilst minimising drift. In steep, fragmented and poorly mechanisable vineyards, the knapsack mistblower (Fig. 8.26) is the simplest type of air-assisted sprayer.

8.3.5.2   Air Shear Pneumatic Droplet Generation

In pneumatic air-shear sprayers (Fig. 8.27), the spray mixture circulates at low pressure (1–2 bar) and the liquid is broken up by a small volume of air driven at very high speed (up to 100 m/s or >300 km/h) that generates a cloud of very fine droplets (<100 µm). The further away these droplets are produced from the target, the greater the risk of drift. Different air- and spray-mixture supply systems have been developed to avoid this drawback. Mounted on large self-propelled machines that operate

Fig. 8.24 Air-blast axial sprayers spread droplets generated by the nozzles to the foliage via airflow

Fig. 8.25 Tangential or crossflow fans blow air horizontally onto the leaf canopy

Fig. 8.26 Knapsack motorised air-assisted sprayers or mistblowers is adapted for small, poorly mechanisable vineyards

Fig. 8.27 With pneumatic air-shear sprayers, a powerful airflow shatters the spray into very small drift-prone droplets

on several vine rows simultaneously, this system in among the most efficient. It requires low volumes of spray mixture, provided that each side or 'face' of the canopy is covered. In this case, we speak of air-assisted face-by-face spraying.

8.3.5.3 Nozzles and Droplet Generation

Nozzles are one of the smallest components of a sprayer, but they play an essential role in the quality of the application. The type and positioning of the nozzles, their optimal working pressure and the speed at which the device advances directly affect droplet formation, the quantity of active ingredient sprayed, and distribution of the spray on the surface of the vines' organs as opposed to spray particles drifting in the air or on the ground. The nozzles determine the volume of spray mixture depending on the size of the orifice, shatter the liquid into droplets of varying size and propel it towards the target (organs of the vine). The smaller the nozzle orifice, the lower its output (m s^{-2}), and the higher the working pressure, the finer the droplets.

Droplet size defines the degree of cover of a surface. The finer the droplets, the greater their number must be to evenly cover one square centimetre of leaf surface, and vice-versa (Fig. 8.22). Nozzles can produce a flat-fan, solid or hollow-cone spray with an angle of between 80° and 110°. They generate a population of more-or-less uniform droplets of varying size expressed by the Volume Median Diameter (VMD) which ranges between 50 and 200 μm. The VMD (Fig. 8.28) indicates the relative size of a set of droplets of a volume sprayed by a nozzle. A VMD of 100 μm signifies that half of the sprayed volume consists of droplets bigger than 100 μm and the other half of droplets smaller than this size. The composition of the droplet cloud may also be expressed by the NMD (Number Median Diameter), which represents 50% of the small droplets in absolute terms compared to the absolute number of 50% of the large droplets (Fig. 8.28).

The finer the droplets (VMD <150 μm), the greater the risk of drift. Droplets under 80 μm in diameter are invisible to the naked eye. Air-assisted sprayers generate droplets with a VMD of under 150 μm. When these types of sprayers operate over several rows width with fully developed canopy, it is speculated that droplets will cross the sprayed row, resulting in insignificant coverage of leaves on adjacent vines (Viret et al. 2003). The advantage of air-assisted sprayers is that they use low volumes of water (100–200 L/ha) whilst ensuring good distribution over the leaf wall, thanks to the very fine droplets and the movement of the foliage by the airflow. Face-by-face air-assisted sprayers can combine droplet generation by anti-drift or conventional nozzles (Fig. 8.29) and their transfer to the organs of the vine through the air.

Drift-reducing air-induction nozzles (Fig. 8.30) have a side port that creates a vacuum effect (Venturi effect), reducing the pressure inside the nozzle. The liquid is thus mixed with air, charging the droplets with air bubbles and making them bigger (VMD may exceed 300 μm) and less subject to drift (Fig. 8.31). These nozzles require a higher working pressure (8–15 bars) than conventional nozzles to ensure the incorporation of the air into the liquid. The heavier droplets produced are transported over shorter distances and in optimal conditions of use allow the reduction of drift by over 50%. However, these nozzles are not suitable for the simultaneous treatment of several vine rows, even at the start of the vegetation period, unless a face-by-face sprayer is available.

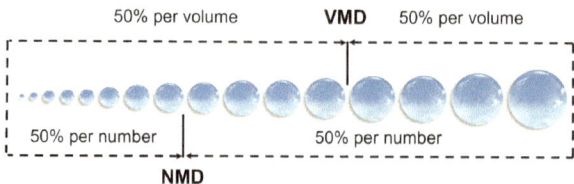

Fig. 8.28 Sprayer nozzles produce a population of droplets of various sizes expressed as VMD (Volume Median Diameter) or NMD (Number Median Diameter)

Fig. 8.29 Pneumatic sprayers (left) can combine the presence of nozzles and a directed airflow to the canopy (right)

Fig. 8.30 Air-induction nozzles generate big droplets charged with air bubbles, which are less susceptible to drift

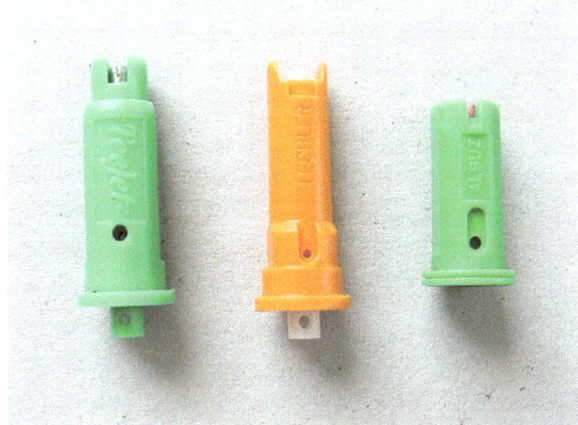

For contact fungicides, the number of medium-sized droplets (250–350 μm) required for optimal plant coverage is around 50–70 droplets per cm², whilst for penetrating or systemic fungicides 20–30 droplets suffice. The lifespan of a droplet ultimately depends on its size, and above all on the temperature and relative humidity. The finer the droplets and the higher the temperatures at low relative humidities, the less likely the droplets are to reach their target (Table 8.2).

8.3.5.4 Application Techniques in Steep Vineyards

In steeply sloping, fragmented and difficult-to-access vineyards, spraying equipment is still often worn on the back (Fig. 8.26). A laborious task, spraying has been made easier by using high-pressure (40-bar) lances using up to 2000 L of water per

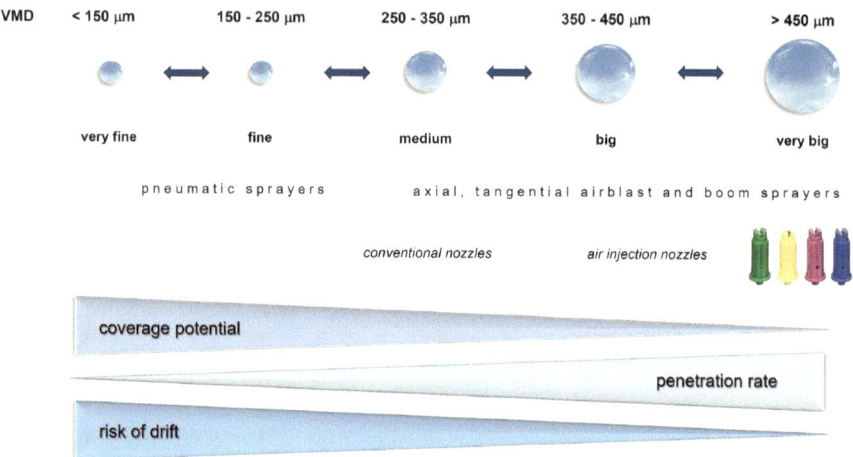

Fig. 8.31 Coverage potential, canopy penetration rate and risk of drift in relation to the droplet size produced by different sprayer types

Table 8.2 Temperature and relative humidity influence the evaporation and extinction time of droplets in relation to their size

Droplet size	20 °C—80% RH		30 °C—50% RH	
	Time before evaporation [s]	Distance before extinction [m]	Time before evaporation [s]	Distance before extinction [m]
50 μm	12.5	0.13	3.5	0.03
100 μm	50	6.7	14	1.8
200 μm	200	81.7	56	21

hectare spread remotely by dragging hoses through the vines from a tank of spray mixture and a pump. A more streamlined approach is application by helicopter. The ramp placed under the aircraft produces large droplets whipped into the foliage by the turbulence of the rotor (Fig. 8.32). To reduce drift, the length of the spraying ramp must be less than the diameter of the rotor, so as to minimise vortex phenomena. The helicopter carries a tank with a capacity of approximately 600 L and is equipped with a spray boom containing several dozen nozzles (50). It flies at a minimum speed of 50 kmh^{-1}, applying low volumes of spray mixture (100 Lha^{-1}), and can cover 1 ha of vineyard in 2 min. This process, which would take 5–6 h by hand, is significantly expedited (Viret et al. 2003). Application quality is satisfactory at the time of the first treatments when the vegetation is not very dense, but in full vegetation this technique has the disadvantage of imperfect coverage of the undersides of the leaves, the leaves at the bottom of the canopy and the bunches inside the canopy (Viret et al. 2003). Spraying by helicopter has been banned in the European Union since 2009 (Directive 2009/128/EC). However, in Switzerland, it is allowed under strict regulations and controls.

Fig. 8.32 Although helicopters are very efficient at spraying vineyards they generate significant drift, and deposit and coverage inside the vine canopy is limited. Old Aerospatiale SA315B Lama (top left and right) generates noticeably higher drift than the modern Eurocopter Ecureuil AS350 B2 (bottom)

Although drone spraying (Fig. 8.33) is very widespread in Asian rice fields (Yan et al. 2021; Xiongkui et al. 2017), it represents a very recent development in viticulture. In their current configuration, drones and all other unmanned aerial spraying systems (UASS) have limited transport capacity and require know-how to fly in automatic mode over rugged terrain. Flight planning based on 3D modelling of the terrain and GPS centimetre-level positioning accuracy are essential to enable a sufficiently high-quality application with complete safety. Drones equipped for crop spraying consist of the same components as traditional sprayers, viz., a 10–50-L-capacity tank, one or more pumps, hoses and nozzles, in addition to a battery and varying numbers of propellers. In steeply sloping vineyards, the drone sprays around 75–140 L ha^{-1} and habitually flies at a speed of 3–10 km h^{-1} at a

Fig. 8.33 Unmanned aerial spraying systems (UASS) are used in steeply sloping vineyards

height of 1.5–3 m above the leaf wall. Application quality of the current systems is similar to that of the helicopter, with the same limitations in terms of coverage of the underside of the leaves, of the leaves at the bottom of the leaf wall and of the bunches. To guarantee satisfactory efficacy, additional ground applications are necessary. The very rapid development of spray drones observed recently points to significant improvements in the coming years.

Creating terraces or benches in steeply sloping vineyards provides access to axial or pneumatic sprayers mounted on motorised tracked vehicles allowing for more efficient working (Fig. 8.34).

Fig. 8.34 In steep vineyards, terraces allow the use of small tracked-vehicle-mounted axial fan sprayers (left). However, there are challenges in adapting the angle of the nozzles and air flow to reach the foliage hedge when the terrace slope is too steep (right)

8.3.6 Calibration of Spraying Device, Application Quality and Drift

The efficacy of fungal disease control is directly affected by the quality of application, which depends on sprayer calibration and adaptation to the leaf-wall structure. The technical performance of spraying equipment is guaranteed provided that the equipment used is maintained and checked on a regular basis. Nozzle wear as a function of the materials (ceramic, polymer, stainless steel) may lead to significant differences in flow rate due to abrasion and deformation of the opening resulting from frequency of use, type of spray mixture and water hardness. Ceramic nozzles have the highest abrasion resistance. Proper calibration of the sprayer is one of the basic principles of precision viticulture. It ensures effective control, minimises drift, and preserves the environment.

In Europe, certain countries require an official technical inspection of sprayers at regular intervals (Koch 1996). This procedure allows malfunctions to be detected and ensures that the equipment components are in good running order. This inspection does not, however, replace the regular calibration of equipment before use, or the adaptation of airflow and spray flow to the canopy.

8.3.6.1   Calibration of Spraying Devices and Adaptation of Air and Spray Flows to the Canopy

Before each spraying campaign and prior to the initial application of plant protection treatments, it is essential to inspect and verify all functions of the sprayer. Sprayer calibration is an operation of the utmost importance in which each technical parameter plays a key role in ensuring the quality of the application and the success of pathogen control. Before each control campaign and before the first plant-protection treatments all the functionalities of the sprayer must be inspected and checked before each use. In this respect, on-board computers are standard feature of modern sprayers, enabling continuous monitoring of spraying performance. Cleaning the equipment after each use also allows any defects to be noted and proper functioning to be monitored. Whatever the type of equipment, the sprayer can be calibrated easily by carrying out the following steps:

1. **Travel speed**

    As a rule, travel speed cannot be read reliably from the odometers of agricultural vehicles unless the vehicles are equipped with a high-performance navigation system. The effective speed can be calculated by timing how long is needed to travel a previously measured distance of 50–100 m while advancing with the same number of revolutions per minute of the engine and with the same gear ratio as during spraying. For plots on sloping ground, the reliability of the measurement can be ensured by bearing in mind the difference in speed between travelling down and up the plot, which may require an adjustment.

$$\text{Travel speed}\left(\text{kmh}^{-1}\right) = \frac{\text{Distance travelled}\left[\text{m}\right]}{\text{time taken}\left[\text{s}\right]} \times 3.6$$

2. **Nozzle flow rate**

    Nozzle flow rate can be measured directly at each of the nozzle outlets (Fig. 8.35), or by the total flow of all the nozzles. With the first method, the nozzles are covered with a rubber sleeve that allows the water to be collected in a measuring cylinder after running the pump without the blower at the number of revolutions per minute and working pressure used for spraying. This operation must be performed for each nozzle measuring for a fixed time, generally 30–60 s. The values obtained are compared to the tables of nozzles complying with international ISO colour coding and standards (ISO 10625:2018 Equipment for crop protection—Sprayer nozzles—Colour coding for identification) supplied by the manufacturers in relation to the working pressure. This check allows any differences in flow rate between one nozzle and another to be identified, so that, where applicable, the appropriate measures may be taken (changing of the nozzles, cleaning of the opening or filter). The flow-rate tolerance between one nozzle and another is ±5%. If the difference is greater than 5% adjusting the working pressure or travel speed may suffice, whilst the nozzles must be replaced

Fig. 8.35 Flow rate is measured by placing rubber tubes over the nozzles to collect the spray in a graduated cylinder or container

if the difference exceeds 10%. Measuring nozzle flow rate allows the initially defined volume of spray mixture to be calculated.

Required nozzle output:

$$\text{Flow rate of each nozzle}\left[\text{L min}^{-1}\right] = \frac{\text{speed}\left[\text{km h}^{-1}\right] \times \text{row width}\left[\text{m}\right] \times \text{application rate}\left[\text{L ha}^{-1}\right]}{600 \times \text{number of operating nozzles}}$$

$$\text{Application rate}\left[\text{L ha}^{-1}\right] = \frac{600 \times \text{number of operating nozzles} \times \text{flow rate of nozzles}\left[\text{L min}^{-1}\right]}{\text{speed}\left[\text{km h}^{-1}\right] \times \text{row width}\left[\text{m}\right]}$$

The 2 min test allows to determine the flow rate of all the nozzles, by filling the water tank to maximum capacity with water and by spraying for 2 min under the conditions of application. The volume of water sprayed is determined by refilling the water tank to the maximum, measuring the amount added in a graduated bucket.

$$\text{Flow rate of nozzles}\left[\text{L min}^{-1}\right] = \frac{\text{speed}\left[\text{km h}^{-1}\right] \times \text{row width}\left[\text{m}\right] \times \text{application rate}\left[\text{L ha}^{-1}\right] \times 2\text{min}}{600}$$

3. **Adjusting the flow of liquid and air to the leaf wall**

By directing the flow of air and spray mixture, the droplets are projected in a more or less targeted manner towards the organs of the vine. When calibrating, this factor is essential for optimising the application. Flow is adjusted to the leaf wall by placing the sprayer in the vineyard and adjusting the angles of the nozzles and deflectors to the maximum height of the leaf wall. To orient the deflectors, thin tapes may be attached to the nozzles to check the angle of the airflow transporting the droplets towards the foliage when the turbine is running (Fig. 8.36). If the airflow clearly exceeds the maximum height of the leaf wall, the angle of the deflectors and nozzles should be modified. For convenience's sake this point is very often disregarded when spraying young vines, with all nozzles kept open and the turbine kept running at full throttle. In this case, closing the uppermost nozzles of the equipment individually is generally sufficient to prevent non-point pollution. This modification is also useful for spraying vines in production during the pre-flowering period, when the leaf canopy is not yet completely developped.

The air produced by the sprayer serves to transport the droplets and distribute them well in the foliage by creating turbulence in the latter so as to cover the top side and underside of the leaves. In general, air-assisted sprayers produce excessively large volumes of air. The optimal air volume should allow the spray mixture to slightly pass through the leaf wall of the treated vine rows. If a large cloud of droplets is visible beyond the treated rows, the air volume is too large or the equipment is not calibrated properly (Fig. 8.37a). In this case, the fine particles inevitably drift and are lost. Conversely, an insufficient volume of air prevents proper penetration of the spray mixture inside the leaf wall. That is why air volume and travel speed (not to exceed 5–6 km h^{-1}) must be adjusted according to the crop parameters. The required volume of air which the turbine must produce can be calculated by multiplying the interline distance by the height of the leaf wall and by the travel speed and dividing the result by a foliage density factor (factor 2 for a high density, 3 for an average density and 4 for a low density) (Fig. 8.38).

Fig. 8.36 Properly oriented air- and spray flows are essential for optimising spray application. The angles of nozzles and deflectors should be precisely adjusted to the leaf canopy using thin tapes fixed on the nozzles

Fig. 8.37 Excessive airflow and improperly adjusted air- and spray flows inevitably lead to drift (**a**). By contrast, air and spray flows directed at the canopy allow droplets to travel through the sprayed vine row with almost no drift (**b**, center part of the image, sprayer operating in the right-hand row beside)

Airflow [m³h⁻¹] = S [m] x L [m] x mean air speed [m s⁻¹] x 3600 x canopy density*

Airflow (m³h⁻¹) = $(\frac{D}{2})^2$ x π x mean air speed [m s⁻¹] x 3600 x canopy density*

*canopy density: factor 2 for high, 3 for medium, 4 for low density

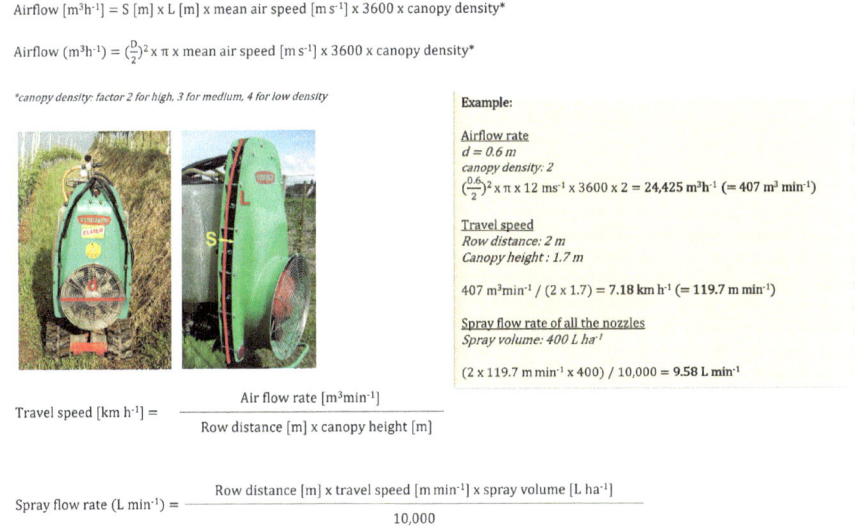

Example:

Airflow rate
d = 0.6 m
canopy density: 2
$(\frac{0.6}{2})^2$ x π x x 12 ms⁻¹ x 3600 x 2 = 24,425 m³h⁻¹ (= 407 m³ min⁻¹)

Travel speed
Row distance: 2 m
Canopy height : 1.7 m

407 m³min⁻¹ / (2 x 1.7) = 7.18 km h⁻¹ (= 119.7 m min⁻¹)

Spray flow rate of all the nozzles
Spray volume: 400 L ha⁻¹

(2 x 119.7 m min⁻¹ x 400) / 10,000 = 9.58 L min⁻¹

$$\text{Travel speed [km h}^{-1}\text{]} = \frac{\text{Air flow rate [m}^3\text{min}^{-1}\text{]}}{\text{Row distance [m] x canopy height [m]}}$$

$$\text{Spray flow rate (L min}^{-1}\text{)} = \frac{\text{Row distance [m] x travel speed [m min}^{-1}\text{] x spray volume [L ha}^{-1}\text{]}}{10,000}$$

Fig. 8.38 An airflow optimally adjusted to the vine enables good penetration of the spray into the canopy and minimises drift. Airflow can be calculated using a handheld anemometer to measure wind speed at the flow outlet and applying the above formulae. Travel speed must be adjusted to the air volume produced, and can be calculated using the airflow rate, as well as the spray flow rate of all nozzles given the travel speed obtained, which can be checked at the sprayer nozzles (Fig. 8.35). *d* diameter of the fan (m), *S* air-output width (m), *L* air-output length (m)

The volume of air actually produced by the turbine (in m³ h⁻¹ or in m³ min⁻¹) is measured by means of a wind gauge placed near the nozzles. This measurement, taken in different places, allows the mean airflow speed (m s⁻¹) to be calculated. Air volume (in m³ s⁻¹ or m³ h⁻¹) is calculated by measuring the area of the space from where the air is projected (= width × height in m²), multiplied by the mean air speed measured in m s⁻¹. Another approach is to consider the area of the turbine by multiplying the square of the radius by π (3.141), then multiplying the result by the mean air speed (m s⁻¹). The volume of air produced by the turbine is also dependent on the number of revolutions per minute of the tractor engine. This gives an additional adjustment option of reducing the engine speed. Some sprayers have several turbine rotation speeds that enable the forced air to be adjusted according to the vegetative development of the vine.

4. **Checking spray-mixture distribution**

Securing water-sensitive papers and strips, which turn blue upon contact with water, to a wooden stand or vine leaves provides an easy method for checking droplet deposition and distribution in the canopy. This straightforward yet highly illustrative procedure allows for fine-tuning the angle of the nozzles and deflectors. The droplet size and extent of leaf coverage are directly observable on the water-sensitive paper (Fig. 8.39). These papers can also be utilised to assess off-target particle drift above the canopy or beyond the vineyard, as well as the penetration of droplets into the canopy and their impact on adjacent vine rows. Additionally, visualisation of spray deposit and distribution can be achieved using fluorescent tracers, which are visible under ultraviolet light in dark conditions.

Fig. 8.39 Water-sensitive paper allows the direct display of spray deposit on the target surface, of the distribution, size and number of droplets in the canopy and of off-target particle drift

8.3.6.2 Drift

Drift consists of the set of droplets in a cloud of spray mixture that fail to reach their intended target (Fig. 8.40). Even when optimally calibrated, all sprayers generate a form of drift through the essentially unavoidable evaporation of a tiny percentage of the spray mixture. A portion of the very fine droplets disperse into the atmosphere, generally in a very low concentration, and dry before reaching the organs of the vine. Conversely, particle drift due to poor adjustment of the sprayer can be very largely avoided by following the principles of good practice regarding the application of plant-protection products.

In addition to the use of anti-drift nozzles, following specific optimisation measures can be implemented to reduce the unintentional impact of plant-protection products on the environment.

Tunnel Sprayers and Recycling Technologies

Tunnel sprayers are equipped with side panels that surround the leaf wall, thereby nearly completely avoiding particle drift. The spray liquid recovered in this way is pumped into the tank and reintroduced into the sprayer circuit (Fig. 8.41). This type of sprayers is generally towed or self-propelled and can treat several vine rows in a single pass. The recycling rate is high up until flowering, possibly accounting for over 30% of the spray mixture at the time of the first treatments when the canopy is still poorly developed. This figure decreases to less than 5% once the vine is in full vegetation (Viret et al. 2003). Tunnel-sprayers are suitable for large, easily mechanisable vineyards. The new generations are crossflow air-assisted sprayers improving the rate of leaf coverage. These systems can also be adapted for use on small tracked devices, but particular attention must be paid to their handling.

Fig. 8.40 Drift refers to all particles which fail to reach their target and are either lost through evaporation or end up on other surfaces, resulting in undesirable environmental contamination

Fig. 8.41 Tunnel sprayers with recycling techniques, which combine crossflow air assistance and nozzles (**a**, **b**), or those without air assistance (**c**), enable up to 30% of the spray volume to be recycled. Over-row tunnel sprayers are designed to simultaneously spray four rows of densely planted vineyards, as depicted in the Champagne region of France (**d**)

Sensor Techniques

Infrared sensors detect leafless zones in the leaf wall and automatically control the magnetic opening of the nozzles. Used primarily for herbicide application, these technologies should allow drift to be reduced by 25–30% and application to be restricted to the intended target.

Hedges at the Edge of the Plot and Respect of the Surroundings

A hedge at the edge of a plot can serve as an obstacle to drift spread, particularly near surface waters or housing. Another approach for preventing drift beyond the plot consists in treating the rows at the edge only towards the inside of the plot.

Factors Influencing Drift

- **Size of droplets and volume of spray mixture**: The application of a low volume of spray mixture per hectare involves the formation of very fine droplets, which are the most prone to drift. Each nozzle has an optimal working-pressure range given by the manufacturer. The pressure dictates the flow rate of the spray mixture through the nozzles in litres per minute. The greater the pressure, the higher the proportion of fine droplets and the higher the flow rate, whilst the opposite holds true when the pressure is below the optimal values. In a pressure range of 5–10 bar for standard nozzles and 8–15 bar for air-injection anti-drift nozzles, all of the droplets produced are the ideal size for good distribution over the target.
- **Distance between the nozzles and the target (working width)**: The further the nozzles are from the target, the greater the risk of drift. When treating several rows of width for the sake of efficiency, the distance between the nozzles and the adjacent vine rows is high and favours drift. The use of either canons (Fig. 8.42) operating from access roads over large distances or of helicopters in steep and poorly accessible vineyards, is rational but generates significant drift owing to the great distance between the equipment and the vine (Viret et al. 2003).
  The given angle of the nozzles, as specified by the manufacturers, which defines the angle of the jet of spray mixture, can also lead to drift if the distance to the target is unsuitable. For example, flat-jet nozzles with an angle of 80° should not operate at over 75 cm distance from the leaf wall, whilst the same nozzle with an angle of 110° should be placed at the closer distance of 50 cm.
- **Wind**: Wind is one of the main factors influencing drift. Plant-protection treatments must be carried out in calm weather. Starting from a current of 3 m s^{-1} (10.8 km h^{-1}), particle drift increases linearly with wind speed as a function of the proportion of fine droplets. At 3 m s^{-1}, application is still possible with some drift. From 5 m s^{-1} (18 km h^{-1}) when constant leaf movement is visible, treatment can no longer be carried out without sizable losses owing to drift.
- **Temperature and relative humidity**: These two parameters can reinforce drift due to evaporation. Above 25 °C, when the relative humidity is low (<50%), the proportion of fine droplets evaporating into the atmosphere and missing the target increases significantly. Conversely, applying plant-protection products on leaves that are damp from dew or after rainfall leads to spray-mixture run-off. This loss also accounts for a percentage of drift.
- **Travel speed of vehicle**: Vehicle travelling speed should not exceed 6 km h^{-1}. Above this speed, for airblast sprayers, turbulence forms behind the equipment, which concentrates the droplet cloud behind the sprayer in a sort of siphon effect, causing significant off-target deposits and drift.

Fig. 8.42 In poorly accessible steep vineyards, the use of large cannons operating up to a distance of 50 m produces high air volumes that generate small, drift-prone droplets

8.3.7   Storage, Handling and Management of Plant Protection Products

8.3.7.1   Handling

Plant protection products are toxic products whose harmfulness depends on the chemical structure of their molecules and on product formulation. Internationally valid risk symbols are defined by the United Nations for chemical products under the abbreviation GHS ('Global Harmonized System') and are found on the packaging. These pictograms illustrate the level of hazard to which the user is exposed.

Surveys on the source of pollution incidents associated with plant protection products reveal that the majority of these incidents stem from handling before and after the treatments. If the sprayer is perfectly calibrated and suited to the canopy, accidental diffuse contaminations occurring during treatment can be minimized or avoided. Effluents causing pollution incidents are returned unused spray mixture, tank-bottom residues, the rinse water from the spraying system and tanks, as well as the water for cleaning the outside parts.

Once all the vine treatment optimisation parameters have been considered, at the end of application there should be no more than a technical residue in the tank, pump and hoses of the sprayer. This residue can be heavily diluted with clean water and either sprayed on the vine or recovered in a specific rinsing and recycling area.

8.3.7.2   Storage of Plant Protection Products

Broadly speaking, plant-protection product stocks must be reduced to the minimum. Unused products can generally be returned to the suppliers. Plant protection products must be stored in their original packaging in well aired or ventilated premises specifically equipped for this purpose, locked and inaccessible to children and animals. The packaging must be sealed, stored by category on shelving or grating and

Fig. 8.43 The risk of end-user contamination by plant protection products requires the wearing of protective suiting and accessories throughout the entire handling process, including spraying in the vineyard, especially for all equipment without a closed driving cabin

kept away from moisture and frost. These premises are generally also used to weigh or measure the products and store personal protective equipment. For proper stock management, it is recommended that an up-to-date product inventory be kept.

8.3.7.3 Protective Equipment

Handling plant protection products involves a risk of contamination for the user during weighing, preparation and application of the spray mixture and cleaning of the equipment (Fig. 8.43). To prevent occasional or chronic undesirable effects users must avail themselves of appropriate protection and refrain from smoking, eating or drinking when handling the products. Protective equipment (jumpsuit, gloves, mask, protective goggles) reduces user contamination by 98% compared to working with no protection. Wearing a jumpsuit, waterproof gloves, waterproof boots or footwear, goggles, headgear or a full mask fitted with filters protects the skin, eyes, face and airways. After spraying, the protective equipment must be removed and washed with care. The hands, face and other exposed parts of the body must be rinsed and washed with soap and water.

8.3.7.4 Washing Area and Handling of Spraying Residues

The sprayer must be washed in an area provided for this purpose that allows recovery of the contaminated water. In some winegrowing regions, individual or collective washing stations ensure proper management of rinse residues (Fig. 8.44). The contaminated substrate can either be evaporated, followed by incineration of the solid residues, or recycled by successive filtrations over activated charcoal or

Fig. 8.44 Collective sprayer cleaning and recycling areas in or next to vineyards allow the recovery of residues, waste and contaminated cleaning water (**a**), which can then be properly recycled in biologically active systems (**b**)

Fig. 8.45 Plant-protection product containers are deemed to be special waste. They may not be stored in nature (**a**), reused, or burnt on the field, but must be collected and properly destroyed (**b**)

according to other principles such as coagulation, flocculation, or reverse osmosis. The wash bay is equipped with a high-pressure plant, a sloping concrete area, a tunnel for rinsing the nozzles, a tank for recovering the contaminated water and a settling tank. The settled wastewater can be channelled to a water-treatment plant or treated separately. Biological decontamination on a substrate enables treatment effluents to be decontaminated using microorganisms. Various closed or open systems have been developed based on the principle of gravity-fed settling followed by passage through a biologically active substrate (Fig. 8.44).

8.3.7.5   Handling of Empty Packaging

End-of-life plant protection products and their empty packaging are deemed to be hazardous special waste. They must not be left on-site (Fig. 8.45), buried or burned, or used for other purposes. Empty paper bags, plastic containers and all other contaminated packaging must be transported to dedicated collection centres or waste sorting centres. Some suppliers offer to take back the empty packaging in order to treat it appropriately. End-of-life, out-of-date or no-longer-authorised plant protection products must be returned to the supplier or to the specialist toxic-product collection services.

References

Anant JK, Inchulkar SR, Bhagat S (2018) An overview of copper toxicity relevance to public health. Eur J Pharm Med Res 5(11):232–237

Anke T, Oberwinkler F, Steglich W, Schramm G (1977) The strobilurins—new antifungal antibiotics from the basidiomycete *Strobilurus tenacellus*. J Antibiot 30(10):806–810

Baggiolini M (1990) Production Intégrée en Suisse: I. aperçu historique de la "production agricole intégrée". J Swiss Entomol Soc 63:493–500. https://doi.org/10.5169/seals-402422

Baker RE, Peña J-M, Jayamohan J, Jérusalem A (2018) Mechanistic models versus machine learning, a fight worth fighting for the biological community? Biol Lett 14:20170660. https://doi.org/10.1098/rsbl.2017.0660

Basso TC, Berti N, Leoni S, Schnee S, Hewison S, Fabre AL, Kasparian J, Dubuis PH, Wolf JP (2023) Digital holography and artificial intelligence for real-time detection and identification of pathogenic airborne spores. 44th World Congress of Vine and Wine, BIO Web of Conferences 68:01020. https://doi.org/10.1051/bioconf/20236801020

BBCH (2001) Growth stages of mono-and dicotyledonous plants, BBCH monograph second edition by Uwe Meier. Federal Biological Research Center for Agriculture and Forestry, Berlin

Becker WF, von Jagow G, Anke T, Steglich W (1981) Oudemansin, strobilurin A, strobilurin B and myxothiazol: new inhibitors of the bc1 segment of the respiratory chain with an E-β-methoxyacrylate system as common structural element. FEBS Lett 132(2):329–333. https://doi.org/10.1016/0014-5793(81)81190-8

Beffa T, Pezet R, Turian G (1987) Multiple-site inhibition by colloidal elemental sulfur (S°) of respiration by mitochondria from young dormant α spores of *Phomopsis viticola*. Physiol Plant 69(3):443–450. https://doi.org/10.1111/j.1399-3054.1987.tb09222.x

Bleyer G, Kassemeyer HH, Krause R, Viret O, Siegfried W (2008) VitiMeteo *Plasmopara*-Progronemodell zur Bekämpfung von *Plasmopara viticola* (Rebenperonospora) im Weinbau. Gesunde Pflanz 60:91–100. https://doi.org/10.1007/s10343-008-0187-1

Blum M, Waldner M, Gisi U (2010) A single point mutation in the novel PvCesA3 gene confers resistance to the carboxylic acid amide fungicide mandipropamid in *Plasmopara viticola*. Fungal Genet Biol 47:499–510

Boller EF, Avilla E, Jörg E, Malavolta C, Wijnands F, Esbjerg P (2004) Integrated production: principles and technical guidelines, 3rd edition. IOBC WPRS Bull 27(2):50

Brent KJ, Hollomon DW (2007) Fungicide resistance: the assessment of risk (2nd revised edition). FRAC monograph no 2, Ed. Fungicide Resistance Action Committee, Croplife International, Brussel, Belgium, 52 p

Caffi T, Legler SE, Rossi V, Bugiani R (2012) Evaluation of a warning system for early-season control of grapevine powdery mildew. Plant Dis 96(1):104–110. https://doi.org/10.1094/PDIS-06-11-0484

Carisse O, Van der Heyden H, Tremblay DM, Hébert PO, Delmotte F (2021) Evidence for differences in the temporal progress of *Plasmopara viticola* clades *riparia* and *aestivalis* airborne inoculum monitored in vineyards in eastern Canada using a specific multiplex quantitative PCR assay. Plant Dis 105(6):1666–1676. https://doi.org/10.1094/PDIS-06-20-1164-RE

Davy A, Claverie M, Codis S, Bernard FM, Colombier M, Davidou L, Girard M, Mornet L, Perraud J-P, Rives C, Vergnes D (2010) Trials results of the "Optidose" method using an adjustment of the pesticide dose for control of downy and powdery mildew. In: Proc. 6th Int. workshop on downy and powdery mildew, Bordeaux (France)

Delucchi V (1987) La protection intégrée des cultures—Integrated pest management in cultivated systems. In: Integrated pest management—protection intégrée: quo Vadis? Proceedings of a symposium held in Geneva from 9 to 11 December 1986 within the frame of the exhibition "Parasitis 86", 7–23

Douillet A, Laurent B, Beslay J, Massot M, Raynal M, Delmotte F (2022) LAMP for in-field quantitative assessments of airborne grapevine downy mildew inoculum. J Appl Microbiol 133(6):3404–3412. https://doi.org/10.1111/jam.15762

EPPO (2017) Efficacy evaluation of plant protection products PP 1/214 (4), Principles of acceptable efficacy. EPPO Bull 47(3):293–296. https://doi.org/10.1111/epp.12395

FAO (2016) Manual on development and use of FAO and WHO specifications for pesticides, first edition, third revision, FAO plant production and protection paper 228

Fritz R, Lanen C, Chapeland-Leclerc F, Leroux P (2003) Effect of the anilinopyrimidine fungicide pyrimethanil on the cystathionine β-lyase of *Botrytis cinerea*. Pestic Biochem Physiol 77(2):54–65

Fujinami A, Ozaki K, Yamamoto S (1971) Studies on biological activity of cyclic amide compounds. Part 1. Antimicrobial activity of 3-phenyloxazolidine-2,4-diones and related compounds. Agric Biol Chem 35:1707–1719

Gadoury DM, Sapkova S, Cadle-Davidson L, Underhill A, McCann T, Gold KM, Gambhir N, Combs DB (2023) Effects of nighttime applications of germicidal ultraviolet light upon powdery mildew (*Erysiphe necator*), downy mildew (*Plasmopara viticola*), and sour rot of grapevine. Plant Dis 107(5):54–65. https://doi.org/10.1094/PDIS-04-22-0984-RE

Gaetke LM, Chow-Johnson HS, Chow CK (2014) Copper: toxicological relevance and mechanisms. Arch Toxicol 88(11):1929–1938. https://doi.org/10.1007/s00204-014-1355-y

Gil E, Llorens J, Landers A, Llop J, Giralt L (2011) Field validation of Dosaviña, a decision support system to determine the optimal volume rate for pesticide application in vineyards. Eur J Agron 35:33–46. https://doi.org/10.1016/j.eja.2011.03.005

Gil E, Campos J, Ortega P, Llop J, Gras A, Armangol E, Salcedo R, Gallart M (2019) DOSAVIÑA: tool to calculate the optimal volume rate and pesticide amount in vineyard spray applications based on a modified leaf wall area method. Comput Electron Agric 160:117–130. https://doi.org/10.1016/j.compag.2019.03.018

Kast WK (1997) A step by step risk analysis (SRA) used for planning sprays against powdery mildew (OiDiag-system). Vitic Enol Sci 52:230–321

Keller M, Mills LJ (2021) High planting density reduces productivity and quality of mechanized Concord juice grapes. Am J Enol Vitic 72(4):358–370. https://doi.org/10.5344/ajev.2021.21014

Kennelly MM, Gadoury DM, Wilcox WF, Seem RC (2007) Vapor activity and systemic movement of mefenoxam control grapevine downy mildew. Plant Dis 91(10):1260–1264. https://doi.org/10.1094/PDIS-91-10-1260

Keon JPR, White GA, Hargreaves JA (1991) Isolation, characterization and sequence of a gene conferring resistance to the systemic fungicide carboxin from the maize smut pathogen *Ustilago maydis*. Curr Genet 19:475–481

Kirchmann H, Thorvaldsson G, Bergstrom L, Gerzabek M, Andren O, Eriksson LO, Winninge M (2008) Fundamentals of organic agriculture—past and present. In: Kirchmann H, Bergström L (eds) Organic crop production—ambitions and limitations. Springer, Dordrecht, pp 13–37. https://doi.org/10.1007/978-1-4020-9316-6_2

Koch H (1996) Periodic inspection of air-assisted sprayers. EPPO Bull 26(1):79–86. https://doi.org/10.1111/j.1365-2338.1996.tb01531.x

Kumar Anant JK, Inchulkar SR, Bhagat S (2018) An overview of copper toxicity relevance to public health. Eur J Pharm Med Res 5(11):232–237

Labite H, Butler F, Cummins E (2011) A review and evaluation of plant protection product ranking tools used in agriculture. Hum Ecol Risk Assess 17(2):300–327. https://doi.org/10.1080/10807039.2011.552392

Landers AJ (2010) Effective vineyard spraying a practical guide for growers. Cornell University Digital Print Services, Ithaca, NY

Linder C, Kehrli P, Viret O (2016) La Vigne (vol 2), ravageurs et auxiliaires. Ed. AMTRA, Lausanne, 394p

Lu W, Newlands NK, Carisse O, Atkinson DE, Cannon AJ (2020) Disease risk forecasting with Bayesian learning networks: application to grape powdery mildew (*Erysiphe necator*) in vineyards. Agronomy 10:622

Marès H (1856) Mémoire sur la maladie de la vigne, Bull. Soc. Centrale d'Agr. du Départ. de l'Hérault, p 165

Marti G, Schnee S, Andrey Y, Simoes-Pires C, Carrupt PA, Wolfender JL, Gindro K (2014) Study of leaf metabolome modifications induced by UV-C radiations in representative *Vitis*, *Cissus* and *Cannabis* species by LC-MS based metabolomics and antioxidant assays. Molecules 19(9):14004–14021. https://doi.org/10.3390/molecules190914004

Millardet A (1887) Nouvelles recherches sur le développement et le traitement du mildiou et de l'anthracose. G. Masson, Paris

Molitor D, Junk J, Evers D, Hoffmann L, Beyer M (2014) A high-resolution cumulative degree day-based model to simulate phenological development of grapevine. Am J Enol Vitic 65(1):72–80

Molitor D, Augenstein B, Mugnai L, Rinaldi PA, Sofia J, Hed B, Dubuis P-H, Jermini M, Kührer E, Bleyer G (2016) Composition and evaluation of a novel web-based decision support system for grape black rot control. Eur J Plant Pathol 144:785–798. https://doi.org/10.1007/s10658-015-0835-0

Müller K, Rabanus A, Kotte W (1923) Biologische Versuche mit der Rebenperonospora zur Ermittlung der Inkubationszeit. Weinberg und Keller 2:65–71

Nave BT, Koehle H, Kogel KH, Opalski K (2008) The mode of action of metrafenone. In: Modern fungicides and antifungal compounds V: 15th international Reinhardsbrunn symposium, Germany, pp 39–43

OIV (2016) Résolution OIV-CST 518–2016, Principes généraux OIV de la vitiviniculture durable, aspects environnementaux sociaux, économiques et culturels. https://www.oiv.int/public/medias/4973/oiv-cst-518-2016-fr.pdf

Onofre RB, Gadoury DM, Stensvand A, Bierman A, Rea MS, Peres NA (2021) Use of ultraviolet light to suppress powdery mildew in strawberry fruit production fields. Plant Dis 104. https://doi.org/10.1094/PDIS-04-20-0781-RE

Parker JE, Warrilow AGS, Price CL, Mullins JGL, Kelly DE, Kelly SL (2014) Resistance to antifungals that target CYP51. J Chem Biol 7:143–161. https://doi.org/10.1007/s12154-014-0121-1

Pasteris RJ, Hanagan MA, Bisaha JJ, Finkelstein BL, Hoffman LE, Gregory V, Andreassi JL, Sweigard JA, Klyashchitsky BA, Henry YT, Berger RA (2016) Discovery of oxathiapiprolin, a new oomycete fungicide that targets an oxysterol binding protein. Bioorg Med Chem 24(3):354–361. https://doi.org/10.1016/j.bmc.2015.07.064

Paull J (2011a) The secrets of Koberwitz: the diffusion of Rudolf Steiner's agriculture course and the founding of biodynamic agriculture. J Soc Res Policy 2(1):19–29

Paull J (2011b) Biodynamic agriculture: the journey from Koberwitz to the world, 1924-1938. J Organ Syst 6(1):27–41

Pertot I, Caffi T, Rossi V, Mugnai L, Hoffmann C, Grando MS, Gary C, Lafond D, Duso C, Thierry D, Mazzoni V, Anfora G (2016) A critical review of plant protection tools for reducing pesticide use on grapevine and new perspectives for the implementation of IPM in viticulture. Crop Prot 97:70–84. https://doi.org/10.1016/j.cropro.2016.11.025

Pezet R, Pont V (1977) Elemental sulfur: accumulation in different species of fungi. Science 196(4288):428–429. https://doi.org/10.1126/science.850786

Planas S, Román C, Sanz R, Rosell-Pablo JR (2016) A proposal for dose expression and dose adjustment in the EU-Southern zone (DOSA3D system). In: Proc. workshop on harmonized dose expression for the zonal evaluation of plant protection products in high growing crops. EPPO, Vienna

Planas S, Román C, Sanz R, Rosell-Pablo JR (2022) Bases for pesticide dose expression and adjustment in 3D crops and comparison of decision support systems. Sci Total Environ 806:150357. https://doi.org/10.1016/jscitotenv.2021.150357

Román C, Peris M, Esteve J, Tejerina M, Cambray J, Vilardell P, Planas S (2022) Pesticide dose adjustment in fruit and grapevine orchards by DOSA3D: fundamentals of the system and on-farm validation. Sci Total Environ 808:152158. https://doi.org/10.1016/j.scitotenv.2021.152158

Rossi V, Caffi T, Giosuè S, Bugiani RA (2008) Mechanistic model simulating primary infections of downy mildew in grapevine. Ecol Model 212:480–491

Rüegg J, Siegfried W, Raisigl U, Viret O, Steffek R, Reisenzein H, Persen U (2001) Registration of plant protection products in EPPO countries: current status and possible approaches to harmonization. Bull OEPP/EPPO 31:143–152. https://doi.org/10.1111/j.1365-2338.2001.tb00983.x

Sano S, Kasahara I, Yamanaka H (2007) Development of a novel fungicide, cyflufenamid. J Pestic Sci 32(2):137–138

Schultz HR (1992) An empirical model for the simulation of leaf appearance and leaf development of primary shoots of several grapevine (Vitis vinifera L.) canopy-systems. Sci Hortic 52:179–200

Siegfried W, Viret O, Huber B, Wohlhauser R (2007) Dosage of crop protection products adapted to leaf area index in viticulture. Crop Prot 26(2):73–82. https://doi.org/10.1016/j.cropro.2006.04.002

Strizyk S (1983) Grape downy mildew: data of the EPI model (*Plasmopara Viticola*; simulation model; "Etat Potentiel d'infection" infection potential position). Phytoma 350:14–15

Syngenta (ed) (2004) Manual of field trials in crop protection, 4th edn. Syngenta International, Basel, pp 91–94

Taylor AA, Tsuji JS, Garry MR, McArdle ME, Goodfellow WL, Adams WJ, Menzie CA (2020) Critical review of exposure and effects: implication for setting regulatory health criteria for ingested copper. Environ Manag 65:131–159. https://doi.org/10.1007/s00267-019-01234-y

Toquin V, François B, Sirven C, Gamet S, Latorse M-P, Zundel J-L, Schmitt F, Beffa R (2006) A new mode of action for fluopicolide: modification of the cellular localization of a spectrin-like protein. Pflanzenschutz-Nachrichten Bayer 59(2–3):171–184

Van der Heyden H, Dutilleul P, Charron JB, Bilodeau GJ, Carisse O (2021) Monitoring airborne inoculum for improved plant disease management a review. Agron Sustain Dev 41:20. https://doi.org/10.1007/s13593-021-00694-z

Velásquez-Camacho L, Otero M, Basile B, Pijuan J, Corrado G (2023) Current trends and perspectives on predictive models for mildew diseases in vineyards. Microorganisms 11:73. https://doi.org/10.3390/microorganisms11010073

Velasquez-Camacho L, Oterto M, Basile B, Pijuan J, Corrado G (2023) Current trends and perspectives on predictive models for mildew diseases in vineyards. Microorganisms 11(1):73. https://doi.org/10.3390/microorganisms11010073

Viret O, Siegfried W, Holliger E, Raisigl U (2003) Comparison of spray deposits and efficacy against powdery mildew of aerial and ground-based spraying equipment in viticulture. Crop Prot 22(8):1023–1032. https://doi.org/10.1016/S0261-2194(03)00119-4

Viret O, Siegfried W, Wohlhauser R (2005) Crop adapted spraying in viticulture for precise and ecological application. In: VIII Work Spray Appl Tech Fruit Grow. Suprofruit. Barcelona (ES)

Viret O, Dubuis P-H, Fabre A-L, Bloesch B, Siegfried W, Naef A, Hubert M, Bleyer G, Kassemeyer H-H, Breuer M, Krause R (2011). www.agrometeo.ch an interactive platform for a better management of grapevine diseases and pests. IOBC/WPRS Bull 67:85–91

Wijnands FG, Baur R, Malavolta C, Gerowitt B (2012) Integrated pest management—design and application of feasible and effective strategies. IOBC/WPRS, Brussels

Xiongkui H, Bonds J, Herbst A, Langenakens J (2017) Recent development of unmanned aerial vehicle for plant protection in East Asia. Int J Agric Biol Eng 10(3):18–30

Yamaguchi I, Fujimura M (2005) Recent topics on action mechanisms of fungicides. J Pestic Sci 30:67–74

Yan X, Zhou Y, Lui X, Yang D, Yuan H (2021) Minimizing occupational exposure to pesticide and increasing control efficacy of pests by unmanned aerial vehicle application on cowpea. Appl Sci 11(29):9579. https://doi.org/10.3390/app11209579

Zaker M (2016) Natural plant products as eco-friendly fungicides for plant diseases control—a review. Agriculturists 14(1):134–141

Zufferey V, Gindro K, Verdenal T, Murisier F, Viret O (2022) La Vigne: anatomie et physiologie. AMTRA. ISBN: 978-3-85928-112-7

Index

A

Active substances, 6, 135, 215, 230, 249, 258, 259, 266, 269, 270, 401–411, 413–415, 418, 420

Anatomy, 17, 22, 23, 25, 27–31, 34, 36, 37, 39, 44, 45, 49, 52, 55, 57, 356, 384

Angular leaf spot, 287, 288

Anthracnose, 3, 72, 129, 181, 252, 259–265, 284, 409

Armillaria root rot, 353, 354, 356, 359–361

Ascomycota, 122, 141, 144–146, 150, 152, 155, 157, 163, 166, 167, 216, 231, 249, 259, 265, 266, 270–272, 276, 283, 284, 287, 292, 293, 313, 315, 316, 325–327, 330, 336, 342, 343, 359, 362, 364–366

B

Basidiomycota, 141, 144–146, 148, 150, 151, 163, 166, 167, 289, 290, 313, 316, 353, 354, 359

Berries, 3, 14, 17, 18, 23, 38, 44, 45, 52, 55, 58, 63, 64, 66, 67, 76, 77, 82, 87, 201, 203, 204, 210–212, 216, 217, 223–226, 228, 232, 238, 240–243, 245–252, 254–258, 260–276, 278, 280, 283–286, 293–298, 319, 322, 336, 338, 339

Berry infection, 232, 238, 239, 242, 245, 255, 270

Biodynamic disease management, 412, 416

Bioluminescence, 358, 359

Biotrophs, 142, 143, 181

Black dead arm, 325–327, 366

Black foot disease, 318, 342

Black rot, 3, 58, 60, 61, 67, 68, 71, 76, 96, 97, 130, 152, 212, 216, 249–258, 261, 262, 265, 266, 268, 339, 401, 405, 406, 409, 410, 419, 420, 425

Breeding, 6, 12, 15, 21, 60, 62–64, 67, 68, 71–73, 75, 76, 78, 80, 82–84, 86, 88, 89, 91, 92, 94, 96, 97, 198, 258, 419

Breeding for resistance, 60, 61, 63, 64, 67, 68, 72–78, 80, 83, 84, 86, 88, 89, 91, 92, 95, 97

Buds, 17, 18, 20, 30–32, 34, 36, 51, 53, 121, 128, 203, 209, 219, 222, 224, 226, 227, 242, 243, 245, 254, 258, 265, 271, 286, 318, 328, 338, 340, 341, 343, 375–378, 387, 391, 393, 431, 432

Bunch, 18, 23, 32, 36–38, 40, 44, 45, 49, 50, 54, 56, 57, 60–64, 92, 128, 144, 198, 201, 209–212, 218, 228–231, 236, 238, 239, 243, 247–249, 252, 255–260, 266–269, 272, 273, 276, 278, 280, 286, 294, 296, 297, 319, 331, 333, 337, 339, 387, 403, 412, 420, 426, 427, 442, 443